# POLLUTANT TRANSPORT AND FATE IN ECOSYSTEMS

# POLLUTANT TRANSPORT AND FATE IN ECOSYSTEMS

## SPECIAL PUBLICATION NUMBER 6 OF THE BRITISH ECOLOGICAL SOCIETY

EDITED BY

P.J. COUGHTREY
Associated Nuclear Services,
Eastleigh House, East Street,
Epsom, Surrey

M.H. MARTIN
Department of Botany,
University of Bristol,
Woodland Road, Bristol

M.H. UNSWORTH
Institute of Terrestrial Ecology,
Bush Estate, Penicuik, Midlothian

BLACKWELL SCIENTIFIC PUBLICATIONS
OXFORD LONDON EDINBURGH
BOSTON PALO ALTO MELBOURNE
1987

Blackwell Scientific Publications
Osney Mead, Oxford, OX2 0EL
8 John Street, London, WC1N 2ES
23 Ainslie Place, Edinburgh, EH3 6AJ
52 Beacon Street, Boston
Massachusetts 02108, USA
667 Lytton Avenue, Palo Alto
California 94301, USA
107 Barry Street, Carlton
Victoria 3053, Australia



First published 1987.



Set by Poole Typesetting, Bournemouth, Dorset
and printed in Great Britain.

DISTRIBUTORS
USA and Canada
Blackwell Scientific Publications Inc
P O Box 50009, Palo Alto
California 94303

Australia
Blackwell Scientific Publications
Australia (Pty) Ltd
107 Barry Street
Carlton, Victoria 3053

British Library
Cataloguing in Publication Data
Pollutant transport and fate in ecosystems.
– (Special publication number 6 of the
British Ecological Society)
1. Pollution – Environmental aspects
I. Coughtrey, P.J. II. Martin, M.H.
III. Unsworth, M.H. IV. Series
574.5'222 QH545.A1

ISBN 0-632-01627-2

Library of Congress
Cataloging-in-Publication Data
Pollutant transport and fate in ecosystems.
(Special publication no. 6 of the British Ecological Society)
"A selection of the papers that were presented at a meeting of the Industrial Ecology Group of the British Ecological Society, held at the University of Bristol, 1-4 April 1985" – Pref.
Bibliography: p.
Includes index.
1. Pollution — Environmental aspects – Congresses.
2. Pollutants – Congresses.
3. Biogeochemical cycles – Congresses.
I. Coughtrey, P.J. II. Martin, M.H.
III. Unsworth, M.H. IV. British Ecological Society. Industrial Ecology Group.
V. Series: Special publication . . . of the British Ecological Society; no. 6.
QH545.A1P647 1987 574.5'222 86-24501

# Contents

Preface vii

**F.T. Last.** Introduction 1

**E.I. Hamilton.** The Periodic Table of the elements: geochemical and biochemical mechanisms 5

**K.R. Dyer.** Flushing and dispersal mechanisms in estuaries 35

**R.E. Lewis.** Transfer mechanisms for dissolved pollutants in estuaries 47

**D.I. Little, S.E. Howells, T.P. Abbiss & Dale Rostron.** Some factors affecting the fate of estuarine hydrocarbons and trace metals in Milford Haven 1978–82 55

**P.A. Cawse.** Trace and major elements in the atmosphere at rural locations in Great Britain 1972–81 89

**H.M. ApSimon, A.J.H. Goddard, P.M. Manning & K. Simms.** Facts and fallacies in wet deposition modelling 113

**M.H. Unsworth & A. Crossley.** The capture of wind-driven cloud by vegetation 125

**R.A. Skeffington.** Transport of acidity through ecosystems 139

**J.N. Cape, D. Fowler, J.W. Kinnaird, I.A. Nicholson & I.S. Paterson.** Modification of rainfall chemistry by a forest canopy 155

**H.G. Miller, J.D. Miller & J.M. Cooper.** Transformations in rainwater chemistry on passing through forested ecosystems 171

**M. Wainwright & W. Nevell.** The microbial sulphur cycle in atmospheric polluted soils 181

**A.W. Davison.** Pathways of fluoride transfer in terrestrial ecosystems 193

**D.C. Seel, A.G. Thomson & R.E. Bryant.** Bone fluoride in four species of predatory bird in the British Isles 211

**M. Hutton & C. Symon.** Sources of cadmium discharge to the UK environment 223

**R.M. Harrison.** Physico-chemical speciation and chemical transformations of toxic metals in the environment 239

**D.A. Jenkins.** Trace elements in saxicolous lichens 249

**B.E. Davies.** Plant-availability of heavy metals in soils 267

**R.M. Harrison & W.R. Johnston.** Experimental investigations on the relative contribution of atmosphere and soils to the lead content of crops 277

**N.W. Lepp & N.M. Dickinson.** Partitioning and transport of copper in various components of Kenyan *Coffea arabica* stands 289

**S.P. McGrath.** Long-term studies of metal transfers following application of sewage sludge 301

**M.H. Martin & P.J. Coughtrey.** Cycling and fate of heavy metals in a contaminated woodland ecosystem 319

**P.R. Evans, J.D. Uttley, N.C. Davidson & P. Ward.** Shorebirds (S.Os Charadrii and Scolopaci) as agents of transfer of heavy metals within and between estuarine ecosystems 337

**H.A. Grogan, N.G. Mitchell, M.J. Minski & J.N.B. Bell.** Pathways of radionuclides from soils to wheat 353

**B.J. Howard.** $^{137}Cs$ uptake by sheep grazing tidally-inundated and inland pastures near the Sellafield reprocessing plant 371

**D. Jackson & A.D. Smith.** Generalized models for the transfer and distribution of stable elements and their radionuclides in agricultural systems 385

**Subject index** 403

# Preface

This publication contains a selection of the papers that were presented at a meeting of the Industrial Ecology Group of the British Ecological Society, held at the University of Bristol 1–4 April 1985. The aim of the meeting was to discuss the processes and mechanisms underlying the transfer of pollutants and contaminants in ecological systems. The discussion of the impact of pollutants on individual organisms, populations and communities was specifically excluded. The contents of this volume reflect a diverse range of studies on various pollutants, contaminants and ecological systems. Participants were requested to identify parallels between transfer, distribution and fate of a wide range of materials and this range of interests was reflected in the contributions received. The industrial ecologist is often requested to provide responses to questions raised by members of the public as well as by industry and government authorities. Responses to these questions need to be based on sound chemical, physical and biological principles and to contain both a qualitative and quantitative element. The papers presented at the meeting provided examples of mechanisms and processes involved in pollutant transport through ecosystems as well as of the significance of long-term or widespread investigations in the identification of temporal or geographical trends. In addition, examples were also provided of studies involving complex systems and diverse materials with the aim of identifying underlying principles. The contents of this volume tend to reflect topics of current environmental concern, e.g. acid deposition, heavy metals, radioactivity, etc. for which information is being collated in order to provide a basis for assessments concerning future impact. Such assessments will require a combination of the information on transport and fate within ecosystems with knowledge of the effects of pollutants on the system. Nevertheless, the interpretation of data concerning effects of a pollutant needs to be placed in the wider context of the occurrence, distribution and fate of that pollutant. The purpose of this publication is to provide that wider context. The integration of the many aspects of pollutant behaviour and effects is a procedure with which the ecologist should be familiar. In the future we may be faced with aspects of pollutant distribution, fate and impact which, today, are unforeseen, or which seem to have little practical significance. In order to prepare for these events it is pertinent that the experiences of various groups and individuals be combined and that industrial ecologists develop their ability to assimilate information from a wide range of subject areas.

Only a selection of the papers presented at the meeting were subsequently submitted for publication and included in this volume. We should like to thank all contributors to the meeting and particularly the chairmen of the various sessions for directing the stimulating discussions. We are also grateful to the authors of the papers in this volume for their perseverance during the editing procedure.

# Introduction

F. T. LAST
*Institute of Terrestrial Ecology, Bush Estate, Penicuik, Midlothian, Scotland EH26 0QB*

Just over 20 years ago the British Ecological Society arranged a symposium concerned with 'Ecology and the Industrial Society.' In his Address, Professor Clapham (1965), who took the opportunity to air many of his hobby-horses, started by lending support to Alvin Weinberg's forthright statement that 'scientists cannot expect society to support science because scientists find it an enchanting diversion'. Professor Clapham then proceeded to identify three problems. They were associated with:

(a) the need 'to increase the efficiency of land-use in growing plants and animals for food and a wide variety of other purposes;

(b) the problems involved in creating new ecosystems or re-creating old ones;

(c) ecological aspects of the problem of environmental deterioration arising from increasing populations and intensified industrialization'.

In discussing the third problem, he wrote 'that man has recently brought about the changes in the chemical composition of the soil, of the water of ditches, streams, lakes and even of the sea; changes too of the air that collectively constitute a massive modification of the environment'. These perceptive statements have a familiar ring and immediately bring to mind the problems confronting those of us concerned with pollution, 'agriculture and the environment', 'forestry and the environment', etc. In tackling these problems we have come to realize that Professor Clapham's plea for 'repeated quantitative assessments of different ecosystems, so as to have bases against which to measure change or detect perturbation', has passed largely unheeded.

But what has happened since the earlier symposium? Undoubtedly the number of experts has greatly increased, a change which has added strength and exposed weaknesses. As Nicholas Butler of Columbia University, USA suggested, an expert is a person who knows more and more about less and less. Fortunately, there are now indications that the prime advantage of expert knowledge, namely the ability to predict, is being increasingly balanced by a healthy and increasing interest in wider perspectives, the strength of ecologists. The individual concerned with atmospheric pollutants has come to realize that the atmospheric phase is only the beginning of a potentially long sequence of events subsequently affecting, directly and indirectly, terrestrial, freshwater and marine ecosystems.

To return to my question: What has happened in the pollution field during the last 20 years?

(a) Undoubtedly we are much more aware of the often vast distances over which airborne pollutants are dispersed. While they may still pose some locally intense problems, despite the changes enforced by the Clean Air Acts (1956, 1968), many of them create greater concern regionally, nationally and internationally *vide* acid rain (*sensu stricto*).

(b) With this change from local to international concern, the rather naive concept of damage, held in the 1950s, has been totally overturned and, as a result, we now tend to be less certain of 'effects' than in earlier years.

(c) While new pollutants have undoubtedly been identified, there has been a greatly increased awareness of the potential importance of substances which have been known for possibly hundreds of years. Some have changed from contaminants to being regarded as pollutants, e.g. nitrogen dioxide, $NO_2$, which (i) in mixtures can greatly exacerbate the damage done to some plants by sulphur dioxide but rarely, if ever, occurs by itself in damaging amounts and (ii) is one of the inputs in the photochemical reaction producing photoxidants including phytotoxic ozone.

But, at this stage, we should ask ourselves about the definition of pollution: there is a surprising diversity of opinions. Attempts have been made to attach separately identifiable meanings to 'pollution' and 'contamination'. But is this wise and necessary, remembering that their verbs both mean 'to make physically impure'? In its Tenth Report, the Royal Commission on Environmental Pollution (1984) referred to contamination/contaminant as 'the introduction or presence in the environment of alien substances or energy' when it did not wish to pass judgement on whether they cause, or are liable to cause, damage or harm. This definition is supposed to act as a foil to the suggestion made by the National Research Council, USA in its Report on the '*Troposphere transport of pollutants and other substances*' (1978): 'substances introduced into the environment become pollutants only when their distribution, concentration or physical behaviour are such as to have undesirable or deleterious consequences'. This point of view was upheld by Mellanby (1967) and elaborated by Holdgate (1979) who defined pollution as 'the introduction by man into the environment of substances or energy liable to cause hazards to human health, harm to living resources and ecological systems, damage to structures or amenity, or interference with legitimate uses of the environment'. Where do sulphur dioxide and nitrogen dioxide fit? Both supply essential plant nutrients and have been shown to be beneficial in some circumstances, usually where intensively managed crops are exposed to small atmospheric concentrations. Thus, they may be beneficial in small amounts (contaminants?) and harmful in larger (pollutant?) doses. It is also possible that with new knowledge, today's contaminants may become tomorrow's pollutants; a proposition, based on value judgements, which I regard as unsound and likely to lead to complacent attitudes. Because the search for knowledge about 'pollutants' is accorded, for more or less understandable social reasons, greater priority than that about contaminants, it is possible that the damage done to the assemblages of trees in the San Bernardino Valley, near Los Angeles, might have been avoided if the increased concentrations of ozone, attributable to the activities of man, had been regarded as pollutant instead of contaminant. The same may prove to be the case with forest decline in parts of continental Europe. However, on balance, I prefer to restrict myself to the use of pollution and pollutant, terms that can be applied with as much facility to concentrations of naturally-occurring substances, which have been augmented by the activities of man, as to substances which are never formed by natural processes.

Setting aside the niceties of the English language, what should a symposium about pollutants be discussing?

(a) Atmospheric pollutants: smoke, $SO_2$, NO and $NO_2$ (including their formation during, and after, straw burning), particulate $SO_4^{2-}$ and $NO_3^-$ and acid rain, hydrocarbons, ozone, ammonia (from animal wastes), lead, fluoride, etc.

(b) Freshwater pollutants: sewage, industrial effluents, products of agriculture including excess nitrate, acid rain (direct and indirect as in drainage and surface run-off).

(c) Thermal pollutants.

(d) Radiation.

(e) Marine pollutants: oil, sewage and other substances flowing from industrial outflows.

(f) Pesticides: herbicides, fungicides and insecticides.

Each of these forms of pollution might warrant a symposium of its own but, in the event, their study can be apportioned to the same five subdivisions:

(i) Sources.
(ii) Transport.
(iii) Deposition and/or Uptake.
(iv) Effects (direct and indirect).
(v) Fates.

To gain the greatest benefit from the series of papers that constitute this book, I suggest that the reader attempts to compare and contrast. Think of the laws that govern the dispersion of gases (air) and liquids (freshwater and marine); also the influences of weather and tides. Is the atmospheric transport of radioactive particles and particulate fluoride governed by the same laws as the transport of $SO_4^{2-}$ and $NO_3^-$ in air, aerosols (occult deposition) and precipitation (rain, hail and snow)? How do these laws differ from those controlling the movement of gases and what effects do they have on residence time and possible physico-chemical modifications? What is the significance of mean concentrations in relation to those occurring in episodes? Does the resistance analogue identifying atmospheric, canopy and soil resistances (Fowler & Unsworth 1974), and developed to aid the understanding of pollutant uptake in terrestrial ecosystems, have relevance to the uptake and dose perceived by terrestrial animals and aquatic biota? Can the resistance analogue be used to identify the importance of different processes at the cellular level, in addition to those occurring among assemblages of vegetation? Can parallels be usefully sought among the biochemical reactions of plants and animals to intruding pollutants, whether they are 'inactivated' by being (i) bound to proteins, e.g. cadmium to metallothionein, a sulphur-containing amino acid, or (ii) reduced or oxidized to less damaging forms. In short, are the pathways of different pollutants unique as we tend to suggest by our overriding concern for one at a time? For that matter, I suggest that pollution research could gain considerably from a greater appreciation of the studies defining the flow of nutrients through complex natural and semi-natural ecosystems with the development of (i) transfer functions between compartments, (ii) a relatively sophisticated understanding of the interplay between herbivores, omnivores and carnivores and (iii) knowledge of outputs as well as inputs so as to enable budgets to be prepared which enable major omissions to be identified.

For the future, pollution research must be based upon a more apposite balance of description and prediction, with both statistical and chemical sophistication — an investigation of acidic deposition must focus on ionic balances and not be restricted to a consideration of $SO_4^{2-}$ and or $NO_3^-$. We should recognize the parallels with flows of nutrients, both in plants and animals, accepting that pollutants may sometimes be damaging, sometimes beneficial and, on other occasions, seemingly without effect. But, whatever their impact, we should know their pathways and fates, remembering that a pollutant such as fluoride, although accumulated, may have seemingly no effect on foxes while severely impairing cattle, a possible distinction between carnivores and herbivores.

## REFERENCES

**Clapham, A. R. (1965).** Symposium Address. *Ecology and the Industrial Society.* (Ed. By G. T. Goodman, R. W. Edwards & J. M. Lambert), pp. 1-9. Symposium of the British Ecological Society 1964. Blackwell Scientific Publications, Oxford.

**Clean Air Act (1956).** Her Majesty's Stationery Office, London.

**Clean Air Act (1968).** Her Majesty's Stationery Office, London.

**Fowler, D. & Unsworth, M. H. (1974).** Dry deposition of sulphur dioxide on wheat. *Nature (London)*, 249, 389–390.

**Holdgate, M. W. (1979).** *A Perspective of Environmental Pollution.* The Cambridge University Press, Cambridge.

**Mellanby, K. (1967).** *Pesticides and Pollution.* Collins, London.

**Royal Commission on Environmental Pollution** (1984). *Tenth Report. Tackling Pollution - Experience and Prospects.* Her Majesty's Stationery Office, London.

# The Periodic Table of the elements: geochemical and biochemical associations

E. I. HAMILTON
*Institute for Marine Environmental Research, Prospect Place, The Hoe, Plymouth, PL1 3DH*

## SUMMARY

**1** The relative abundances and correlations between a number of elements of the Periodic Table for geochemical and biological natural systems are considered.
**2** Three element associations can be identified in biological materials: the first can be understood in terms of basic inorganic chemistry and is typical of elements which are not involved in biochemical processes; the second reflects interactions between elements and the surfaces of biochemical compounds; and the third is an association with seawater.
**3** The manner in which the abundance of the elements in the natural environment are reflected in man is discussed.

## INTRODUCTION

This paper, prompted by scientific curiosity, concerns relationships between the abundance of the elements in the natural environment and those in man. An examination of the literature for the abundance of the chemical elements in the geosphere and biosphere shows that, with respect to relative enrichment and depletion factors, chemical compositions are similar. Before the agricultural revolution (6000 BC) the transfer of elements from the geosphere to the biosphere (for example, rocks→ soil→ plants→ animals→ man), and the relative abundances of the elements in living organisms must, to a degree, have reflected the geochemistry of the natural environment modified by the biological uptake and recycling of the elements. Simple relationships are not expected and only certain types of living organisms can proliferate in a particular geochemical environment. Although plants and animals tolerant to certain elements do exist and colonize particular chemical niches, overall those which provide the main source of food to most forms of animal life have a wide distribution and are adapted to a tolerable range of chemical environments. After the Agricultural and Industrial Revolution (*c.*1800) the natural environment became altered as a result of man's cultural and technological advances. The association between elements of the Periodic Table, illustrated in Fig. 1, is based upon similarities in the chemical properties of the elements which are, in general, maintained in the geosphere and biosphere. However, following the availability of one element to an organism in an unusual amount there may also be an abnormal degree of uptake, such that the levels which are normally acceptable are exceeded and toxic effects may accrue. Some plants and animals are highly enriched or are tolerant to certain elements, the origins of which lie in the process of organic evolution.

Today, many studies of environmental pollution concern acute events, such as those

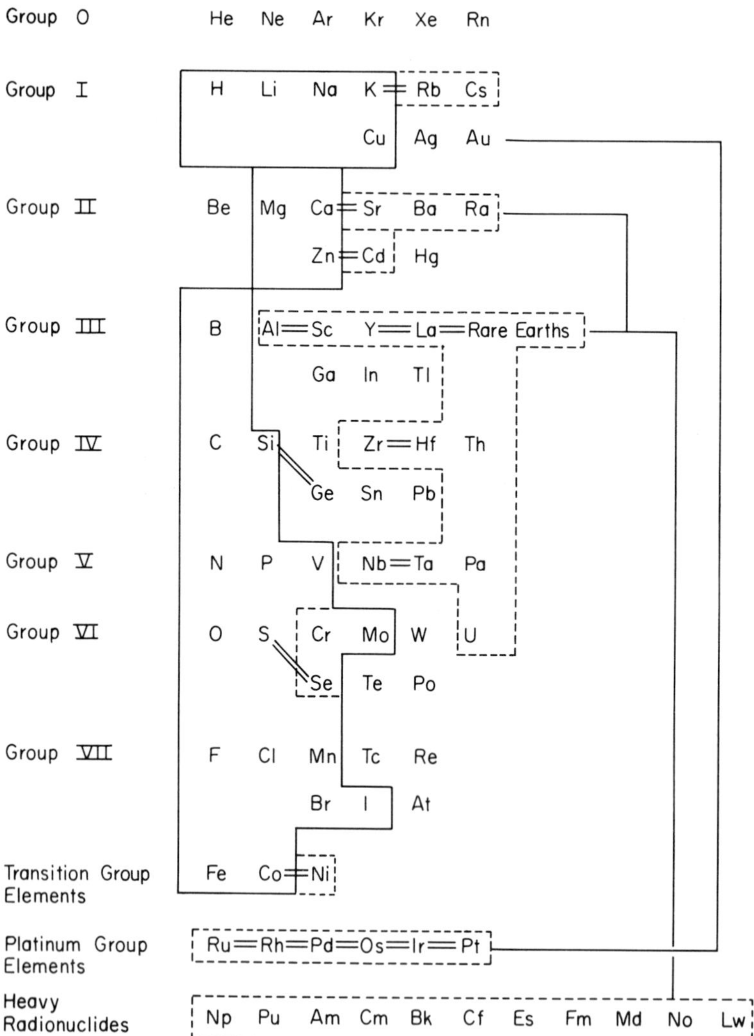


FIG. 1. Element groupings based upon the Periodic Table of the elements (Hamilton 1979). Elements enclosed by a continuous line are considered to be essential, those enclosed by a broken line represent elements having similar chemistry and amendable to isomorphous substitution. The double continuous lines represent major links between a major and a minor element such that associations are retained in the passage of these elements from the natural environment to man.

following accidental release of an element into the environment or the proximity to a source, e.g. an effluent outfall or a smelter. It is acknowledged that sub-toxic effects also occur and that there is synergism and antagonism between elements which can result in either beneficial or detrimental effects to organisms. The hypothesis that uniformity in chemical composition should prevail in major components of the biosphere is examined in this paper through associations which exist between elements in geochemical and biological systems. An objective is to identify those trends of use in identifying changes which have taken place since man first began to modify the chemical composition of his environment, and also to determine whether or not patterns in the abundances of the chemical elements can eventually be related to those which are characteristic of various types of morbidity in both man and components of his environment. For several decades there has been little attention to explaining relationships between the chemical elements in the geosphere and biosphere. The pioneering work of Vinogradov (1933), Vernadsky (1939), Frey-Wyssling (1935) and Steinberg (1938 remains to be exploited; more recently the subject has been discussed by Hutchinson (1943) and Shaw (1960), while Williams (1953) has pioneered research in the interaction of metal ions in biological systems.

The discipline of biogeochemistry provides an interface between inorganic and organic processes. Williams (1983) stated that 'Very many elements are not inorganic and much of organic chemistry is inorganic'. In considering any uniformity in the relative abundances of elements in living organisms (in relation to their inorganic source-materials) interest centres upon the extent to which an element entering a living organism is actually involved in biological processes, as distinct from one which is not recognized and is 'simply passing through'. The chemical composition of the atmosphere and seawater is directly involved in organic evolution with the eventual emergence of living organisms some 2000 million years ago. Holland (1984) considered that the constancy in the chemical evolution of the atmosphere and the oceans throughout geological time is a consequence of chemical and physical forces; Lovelock (1979) visualized the constancy as a homeostasis maintained for and by the biota. If constancy is controlled initially by geological forces, then uniformity in element abundances in biota would be expected and this would give rise to characteristic regional element abundance patterns. If biological control is dominant, then biochemical selection and adaptation would operate and should also result in uniformity in compositions (together with local differences as influenced by biological adaptation to the geochemical characteristics), but under the control of biochemical processes which need not be the same as those which are influenced by geochemical processes. The fundamental nature of associations between elements in the geosphere are influenced by the degree to which elements form stable compounds with silicon and oxygen (i.e. $SiO_4$ tetrahedra) and in the biosphere with hydrogen, carbon, oxygen and nitrogen.

In the approach that is adopted here, man is central to the discussion as the main cause for altering any natural balance between elements (local, regional or global). Agricultural practices are also a central theme as they affect major pathways whereby elements in non-natural associations are made directly available to food chains (e.g. fertilizers and use of the blanket-approach in providing element additives irrespective of whether or not they are really required). As a result of these processes, element fluxes are delivered to all components of the natural environment and are most noticeable in proximity to major centres of human populations, lakes, rivers, estuaries and coastal waters.

If it is possible to identify associations between elements which can be considered as natural, as distinct from those which can be attributed to man's influences,then a basis can be established in order to determine to what extent, if any, man has altered the composition and possibly the stability of natural ecosystems in relation to biological processes. In recent years the improvements which have been made in analytical techniques now show that the most significant alterations to element budgets and distributions in the natural environment are associated with terrestrial and nearshore ecosystems; the open oceans are still relatively uncontaminated.

In this paper attention is centered, using a multi-element approach, on man. Some of the element compositions of major systems such as those of rocks, soils, seawater, air and foods are considered; an extension of this approach which considers the transformation of elements to sub-cellular components of living organisms will be presented elsewhere (E. I. Hamilton, unpublished); here, only preliminary data for the abundance of elements in DNA from herring tissue is provided in order to illustrate that, at this level of investigation, some major differences between the abundance of elements in the geosphere and biospheres appear to exist.

Bowen (1966, 1979) noted a similarity between the chemical composition of marine and non-marine plants. John (1983) using data published by Hamilton & Minski (1972–73a) noted that the correlations between the abundance of elements in human blood and the atmospheric aerosol were much stronger than those between crustal rocks and seawater. Hamilton also noted that the shape of the abundance curve for the distribution of elements in crustal rocks, seawater and human blood were very similar, thus indicating some type of uniformity in the relative abundance of the elements. Li (1984) considered the matter further and proposed that the partition of elements between seawater and marine solids is basically controlled by the same surface-adsorption mechanisms as those which exist between soil pore-water and land solid-phases. He further separated, on the basis of linear correlation relationships, the elements in marine organisms, plants, red clay, seawater and soils into 'biophilic' and 'biophobic' groups.

The studies of Bowen, John and Li have all relied, in part, upon the use of average values for the concentrations of elements in a variety of materials as derived from literature surveys ignoring any influence caused by local geochemical abundance patterns, and not considering the quality of the data. A more closely-defined association between elements for a defined geography is considered here in order to determine whether or not any significant differences to the 'broad-brush' approach exist.

## METHODS

In 1964 the UK Ministry of Health and the Medical Research Council Radiological Protection Service at Sutton, Surrey, established an ultra-clean laboratory for the specific purpose of investigating the relative abundance of the elements in man and major components of his environment. Such data were required in order to understand the processes through which radionuclides in the environment are transferred to man. Following government reorganization, research was terminated in 1971, but some of the initial data were published by Hamilton *et al.* (1973a) and summarized by Hamilton (1979); these data are considered further here. Details of the methods and procedures used in the study have been described by Hamilton, Minski & Cleary (1973a, b) and Hamilton & Minski (1972), and Hamilton (1979). Briefly, the analytical techniques of spark-source mass-spectrometry, X-ray fluorescence, atomic absorption and neutron-activation analysis were used; all procedures were carried out under ultra-clean controlled conditions; a comprehensive programme of contamination control was used; and the various analytical methods were validated by analysis of a variety of Standard Reference Materials (SRMs) and cross-standardization of methods with defined chemical matrices. As the quality of the data and the origin of the source materials are defined geographically, realistic comparisons between data obtained for different types of samples by one laboratory can be made with some confidence. The study involved a multi-element approach (*c*. 30 elements) but the accuracy of all data are not comparable, especially for those elements which occur at very low levels even when the precision of the analyses is high, hence a comparison of levels of elements in different samples is considered to be valid.

### *Data quality*

When the data were originally obtained there were very few similar data for purposes of comparison. During the last decade there have been considerable improvements in both the quality and quantity of data. But for many elements the quality is still poor, which is important to this study as many of the potentially useful pairs of elements involve a consideration of major and trace elements, e.g. K–Rb, K–Cs, Si–Ge, Al–Ga, Zr–Hf and Ba-rare-earth-elements. The following examples illustrate the quality of the data used in this study.

*Coal.* The concentration of some elements in coal which is representative of the UK coalfields is given in Table 1. The Sutton data (i.e. Hamilton 1974) compare well with the more limited data obtained by two other laboratories.

*Human liver and kidney.* Unlike coal, which only involved sampling a powder from a bottle, the analytical procedures for these samples involve excision from the corpse, preparation of a representative sample based upon the biological structure of the organ,

TABLE 1. Comparative data for the concentration of various elements in coal representative of all major UK coalfields. Mean value ± standard deviation.

| Element | Concentration (mg $kg^{-1}$) | | |
|---|---|---|---|
| | Hamilton (1974) | Goetz *et al.* (1979) | Salmon *et al.* (1984) |
| K | 4000 ± 3000 | – | 6000 |
| V | 41·5 ± 23·4 | 47·9 ± 26·8 | — |
| Cr | 35·2 ± 4·2 | 31·9 ± 7·9 | — |
| Mn | 100·4 ± 43·6 | 129·3 ± 51·6 | — |
| Co | 9·3 ± 3·2 | 9·9 ± 4·1 | — |
| As | 18·5 ± 14·3 | 16·8 ± 15·9 | — |
| Mo | 3·0 ± 1·8 | 3·1 ± 1·5 | — |
| Cd | 0·2 ± 0·1 | 0·3 ± 0·2 | — |
| La | 15·2 ± 8·4 | 18·9 ± 7·4 | — |
| Pb | 15·8 ± 7·4 | 21·7 ± 10·2 | — |
| Th | 4·3 ± 3·7 | 3·8 ± 1·1 | 3·2 |
| U | 1·4 ± 0·5 | 2·0 ± 0·7 | 1·2 |
| *n* = | 120 | 13 | 20 |

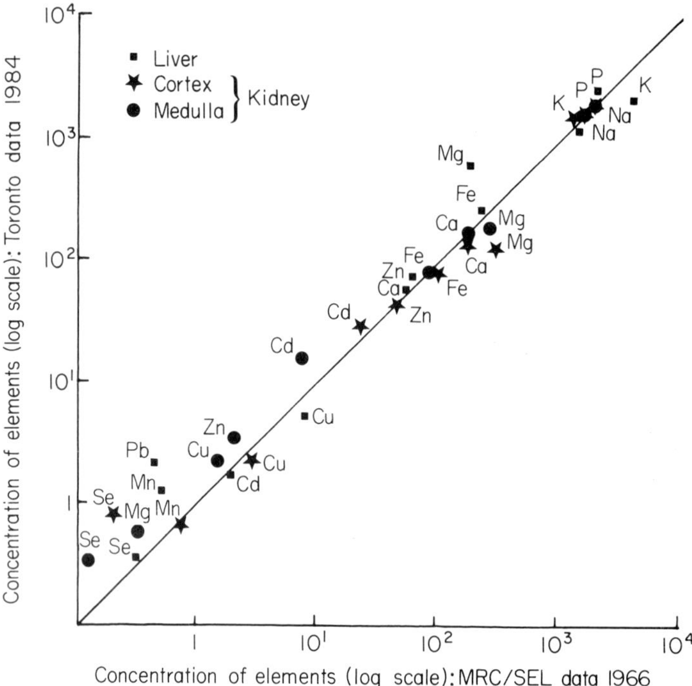


FIG. 2. A comparison between the concentration of various elements in human liver and kidney (cortex and medulla) obtained by Subramanian *et al.* (1985) for samples from Toronto, Canada, and those of Hamilton *et al.* (1973b) for the UK.

freeze-drying, homogenization and other processing as required by a particular method of analysis. For kidney there is an additional need to remove the external capsule membrane, extraneous urethra and to dissect the organ into medulla and cortex. All of these procedures are susceptible to various sources of contamination. The analytical data obtained at Sutton are compared with those obtained by Subramanian, Meranger & Burnett (1985), who used inductive-coupled plasma-spectrometry, on tissues from unembalmed cadavers taken from normal subjects (residents of Toronto, Canada) as identified by the absence of histopathological changes. Both laboratories checked the quality of their measurements against the US National Bureau of Standards (NBS) SRM 1577/bovine liver and the Sutton data were also checked against Bowen's kale (Bowen 1973). If some allowance is made for the differences in the geochemical environment of Sutton and Toronto, the similarity of the values obtained for all three samples is remarkable. These values are supported by comparison with literature values for the same organs from many countries. If the quality of the data is accepted, then it is surprising that, in selecting two organs which are associated with the excretion of elements, much larger differences are not observed. Although not considered here, the outlying values (Pb and Mn in liver; Zn, Cu, Mn, Se in kidney medulla and Se and Mg in kidney cortex) are potentially useful indicators for local element-uptake patterns in relation to regional patterns of morbidity.

*Aerosol.* The abundance of the elements in the urban Sutton aerosol is compared in

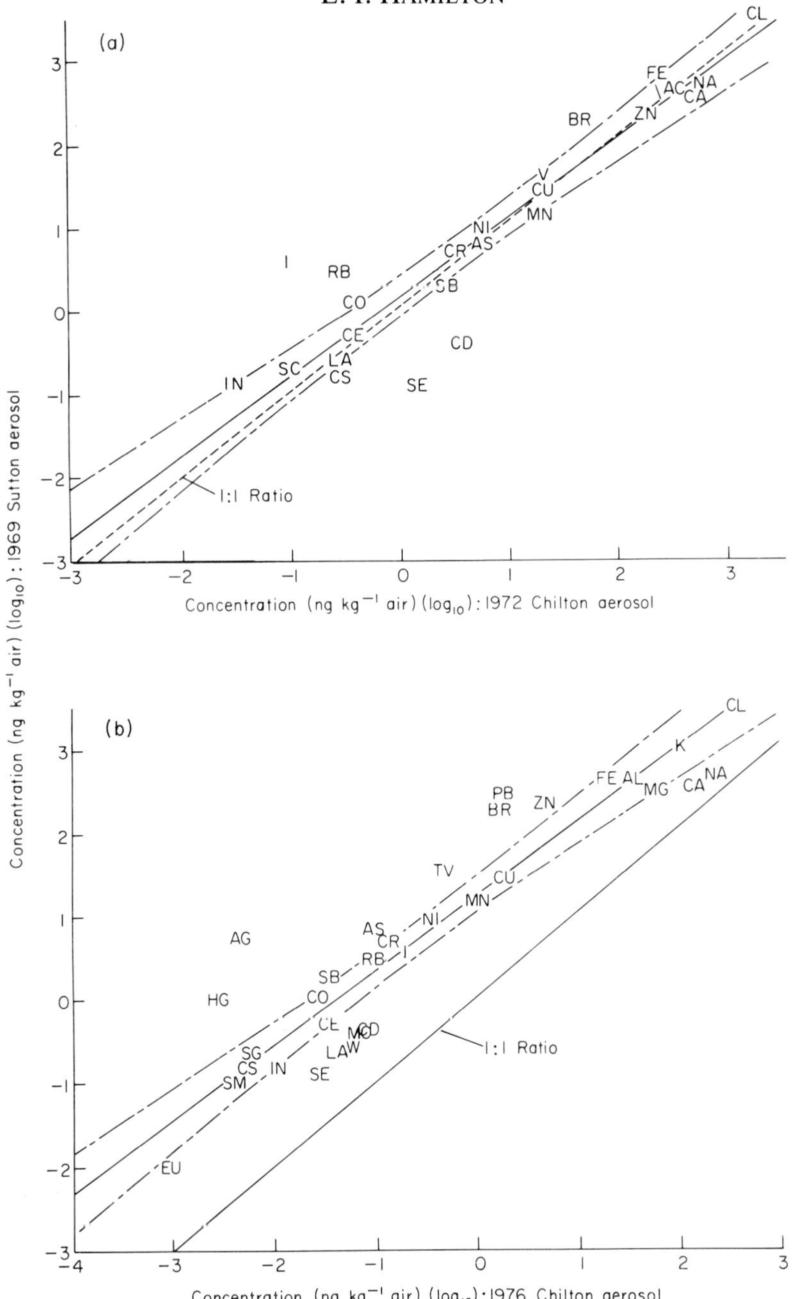


FIG. 3. A comparison between the concentration of various elements in air particulate debris sampled at Sutton (Hamilton 1979) and Chilton (Salmon *et al.* 1978) for the years 1969, 1972 and 1976.

Fig. 3a and b with that for a rural site at Chilton some 60 miles to the north. Data for Sutton for 1969 are compared with those for Chilton in 1972 (Fig 3a) and Sutton for 1969 with Chilton for 1976 (Fig. 3b) which may be considered further in relation to retrospective trend analysis of the concentration of elements in UK air particulate-material (1957–74) as described by Salmon *et al.* (1978). The observed high correlation between the concentration of elements in the two different localities reflects considerable similarities in the overall element composition of the global aerosol as observed by Rahn (1976).

In the Sutton work the US Geological Survey Standards, G1, W1, G2, AGV-1, DTS-1, NBS 69a Bauxite, 1b Limestone (argillaceous), 70a Feldspar (potash), and 97a Clay and the Atomic Energy Agency, Vienna interlaboratory comparison samples A2–Animal blood and A3/1 Calcined Bone were also used.

From the evidence presented above the overall quality of the Sutton data is considered to be sufficient for the purpose of comparing the abundance of elements in a variety of different samples. Other data used in this study, which were obtained from the literature are for well-authenticated materials, e.g. the average composition of seawater, oceanic pelagic clay and the solar abundances of the elements.

### *Selection of samples and treatment of data*

For the purpose of this paper attention is directed towards only a part of the data bank. Where appropriate, for purposes of illustrating certain features, the data are presented either in their original form, or as an alternative such as ash weight; for example, it is more appropriate to consider the element composition of seawater v. crustal abundances in terms of ashed weight rather than 'wet' seawater; the aerosol data are treated as dry weight, but the original samples are known to contain between 15 and 30% organic matter and hence an appropriate correction needs to be applied when relating the relative abundances of elements in air to those in different types of sample.

For this study, multi-element comparisons between different samples are made by linear regression correlation analysis using logarithmically-transformed data, and therefore interest is focused only upon a slope of one for the regression line or the extent by which the slope differs from unity. The data have also been subjected to principal-component analysis.

## RESULTS

The equations for the various regressions which have been examined in this study are given in Table 2. The various pairs of samples which have been selected are considered to be logical choices in an examination of element transfer patterns, albeit that some are rather extreme. The manner in which the correlation coefficient ($r$) is related (sample suite to sample suite) is illustrated between the abundances of the elements in crustal rocks, oceanic pelagic clay and various biological samples and between seawater and biological tissues is poor; the lowest correlation coefficient is observed for seawater v. herring DNA. The intra-correlation coefficients for either biological or geochemical samples are high, e.g. crustal abundances v. Esk sediments and seal blood v. human blood.

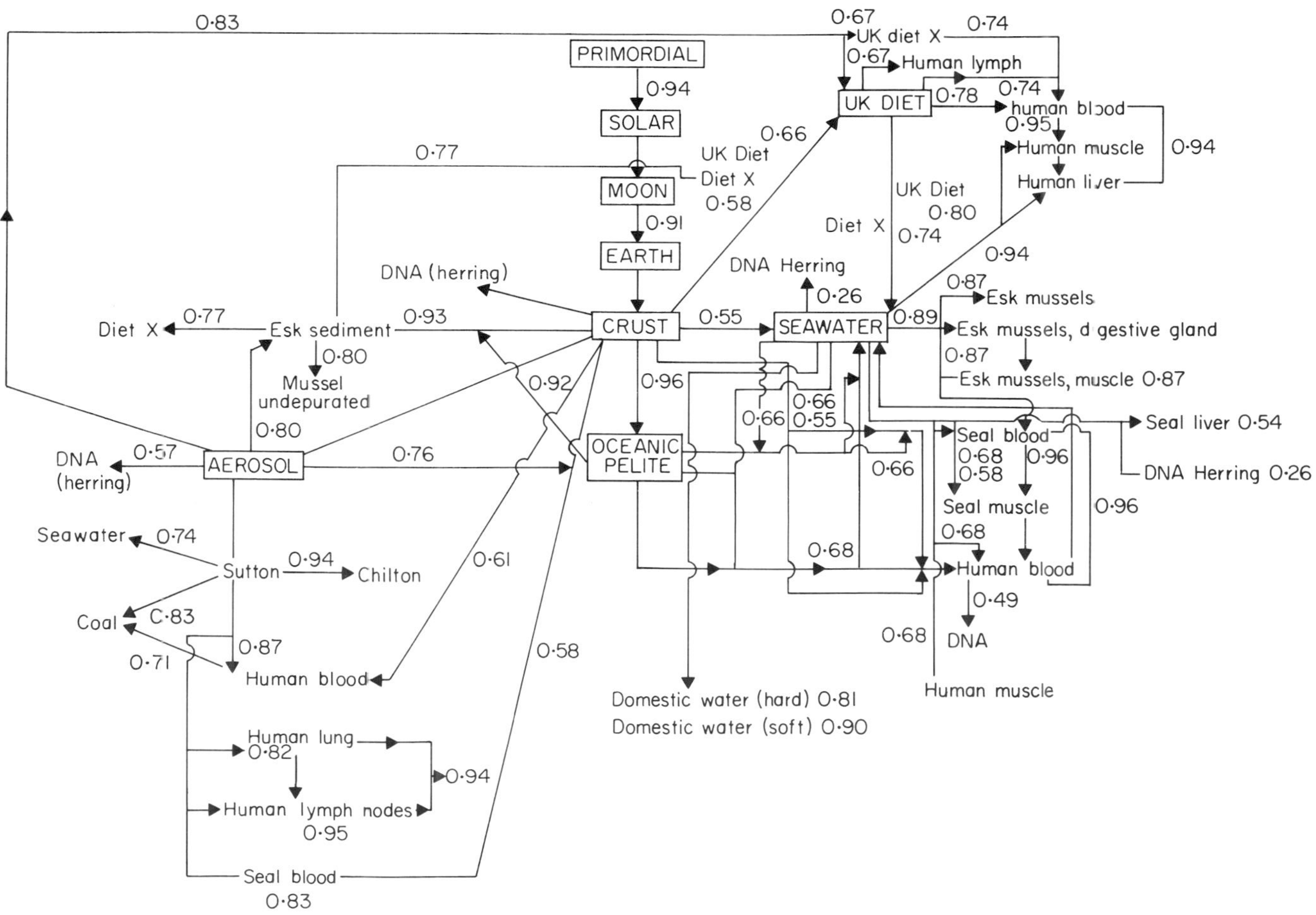


Fig. 4. Correlation coefficients between various biotic and abiotic systems (see Table 2 for reference sources).

TABLE 2. Least squares linear regression analysis for the abundance of elements in various groupings. All data are logarithmically transformed. $n$ = number of elements, $R$ = percentage variability accounted for by the regression.

| Components of regression line | ($n$) | Regression line | Corr coeff. ($r$) | $R$ | Foot notes |
|---|---|---|---|---|---|
| Primordial abundances v. Solar abundances | 57 | $Y = -0{\cdot}08 \pm 0{\cdot}07 + (0{\cdot}97 \pm 0{\cdot}07)X$ | 0·98 | 96·4 | 1 |
| Earth v. Moon | 33 | $Y = 0{\cdot}31 \pm 0{\cdot}21 + (0{\cdot}81 \pm 0{\cdot}06)X$ | 0·89 | 78·6 | 2 |
| Crustal abundance v. Oceanic pelagic clay | 60 | $Y = 0{\cdot}25 \pm 0{\cdot}08 + (1{\cdot}05 \pm 0{\cdot}04)X$ | 0·96 | 92·6 | 3 |
| Crustal abundances v. Seawater | 60 | $Y = -4{\cdot}76 \pm 0{\cdot}42 + (1{\cdot}01 \pm 0{\cdot}20)X$ | 0·55 | 30·1 | 4 |
| Crustal abundances v. Esk (mud) sediment | 39 | $Y = 0{\cdot}78 \pm 0{\cdot}09 + (0{\cdot}66 \pm 0{\cdot}04)X$ | 0·93 | 86·0 | 5 |
| Crustal abundances v. Seal blood | 35 | $Y = -1{\cdot}60 \pm 0{\cdot}40 + (0{\cdot}61 \pm 0{\cdot}16)X$ | 0·55 | 30·3 | 6 |
| Crustal abundances v. Human blood | 33 | $Y = -1{\cdot}72 \pm 0{\cdot}39 + (0{\cdot}70 \pm 0{\cdot}17)X$ | 0·61 | 36·9 | 7 |
| Crustal abundances v. UK Total diet | 39 | $Y = 2{\cdot}01 \pm 0{\cdot}27 + (0{\cdot}63 \pm 0{\cdot}12)X$ | 0·66 | 44·0 | 8 |
| Crustal abundances v. UK diet GI transfer | 33 | $Y = 1{\cdot}04 \pm 0{\cdot}46 + (0{\cdot}69 \pm 0{\cdot}19)X$ | 0·54 | 28·9 | 9 |
| Crustal abundances v. DNA (herring) | 29 | $Y = -0{\cdot}47 \pm 0{\cdot}24 + (0{\cdot}33 \pm 0{\cdot}11)X$ | 0·51 | 26·3 | 10 |
| Oceanic pelite v. Esk (Mud) sediment | 39 | $Y = 0{\cdot}65 \pm 0{\cdot}11 + (0{\cdot}61 \pm 0{\cdot}04)X$ | 0·92 | 84·7 | 4,3 |
| Oceanic pelagic clay v. Seawater | 61 | $Y = -5{\cdot}22 \pm 0{\cdot}40 + (1{\cdot}14 \pm 0{\cdot}17)X$ | 0·66 | 43·9 | 3 |
| Oceanic pelagic clay v. Seal blood | 36 | $Y = -2{\cdot}05 \pm 0{\cdot}40 + (0{\cdot}71 \pm 0{\cdot}13)X$ | 0·66 | 43·8 | 3,7 |
| Oceanic pelagic clay v. Human blood | 34 | $Y = -2{\cdot}08 \pm 0{\cdot}41 + (0{\cdot}75 \pm 0{\cdot}14)X$ | 0·68 | 43·3 | 3,7 |
| Esk (mud) sediment v. Esk mussel/digestive gland | 25 | $Y = -1{\cdot}60 \pm 0{\cdot}58 + (0{\cdot}25 \pm 0{\cdot}24)X$ | 0·80 | 63·4 | 11 |
| Esk (mud) sediment v. Esk mussel/muscle | 21 | $Y = -1{\cdot}57 \pm 0{\cdot}55 + (1{\cdot}37 \pm 0{\cdot}23)X$ | 0·80 | 64·3 | 12 |
| Esk (mud) sediment v. UK Total diet | 28 | $Y = 1{\cdot}20 \pm 0{\cdot}39 + (1{\cdot}00 \pm 0{\cdot}16)X$ | 0·77 | 58·5 | 5,8 |
| Esk (mud) sediment v. UK Diet GI transfer | 25 | $Y = 0{\cdot}35 \pm 0{\cdot}62 + (1{\cdot}07 \pm 0{\cdot}26)X$ | 0·66 | 42·8 | 4,8 |
| Seawater v. Esk mussels/digestive gland | 21 | $Y = 2{\cdot}30 \pm 0{\cdot}23 + (0{\cdot}23 \pm 0{\cdot}06)X$ | 0·89 | 79·5 | 3,11 |
| Seawater v. Esk mussels/muscle tissue | 21 | $Y = 2{\cdot}26 \pm 0{\cdot}25 + (0{\cdot}46 \pm 0{\cdot}06)X$ | 0·87 | 75·7 | 3,12 |
| Seawater v. Seal blood | 36 | $Y = 0{\cdot}50 \pm 0{\cdot}28 + (0{\cdot}43 \pm 0{\cdot}08)X$ | 0·68 | 46·6 | 3,6 |
| Seawater v. Seal muscle | 37 | $Y = 0{\cdot}47 \pm 0{\cdot}31 + (0{\cdot}37 \pm 0{\cdot}09)X$ | 0·59 | 34·2 | 3,6 |
| Seawater v. Seal liver | 36 | $Y = 0{\cdot}77 \pm 0{\cdot}33 + (0{\cdot}34 \pm 0{\cdot}09)X$ | 0·53 | 28·7 | 3,6 |
| Seawater v. Human blood | 34 | $Y = 0{\cdot}50 \pm 0{\cdot}30 + (0{\cdot}46 \pm 0{\cdot}08)X$ | 0·72 | 51·7 | 3,15 |
| Seawater v. Human muscle | 31 | $Y = 0{\cdot}49 \pm 0{\cdot}34 + (0{\cdot}48 \pm 0{\cdot}10)X$ | 0·68 | 46·1 | 3,15 |
| Seawater v. DNA (herring) | 29 | $Y = 0{\cdot}44 \pm 0{\cdot}37 + (0{\cdot}14 \pm 0{\cdot}10)X$ | 0·26 | 7·0 | 3,15 |
| Seawater v. UK diet GI transfer | 34 | $Y = 3{\cdot}57 \pm 0{\cdot}27 + (0{\cdot}67 \pm 0{\cdot}07)X$ | 0·85 | 72·2 | 3,8 |
| UK Diet v. Crustal abundances | 39 | $Y = -0{\cdot}56 \pm 0{\cdot}43 + (0{\cdot}69 \pm 0{\cdot}13)X$ | 0·66 | 44·0 | 8,3 |
| UK Diet v. Seawater | 40 | $Y = -6{\cdot}30 \pm 0{\cdot}59 + (1{\cdot}36 \pm 0{\cdot}17)X$ | 0·80 | 63·3 | 8,3 |
| UK Diet v. Human blood | 33 | $Y = -3{\cdot}88 \pm 0{\cdot}24 + (1{\cdot}08 \pm 0{\cdot}06)X$ | 0·95 | 90·2 | 8,15 |
| UK Diet GI transfer v. Human blood | 30 | $Y = -1{\cdot}70 \pm 0{\cdot}36 + (0{\cdot}60 \pm 0{\cdot}10)X$ | 0·74 | 55·0 | 8,15 |
| UK Diet v. UK GI transfer | 34 | $Y = -1{\cdot}23 \pm 0{\cdot}52 + (1{\cdot}10 \pm 0{\cdot}14)X$ | 0·80 | 64·1 | 8 |
| Aerosol Sutton (1969) v. Aerosol Chilton (1976) | 35 | $Y = -1{\cdot}25 \pm 0{\cdot}11 + (0{\cdot}97 \pm 0{\cdot}11)X$ | 0·93 | 86·1 | 13 |
| Sutton aerosol v. Oceanic pelagic clay | 59 | $Y = 1{\cdot}40 \pm 0{\cdot}16 + (0{\cdot}86 \pm 0{\cdot}10)X$ | 0·76 | 58·0 | 13,3 |
| Sutton aerosol v. Esk (mud) sediment | 39 | $Y = -1{\cdot}32 \pm 0{\cdot}14 + (0{\cdot}65 \pm 0{\cdot}08)X$ | 0·80 | 63·7 | 13,4 |
| Sutton aerosol v. Seawater | 59 | $Y = -3{\cdot}73 \pm 0{\cdot}28 + (1{\cdot}43 \pm 0{\cdot}17)X$ | 0·74 | 55·2 | 13,3 |
| Sutton aerosol v. Human lung | 33 | $Y = -1{\cdot}12 \pm 0{\cdot}24 + (1{\cdot}05 \pm 0{\cdot}14)X$ | 0·81 | 65·4 | 13,1 |
| Sutton aerosol v. Human lymph nodes | 35 | $Y = -0{\cdot}84 \pm 0{\cdot}20 + (1{\cdot}00 \pm 0{\cdot}12)X$ | 0·83 | 69·3 | 13,1 |
| Sutton aerosol v. Human blood | 34 | $Y = -1{\cdot}78 \pm 0{\cdot}22 + (1{\cdot}30 \pm 0{\cdot}13)X$ | 0·87 | 74·9 | 13,1 |
| Sutton aerosol v. Seal blood | 36 | $Y = -1{\cdot}66 \pm 0{\cdot}23 + (1{\cdot}77 \pm 0{\cdot}13)X$ | 0·83 | 69·4 | 13,6 |
| Sutton aerosol v. Coal (UK) | 46 | $Y = 0{\cdot}96 \pm 0{\cdot}13 + (0{\cdot}75 \pm 0{\cdot}08)X$ | 0·83 | 68·4 | 14 |
| Sutton aerosol v. DNA (herring) | 29 | $Y = -0{\cdot}46 \pm 0{\cdot}22 + (0{\cdot}59 \pm 0{\cdot}16)X$ | 0·57 | 32·4 | 13,1 |
| Sutton aerosol v. UK Diet | 40 | $Y = 2{\cdot}12 \pm 0{\cdot}19 + (1{\cdot}04 \pm 0{\cdot}11)X$ | 0·83 | 69·3 | 13,8 |
| Sutton aerosol v. UK Diet GI transfer | 34 | $Y = 1{\cdot}10 \pm 0{\cdot}39 + (1{\cdot}12 \pm 0{\cdot}22)X$ | 0·67 | 44·4 | 13,8 |
| Human lymph node v. UK Diet | 32 | $Y = 0{\cdot}22 \pm 0{\cdot}10 + (0{\cdot}89 \pm 0{\cdot}06)X$ | 0·94 | 89·1 | 15,8 |

| | | | | | |
|---|---|---|---|---|---|
| Human lung v. Human lymph node | 32 | $Y = 0{\cdot}22 \pm 0{\cdot}10 + (0{\cdot}89 \pm 0{\cdot}10)\,X$ | 0·94 | 89·1 | 15 |
| Whole human body v. Crustal abundances | 33 | $Y = 1.94 \pm 0{\cdot}25 + (0{\cdot}52 \pm 0{\cdot}12)\,X$ | 0·61 | 36·9 | 15,3 |
| Whole human blood v. Seawater | 34 | $Y = -1{\cdot}55 \pm 0{\cdot}40 + (1{\cdot}12 \pm 0{\cdot}19)\,X$ | 0·72 | 51·7 | 15,3 |
| Whole human blood v. Seal blood | 28 | $Y = -0{\cdot}01 \pm 0{\cdot}10 + (0{\cdot}93 \pm 0{\cdot}05)\,X$ | 0·96 | 92·7 | 15 |
| Whole human blood v. Human liver | 27 | $Y = 0{\cdot}25 \pm 0{\cdot}14 + (0{\cdot}94 \pm 0{\cdot}07)\,X$ | 0·94 | 88·3 | 15 |
| Whole human blood v. Human muscle | 26 | $Y = -0{\cdot}09 \pm 0{\cdot}13 + (0{\cdot}97 \pm 0{\cdot}07)\,X$ | 0·95 | 90·1 | 15 |
| Whole human blood v. UK Diet | 33 | $Y = 3{\cdot}57 \pm 0{\cdot}10 + (0{\cdot}84 \pm 0{\cdot}05)\,X$ | 0·95 | 90·2 | 15,8 |
| Whole human blood v. Diet GI transfer | 30 | $Y = 2{\cdot}64 \pm 0{\cdot}32 + (0{\cdot}92 \pm 0{\cdot}16)\,X$ | 0·74 | 55·0 | 15,8 |
| Whole human blood v. Coal (UK) | 27 | $Y = 1{\cdot}94 \pm 0{\cdot}20 + (0{\cdot}52 \pm 0{\cdot}10)\,X$ | 0·74 | 54·5 | 15 |
| Whole human blood v. DNA (herring) | 20 | $Y = 0{\cdot}44 \pm 0{\cdot}31 + (0{\cdot}44 \pm 0{\cdot}18)\,X$ | 0·49 | 24·1 | 16 |
| UK domestic water: hard (am flow) v. Seawater | 34 | $Y = -4{\cdot}19 \pm 0{\cdot}45 + (1{\cdot}42 \pm 0{\cdot}19)\,X$ | 0·80 | 64·6 | 16,3 |
| UK domestic water: hard (am) v. UK domestic water: soft (am) | 30 | $Y = -0{\cdot}43 \pm 0{\cdot}25 + (1{\cdot}00 \pm 0{\cdot}10)\,X$ | 0·89 | 79·1 | 16 |
| UK domestic water: hard (am) v. Diet GI transfer | 25 | $Y = 0{\cdot}49 \pm 0{\cdot}50 + (1{\cdot}10 \pm 0{\cdot}19)\,X$ | 0·77 | 59·0 | 16,8 |

FOOTNOTES

1. Primordial abundances. Represented by carbonaceous chondrites which may reflect the initial composition of all meteorites and the unfractionated sample of the non-volatile component of primordial matter; data from Cameron (1973) and for boron from Curtis *et al.* (1980). Solar abundances determined from spectral analysis of the solar atmosphere; data from Ross & Aller (1976).
2. Abundance of elements in the Earth according to estimates by Ganapathy & Anders (1974). Composition of the Moon data from Ganapathy & Anders (1974).
3. Abundance of the elements in the Earths crust from Taylor (1964).
4. Oceanic pelagic clay and seawater composition from Li (1982).
5. Fine-grained sediment of the Esk estuary, Cumbria, UK; E. I. Hamilton (unpublished data).
6. Seals from NW Scotland and the Farne Islands, North Sea; Hamilton (1976).
7. Whole human blood representative for the population of the UK, 1969; Hamilton *et al.* (1973b).
8. UK total diet obtained from a national survey for dietary intake of elements; data from Hamilton & Minski (1972–73b).
9. Data corrected for element transfer across the gastro-intestinal tract of man using data provided by ICRP (1975).
10. Data for DNA herring tissues obtained from a commercial product; Hamilton (1979).
11. Data obtained for digestive gland of the mussel (*Mytilus edulis* L.) from the Esk estuary, Cumbria, UK. Data have been corrected for residual sediment burden; Hamilton (unpublished data).
12. Muscle tissue from Esk animals corrected for detrital sediment contamination.
13. Airborne particulate debris collected on filters. The Sutton data are for an urban area (see Hamilton 1974) and those for Chilton from a rural community (see Cawse 1975 and Peirson *et al.* 1974).
14. Samples of coal representative for the total UK consumption for generation of electric power (see Salmon *et al.* 1984).
15. Samples of human tissue (Sutton, Surrey, UK) from healthy individuals (see Hamilton 1973, 1979).
16. Hard and soft water sampled from domestic taps for the first flow in the morning (see Hamilton *et al.* 1979b).

The correlations between UK diet (both forms, see footnotes to Table 2) and seawater and human blood are also high. The main feature of Table 2 is the consistent high correlation (except for herring DNA) between the Sutton aerosol and most other parameters. A poor correlation between geochemical indices and those of a biological nature is perhaps to be expected; the poor correlation between seawater and most biological samples (except human blood) is a little unexpected if the common belief that life originated from the seas, and that human plasma can be considered as 'relict' seawater is correct. An exhaustive treatment of the data given in Fig. 4 and Table 2 is not given here, but the following examples identify some of the more interesting features of the data.

(a) The inter-element relationships for solar elemental abundances, their accretion in the form of carbonaceous chondrites and subsequent fractionization to ultimately give rise to the crust of the Earth, together with the survival of the most resistant phases of crustal weathering, to give rise to the oceanic pelagic clay, is illustrated in Figs 5 and 6. Apart from

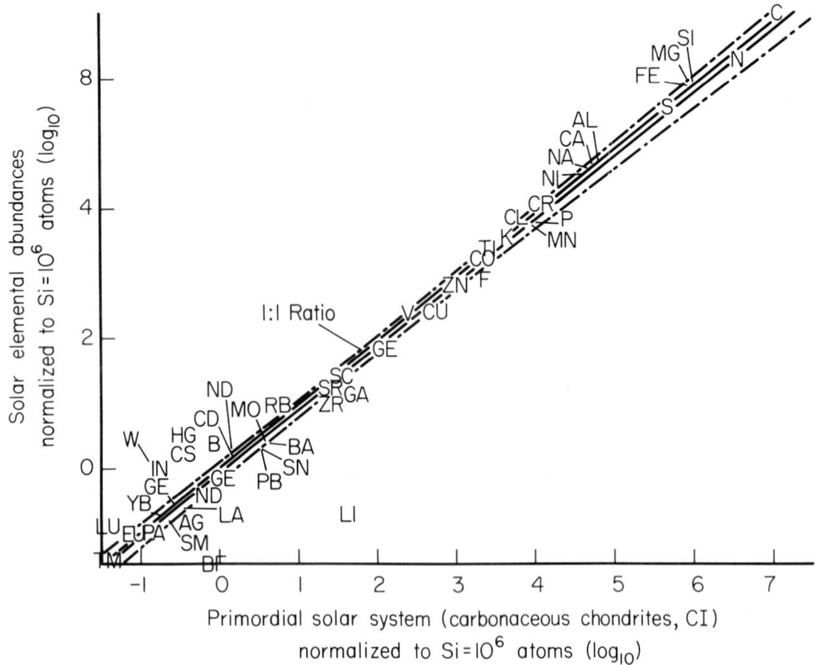


FIG. 5. Solar element abundances (Ross & Aller 1976) v. the composition of the Primordial Solar System (Cameron 1973; Curtis *et al.* 1980).

the lithium anomaly (see Burbidge *et al.* (1957) and Clayton (1968) for an explanation) the overall correlation coefficients for the regressions are high. In Fig. 6 an enrichment in soluble elements present in oceanic pelagic clay is observed. Fig. 7 indicates little correlation between the abundances of the elements in seawater (ash weight) with those of the Earth's crust; it should be noted that an examination of the regressions between soluble elements is high, and that the ratio is *c.* 1. Fig. 8 compares whole human blood (ashed) with continental crust; the correlation coefficient is low, but most of the conservative elements show a regression with a slope of *c.* 1. The correlation between elements in whole human blood (ashed) and seawater is poor, but as illustrated in Fig. 9 the regression line for conservative elements approaches a slope of *c.* 1. Fig. 10 illustrates a significant correlation between human blood (ashed) and the Sutton aerosol (dry weight) which is maintained in Fig. 11 for lung (ashed) v. Sutton aerosol and in Fig. 12 for lung (wet weight) v. lymph nodes (wet weight). A proportion of the particulate debris which enters the lung of man is removed to the lymph nodes through the action of macrophages which provide a pathway

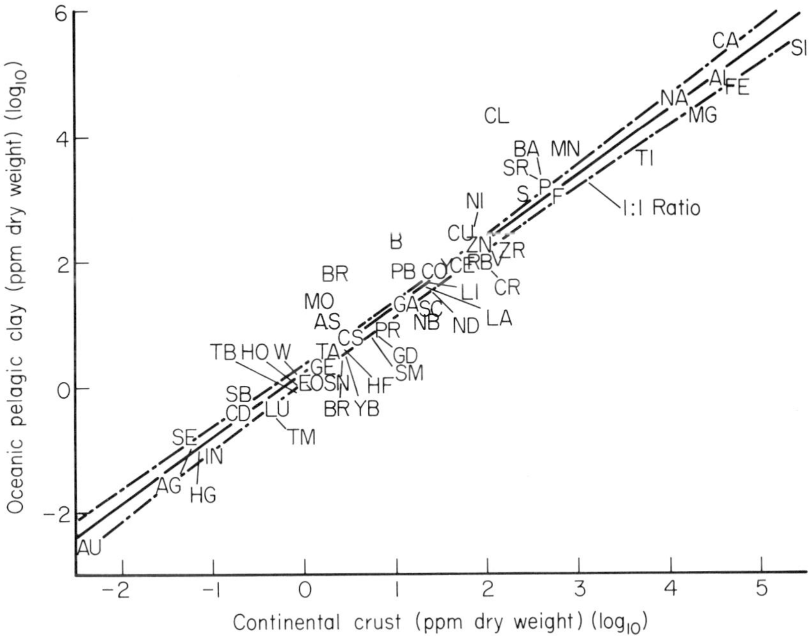


FIG. 6. Abundances of the elements in oceanic pelagic clay (Li 1982) v. continental crust (Taylor 1964).

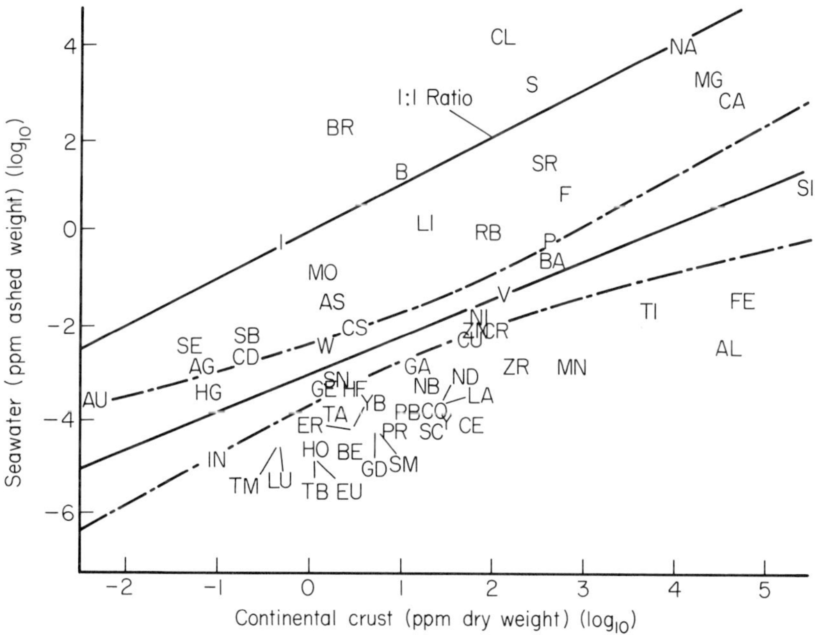


FIG. 7. Abundances of the elements in oceanic seawater (Li 1982) v. continental crust (Taylor 1964).

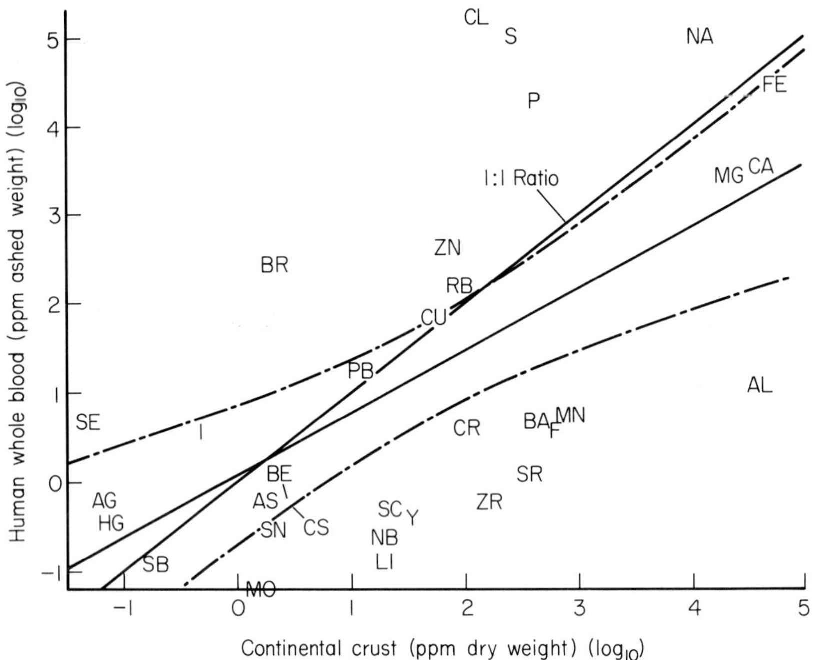


FIG. 8. Abundance of the elements in human whole blood (Hamilton *et al.* 1973b) v. continental crust (Taylor 1964).

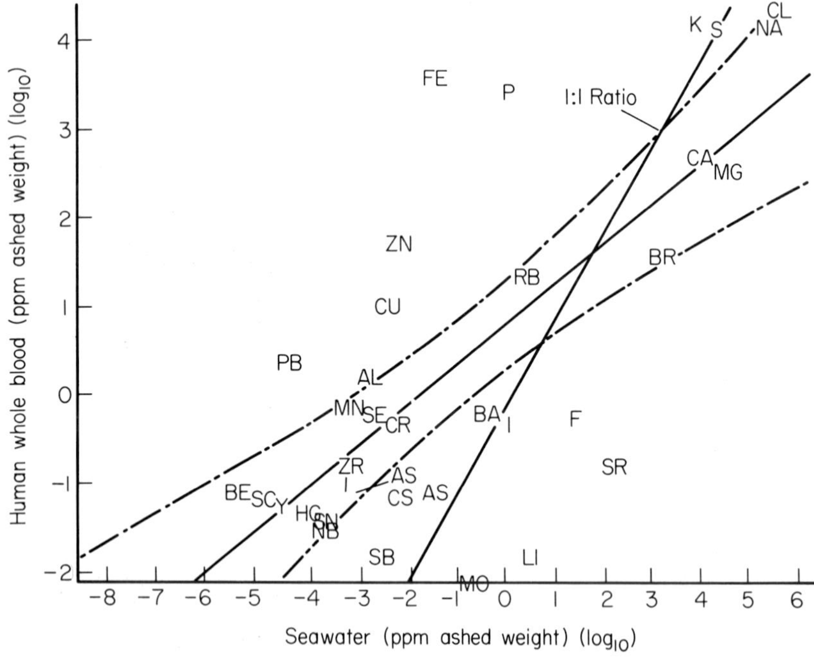


FIG. 9. Abundance of the elements in human whole blood (Hamilton *et al.* 1973b) v. seawater (Li 1982).

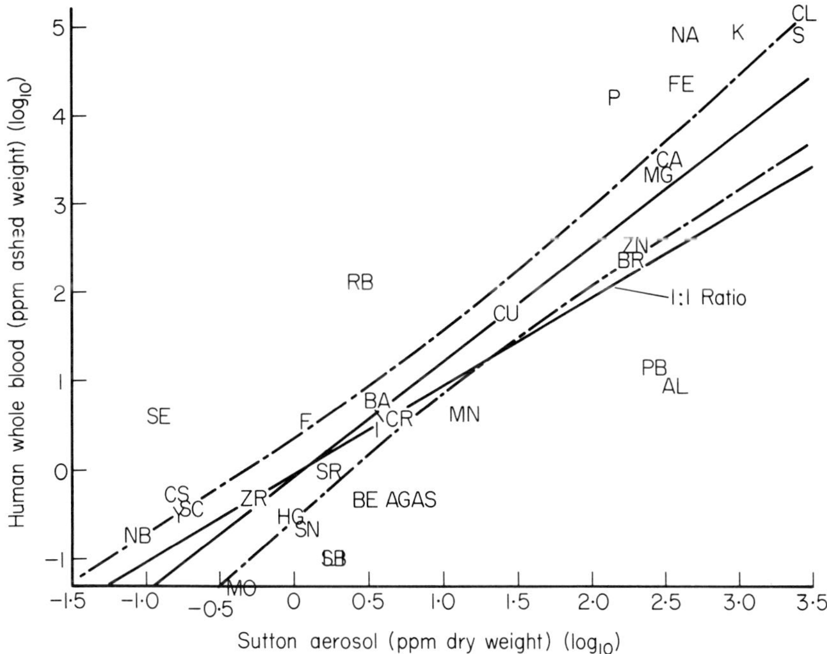


FIG. 10. Abundance of the elements in human whole blood (Hamilton *et al.* 1973b) v. the Sutton aerosol (Hamilton 1979).

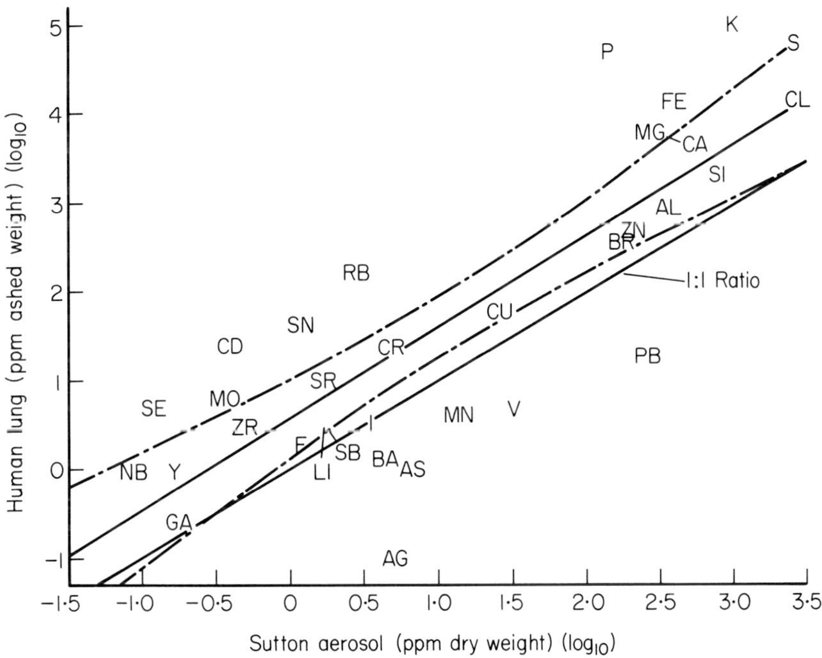


FIG. 11. Abundance of the elements in human lung (Hamilton *et al.* 1973b) v. the Sutton aerosol (Hamilton 1979).

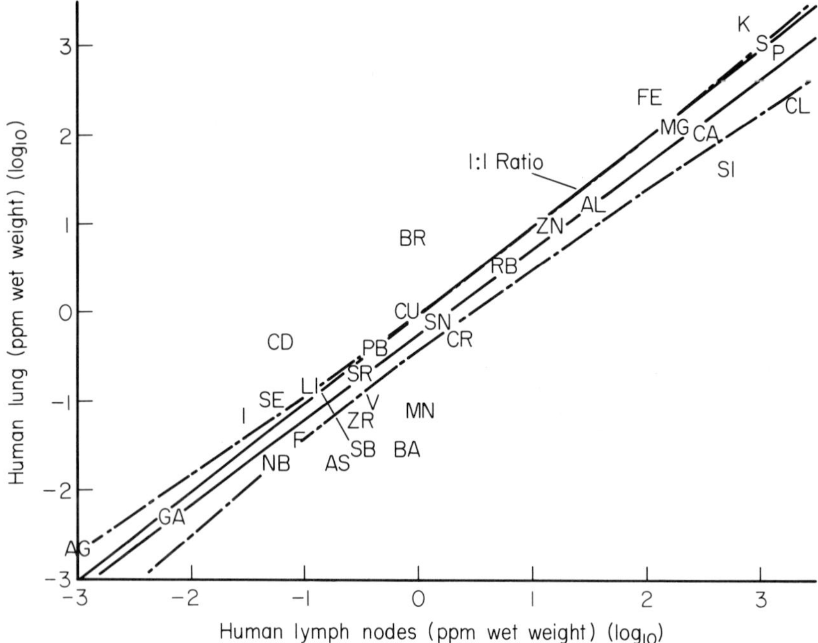


FIG. 12. Abundance of the elements in human lung (Hamilton *et al.* 1973b) v. human lymph nodes (Hamilton *et al.* 1973b).

for entry of geochemically-related elements, i.e. particulate material. Finally, in Fig. 13 the high correlation between the blood of two different mammals, seal and man is illustrated; closer examination of the data can be used to illustrate well known physiological differences between the blood of both species.

Apart from the linear correlation coefficients, an examination of the relative position of elements along a regression line is also of interest but is not discussed here.

An analysis of the data set by principal component analysis is illustrated in Fig. 14 which supports, with the exception of DNA from the herring, the high correlation between the various samples. After the first component is removed the residuals, as illustrated in Fig. 15, show various sample groupings, e.g. sediments, mammalian tissues and invertebrate tissues.

## DISCUSSION

The relative enrichment and depletion of the elements of the Periodic Table for the geochemical and biological pairs of samples considered here are found to be of a similar magnitude. In general they are weakly related to the chemical composition of seawater and more strongly correlated with the composition of the atmospheric aerosol as described by John (1983).

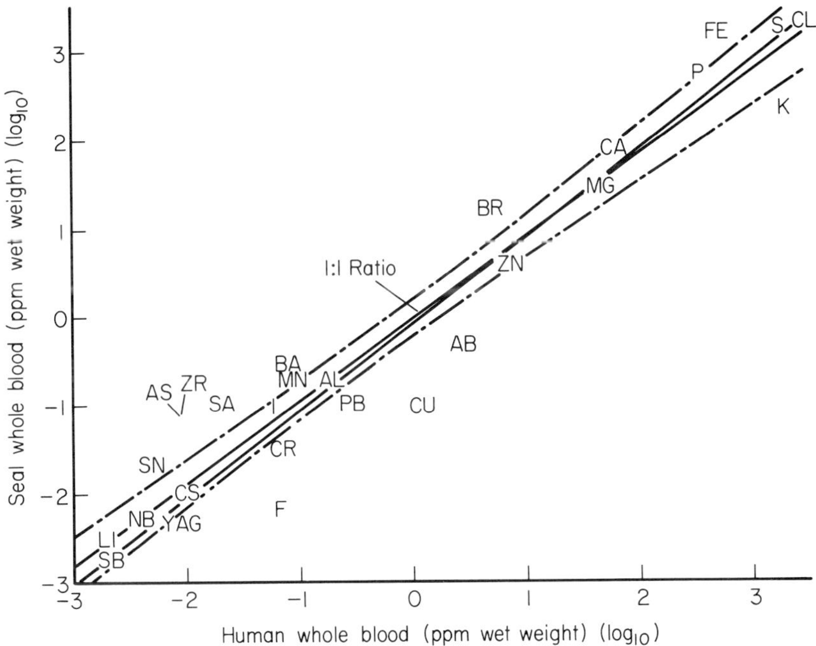


FIG. 13. Abundance of the elements in whole seal blood (Hamilton 1979) v. human whole blood (Hamilton *et al.* 1973b).

A comparison between the correlation coefficients for human blood v. crustal rocks, seawater, and aerosol obtained by John (1983), and those determined in this study are given in Table 3; the relative order of correlation for both sets of data is the same, but the data used in this paper are of more recent origin than those used by John. Li (1984), pursuing a similar line of investigation concluded that the partition of elements between seawater and marine solid phases (organisms and pelagic clays) and soil pore-water and land solid-phases (plants and soils) is basically controlled by the same surface adsorption mechanisms. Li also noted that the elements in living organisms are separated into 'biophile' and 'biophobe' groups as illustrated in Fig. 16 for the elemental concentration ratio for algae over seawater.

Using more recent data, but for a more limited data base (Eisler 1981), the concentration of elements in algae shows the same type of separation between the elements as illustrated in Fig. 17. The 'biophobe' group of Li includes several elements which are essential for living organisms, e.g. manganese, cobalt, iron, nickel, chromium, vanadium, molybdenum and fluorine. With a few exceptions, the 'biophilic' group of elements share a common property, namely their presence in saline systems in the dissolved state. Li (1984) chose algae as a reference material 'simply because it represents a primitive life form'. However, marine algae are very advanced forms of life which have adapted to a special environment. Data for the algae are likely to include contamination by anthropogenic inputs and thus the separation into 'biophobic' and 'biophilic' groups may

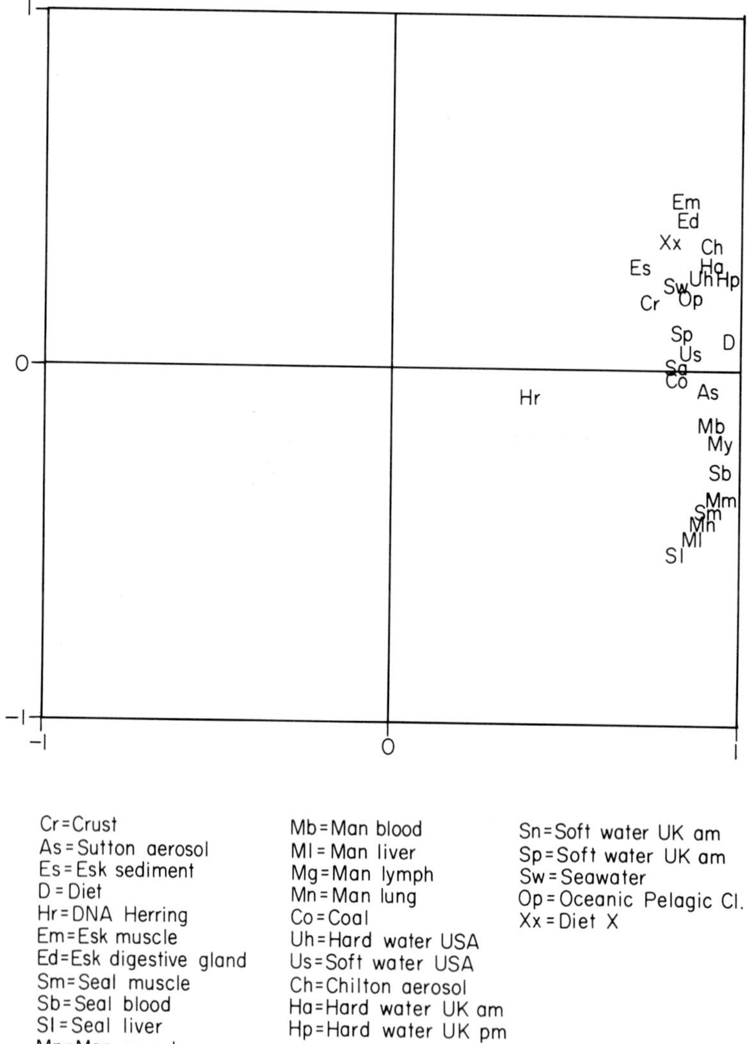


FIG. 14. Principal component analysis for the sample pairs given in Table 2.

simply reflect a high loading of contaminants by marine algae (see Shubert 1984). Furthermore, the data may also reflect adsorption of elements onto external surfaces of algae rather than transport into cells. If the approaches of John and Li are linked, the oceanic pelagic clay can be replaced by the Sutton aerosol and the algae by UK human blood. Fig. 18 illustrates that, in this comparison, the two groups identified by Li are not apparent, but that, with the exception of iron, some of the 'biophilic' elements are still linearly correlated with a slope of *c.* 1. Nevertheless, the concept of adsorption processes as being of some importance in the partitioning of the elements is still applicable to the atmospheric aerosol. Indeed many of the elements present in aerosols are in soluble forms

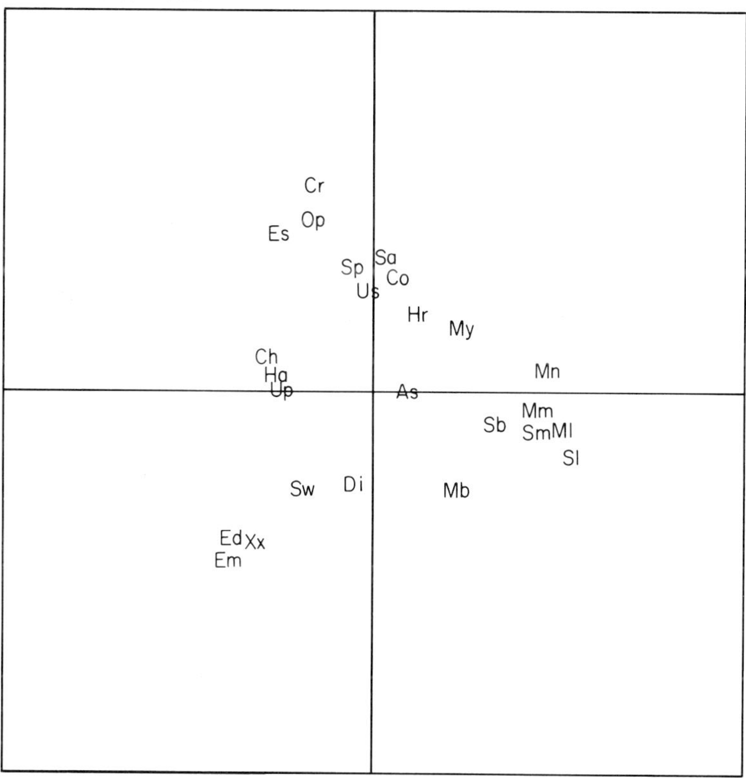


FIG. 15. Principal component analysis for residuals after removal of the first component.

TABLE 3. Comparative data for $r^2$ for selected sample pairs

| Sample pair | $r^2$ John 1984 | This work |
|---|---|---|
| Crustal rocks v. Human blood | 0·29 | 0·37 |
| Seawater v. Human blood | 0·41 | 0·52 |
| Sutton aerosol v. Human blood | 0·62 | 0·79 |

(see Cawse 1978) and therefore when they are associated with the surfaces of solid phases they are more available for transfer into biological materials by solution processes.

There is clearly a close association between elements in geochemical and biological systems which may simply reflect the general availability of elements and the general manner in which they have become incorporated into the process of organic evolution because of their universal occurrence. Nevertheless, from a consideration of the relative abundances of the elements, the overall picture seems to indicate that the biological components contain 'particles of inorganic materials' rather than the incorporation of the elements into biota through dissolved forms. A process of reconstitution of the characteristic geochemical abundances in biological matrixes seems unlikely. It is easy to accept that all living organisms can contain some inorganic solids, and that uptake of these will be by pinocytosis in lower forms of life and by macrophage scavenging in higher forms. The extent to which these processes can result in a discernible inorganic signal being transferred and detected in a biological system is at present under examination by studying the marine mussel. This is a filter feeder in which transport of elements in both dissolved and solid states does take place. Until studies have been completed for the distribution of elements in various constituents of cells (e.g. cytosol, mitochondria, lysosomes and nuclei) it is not possibe to comment upon whether or not the inorganic pattern of elements does extend to within cells. Lovelock (1979) is of the opinion that, since life began, the salinity of the oceans and hence the biological availability of the elements to living organisms has been under biological control. If life began in the seas then, from the studies reported here, and supported by those of John and Li, the abundances of the elements in seawater as a

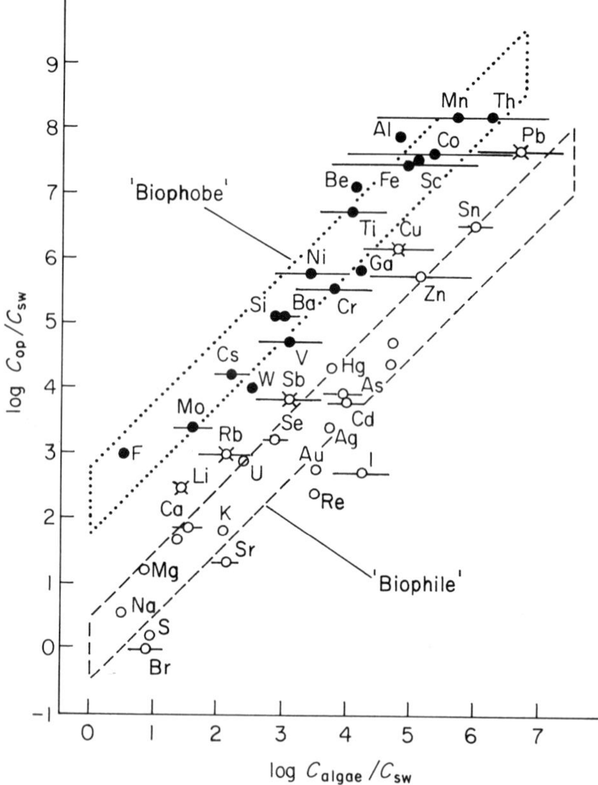


FIG. 16. Plot of the elemental concentration ratios for algae concentration in algae/concentration in seawater against that for oceanic pelagic clays (concentration in pelagic clays/concentration in seawater) (Li 1984).

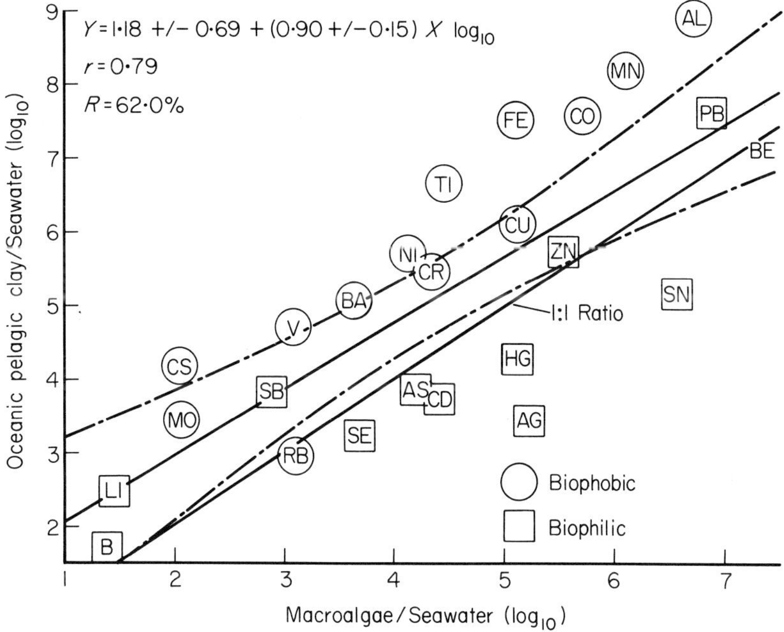


FIG. 17. Plot of the elemental concentration ratios of macroalgae and seawater v. oceanic pelagic clay and seawater (data from Eisler 1981 and Li 1984 respectively).

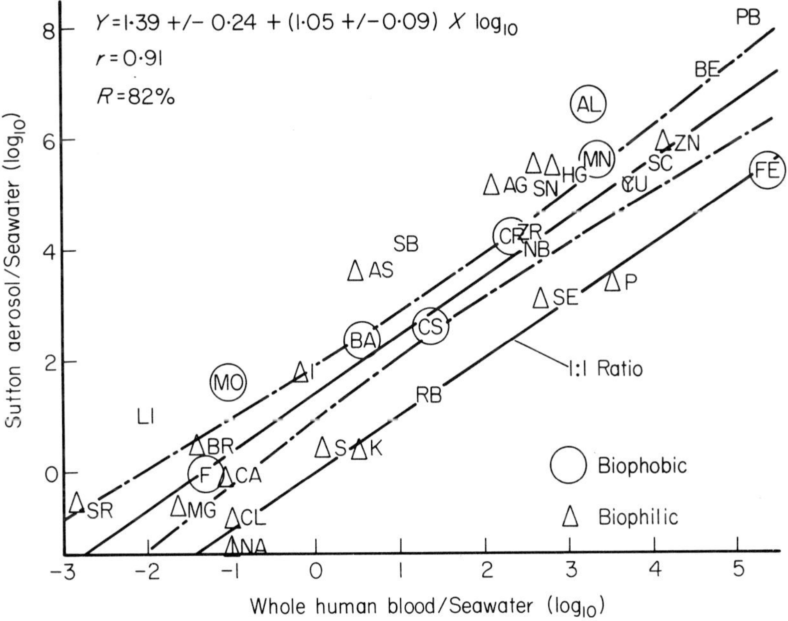


FIG. 18. Plot of the elemental concentration ratios of whole human blood and seawater v. Sutton aerosol and seawater (Hamilton 1979; Li 1984).

major influence do not appear to be related to those in living organisms. Animals and plants themselves exert a control over uptake and loss of many cations and anions by homeostasis, a word which was invented by the American physiologist Walter Cannon and refers to the remarkable state of constancy in which living things maintain themselves when their environment changes. Except over very short times, and with very few exceptions, the salinity of living cells cannot survive a change of >6‰.

The living cell is a dissipated structure in which control of inorganic processes is readily accomplished by kinetic processes; it is energized in a variety of ways and is a system which is concerned with making chemical and physical gradients. It is possible that, when organic evolution first started, it was based upon a geochemical element matrix which has been maintained as an essential framework upon which organic evolution developed, but which is now a structural component of life rather than one which interacts with the biology of life. Element interactions and bioavailability are associated with physical barriers, the attraction and repulsion between ions and compounds, diffusion processes and active ion pumping. Some forms of chemical binding give rise to stable end products, others are gated and are especially associated with chelates; kinetic stability is maintained, but there is often thermodynamic instability. Through processes of precipitation (no energy used) unwanted elements and compounds can be removed (element stores or for purposes of detoxification) from the organic processes. The interaction between elements is through programmed sequences of exchange, e.g. the interaction of elements such as Cu, Zn, Ni, Mo, Si, B, Fe(+3) and Mg with polymers; the concentration of K, Mg, Fe, Cu and Zn in particular biochemical compounds within cells compared with the rejection of Na, Ca and A1 and the neutrality of others such as Mg and Mn. In biological processes, monomers move in and out of cells without any hormone control while polymers become locked up in buffer systems, e.g. the $CaPO_4$ systems in vertebrates.

Under homeostatic control most Na is contained in extracellular sites while K is intracellular; Ca is associated with few proteins and is pumped into cells against the homeostatic gradient ($10^{-8}$ in cells and $10^{-3}$ outside); extracellular Ca is buffered by the precipitation of $CaPO_4$ and requires solubilization by the Ca binding protein (CaBP) while all contractile processes require Ca (calmodulin). Monomers can pass freely through biological membranes but there is a restricted flow of polymers, CaBP solubilizes Ca but, although homeostasis is excellent in extracellular areas it is is not good in intracellular areas. In the human body, Fe should precipitate in plasma, but it is kept in solution by the transfer of the glycoprotein transferrin, which provides a shuttle service between sites of adsorption and sites of utilization. However, only *c.* 30% of the available transferrin sites are occupied by ferric iron. Mitochondria recognize Fe not as the element but rather by its chelate, while electron-transfer processes do not recognize Fe but rather the protein with which it is associated. Most of the transitional elements exist in a steady state and in a stable form associated with small labile components and aquated ions at very low concentrations. As man requires a dietary intake of *c.* 1 mg Fe per day, which is about the same amount that he excretes daily, the balance is precarious and, indeed, iron deficiency is a major form of morbidity which affects *c.* 500 million individuals. Letendre (1985) proposed that, with the exception of certain lactobacilli, iron is an essential component of electron-transport chains where it exists as insoluble hydroxy-aquo complexes bound to carrier proteins. In the absence of any utilizable Fe, bacteria cannot survive and proliferate. In instances of hyperferremia (e.g. haemochromatosis, siderosis, malaria) an increase in circulating iron is associated with an increase in the severity of bacterial infections. However, as illustrated in Table 4, the abundance of many elements in total national diets are often very similar and it is surprising that the Indian diet contains so much iron considering that, for a large number of the population, iron deficiency diseases are known and that the stable food is cereal rather than meat which is a major source of iron.

Copper, like iron, is mainly contained in stores, and that which circulates in human plasma is in the form of labile complexes in equilibrium with aquated ions and low

TABLE 4. Some examples of the daily intake of elements by man for different countries

| Element | UK Survey* 1966–67 | ICRP average diet† | Indian diet‡ | German diet§ | USA diet (adults) ¶ |
|---|---|---|---|---|---|
| Al | 2·3 ± 1·1 mg | 45 mg | — | — | — |
| Ag | 27 ± 17 μg | 70 μg | — | — | — |
| Ba | 603 ± 225 μg | 750 μg | — | 440 μg | — |
| Be | 15 ± μg | 12 μg | — | — | — |
| Br | 8·4 ± 0·9 μg | 7·5 mg | — | 2·5 mg | — |
| Cd | 64 ± 30 μg | 150 μg | — | <50 μg | — |
| Ca | 1·4 ± 0·02 g | 1·1 g | 0·71 g | 0·38 g | 1·1 ± 0·2 g |
| Cs | 13 ± 7 μg | 10 μg | — | 13 μg | — |
| Cl | 5·4 ± 0·06 g | 5·25 g | — | 4·5 g | — |
| Cr | 320 ± 162 μg | 150 μg | 150 μg | 52 μg | — |
| Cu | 3·1 ± 0·8 mg | 3·5 mg | 5·8 mg | 2·7 μg | 1·6 ± 0·7 mg |
| I | 220 ± 51 μg | — | — | — | 319 ± 316 |
| K | 2·8 ± 0·03 μg | 3 g | 4·1 g | 2·4 g | 3·5 ± 0·2 g |
| Li | 107 ± 5 μg | 2 mg | 100 μg | — | — |
| Mg | 0·25 ± 0·02 g | 0·27–0·34 g | 0·74 g | — | 0·34 ± 0·03 g |
| Mn | 2·7 ± 0·8 mg | 3·7 mg | 8·3 mg | 2·7 mg | 3·6 ± 0·6 mg |
| Mo | 128 ± 34 μg | 300 μg | — | 70 μg | — |
| Na | 4·6 ± 0·4 μg | — | — | — | 4·7 ± 0·4 g |
| P | 1·9 ± 0·03 g | 1·8 –2·6 g | 1·5 μg | — | 1·7 ± 0·2 g |
| Pb | 320 ± 150 μg | 440 μg | — | — | — |
| Rb | 4·4 ± 1·5 mg | 2·2 mg | 2·7 mg | 1·9 mg | — |
| S | 0·94 ± 0·06 g | 0·85 g | 1·2 g | — | — |
| Sb | 34 ± 27 μg | 50 μg | — | 23 μg | — |
| Sr | 858 ± 144 μg | 1·9 mg | 4 mg | — | — |
| Tl | <2 ± μg | 1·5 μg | — | — | — |
| Th | <0·05 ± μg | 3·0 μg | — | — | — |
| Ti | 800·0 μg | 850 μg | — | — | — |
| U | 0·99 μg | 1·9 μg | — | — | — |
| Zn | 14·3 ± 1·2 mg | 13·0 mg | 16·1 mg | 11·8 mg | 12·8 ± 1·1 mg |

*Hamilton & Minski (1972–73b)
†ICRP (1975)
‡Soman *et al.* (1969)
§Schelenz (1977)
¶ Pennington *et al.* (1984).

molecular weight complexes. e.g. ceruloplasmin. Although there are very few quantitative data for the abundance of many elements in the diet of man, the data given in Table 4 do indicate that the abundances (daily intake) are very similar; hence, the importance of the food chain in reducing the large differences which occur in the geochemical environment from which the food originates. The data in Table 4 may be compared with the considerable differences in the daily intake of major foods as shown in Table 5.

TABLE 5. Per caput estimates of food supplies (g $day^{-1}$) by geographical region

| Food groups | Far East* | India† | Africa* | Europe* | N. America* |
|---|---|---|---|---|---|
| Cereals | 404 | 470 | 330 | 375 | 185 |
| Starchy roots | 156 | — | 473 | 377 | 136 |
| Sugar | 22 | 20 | 29 | 79 | 113 |
| Pulses, nuts | 56 | 70 | 37 | 15 | 19 |
| Vegetables, fruits | 128 | 116 | 215 | 316 | 516 |
| Meat | 24 | — | 40 | 111 | 248 |
| Eggs | 3 | 15 | 4 | 213 | 55 |
| Fish | 27 | — | 16 | 38 | 26 |
| Milk | 51 | 80 | 96 | 494 | 850 |
| Fats, oils | 9 | 15 | 19 | 44 | 56 |

*Sukhatme (1961).
†ICMR (1964).

Compared with the complexity of the behaviour of elements in the living cell, those in inorganic geochemical systems can be accounted for by fairly simple processes which are mainly controlled by temperature and pressure through which the elements become partitioned between various mineral phases. The weathering processes, through which elements are returned to aqueous systems is more complicated, but the processes are generally understood. There seems to be no obvious pathway through which the relative abundance of geochemically associated groups of elements in rocks and minerals can be transferred without some fractionation or differentiation to living organisms except by the actual transfer of the elements in solid mineral forms. Although the manner in which biological processes accommodate an element is very specific, many of the elements,and especially those at high concentrations, are sequestrated or act as buffer reservoirs. Because of the similarity which exists between elements of the same group of the Periodic Table, the presence of one element as being essential for a particular biological process will inevitably mean that it will become associated (in tissues) with other elements of the same and related chemical groups; hence, an apparent geochemical association will be maintained, e.g. the analogy of haemoglobin in human blood and mafic minerals, and calcium carbonate of phosphate deposits as scavengers or adsorbers of many elements. Elements can be transported in and out of biological systems when their particular chemical and physical characteristics are not recognized, hence they do not interact with the various internal element reservoirs. In any gross chemical analysis of a living organisms the importance of the dynamic processes is lost and the analysis will be biased towards those elements which have accumulated at a particular site. It is at these sites that the expected chemical associations between elements will be preserved on a basis of similarity in chemical behaviour rather than of geochemical associations. The apparent association between aerosol and human blood may be fortuitous but, if groups of characteristic geochemical element abundance ratios can be found in biological materials, then this would provide evidence to support the occurrence of many elements with rock detritus. The matter is not simple, as, apart from requiring exceptionally high-quality analysis on representative samples, it is inevitable that the rock detritus will become highly fractionated in transfer from a geochemical to a biological environment. For example, the Si/Al ratio could be useful, but would require the ingestion and inhalation by man of feldspar and quartz in natural proportions. In practice, as quartz is more resistant, it is likely to be more available for intake by man. In the fractionation process aluminium will become enriched in the finely divided clay fraction (weathering of minerals), hence the crustal Si–Al ratio is not likely to be retained. In the case of the characteristic K–Rb ratio (as neither element is fractionated by most normal geochemical or biochemical processes) the ratio should not vary irrespective of what the materials are.

A large number of possible geochemical and biochemical indicators are being examined and some data for K, Rb, Si and Al are given in Table 6. It is interesting to note

TABLE 6. The K–Rb and Si–Al ratio for varius human tissues and environmental samples

| Sample | K–Rb | Si–Al |
|---|---|---|
| Blood (human) | 1079 | 17 |
| Brain (human) whole | 625 | 460 |
| Brain (human) frontal lobe | 766 | 276 |
| Brain (human) basal ganglia | 766 | 346 |
| Muscle (human) | 570 | 10 |
| Ovary (human) | 270 | 19 |
| Testes (human) | 79 | 8 |
| Liver (human) | 343 | 3 |
| Kidney (human) whole | — | 28 |
| Kidney (human) cortex | 462 | 93 |
| Kidney (human) medulla | 727 | 11 |
| Lung (human) | 571 | 2 |
| Lymph nodes (human) | 125 | 15 |
| Sutton aerosol | 350 | 2 |
| Average crustal rocks | 233 | 3 |
| Oceanic pelagic clay | 227 | 3 |
| Seawater | 3167 | 2000 |
| UK diet | 643 | — |

that the K–Rb ratio is not constant. This is surprising and could be accounted for if the Rb measurements are not accurate, or alternatively the data may indicate that fractionation between K and Rb occurs in biological systems. Of all the samples examined, blood has a

K–Rb ratio nearest to seawater and must reflect the transport of soluble constituents across the gastro-intestinal tract of man and may reflect the inclusion of a component of relict seawater in man. For the Si–Al ratio the complication of excess silicon (quartz) is a problem, but the data do illustrate that, for the lung at least, the characteristic geochemical ratio is observed. However, it seems probable that the quantitative transfer of rock debris from the lung to the systemic circulation is not likely to occur. Neither is the debris which accumulates in the lung with age all likely to reach the deep lung where macrophage transport would presumably not differentiate between different chemical substances, while the chemical solution of finely divided quartz is likely to be more rapid than that for feldspar or detrital clay minerals. If the global aerosol is involved in the transfer of element associations characteristic of biological materials, then the processes indicated by Li (1982, 1984) seem reasonable; in particular, the role of surface adsorption in providing a mechanism suitable for the adsorption of elements at a site for potential catalytic chemical processes.

Ignoring for the present the question of the origin of life, the following scenario seems suitable for the element interaction between abiotic and biotic systems. Because of the probable need for life to develop under constant environmental conditions, and over a long period of time, the sea, or shore line, seems the most appropriate site;the latter is preferred as it is associated with rapid changes in environmental conditions against which the unstable changes arising from exposure to the air could stabilize. The physical formation of natural liposomes along the shoreline would encapsulate fine-grained sediment together with seawater; the inclusion of clay debris is more acceptable than quartz grains as the surface would contain elements which are associated with biological catalytic processes, many of which are associated with enzymes (e.g. Mn, Fe). During aerial exposure, the clay debris would absorb photon energy and thus provide a potential for the production of electro-chemical cells within the liposomes in which electron transport processes could take place. There is abundant evidence that organic compounds were present in the so-called primeval sea, and, if modern analogues are any guide, they would coat the surface of the mineral debris and also the external surface of the liposomes. Ageing and ordering processes would then provide a coating through which the lifetime of the liposomes could be extended and thus allow time for the process of organic evolution to develop. Those products which survived would increase in numbers, and perhaps provide a suitable matrix from which further potentially biotic species could develop. With the passage of geological time it is not impossible for various types of membranes to be developed which would have selective uptake for some elements and also provide mechanisms for cation pumps; for example, the interfacing and coupling of organic ‘coatings’ with inorganic substrates. Recently Kimura *et al.* (1985) produced laboratory preparations of lithium ionophores with the property of proton driven cation transport, and for which the selectivity for lithium over sodium was very high. There appears to be no reason why similar structures should not be present in the natural environment and, when coupled to catalytic properties of the transition group of elements, that quite sophisticated cation selective processes should be produced that would simulate common cation transport processes and mechanisms which are found in living cells.

Throughout the world today vast sums of money are being spent to provide data in order to evaluate the extent to which the natural environment has been contaminated by man; many studies by nature are simply catalogues for levels of elements in a wide variety of materials. If it can be shown that living organisms can tolerate, without any harm, a certain increase or decrease in the abundance of an element then baselines for acceptable contamination of the natural environment can be established. Today it is not possible to take such decisions because we do not know whether or not the stability or essential diversity of natural ecosystems has been altered by the activities of man. An extension of the study presented here may provide some useful clues, but, even if not successful, will at least provide a framework for interdisciplinary studies which is an essential requirement in considering the long-term disposal of contaminants to the natural environment.

## ACKNOWLEDGMENTS

I thank J. M. Colebrook for the principal component analysis and H. E. Stevens for technical assistance.

## REFERENCES

**Bowen, H.J.M. (1966).** *Trace Elements in Biochemistry.* Academic Press, London.

**Bowen, H.J.M. (1973).** The use of reference material in the elemental analysis of biological samples. *Atomic Energy Review,* **13,** 451–457.

**Bowen, H.J.M. (1979).** *Environmental Chemistry of the Elements.* Academic Press, London.

**Burbidge, E.M., Burbidge, G.R., Fowler, W.A. & Hoyle, F. (1957).** Synthesis of the elements in stars. *Review of Modern Physics,* 650.

**Cameron, A.G.W. (1973).** Abundances of the elements in the solar system. *Space Science Review,* **15,** 121–146.

**Cawse, P.A. (1975).** *A survey of atmospheric trace elements in the UK: Results for 1974.* UKAEA-AERE Report, R8038. HMSO, London.

**Cawse, P.A. (1978).** UKAEA-AERE Report, R9164. P.80 in Environmental and Medical Sciences Division, AERE-PR-EMS/5 Progress Report January–December 1977 (Ed. by M. Hainge). UKAEA.

**Clayton, D.D. (1968).** *Principles of Stellar Evolution and Nucleosynthesis.* McGraw Hill, New York.

**Curtis, D., Gladney, E. & Jurney, E. (1980).** A revision of the meteorite based cosmic abundances of boron. *Geochimica et Cosmochimica Acta,* **44,** 1945–1953.

**Eisler, R. (1981).** *Trace Metal Concentrations in Marine Organisms.* Pergamon Press, Oxford.

**Frey-Wyssling (1935).** Die unentbehrlichen Elemente der Pflanzennahrung. *Naturwissenschaften,* **23,** 767–769.

**Ganapathy, R. & Anders, E. (1974).** Bulk compositions of the moon and earth, estimated from meteorites. *Proc. 5th. Lunar. Sci.* Conf., pp. 195–206.

**Goetz, L. Sabbioni, E., Springer, A. & Pietra, R. (1979).** Heavy metals: environmental research related to energy production. *International Conference on the Management and Control of Heavy Metals in the Environment.* pp. 275–279. 18–21 Sept. CEP Consultants Ltd., Edinburgh EH1 3QH UK. London.

**Hamilton, E.I. & Minski, M.J. (1972).** Comments on the trace element chemistry of water: sampling a key factor in water quality surveillance. *Environmental. Letters,* **3,** 53–71.

**Hamilton, E.I. & Minski, M.J. (1972–73a).** The abundances of the chemical elements in man's diet and possible relations with environmental factors. *Science of the Total Environment,* 375–394.

**Hamilton E.I. & Minski, M.J. (1972–73b).** Spark sources mass spectrometric sensitivity factors for elements in different matrixes. *International Journal of Mass Spectrometry and Ion Physics,* **10,** 77–84.

**Hamilton, E.I., Minski, M.J. & Cleary, J.J. (1973a).** Problems concerning multi-element assay in biological materials. *Science of the Total Environment,* 1–14.

**Hamilton, E.I., Minski, M.J. & Cleary, J.J. (1973b).** The concentration and distribution of some stable elements in healthy human tissues for the United Kingdom. *Science of the Total Environment,* **1,** 341–374.

**Hamilton, E.J. (1974).** The chemical elements and human morbidity — water, air and places — a study of natural variability. *Science of the Total Environment*, **3**, 3–85.

**Hamilton, E.I. (1976).** Review of the chemical elements and environment chemistry. *Science of the Total Environment*, **5**, 1–62.

**Hamilton, E.I. (1979).** *The Chemical Elements and Man.* Charles C. Thomas, Springfield, USA.

**Hamilton,E.I. (1981).** *An overview: The chemical elements, nutrition, disease and the health of man. Proceedings of the American Institute for Nutrition Research,* pp. 27–61.

**Holland, H.D. (1984).** *The Chemical Evolution of the Atmosphere and Oceans.* Princeton University Press, Princeton, N.J., USA.

**Hutchinson (1943).** The biogeochemistry of aluminium and of certain related elements. *Quarterly Review Biology,* **18**, 1–29.

**Indian Council of Medical Research (1964).** *Diet Atlas of India.* Indian Council of Medical Research. Special Report. Series 48. Hyderabad, India.

**ICRP (1975).** *Reference Man.* ICRP 23. Pergamon Press, Oxford.

**John, W. (1983).** Relationship between trace element concentrations in human blood and atmospheric aerosol. *Science of the Total Environment,* **27**, 21–32.

**Kimura, K., Sakamoto, H., Kitazawa, S. & Shono, T. (1985).** Novel lithium-selective ionophores bearing an easily ionizable moiety. *J. Chem. Soc. Chem. Commu;* 669-670.

**Letendre, E.D. (1985).** The importance of iron in the pathogenesis of infection and neoplasia. *Trends in Biochemical Sciences*, **10**, 166–171.

**Li-Y.H. (1982).** A brief discussion on the mean oceanic residence time of elements. *Geochimica et Cosmochimica Acta.*

**Li-Y.H. (1984).** Why are the chemical compositions of living organisms so similar. *Scweizer Zeit Hydrology,* **46**, 1–8.

**Lovelock, J.E. (1979).** *Gaia.* Oxford University Press, Oxford.

**Pennington, J.A.T., Wilson,D.B., Newell, R.F., Johnson, R.D. & J.E. Van der Veen (1984).** Selected minerals in food surveys, 1974–1981/82. *Journal of the American Dietetic Association.*

**Peirson, D.H., Cawse, P.A. & Cambray, R.S. (1974).** Chemical uniformity of airborne particulate material, and a maritime effect. *Nature (London),* **251**, 675–679.

**Rahn, K. (1976).** *The chemical composition of the atmospheric aerosol.* Tech. Rept. Graduate School of Oceanography, University of Rhode Island, July 1.

**Ross, J.E. & Aller, L.H. (1976).** The chemical composition of the Sun. *Science,* **191**, 1223–1229.

**Salmon, L., Atkins, D.H.F., Fisher, E.M.R., Healy, C. & Law, D.V. (1978).** Retrospective trend analysis of the content of U.K. air particulate material 1957–1974. *Science of the Total Environment,* **9**, 161–200.

**Salmon, L., Toureau, A.E.R. & Lally, A.E. (1984).** The radiological impact of electricity generation by U.K. coal and nuclear systems. *Science of the Total Environment,* **35**, 417–430.

**Schelenz, R. (1977).** Dietary intake of 25 elements by man estimated by neutron activation analysis. *Journal of Radioanalytical Chemistry,* **37**, 539–548.

**Shaw, W.H.R. (1960).** Studies in biogeochemistry. 1. A biogeochemical periodic table. The data. *Geochimica et Cosmochimica Acta,* **19**, 196–207.

**Shubert, L.E. (Ed.) (1984).** *Algae as Ecological Indicators.* Academic Press, London.

**Soman. S.D., Panday, V.K., Joseph, K.T. & Raut, S.R. (1969).** Daily intake of some major and trace elements. *Health Physics,* **17**, 35–40.

**Steinberg, R.A. (1938).** Correlations between biological essentiality and atomic structure of the chemical elements. *Journal of Agricultural Research,* **57**, 851–858.

**Subramanian, K.S., Meranger, J.C. & Burnett, R.T. (1985).** Kidney and liver levels some major, minor and trace elements in two Ontario communities. *Science of the Total Environment,* **42**, 223–235.

**Sukhatme, P.V. (1961).** The world's hunger and future needs in food supplies. *Journal of the Royal Statistical Society. Series A,* **124**, 463–508.

**Taylor, S.R. (1964).** The abundance of chemical elements in the continental crust — a new table. *Geochimica et Cosmochimica Acta,* **28**, 1273–1285.

**Vernadsky, W.I. (1939).** On some fundamental problems of biogeochemistry. *Trudy Biogekhimica Laboratory Akademy.* Nauk. SSSR, 5, 5–17.

**Vinogradov, A.P. (1933).** La composition chemique elementaire des organises vivants et le systeme periodique des elements chemiques. *Compt Rendu Science, Paris,* **197**, 1673.
**Williams, R.J.P. )1983).** Inorganic elements in biological space and time. *Inorganica Chimila Acta,* **79**, 137.
**Williams, R.J.P. (1953).** Metal ions in biological systems. *Biological Reviews,* **28**, 381–415.

# Flushing and dispersal mechanisms in estuaries

K. R. DYER*
*Institute of Oceanographic Sciences,*
*Crossway, Taunton*

## SUMMARY

The flushing and dispersal of water-borne pollutants are governed very strongly by the estuary topography, the presence and strength of secondary currents, and intermittency produced by river discharge variations and the weather. This paper reviews briefly some of the processes involved in these effects.

## INTRODUCTION

The flushing and dispersal of pollutants in an estuary depends strongly on whether they are in solution and travel with the water, or whether they become adsorbed onto particles. Estuaries are normally quite efficient at diluting water-borne pollutants, and transporting them to the sea but, conversely, they are often effective traps for sediment. However, predicting the fate of particle-borne pollutants is especially difficult since the pollutants may not remain bound to the particles in all chemical conditions. They may migrate within the sediment and may be released into the pore water, or the water column, to be adsorbed elsewhere. For instance, it is fairly well known that manganese is frequently cycled between dissolved and particulate forms in response to rapid changes in chemical conditions during mixing.

Whether pollutants are conservative in the water, or are removed by particles can become apparent by consideration of a plot of the dilution of pollutant concentration against salinity. Removal by sediment is facilitated by the strong gradients of ionic concentration, pH, Eh and by the generally high concentrations of suspended particulate matter present in estuaries. Because the chemical interactions take time to come to equilibrium, the durations of constant conditions, or the relative rates of the change of the chemical and physical conditions may become significant. The intensity and persistence of gradients of salinity and suspended sediment concentration are to a great extent controlled by the processes governing turbulent mixing and dispersion. The movement of water-borne pollutants is normally considered to be governed by the same processes that control the movement of fresh water, or inversely, salty water.

Pollutants can be removed from the estuary by being flushed out by the fresh water, and they are diluted by the processes of diffusion and dispersion. Diffusion is the effect of turbulence in mixing adjacent water particles, but where there is also a velocity-shear some particles are travelling faster, and some slower than the mean flow. When coupled with turbulent diffusion this shearing leads to dispersion which tends to extend the pollutant cloud and decrease the gradients in concentration. The processes involved in dispersion

*Present address: Institute of Marine Studies, Plymouth Polytechnic, Drake Circus, Plymouth PL4 8AA.

include shear on the boundaries, topographic effects, secondary flows, and density stratification (Fischer 1972, 1976), and their overall effect is usually represented by a dispersion coefficient. This relates the pollutant flux to the concentration gradient, rather like a coefficient of conductivity. At present, dispersion coefficients cannot be quantified independently despite their importance in models predicting pollutant distributions, though advances have recently been made (e.g. Smith 1980, 1982). However, dispersion coefficients are only useful when the conditions can be considered to be in a steady state, and when there are no transients or patches passing through which persist for longer than the time over which averaging is completed. Consequently, it is easier to predict mean pollutant concentrations over a period than it is to predict the maximum values likely to occur.

The aim of this paper is to review briefly some of the physical mechanisms important in the mixing and dispersal of water in estuaries, placing emphasis on the physical time-scales involved, and on the unsteady nature of the conditions.

## FLUSHING

The flushing of water through an estuary is caused by the river discharge gradually pushing the previous days flow down the estuary. In its passage the fresh water is mixed by various processes with the salt water, so that the eventual discharge through the estuary mouth is a mixture of salt and fresh water. Thus, the concentration of a pollutant discharged into the estuary would become reduced by dilution with the uncontaminated sea water. The total flushing time of the estuary can be thought of as the time taken to replace the accumulated fresh water fraction in the estuary by the river discharge, and can be calculated from the salinity distribution as:

$$T = \frac{V}{R}(1 - \frac{\overline{s}}{S_0}) \qquad (1)$$

where $V$ is the volume of the segment on the estuary, R is the river discharge, $\overline{s}$ is the mean salinity in the segment, and $S_0$ the undiluted sea water salinity. Estuaries such as the Mersey have flushing times of 5–10 days, the Thames has up to 100 days, and the Rhine 3–6 days. The change in flushing time with river discharge for Boston Harbour is shown in Fig. 1 (Ketchum 1952), and this conforms very well to eqn (1). However, in detail the salinity distribution does not vary simply with river discharge. At high river discharge the salt water is pushed down the estuary, and the increased stratification decreases the mixing between the salt and fresh water. Therefore, the river flow reaches the sea more directly, and without diluting the sea water near the bed significantly. At low discharge the salt penetrates further into the estuary, but the reduced stratification allows greater mixing and the rate of exchange of fresh water to the sea is relatively reduced. The upper part of the estuary is likely to respond much faster to changes in the river flow than the lower part, and the changes would be of greater magnitude. Thus, the estuarine circulation buffers the lower estuary from the most drastic temporal changes in salinity. Bowden & el Din (1966) have shown that the best correlation between salinity and river discharge in the lower part of the Mersey estuary was found using the river discharge averaged over the week before sampling.

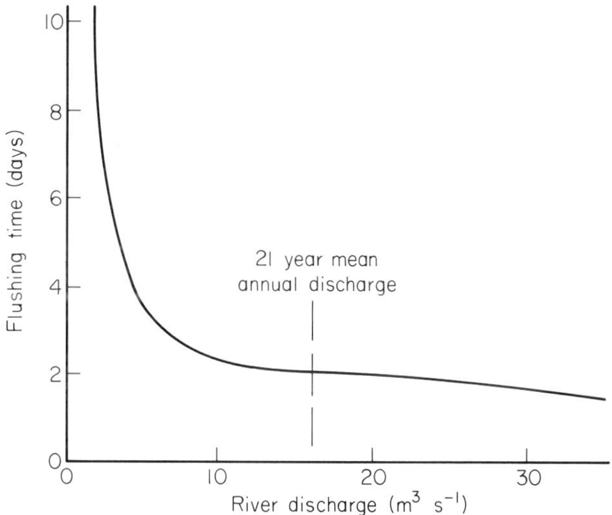


FIG. 1. Variation of flushing time with river discharge for Boston Harbour (after Ketchum 1952).

Additionally, the weather can affect the flushing of an estuary. Elliott (1978) has shown that the Potomac estuary has periods of total inflow, when low barometric pressure and wind effects increases the mean water level in the estuary, as well as periods of total outflow. The total inflow situation occurred for almost a quarter of the time, with durations of the order 2–5 days. It only needs a small increase in water level to accommodate several days of river discharge. Presumably this would initially be held in the upper reaches, and gradually move downstream as the weather changes. Thus, pollutants could be dispersed only intermittently, with the most noticeable effects occurring when the periodicity in the changing weather patterns is similar to the flushing time of the estuary. Weisberg (1976) considered the mean flow in estuaries from a statistical standpoint, and concluded that averaging over a period 5–10 days is required to remove the effects of weather.

## SECONDARY FLOWS

The form and intensity of secondary currents has a significant effect on the rate of dispersion of water-borne material because they transfer water from one side of the estuary to the other and assist the mixing processes. They arise because the water tends to flow in a spiral fashion along the estuary, mainly as a result of the topographic form of the estuary. The secondary flows are manifest as vertical and transverse components of the velocity, but they are difficult to measure as they are caused by only small angular differences between the direction of the current, and the channel direction. There has been considerable interest in secondary flows in rivers (Bathurst, Thorne & Hey 1977; Thorne & Hey 1979) and the dispersion resulting from them (Fischer 1969), but comparatively little work has been carried out in estuaries.

In straight, narrow, well-mixed estuaries (i.e. estuaries less than about 250 m wide) the transverse shear caused by the banks is dominant and the secondary flows form a pair of counter-rotating spirals. On the flood tide the saltier water in the centre of the channel travels faster than that near the sides. Combined with the shear, this leads to a pair of spirals rotating in a sense that produces a convergence at the surface and divergence near the bed. This concept has been confirmed by Nunes (1982) who observed an axial convergence on the flood tide in the Conwy estuary and modelled the flow producing it. West, Knight & Shiono (1984) discussed the flow structure in the Great Ouse and concluded that the secondary circulation reverses on the ebb tide, producing a surface divergence in the centre of the estuary. It is not clear, however, what the tidally-averaged secondary circulation would be, neither is it clear at what stage the circulation would become modified by the presence of bends. Even in comparatively straight rivers there is a tendency for the flow to meander (Einstein & Shen 1964).

In a homogeneous flow the centrifugal force created by the flow round a bend drives the water towards the outside of the bend. As the surface water experiences a larger force than that on the bottom, there will be a downward vertical velocity on the outside of the bend. Near the bed there is a counter-flow towards the inside of the bend and an upward flow near the inner bank. Consequently, on a right-hand bend looking downflow there should be an anti-clockwise secondary circulation, and this circulation is likely to be better developed in wide estuaries where the transverse shear is less dominant. One would expect that the circulation would be in the same sense when the current reverses, because the centrifugal force still acts towards the outside of the bend. Thus, in a homogeneous flow the tidally-averaged secondary circulation should be similar to that in a unidirectional flow.

The presence of density gradients, however, will tend to modify the response, but in a way that is unpredictable at the moment for the instantaneous situation. For the tidally-averaged situation, however, Dyer (1977) proposed from measurements in two estuaries that the secondary circulation in the upper layer of a salt-wedge estuary is the same as that in a homogeneous undirectional flow, with clockwise flow on a left-hand bend (Fig. 2).

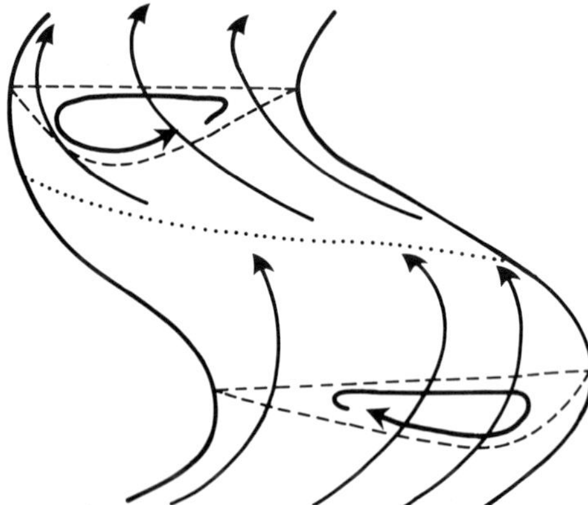

FIG. 2. Schematic diagram of tidally averaged secondary circulation in a salt wedge estuary, showing the surface streamlines, and a convergence line (dotted) between the two secondary circulation cells.

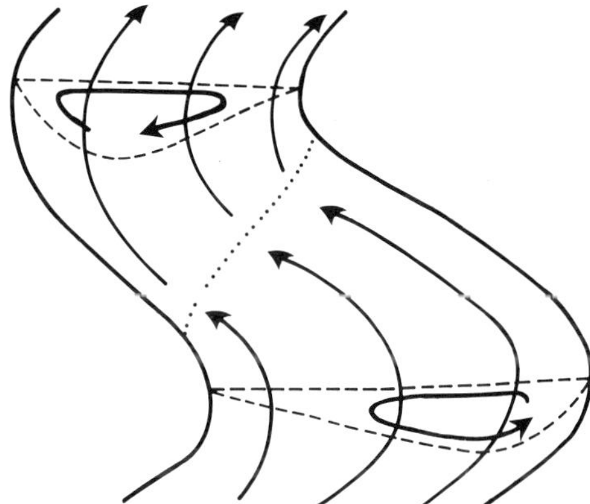

FIG. 3. Schematic diagram of tidally averaged secondary circulation in a partially mixed estuary. Contrast with Fig. 2.

When partially-mixed conditions developed, the circulation reversed and upward flow occurred over the deeper water on the outside of the bend, with a lateral surface flow towards the inside of the bend. Thus, the circulation was anticlockwise on a left-hand bend (Fig. 3). Associated with this reversal was a shift of the maximum velocity away from the outside of the bend, and the fresher surface water tended to flow down the shallower side of the estuary. The vertical velocities involved in the circulation were of the order of $10^{-3}$ cm $s^{-1}$. Since this is similar to the settling velocities of fine inorganic particles and planktonic larvae, settling is likely to be enhanced on the shallower side of the estuary, and limited over the deeper channel. In wide estuaries, however, the secondary circulation may not have time to develop fully before the flow changes. This appears to occur when the estuary is wider than about 200 m (Holley, Harleman & Fischer 1970; Smith 1980, 1982). As a consequence, a pollutant injected on one side of the estuary may not reach the other side of the channel before being carried back again.

Additionally, in wide estuaries there will be a tendency for the ebb and flood flows to preferentially take different channels within the estuary, leading to a horizontal residual circulation pattern. Consequently a pollutant would undergo quicker flushing if discharged into the ebb channel rather than the flood channel.

## PATCHINESS

Because of the intermittency in the river flow, and of the estuarine mixing, it is likely that the salinity distribution does not decrease smoothly towards the upper part of the estuary. There may be zones of high gradient separated by zones of comparative uniformity. There are a number of reasons why such fronts and patches are formed, the most obvious being: discharge events, tributary interactions, and mixing events.

Fig. 4 shows a continuous salinity trace obtained in Southampton Water at a depth of

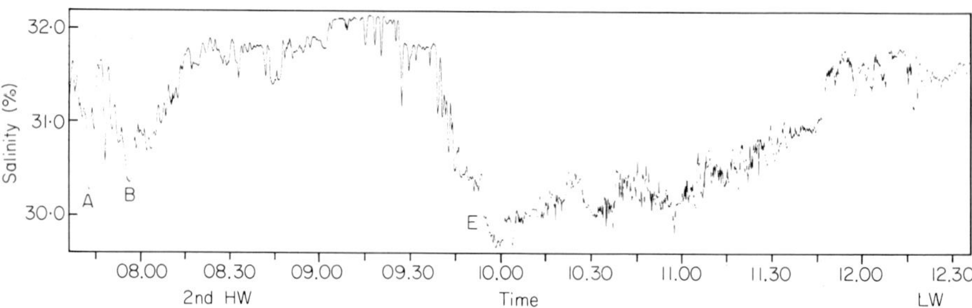


FIG. 4. Continuous salinity trace during the ebb tide at about mid-depth in Southampton Water on 2 April 1978, showing low salinity patches between about 07.45-08.15 and 09.30-11.45 hours. For discussion see text.

4 m, during a period from about high water to low water. Normally one would expect a fairly regular reduction of salinity towards low water, but here the lowest salinity was attained about half way through the ebb tide. Vertical profiles obtained regularly throughout the period showed that between 09.30 and 10.00 hours there was a reduction in salinity by about 1‰ on the bed and 3‰ at the surface. The salinity then gradually rose until about 11.45 hours. It is apparent that a patch of lower salinity water was being advected past the measuring point on the ebb tide. The patch was lower in salinity throughout the water column, had slightly higher stratification than the water on either side, and was bounded by fairly steep frontal structures at either end. Judging by the advection velocities, the leading front would have been about 450 m wide, and the length of the patch was about 2 km.

Downstream of the main patch was another smaller one centred on about 08.00 hours, also with a steep downstream front and a more gradual increase of salinity at the tail end. Particularly distinctive features of this patch were the low salinity events A and B. Event A, for example, must have only been about 50 m wide, and yet it was visible over a 5-day period at two stations 500 m apart on either side of the deep channel of the estuary. Consequently, over that period there must have been little turbulent mixing occurring in that section of the water mass during its oscillation up and down the estuary. On the other hand, between events B and E there was a reduction in salinity at 4 m depth of about 1·5‰ during the 5-day period (Dyer 1982). Therefore, this 1 km length of the water mass must have undergone fairly intense mixing during the tide, presumably due to this part of the water mass encountering shallower water, or a rougher bed at that point in its trajectory where current velocities were at a maximum. Towards spring tides, the length of this zone of mixing would increase as the velocities are higher, but at neap tides it is possible that no mixing would take place. Thus, the mixing appears to be intermittent in time as well as in space.

Although Events A and B were distinctive on every ebb tide, they were not easily distinguishable on the flood tides. Since these events were seen at two stations across the estuary, it is likely that between ebb and flood tides there was a vertical, as well as a possible lateral movement of the internal density field. This may have been a response to the changing secondary circulation pattern.

The origin of the patch was thought to be related to a brief 2-day flood that came down the rivers about a fortnight prior to the beginning of the survey (Dyer 1982). Consequently the overall flushing time for the patches would have been in excess of 20 days, during which time they retained their individuality to a certain extent. It is feasible that the larger upstream patch was the discharge from the River Test, whereas the smaller one could have originated from the River Itchen. In this case, the patches could have had different concentrations of nutrients, trace elements and other constituents despite having similar salinities. Any salinity or temperature-dependent chemical or biological process could thus proceed at different rates within a comparatively restricted area. Within the water patch, processes could readily achieve equilibrium since the rate of mixing is low and there would be many days of virtually constant conditions. For a fixed point in the estuary, however, salinity changes of 2‰ in 30 min are possible as the front of the patch goes past, and even sharper changes are possible for shorter durations.

## TRIBUTARY FLOWS

The interaction of the flows from tributaries with that in the main estuary is an interesting topic that has not received very much attention, though Abraham, de Jong & Van Kruiningen (1986) have recently considered tributary effects in the Rotterdam Waterway. Two important variables that appear to exert a major control on the interaction are the relative river discharges, and the relative phases of the tidal inflow currents. The phases of the currents will depend on the distribution of intertidal volumes. If the subsidiary tributary has a small volume close to low water mark, but extensive intertidal areas close to high water, the main tidal inflow into the tributary is likely to occur late in the rising tide, at a time when the flow in the main channel may be smaller (Fig. 5a). The highest outflow would then occur just after high water. Conversely extensive intertidal areas near low water mark could cause maximum inflow and outflow on either side of low water (Fig. 5b). These two alternative flow states, in combination with the river discharge variations, could produce four tributary junction conditions.

**1** Tributary inflow and outflow dominant near high water (a) low river discharge, (b) high river discharge.

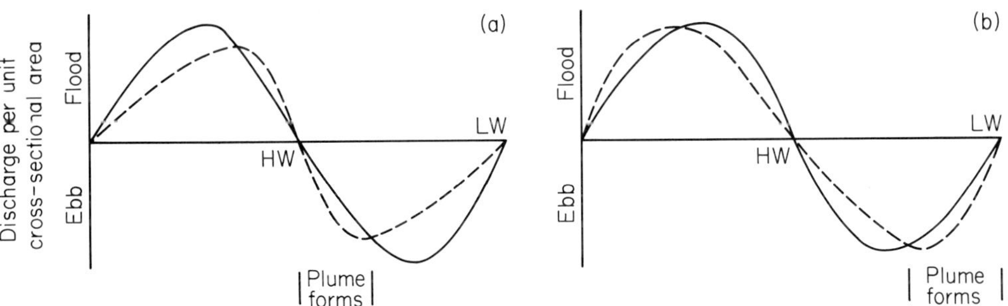


FIG. 5. Schematic diagram illustrating the interaction of a tributary discharge (dashed curved) with that of the main estuary (solid curve).

**2** Tributary inflow and outflow dominant near low water (a) low river discharge, (b) high river discharge.

With little river discharge the water will flow up the tributary and back again with little modification in salinity, except that it might become slightly better mixed in the process. It should not, therefore, differ very much in salinity from the water in the main channel. However, in 2b above, the water issuing from the tributary near low water may be carried up the main channel at the beginning of the flood rather than back into the tributary. Also after high water, in this case, virtually stagnant water of reasonably high salinity may be trapped in the mouth of the tributary.

With relatively high river discharge in the tributary, reasonably high salinity gradients will develop and the discharge into the main channel would be of much fresher water than that prevailing there, so forming a distinct plume. Consequently, a sharp front can quickly become established, early in the ebb tide for 1b and later in the tide for 2b. The salinity contrast is likely to be greatest, however, for 2b, when the most marked front would appear.

Garvine (1977), from observations in the Connecticut River, has shown that a well developed plume will exist during the ebb tide when the ratio of the mean freshwater discharge velocity to the root-mean-squared tidal velocity exceeds about 0·75, and for the flood tide as well when this ratio exceeds 2. Simpson & Nunes (1981) described a front on the flood tide in a small estuary with high river discharge. The front formed about 2 hours after low water, just inside the mouth of the estuary, but once the flood tidal current diminished from its maximum, the front was swept out of the estuary by the river flow.

Normally, however, fronts will exist for a few hours on the ebb tide. They often separate water masses with striking colour differences, and the fronts are marked by concentration of floating debris. The boundary is normally a zone of convergence with the discharged water moving vigorously towards the front, and then sinking at the interface (Fig. 6). The salty and fresher water will be mixed by the shear on the interface, though at a

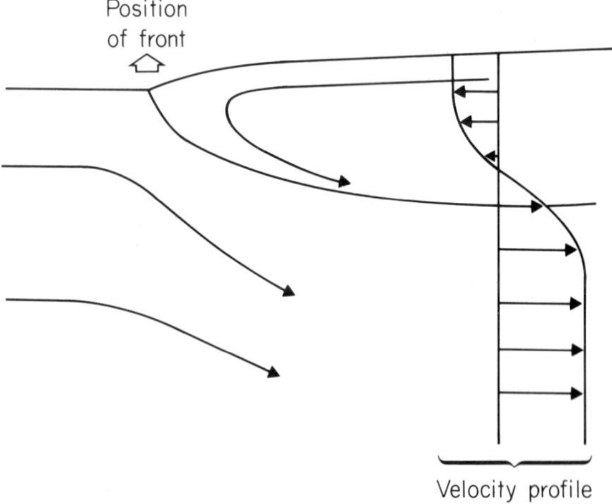


FIG. 6. Schematic diagram of the circulation at a front.

low rate because of the stability of the interface. Because the mixing is limited, once the plume has left the tributary, it could exist for several tidal cycles as a separate entity within the main estuary. Because of their thinness (generally <2m) it is possible that detached plumes may remain undetected by conventional sampling. The dynamics of plumes and their associated fronts have been studied by Garvine (1974), Garvine & Monk (1974) and Kao, Park & Pao (1977).

## MIXING PERIODS

In some estuaries intense mixing periods have been observed on the ebb tide, during which there is a rapid thickening of the surface layer together with an increase in surface salinity. These events were originally described by Partch & Smith (1978) in the Duwamish Estuary. They first observed mixing just after high water, and the mixed water was advected downstream at least 2 km during the ebb tide. Further investigation by Gardner & Smith (1978) suggested that the cause was an internal hydraulic jump which occurred at a sharp change in depth of the estuary. The extent of the mixing seemed to be limited by the fact that the salt wedge was only present at the topographic step for a limited part of the ebb tide. A particularly interesting feature of this was that no remnant of the mixing period was distinguishable returning on the flood tide. Very similar intense mixing periods have been observed in the Tees Estuary (Lewis 1985), and it seems probable that they may result from topographic features or bridge supports (New, Dyer & Lewis 1986).

One possible mixing mechanism involves the creation of standing lee waves on the density interface as the stratified flow passes over a topographic step. If the flow is fast enough the waves grow in amplitude and break, causing mixing. At lower current velocities the waves may not break, but as the current wanes, they will be released and travel against the current, either up or down-estuary depending upon whether they were generated on ebb or flood. Observations have shown that generation of internal waves can occur on both ebb and flood currents, and that these waves can break and cause mixing elsewhere in the estuary. However, breaking at the step will be particularly important near spring tides, and may be the cause of the destratification observed in some estuaries (e.g. Haas 1977; Ruzecki & Evans 1985). This process could result in patches of higher salinity mixed water at the surface that are the converse of the low salinity patches caused by river discharge events, but which may also persist in the estuary for long periods. However, since they have a higher surface salinity than the water upstream and downstream, they would have a tendency to flow lengthways at an intermediate level and become incorporated in the upper halocline, rather than remain as a surface patch.

## CONCLUSIONS

Estuaries are naturally variable since the mixing of salt and freshwater takes place under energetic tidal flow conditions within a complicated topographic situation. Variability of the river flow and of the weather imposes further unsteadiness so that mixing and discharge of pollutants may be very variable and intermittent. One result of this may be the presence within the estuary of water masses with different source characteristics and time histories. Because of the variability and the presence of transient features in the system, prediction of pollutant distribution using dispersion coefficients must rely on careful validation against field data, and the results might be biased by a sampling scheme that does not take the intermittency into account.

At the moment there does not appear to be a sufficiently comprehensive study of any one estuary to be able to judge the relative importances of the various processes. Obviously different processes may be dominant at various times in the same estuary, and different estuaries will show varying responses, especially to those processes dependant on topographic details. Nevertheless, it is important that the effects of secondary flows on dispersion in oscillating stratified flows is better quantified, as well as understanding the movement and alteration of transient events passing through the estuary.

## REFERENCES


**Abraham, G., de Jong, P. & van Kruiningen, F.E. (1986).** *Large-scale Processes in a Partially Mixed Estuary. Physics of Shallow Estuaries and Bays* (Ed. by J. van de Kreeke), pp. 6–21. Springer-Verlag, Berlin.

**Bathurst, J.C., Thorne, C.R. & Hey, R.D. (1977).** Direct measurements of secondary currents in river bends. *Nature (London),* **269**, 504–506.

**Bowden, K.R. & el Din. S.H. (1966).** Circulation, salinity and river discharge in the Mersey estuary. *Geophysical Journal Royal Astronomical Society,* **10**, 383–400.

**Dyer, K.R. (1977).** Lateral circulation effects in estuaries. *Estuaries, Geophysics and the Environment,* pp. 22–29. National Academy of Sciences, Washington.

**Dyer, K.R. (1982).** Localized mixing of low salinity patches in a partially mixed estuary (Southampton Water, England). *Estuarine Comparisons* (Ed. by V. Kennedy), pp. 21–36. Academic Press, New York.

**Einstein, H.A. & Shen, H.W. (1964).** A study on meandering in straight alluvial channels. *Journal of Geophysical Research,* **69**, 5239–5247.

**Elliott, A.J. (1978).** Observations of the meteorologically-induced circulation in the Potomac Estuary. *Estuarine Coastal Marine Science,* **6**, 285–299.

**Fischer, H.B. (1969).** The effect of bends on dispersion in streams. *Water Resources Research,* **5**, 496–566.

**Fischer, H.B. (1972).** Mass transport mechanisms in partially stratified estuaries. *Journal Fluid Mechanics,* **53**, 673–687.

**Fischer, H.B. (1976).** Mixing and dispersion in estuaries. *Annual Review Fluid Mechanics,* **8**, 107–133.

**Gardner, G.B. & Smith, J.D. (1978).** Turbulent mixing in a salt wedge estuary. *Hydrodynamics of Estuaries and Fjords.* (Ed. by J.C.J. Nihoul), pp. 79–106. Elsevier, Amsterdam.

**Garvine, R.W. (1977).** River plumes and estuary fronts. *Estuaries, Geophysics and the Environment,* pp. 30–35. National Academy of Sciences, Washington.

**Garvine, R.W. & Monk, J.D. (1974).** Frontal structure of a river plume. *Journal Geophysical Research,* **79**, 2251–2259.

**Haas, L.W. (1977).** The effect of the spring-neap tidal cycle on the vertical structure of the James, York and Rappahannock Rivers, Virginia,USA. *Estuarine Coastal Marine Science,* **5**, 485–486.

**Holley, E.R., Harleman, D.R.F. & Fischer, H.B. (1970).** Dispersion in homogeneous estuary flow. *Journal Hydraulics Division, ASCE,* **96**, HY8, 1691–1709.

**Kao, T.W., Park, C. & Pao, H.P. (1977).** Buoyant surface discharge and small oceanic fronts: A numerical study. *Journal Geophysical Research,* **82**, 1747–1752.

**Ketchum, B.H. (1952).** *Circulation in Estuaries.* Proceedings of the 3rd Conference on Coastal Engineering, 65–76.

**Lewis, R.E. (1985).** Intense mixing periods in an estuary. *Models of Turbulence and Diffusion in Stably Stratified Regions of the Natural Environment.* Proceedings of the IMA Conference, Cambridge. Oxford University Press, Oxford.

**New, A.L., Dyer, K.R. & Lewis, R.E. (1986).** Predictions of the generation and propogation of internal waves and mixing in a partially stratified estuary. *Estuarine Coastal Shelf Science, zz,* 199–214.

**Nunes, R.A. (1982).** *The dynamics of small-scale fronts in estuaries.* PhD. thesis, University College, North Wales.

**Partch, E.N. & Smith, J.D. (1978).** Time-dependent mixing in a salt wedge estuary. *Estuarine Coastal Marine Science,* **6**, 3–19.

**Ruzecki. E.P. & Evans, D.A. (1985).** *Temporal and spatial sequencing of destratification in a coastal plain estuary.* Proceedings of the Symposium on Circulation Patterns in Estuaries. Virginia, June 1985.

**Simpson, J.H. & Nunes, R.A. (1981).** The tidal intrusion front: an estuarine convergence zone. *Estuarine Coastal Marine Science,* **13**, 257–266.

**Smith, R. (1980).** Buoyancy effects upon longitudinal dispersion in wide well-mixed estuaries. *Philosophical Transactions Royal Society,* **296**, 467–496.

**Smith, R. (1982).** Contaminant dispersion in oscillatory flows. *Journal Fluid Mechanics,* **114**, 379–398.

**Thorne, C.R. & Hey, R.D. (1979).** Direct measurements of secondary currents at a river inflexion point. *Nature (Lond.),* **280**, 226–228.

**Weisberg, R.H. (1976).** A note on estuarine mean flow estimation. *Journal Marine Research,* **34**, 387–394.

**West, J.R., Knight, D.W. & Shiono, K. (1984).** A note on flow structure in the Great Ouse estuary. *Estuarine Coastal Shelf Science,* **19**, 271–290.

# Transfer mechanisms for dissolved pollutants in estuaries

R. E. LEWIS

*ICI PLC, Brixham Laboratory, Freshwater Quarry, Brixham, Devon TQ5 8BA*

## SUMMARY

**1** Various mechanisms control the dispersion of dissolved pollutants in estuaries. However, at any given position there is a tendency for only one or two of these mechanisms to dominate the physical processes.
**2** By using observations of the salinity and velocity distribution, it is possible to estimate the relative contribution of these processes to the overall longitudinal dispersion at different positions along an estuary.
**3** Data obtained on the Tees suggest that the tidal average dispersion is dominated by processes associated with the vertical circulation and the salinity oscillation.

## INTRODUCTION

Many species of the littoral and sub-littoral fauna of an estuary are sedentary or sessile. To a large extent, these animals depend upon the movement of the estuary waters to bring them food and to disperse their larvae to new settlement areas. However, these same water movements may expose the fauna to varying concentrations of dissolved and particulate materials. Since some of these materials may be pollutants, it is important to understand the causes of the water motions, which transport and disperse wastes, if the ecological impact is to be properly assessed.

As a result of the natural fluctuations of salinity, temperature and suspended solids concentration, estuarine animals have evolved adaptations giving a degree of tolerance to extreme conditions. When the inflowing rivers to an estuary are in spate, there may be appreciable decreases in salinity at positions along its length. Many estuarine species can withstand such declines in salinity for relatively short periods, although prolonged exposure to such an extreme could be fatal. However, such prolonged exposure rarely occurs since, even when high freshwater inflows are sustained, the salinity of the estuary readjusts to its 'normal' condition. In seeking to understand the mechanisms which cause this readjustment, a clearer picture of the transport and dispersion mechanisms in estuaries is obtained. Salt is a convenient form of natural tracer so that a study of the salt balance of an estuary is a good method for studying the combined mechanisms which produce transport and dispersion. To a first approximation, it may be assumed that these same mechanisms are responsible for the transport and dispersion of pollutants.

This paper considers three of the mechanisms which cause salt to be dispersed within an estuary and estimates the magnitudes of these processes at different locations along the estuary length. The additive effect of these mechanisms in carrying salt upstream is compared with the downstream advection of salt by the freshwater flow.

## RESIDUAL DRIFT

At the low water stage of tide an estuary can be thought of as empty, leaving only the residual volume of brackish water and the freshwater inflow at the head. As the tidal height rises at the seaward end of the estuary, seawater pushes inwards as a wedge along the bottom due to the greater density of saltwater compared with the less saline brackish waters. If the estuary is shallow and the rise in tidal height rapid, turbulence generated near the bed may be strong enough to render the water column homogeneous. This occurs in estuaries such as the Solway Firth and the Humber. If the tidal energy is not great enough to cause complete vertical mixing, the flow remains stratified or partially stratified during the flood.

In the Tees estuary, the frictional drag of the sides and bed of the estuary slows the flow in the upper reaches so that at 1 hour before high water, the water surface slopes steeply upwards in the downstream direction (Fig. 1). The inertia of the incoming tide causes the upstream movement to continue when high water stage has been reached at the seaward

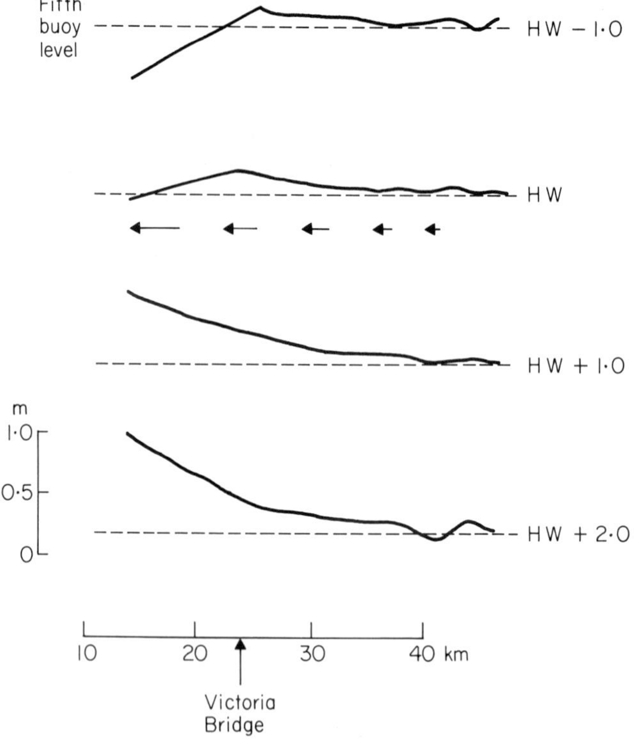


FIG. 1. Longitudinal variation in the surface elevation of the Tees estuary on a tide of mean range. Heights are expressed relative to the level at the fifth buoy, near the estuary mouth.

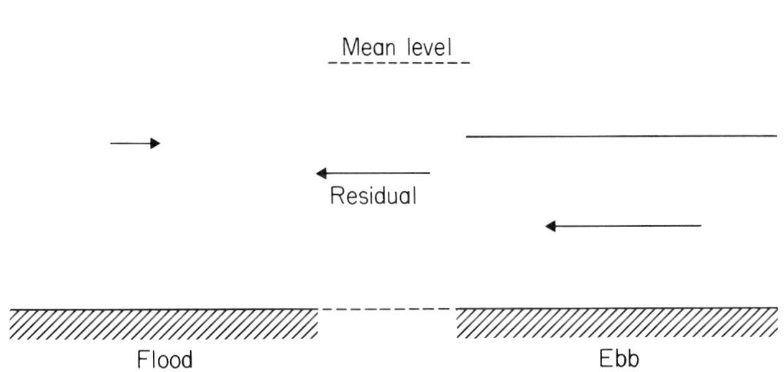


FIG. 2. Generation of a tidal-average residual current.

end of the estuary, thus levelling the surface in the upper reaches. By 1 hour after high water, the water level at the estuary mouth is falling rapidly but the ebb movement in the upper estuary is constrained by frictional drag. This results in a pronounced downward slope of the water surface towards the sea. Near low water, the seaward movement, aided by the freshwater flow, continues and the surface slope decreases until the flooding tide constrains the seaward drift.

The net effect of these processes over a tidal period is summarized in Fig.2. Over the flood tide, the average water depth exceeds the mean tide level and the depth mean velocity is 'slow'. During the ebb tide, the average water depth is lower than the mean tide level and, therefore, to return the flood volume to the sea, the ebb tide current is 'fast'. Thus, the tidal mean residual current is directed towards the sea. It should be noted that this phenomenon, termed the Stokes Drift (Longuett-Higgins 1969), is generated by the tidal oscillation and does not take into account the drift due to the freshwater flow. Observations of the tidal average current in the Tees estuary (Fig. 3) show that it approaches $0 \cdot 2$ m $s^{-1}$ near Victoria Bridge when the current due to the freshwater flow was only about $0 \cdot 01$ m $s^{-1}$. When this residual drift is combined with the tidal fluctuation in salinity, it results in a flux or transport of salt towards the sea.

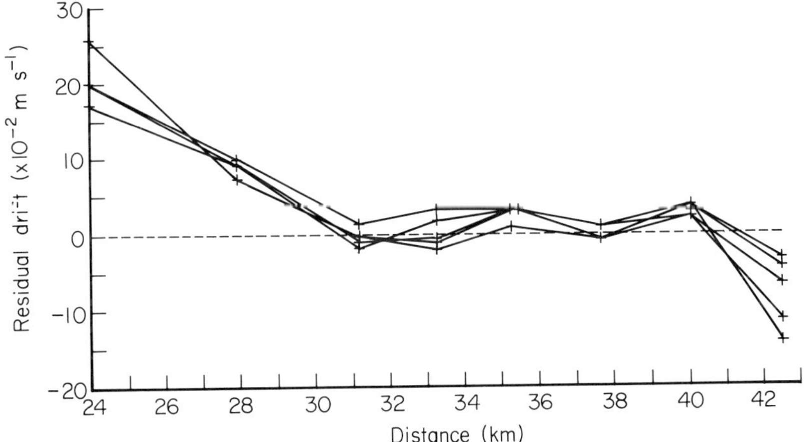


FIG. 3. Observed residual currents in the Tees estuary on five successive days over a period of spring tides.

## VERTICAL CIRCULATION

As the tide floods, the slope of the water surface remains relatively level in the middle and lower reaches of the Tees estuary (Fig. 1). However, on the ebb, the movement towards the sea is resisted by the longitudinal pressure gradient. The latter occurs because the total pressure increases from the head to the mouth of the estuary due to the channel divergence. The restriction on the ebb movement leads to a pronounced slope in the water surface on the ebb tide (Fig. 1). Thus, taken as an average over the whole tidal period, the water surface slopes downwards towards the sea. The longitudinal pressure gradient force due to this slope is a constant function of the depth. By contrast, the counteracting force due to the horizontal density gradient increases linearly with depth (Officer 1976). Therefore, in the near-surface waters the surface slope drives the flow towards the sea since it dominates the density gradient force. At depth, this situation is reversed because a point is reached at which the density gradient force outgrows that due to the surface slope. Consequently, the tidal mean movement in an estuary often displays a characteristic vertical circulation with surface water being taken seawards and the deeper portion of the water column being carried landwards (Fig. 4).

Fig. 4 also shows the vertical salinity distribution typical of a partially stratified estuary. The nature of the circulation is such that the seaward and landward volume flow rates balance. Thus, the volume of water flowing towards the sea carries less salt, due to the lower salinity of the upper layer, compared with the landward transport of salt in the lower layer. The schematized indication of this vertical variation in the salt flux is shown in the figure. Therefore, when averaged over depth, the net effect of the vertical circulation transport is to carry salt towards the land.

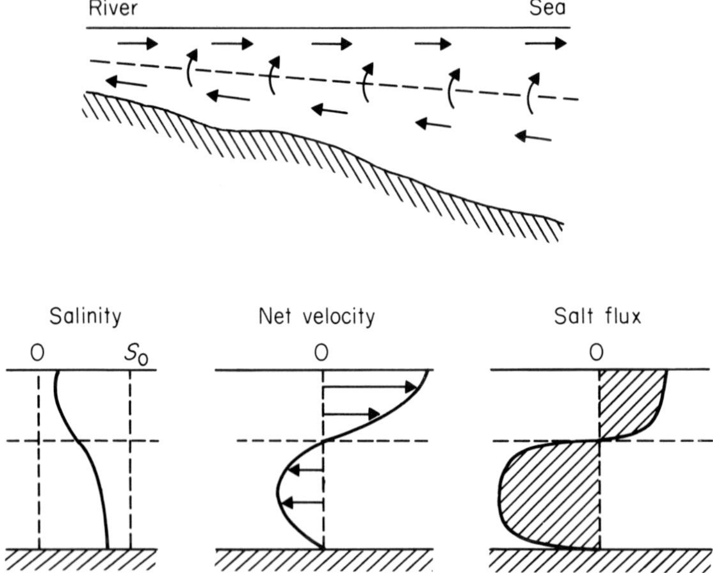


FIG. 4. Tidal-average vertical circulation in an estuary. The flux of salt towards the sea is less than the landward transport.

## VERTICAL MIXING

In an estuary which has no vertical variation of salinity or temperature, small-scale turbulent eddies are principally responsible for the vertical transport of pollutants. However, in estuaries which display some degree of stratification, the turbulence becomes damped and vertical transfer is very limited. For example, the coefficient of vertical diffusion is of the order of 100 $cm^2 s^{-1}$ in a homogeneous estuary and only about 1 $cm^2 s^{-1}$ in a stratified system.

The presence of a stable interface at an intermediate depth can result in some interesting modifications to the fluid flow. Although there may be no signs of disturbance of the estuary surface, waves can form on the interface separating the fresher surface layer from the more saline waters beneath. That such 'internal' waves can exist was first suggested by the Norwegian mathematician, Vilhelm Bjerknes (Gill 1982). Under light wind conditions, the presence of internal waves may be seen as slicks on the sea surface where natural oils are forced to converge by the motion of the internal oscillations. These natural oils damp out the small capillary waves resulting in the smooth appearance of the slicks.

When water flows over an irregular bottom topography, formed by scour holes or sandwaves for example, the pressure fluctuations result in the formation of waves on the density interface. These may be stationary lee waves which only become free to move when the flow slows down, as at the end of a flood or ebb tide. If the roughnesses on the bottom are large and the flow is strong, the waves which try to form on the interface may immediately become unstable and a large amount of turbulence is generated. Such an event is termed an 'internal hydraulic jump'. When it occurs, internal waves may also form on the interface (Fig. 5).

Internal waves which are free to travel are likely to move into regions where they become unstable and break, just as waves break on a beach. Whatever the cause of the instability, the formation of internal hydraulic jumps and breaking internal waves can produce vigorous vertical mixing with the concomitant rapid vertical transport of

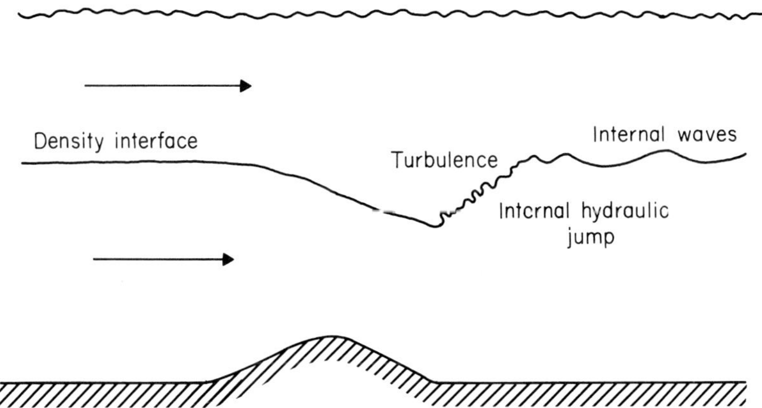


FIG. 5. An internal hydraulic jump forms on a density interface downstream of an obstacle on the estuary bed. When the jump is weak, energy may be radiated away as internal waves.

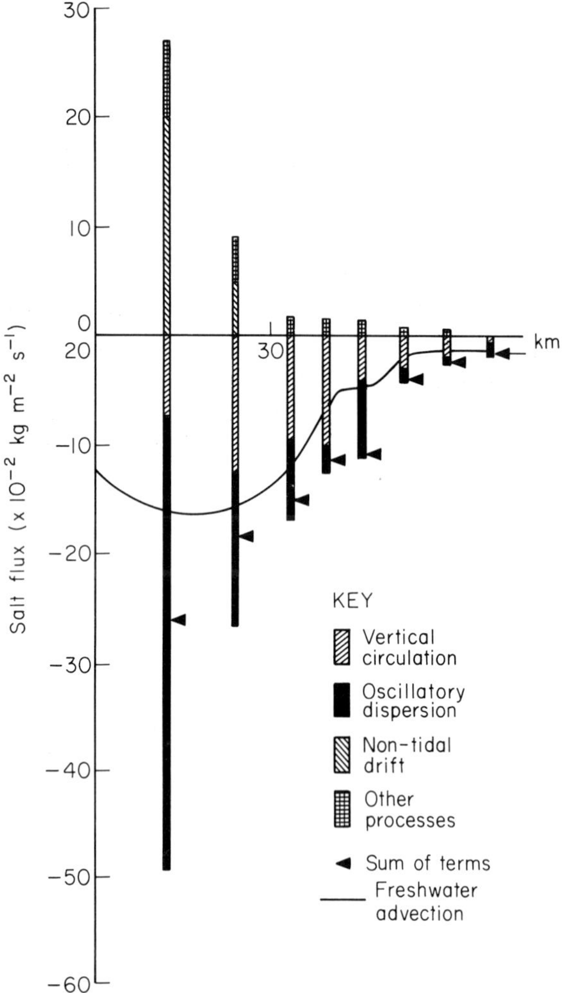


FIG. 6. Components contributing to centre channel flux of salt compared with the cross-sectionally averaged transport necessary to counter advection by the freshwater flow.

material. Such events also influence the longitudinal transport of salt in that the flood and ebb movements of salt are modified by the changes in the vertical salinity distribution.

This results in an overall flux of salt, either upstream or downstream. In addition to the effects of vertical mixing, this oscillatory component of the salt flux can be due to other mechanisms such as transverse secondary circulations and trapping in small embayments along the estuary.

## MAGNITUDES OF DISPERSION MECHANISMS

To determine the significance of the dispersion mechanisms, it is possible to estimate their magnitudes from observational data. This evaluation was undertaken using data for the Tees estuary (Lewis & Lewis 1983). Fig. 6 compares the magnitudes of the dispersive fluxes arising from residual drift, vertical circulation and salinity oscillations with that necessary to counter the seaward advection of salt by the freshwater flow. Allowing for the fact that the observational data were obtained in the centre channel of the Tees and, therefore, give rather higher fluxes than the cross-sectional mean, there is an approximate balance between the salt advected seawards and that taken landwards by dispersive processes.

Fig. 6 clearly shows how, on tidal average, different mechanisms control the dispersion of pollutant in an estuary. Over most of the estuary length, the dispersion due to vertical circulation and salinity oscillation dominates. The vertical circulation varies appreciably in magnitude along the estuary but these fluctuations are smoothed out by complementary variations in the oscillatory dispersion term. In the upper estuary the non-tidal drift effect becomes significant, carrying salt seawards to counter the high upstream transport of salt by the vertical circulation and the salinity oscillation. Thus, the individual mechanisms which disperse pollutants in estuaries show a marked positional variation and at any given position they combine together to produce the overall longitudinal dispersion of dissolved material.

## REFERENCES

**Gill, A.E. (1982).** *Atmosphere - Ocean Dynamics,* Chapter 4. Academic Press, London.

**Lewis, R.E. & Lewis, J.O. (1983).** The principal factors contributing to the flux of salt in a narrow, partially stratified estuary. *Estuarine, Coastal and Marine Science,* **16**, 599–626.

**Longuett-Higgins, M.S. (1969).** On the transport of mass by time varying ocean currents. *Deep Sea Research,* **16**, 43–447.

**Officer, C.B. (1976).** *Physical Oceanography of Estuaries (and Associated Coastal Waters),* Chapter 4. John Wiley & Sons, New York.

# Some factors affecting the fate of estuarine sediment hydrocarbons and trace metals in Milford Haven 1978–82

D.I. LITTLE, S.E. HOWELLS,
T.P. ABBISS* AND DALE ROSTRON

*Field Studies Council Oil Pollution Research Unit, Orielton Field Centre, Pembroke, Dyfed. SA71 5EZ and*
**Leonard Wills Field Centre, Nettlecombe Court, Williton, Taunton, Somerset, TA4 4HT*

## SUMMARY

**1** The objects of the study reported here were to examine spatial variability of sediment hydrocarbon and trace metal concentrations in relation to industrial and environmental variables,and to make a preliminary comparison of the behaviour and fate of the contaminants.

**2** Sediments were sampled and analysed for a range of inorganic and organic contaminants, in addition to major element geochemistry, sedimentology and permeability. Data were analysed by bivariate and multiple linear regression.

**3** Results indicated that total hydrocarbon loads were predominantly anthropogenic and ranged from 1 mg $kg^{-1}$ in coarser sediments to 615 mg $kg^{-1}$ in finer sediments, with some indications of a slight increase since 1978. This range is comparable to other moderately polluted estuaries and harbours with fine sediments.

**4** Trace metal levels (extractable Co, Cr, Cu, Ni, V and Zn) although quite low for fine sediment environments, were found to indicate point sources more reliably than hydrocarbon concentrations, GLC trace patterns or total metal concentrations, in spite of the major chronic and acute inputs of oil being comparatively well known.

**5** Multiple regression equations explained over 92% of hydrocarbon variability with combinations of sediment organic matter, total Cu and mean grain diameter as independent variables. These results were interpreted as reflecting both sewage and refinery effluent inputs.

**6** Extractable trace metal equations were dominated by the total metal present in the sediment which accounted for 21–82% of the percentage variation. Aluminium, Fe and ø mean accounted for up to 91% of the variation of the total concentrations of trace metals, reflecting the importance of aluminosilicates in controlling trace metal distributions.

**7** The results suggest that, in spite of the successful management of the oil industry's effluents and shipping, some signs of both localized and general contamination are appearing after 25 years of development. This is attributed to the overwhelming importance of sedimentological factors in trapping contaminants in the estuary.

## INTRODUCTION

Milford Haven (south-west Wales, U.K.) is of ecological importance because of its location in an area of overlap between the ranges of southern and boreal–arctic marine species. The marine flora and fauna are consequently diverse, and extend far up the estuary into the Cleddau rivers. The estuary is also of national importance for shelduck and teal and supports significant resident and migratory wader populations. Biological surveys have been published by Dias (1960) on plankton, Moyse & Nelson-Smith (1963), Nelson-Smith (1964, 1965, 1967) on rocky shore communities, and Nelson-Smith & Case (1984) on rocky shore and sublittoral communities. Crapp (1971) resurveyed twenty-two of the original thirty-six transects established by Moyse and Nelson-Smith. More restricted surveys were carried out by the Field Studies Council in the following 10 years. In 1979 and 1982 twenty-one transects were resurveyed. Results of these studies and of follow-up work after oiling incidents have been reported by Blackman *et al.* (1973), Baker (1976), Dicks & Hartley (1982), and Woodman & Little (1985).

These studies were predominantly carried out on rocky shorelines. Rigorous port management, including the routine use of chemical dispersants for treatment of all but heavy fuel oil spills, may divert the impact of floating oil from the predominantly rocky shoreline to the sedimentary environment. Inputs from refinery effluents and road run-off are already well mixed into the water column. The scavenging of organic and inorganic contaminants by suspended sediments and the natural or chemical dispersion of floating oil perhaps suggest that bottom-sediment communities are more likely to be affected by pollution. This hypothesis led to the initiation of a benthic macrofaunal monitoring programme at twenty-six stations in the lower and central parts of the estuary (Addy 1976).

During these investigations, very little was published on sediment grain size and nothing on sediment hydrocarbon and trace metal concentrations. To fill some of these gaps, the Welsh Office and the Institute of Petroleum financed a project in 1978 to collect and analyse fifty sediment samples. In 1982 similar samples were obtained at 130 stations as part of the continuing biological monitoring programme. The sedimentological and geochemical studies have become an important part of estuarine research in this area, partly because of the lack of large population centres and heavy industry (until 1960), which means that pollution studies are less hampered by the common problems in estuaries of identifying point sources of contamination and separating them from high 'background' levels. The estuary also stands apart from many of those described in the literature because of its comparatively low freshwater inflow and turbidity, and the high proportion of hard substrata. In spite of these characteristics, sedimentological factors have been found to account for a significant proportion of the spatial variability of several contaminants. An examination of the reasons for this variability and comparisons between contaminants are the objectives of this study.

## THE STUDY AREA

Milford Haven is a drowned river valley enclosing over 110 km of coastline with an entrance only 2·5 km wide. The approximate daily freshwater input from the Eastern and

Western Cleddau Rivers is between $1 \cdot 05$ and $1 \cdot 18 \times 10^6$ m$^3$, with other inputs from the Pembroke, Carew, Cresswell and Gann Rivers and Sandyhaven Pill. However, this volume is much less than the volume change from high to low water of approximately $1 \cdot 82 \times 10^8$ m$^3$ between the estuarine mouth and the junction of the Cleddau Rivers at Picton Point (Nelson-Smith 1965). Significantly reduced salinity (less than 33‰) is found, therefore, only in the reaches above the Cleddau Bridge (the Daucleddau), the Pembroke, Carew, Cresswell and Gann Rivers, and in Sandyhaven Pill.

The location, geology and bathymetry of the study area are summarized in Figs 1 and 2. The bathymetry of the estuary is characterized by a main channel flanked by extensive, shallow subtidal banks. At the estuarine mouth the seabed and intertidal areas are largely exposed Old Red Sandstone bedrock, or of coarse boulder or pebble deposits ranging gradually through sands and shell sands to more muddy sediments further upstream (Fig. 2). The deep-water channel provides a depth of water of 12 m at low water spring tides as far east as Wear Point, over 10 km from the mouth of the estuary. With the growth of the

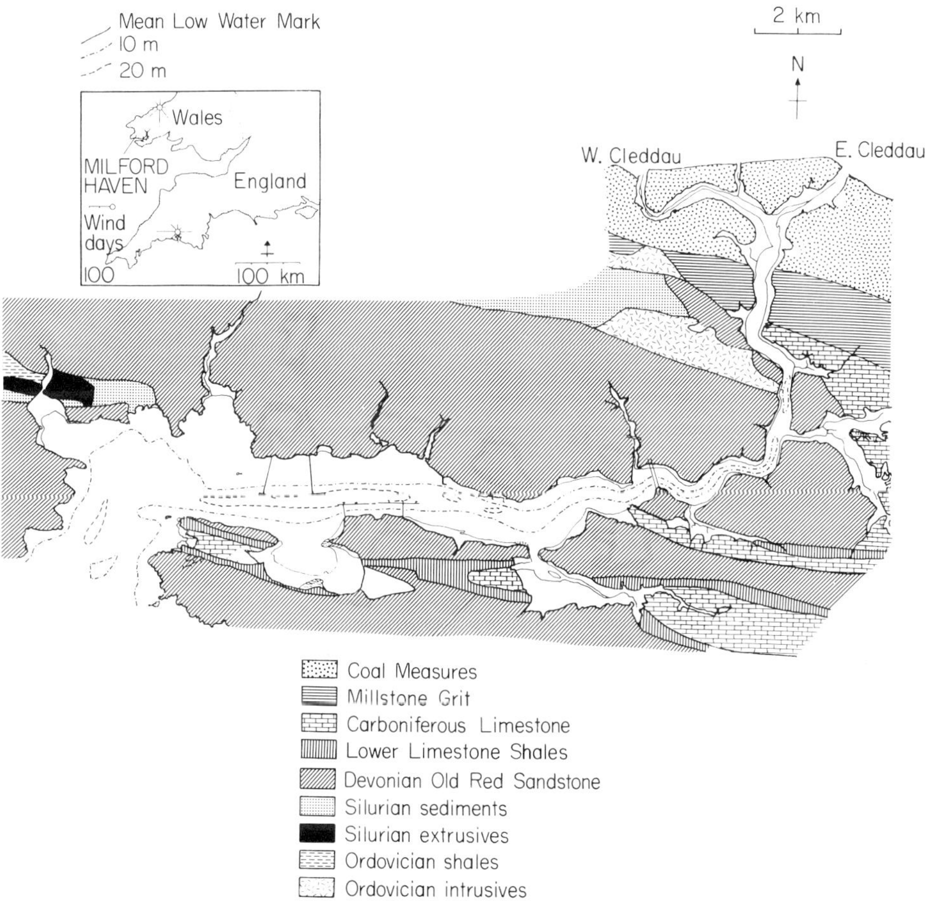


FIG. 1. Location (inset), geology and bathymetry of Milford Haven and the Daucleddau Estuary.

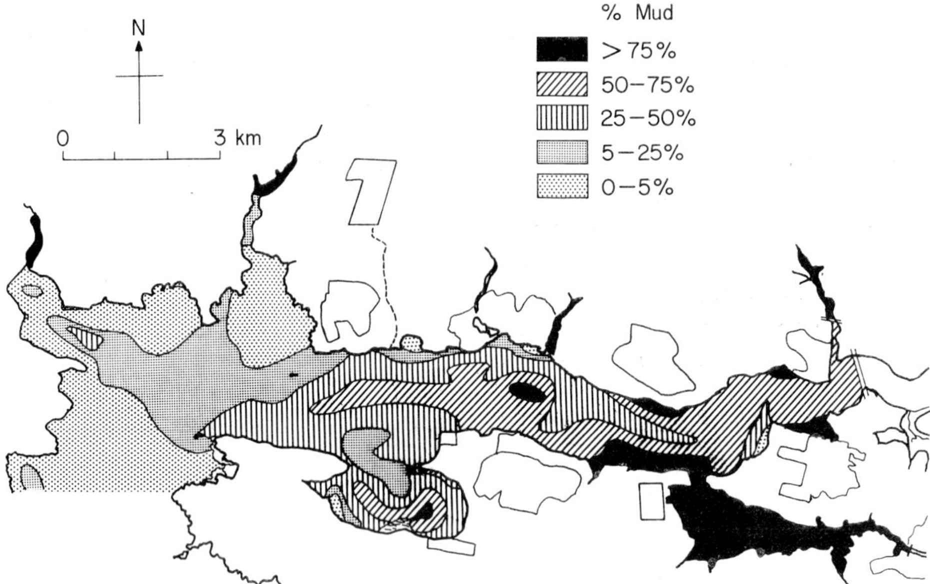


FIG. 2. Percentage mud (<63 μm particle diameter) 1978–79 in Milford Haven sediments.

U.K. oil-refining industry and the growth in tanker sizes, Milford Haven's deep waterway was consequently developed as an oil port.

The potential impact of industrial stresses was investigated by Williams (1971) and Williams & Jolly (1975), who obtained useful data on suspended sediments, freshwater inflow, current, salinity and temperature measurements, in addition to various chemical analyses. The maximum mid-stream suspended sediment concentration in the lower estuary was found to be about 10 mg $l^{-1}$ which is particularly low for an estuary, although the measurements were made in midsummer at the time of the seasonal low. Refinery effluent suspended solid loads were *c.* 50 mg $l^{-1}$. Current measurements indicated that vertical velocity gradients were subordinate to lateral velocity gradients, especially near principal channels. Maximum spring tidal current velocities were recorded at Pembroke Ferry(102 cm $s^{-1}$) with a depth mean velocity of 80 cm $s^{-1}$, compared to 33 cm $s^{-1}$ at Chapel Rocks in the mouth of the estuary. Flood tide depth mean velocities upstream of Milford Haven usually exceeded ebb tide depth mean velocities by *c.* 45 cm $s^{-1}$. However, the reverse was true downstream of the BP jetty in the eastern entrance channel. There is strong evidence to suggest the existence of an ebb-flood channel system in the mouth, since the eastern channel ebbs for much longer than it floods, while the pattern in the western channel is much more symmetrical. Upriver, the flood tide lasts longer and, near the bottom, tends to be more rapid.

Although a salt wedge circulatory pattern is probably restricted to the Daucleddau, density profiles indicated that, at depth, water of a given density is normally found about 2 km upstream of the equivalent density near the surface (Williams 1971). However, this may still only represent a vertical salinity gradient of 1‰ in the lower reaches. Drogue tracking studies revealed that the tidal excursion was *c.* 7 km on a spring tide and about half that on a neap tide. These findings depict a physically-dominated environment which, coupled with a low biological oxygen demand and low ambient heavy metal concentrations, perhaps suggested an estuary well suited to industrialization.

The first refinery was built by Esso Petroleum Company and completed in 1960, with an original design capacity of 4·5 Mt $year^{-1}$, subsequently increased to 8·7 Mt $year^{-1}$, although the refinery closed in 1983. In 1961 BP's Angle Bay Ocean Terminal was opened, with a capacity of 9·2 Mt $year^{-1}$, the crude oil being transferred by pipeline over 100 km to the BP refinery at Llandarcy (Swansea). This installation also closed its crude refining plant in 1985. Texaco opened a refinery in 1964, followed by Gulf in 1968. These two companies, with a joint capacity of 14 Mt $year^{-1}$, completed in 1982 the construction of a catalytic cracker unit to improve and extend their refining abilities. The latest refinery built was that of Amoco (U.K.) Ltd, which came on stream in 1973 with a capacity of 5 Mt

TABLE 1. Milford Haven Conservancy Board oil spill statistics: tankers and oil terminals

| Year | No. of pollution incidents | Tonnage of cargo $10^6$ t | No. of spills per 100 ships | No. of spills per $10^6$ t | Tonnage spilled |
|---|---|---|---|---|---|
| 1961 | 45 | 9·9 | 4·2 | 4·6 | - |
| 1962 | 33 | 11·5 | 2·8 | 2·9 | - |
| 1963 | 28 | 13.0 | 2·3 | 2·1 | 9·8 |
| 1964 | 34 | 17·7 | 2·4 | 1·9 | 8·8 |
| 1965 | 83 | 24·9 | 4·2 | 3·3 | 35·8 |
| 1966 | 72 | 28·9 | 3·0 | 2·5 | 30·5 |
| 1967 | 50 | 28·2 | 1·9 | 1·8 | 267·4[a] |
| 1968 | 52 | 30·0 | 1·9 | 1·7 | 13·6 |
| 1969 | 58 | 39·9 | 1·8 | 1·4 | 16·5 |
| 1970 | 56 | 41·3 | 1·6 | 1·3 | 14·9 |
| 1971 | 50 | 43·1 | 1·4 | 1·1 | 161·0[b] |
| 1972 | 59 | 45·7 | 1·6 | 1·5 | 17·8 |
| 1973 | 50 | 53·1 | 1·3 | 0·9 | 2316·5[c] |
| 1974 | 45 | 59·2 | 1·1 | 0·8 | 14·5 |
| 1975 | 25 | 44·7 | 0·75 | 0·6 | 67·2[d] |
| 1976 | 29 | 43·0 | 0·8 | 0·7 | 63·9[e] |
| 1977 | 28 | 38·5 | 0·8 | 0·7 | 10·1 |
| 1978 | 37 | 40·8 | 1·0 | 0·9 | 10·8 |
| 1979 | 32 | 41·0 | 0·8 | 0·8 | 13·1 |
| 1980 | 26 | 38·8 | 0·7 | 0·7 | 4·9 |
| 1981 | 34 | 32·1 | 0·9 | 1·1 | 18·6 |
| 1982 | 21 | 35·7 | 0·5 | 0·6 | 6·4 |
| 1983 | 21 | 30·6 | 0·6 | 0·7 | 12·8 |
| 1984 | 20 | 31·9 | 0·6 | 0·6 | 109.1[f] |
| | | | | | Total 3224 t |

[a] 250 t: 'Chryssi B. Goulandris'.
[b] 150 t: 'Thuntank 6'.
[c] 2300 t: 'Dona Marika'.
[d] 50 t: 'Golden Endeavour'.
[e] 50 t: 'Tollana'.
[f] 105 t: 'Matco Avon' and 'Agapi'.

year$^{-1}$. This refinery also installed a catalytic cracker unit in 1981. All the refineries and the BP terminal have extensive jetties (Fig. 1) which, with the exception of the Gulf jetty, permit the berthing of tankers up to 250 000 t, the original deep water channel having been dredged to a depth of >18 m. The Central Electricity Generating Board power station, opened in 1970, uses fuel oil supplied by the refineries and has a 2 GW electrical capacity. It produces an effluent of *c.* 4550 m$^3$ day$^{-1}$ containing around 10 mg l$^{-1}$ oil, which is then mixed into the cooling water flow of $1 \cdot 36 \times 10^6$ m$^3$ day$^{-1}$. The refineries have effluents with similar oil content and with flow rates of up to 18 200 m$^3$ day$^{-1}$.

The Milford Haven Conservancy Board, established by government legislation in 1958, is responsible for clearing oil spills and prosecuting offenders when appropriate. Milford Haven was relatively undeveloped before the arrival of the oil industry and, because of the outstanding amenity, scientific and educational interest of the area, and its status as part of a National Park, considerable thought was given to the prevention and treatment of pollution. The Conservancy Board and oil companies, in collaboration with local authorities and conservation organizations, have developed a contingency plan for major spills, and work closely together in the routine control of lesser spills. Table 1 shows the rapid growth in oil cargo and the spillages incurred related to total tonnage handled. It can be seen that the efforts of the Board and the companies have reduced the spill rate and the size of spills considerably over the years.

The Welsh Water Authority is responsible for monitoring effluent quality, but the sheer volume of effluents produced means that *c.* 0·5 t day$^{-1}$ oil enters the estuary from these point sources, an annual total of *c.* 180 t (Table 2). Small quantities of heavy metals, from 0·0015 to 1·82 kg year$^{-1}$ depending on the element, are also discharged. Contamination from oil spills, excluding major incidents, is considerably less than the chronic input, averaging only 14·5 t year$^{-1}$. Allowing for road run-off and sewage inputs, a total of 200–240 t oil, most of it well dispersed in water and already associated with suspended particles, enters the estuary each year.

TABLE 2. Typical coastal oil refinery effluent statistics

| | |
|---|---|
| Total oil | 3 – 65 mg l$^{-1}$ |
| Cr | 11 – 1100 μg l$^{-1}$ |
| Cu | 10 – 130 μg l$^{-1}$ |
| Ni | 0·9 – 82 μg l$^{-1}$ |
| Zn | 10 – 190 μg l$^{-1}$ |
| Suspended sediment | 45 – 60 mg l$^{-1}$ |

## MATERIALS AND METHODS

### *Sample collection and field measurements*

Sediments were obtained at a network of stations designed to cover the intertidal and subtidal sediments downstream of the Cleddau Bridge (Fig. 3). Station location was by two or more intersecting lines of sight backed up by compass bearings to prominent landmarks. Except for the intertidal samples taken in 1979 (obtained using a solvent-rinsed stainless steel trowel), a grab sampler was deployed from an inshore vessel to take a 10 cm deep × 0·1 m² sample of the seabed. In the 1978–79 survey a Shipek grab was used, while in 1982 a Day grab was used. In both surveys cores of sediment (6·5 cm diameter × 10 cm depth) were removed from the grab and stored in pentane-rinsed glass or aluminium containers at −20 °C prior to hydrocarbon analysis. In 1978 sediments were sampled only for grain size and hydrocarbon analyses. In 1982 undisturbed cores were taken from the

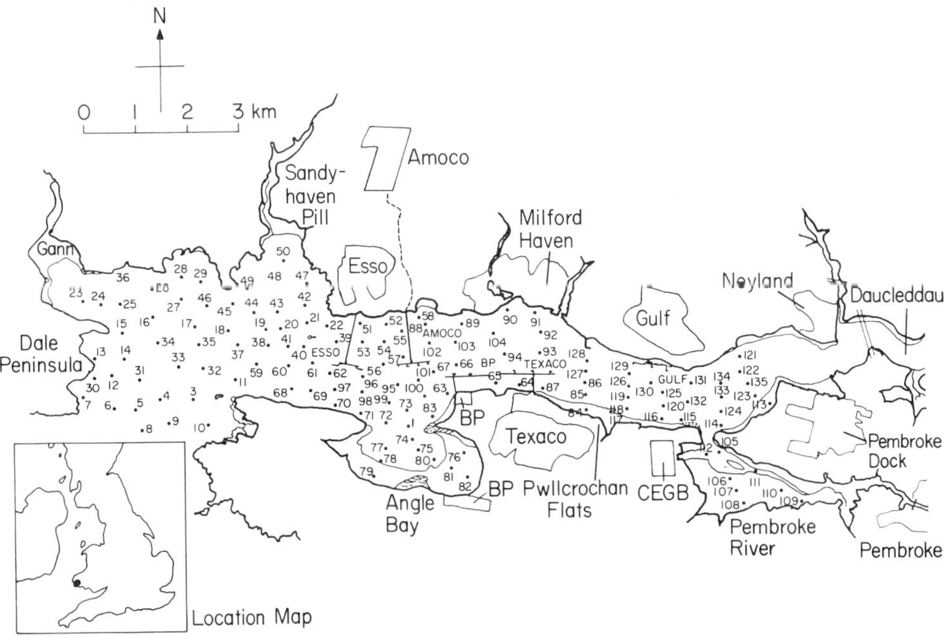


FIG. 3. Sample station layout for the 1982 Milford Haven survey.

centre of the grab at forty-three stations for determination of hydraulic conductivity (water permeability). These were stoppered and stored upright in a field-moist condition. Samples for major and trace element determinations, grain size distribution and organic matter content were taken at all stations in 1982 and stored in clean plastic containers.

Measurements of sediment redox potential at eighty-three stations in 1982 were made in the field on drained sediment using an EIL Model 3030 portable pH/mV meter equipped with a platinum/calomel electrode which had been previously calibrated against a redox buffer (0·003 M potassium ferrocyanide plus 0·003 M potassium ferricyanide). Redox potential was measured at a depth of 2–3 cm in the sediment, and a value of 250 mV added to the field measurement to correct to hydrogen electrode potential (Whitfield 1969).

### *Hydrocarbon analysis*

The sediments were homogenized and extracted twice using methanolic potash and methanol. The residue was then extracted twice with chloroform. This procedure removes the solvent-soluble organic compounds, including hydrocarbons, from the sediment. The methanolic extracts were then shaken in pentane which partitions the hydrocarbons from other organic compounds, the resultant extract being combined with the chloroform extract. These were then separated into aromatic and aliphatic fractions on an alumina:silica gel chromatography column. Solvents were removed and the extracts weighed to give the sediment hydrocarbon concentrations in relation to dry mass of the sediment determined on a separate subsample. Gravimetric analysis provides a simple and standard method for determination of aromatic and aliphatic fractions, the total hydrocarbon concentration (THC) being obtained from their sum.

For the 1978–79 samples a Perkin-Elmer F33 gas chromatograph linked to a Model F56 chart recorder and fitted with a 4 m stainless steel column was used to provide a trace of the analysed fraction. For the 1982 samples this was provided by a Carlo Erba gas liquid chromatograph (GLC) Model FV 4160 under the following conditions: 25 m capillary column, OVI stationary phase, cold on-column injection, 80–300 °C at 5 °C $min^{-1}$, with a 10-minute isothermal period at 300 °C. Methods followed Clarke & Brown (1977) and Baker *et al.* (1984).

### *Major and trace element determinations*

Sediment from the 1982 survey was oven-dried at 80 °C, mixed thoroughly, passed through a 2·0 mm sieve and then ground. This coarse size was selected because of the lack of finer sediments over much of the estuary mouth, and for comparison with hydrocarbon data. 0·1 g (±0·0005 g) of sediment was weighed out and dissolved at *c.* 80 °C in perchloric and hydrofluoric acids. The residue was redissolved in hydrochloric acid. A second subsample of 1·33 g of oven-dry sediment from each station was weighed out and leached in 20 ml of cold 0·5 N HCl by 16 h automatic shaking (Agemian & Chau 1976; Chester & Voutsinou 1981). The supernatant was decanted, the sediment rinsed twice in distilled water, and then re-dried at 105 °C. The weight loss was recorded and the dissolution in $HClO_4$ and HF was carried out on the leached sample as described above.

Simultaneous element determination was carried out using a Phillips PV 8210 inductively-coupled plasma (ICP) atomic emission spectrometer (with 1·5 m air path, PV 8490 power source unit and PV 852 minicomputer). The concentrations of nine major elements were expressed as weight percentage oxides ($Al_2O_3$, $Fe_2O_3$, MgO, CaO, $Na_2O$, $K_2O$, $TiO_2$, $P_2O_5$ and MnO) and concentrations of sixteen trace elements were expressed in mg $kg^{-1}$ (Ba, Ce, Co, Cr, Cu, La, Li, Mo, Nb, Ni, Sc, Sr, V, Y, Zn and Zr). By analysing each sample twice — once using raw sediment and once after the sediment had been leached in dilute HCl — a comparison of residual and extractable element concentrations could be made by subtracting the second series of results from the first, after appropriate correction for weight loss during the extraction. The methods of total sample dissolution and of ICP analysis are summarized by Walsh (1980).

### *Grain size analysis and organic matter content*

Sediment samples from both surveys were split into gravel (>2·0 mm), sand (2 mm–63 μm) and mud (<63 μm) fractions by wet-sieving. The phi notation, where $\varnothing = -\log_2$ particle diameter in mm was used. Sand and gravel fractions were dry-sieved, while the mud fraction was analysed using a Model D industrial coulter counter. Organic material was removed by treatment with hot hydrogen peroxide and sediment dispersed in an ultrasonic bath. The method used is described by Little, Staggs & Woodman (1984).

Data were analysed by microcomputer using the method of moments (Krumbein 1936) which calculates Ø mean particle diameter, Ø standard deviation (a measure of sediment sorting), Ø skewness (symmetry of the size distribution) and Ø kurtosis ('peakedness' of the size distribution). The data were also presented as % mass frequency histograms, % mud and % gravel.

Organic content of the sediment was determined by calculation of the % weight loss from an oven-dry sample on ignition in a muffle furnace for 16 h at 480 °C. This low temperature was used to avoid weight losses due to breakdown of carbonates (Byers, Mills & Stewart 1978) and removal of structural water from the clay fraction (Mook & Hoskin 1982).

### *Hydraulic conductivity and sediment structures*

For cores consisting of sands and sandy silts a constant-head permeameter was used to determine hydraulic conductivity. For finer sediments, a falling-head permeameter was used. These techniques permit the measurement of water flow through the undisturbed core sample under known hydraulic gradients, and thus may help to explain the penetration of contaminants into burrowed but non-depositional sediment areas. The methods used were based on those described by Ackroyd (1969). After testing, the cores were extruded, air-dried, split and photographed.

### *Statistical analysis*

The spatial variability of hydrocarbon and trace metal concentrations was analysed by bivariate and multiple linear regression (Krumbein 1959). Multiple regression allows for the often large number of significant intercorrelations among the independent variables. Some of these variables may be redundant in that their relation to the dependent variable is affected or confounded by their relation to other independent variables. Statistical criteria for the inclusion or exclusion of variables (by analysis of variance) were functions of each variable's $F$ value and its degrees of freedom, and were $P < 0{\cdot}05$ for entry to and $P > 0{\cdot}10$ for removal from the equation. Extensive use was made of *Statistical Package for the Social Sciences* (Nie *et al.* 1975; Hull & Nie 1981). Definitions and abbreviations for the variables are given in Table 3. Coefficients of variation for some of the chemical and granulometric analyses are presented in Table 4.

TABLE 3. Summary of variables used in the statistical analyses with the abbreviations used in the tables

| Variables | Abbreviations |
|---|---|
| Ø mean particle diameter | MEAN |
| Ø standard deviation | SD |
| Percentage mud (<63 $\mu$m) | MUD |
| Percentage gravel (>2 mm) | COA |
| Percentage combustible material (at 480 °C) | COM |
| Percentage $A1_2O_3$ | AL |
| Percentage $Fe_2O_3$ | FE |
| Percentage soluble in 0·5N HC1 | SOL |
| Total Co (mg $kg^{-1}$) | TCO |
| Total Cr (mg $kg^{-1}$) | TCR |
| Total Cu (mg $kg^{-1}$) | TCU |
| Total Ni (mg $kg^{-1}$) | TNI |
| Total V (mg $kg^{-1}$) | TV |
| Total Zn (mg $kg^{-1}$) | TZN |
| Total hydrocarbons (mg $kg^{-1}$) | THC |
| Aliphatic hydrocarbons (mg $kg^{-1}$) | ALI |
| Redox potential (mV) | RED |
| Hydraulic conductivity coefficient (cm $s^{-1}$) | PER |
| Co extractable in 0·5N HC1 (mg $kg^{-1}$) | CO |
| Cr extractable in 0·5N HC1 (mg $kg^{-1}$) | CR |
| Cu extractable in 0·5N HCL (mg $kg^{-1}$) | CU |
| Ni extractable in 0·5N HCL (mg $kg^{-1}$) | NI |
| V extractable in 0·5N HC1 (mg $kg^{-1}$) | V |
| Zn extractable in 0·5N HC1 (mg $kg^{-1}$) | ZN |

TABLE 4. Coefficients of variation derived from replicate analyses

| | |
|---|---|
| Ø mean | 6·34% |
| Ø standard deviation | 4·33% |
| Ø skewness | 3·91% |
| Ø kurtosis | 5·54% |
| Percentage mud | 5·53% |
| Percentage $A1_2O_3$ | 2·99% |
| Percentage $Fe_2O_3$ | 3·49% |
| Percentage Ca0 | 5·46% |
| Co (mg $kg^{-1}$) | 4·43% |
| Cr (mg $kg^{-1}$) | 6·07% |
| Cu (mg $kg^{-1}$) | 4·67% |
| Ni (mg $kg^{-1}$) | 3·18% |
| V (mg $kg^{-1}$) | 3·07% |
| Zn (mg $kg^{-1}$) | 1·37% |
| Aliphatic hydrocarbons (mg $kg^{-1}$) | 19·90% |

# RESULTS

## *Geochemistry*

### *Redox potential*

Corrected values in mV obtained in 1982 are presented in Table 5. Values were not measured at some stations in Angle Bay and in the estuarine mouth. Negative and low positive readings, indicative of reducing sediments, were found to occur at about a third of the stations, notably at stations 56 and 57 close to the Amoco jetty, in the areas of fine sediment between the Texaco jetty and Milford Haven, and near Pembroke Dock. In addition, values were low at Station 84 on Pwllcrochan Flats and at Stations 24–26 on Dale Shelf.

### *Hydrocarbon concentrations*

The aliphatic and THC concentrations measured in 1982 are presented in Table 5. In 1978–79 the range in aliphatic hydrocarbon concentrations was found to be 4–480 mg $kg^{-1}$ (mean value 78 mg $kg^{-1}$). The subtidal sediments ranged in concentration from 4 to 190 mg $kg^{-1}$ with only three of the intertidal oil concentrations exceeding 190 mg $kg^{-1}$. In 1982 the range in aliphatic hydrocarbon concentrations was from $<1$ to 410 mg $kg^{-1}$ (mean value 104 mg $kg^{-1}$) with, again, many of the higher values being obtained from muddy intertidal areas, in particlar Pennar Flats and Pembroke River (Table 5). Generally the lower values in both surveys were obtained from samples taken near the estuarine mouth. Scattered stations with higher oil concentrations were found near some refinery jetties and towns (Fig. 4). In 1982 the range in THC, not determined in 1978–79, was $<1$–615 mg $kg^{-1}$.

The GLC traces of the aliphatic hydrocarbons indicated several common features at many sites: firstly, all the traces showed resolved peaks superimposed on unresolved complex mixture (UCM), with the latter being the major component quantitatively. The UCM spanned a boiling range equivalent to an n-alkane range from n-$C_{11}$ to n-$C_{40}$ and the concentration increased with boiling point, maximizing between n-$C_{26}$ and n-$C_{35}$ (Fig. 5a–f). Many traces also showed a relatively high concentration of UCM over the range equivalent to n-$C_{16}$ to n-$C_{26}$, apparent as a shoulder on the UCM envelope (Fig. 5a, b, c and e). Although the resolved peaks also ranged from n-$C_{11}$ to n-$C_{40}$, there was a variety of distribution patterns. For example, over the range n-$C_{13}$ to n-$C_{20}$, there was frequently a predominance of n-alkanes with an even number of carbon atoms (Fig. 5b), whilst over the range n-$C_{25}$ to n-$C_{34}$ the n-alkanes with an odd number of carbon atoms often predominated (Fig. 5b, e and f). Enhanced concentrations over the range n-$C_{23}$ to n-$C_{36}$ (Fig. 5d), n-$C_{33}$ to n-$C_{40}$ (Fig. 5b and e), or clusters of peaks around n-$C_{17}$ (Fig. 5c) or n-$C_{22}$ (Fig. 5f) were common. The iso-alkanes pristane and phytane were pronounced, with pristane present in greater concentrations than phytane. Similarly, prominent peaks occurred between n-$C_{29}$ and n-$C_{34}$ probably representing triterpanes and steranes (Fig. 5b and e). The aromatic fractions generally showed a similar boiling point range and ratio of resolved peaks to UCM to the respective aliphatic fractions.

TABLE 5. Summary of selected sedimentological and geochemical data (1982) used in the statistical analyses. For variable names, see Table 3

| Station | MEAN | SD | MUD | COA | COM | PER | THC | ALI | RED | AL | FE | SOL | TCO | TCR | TCU | TNI | TV | TZN |
|---|---|---|---|---|---|---|---|---|---|---|---|---|---|---|---|---|---|---|
| 1 | 3·32 | 1·25 | 12·74 | 0·28 | 1·68 | – | 13 | 8 | – | 4·24 | 1·65 | 17·29 | 7 | 26 | 17 | 27 | 30 | 32 |
| 2 | 2·58 | 0·76 | 1·87 | 0·01 | 0·90 | – | – | – | – | 4·07 | 1·99 | 28·57 | 9 | 22 | 20 | 27 | 34 | 17 |
| 4 | −0·84 | 2·28 | 0·84 | 48·78 | 1·31 | – | – | – | – | – | – | – | – | – | – | – | – | – |
| 6 | 0·21 | 3·13 | 6·97 | 4·70 | 2·94 | – | – | – | – | 4·32 | 2·60 | 36·09 | 10 | 18 | 23 | 34 | 43 | 32 |
| 7 | 2·62 | 0·96 | 3·37 | 0·04 | 0·88 | – | 17 | 14 | – | 4·41 | 2·22 | 27·82 | 9 | 24 | 19 | 29 | 37 | 19 |
| 10 | 3·01 | 1·03 | 3·15 | 1·32 | 0·80 | – | 14 | 11 | – | 3·85 | 2·43 | 21·05 | 8 | 32 | 20 | 27 | 40 | 23 |
| 11 | 2·72 | 1·19 | 3·58 | 2·29 | 1·31 | – | – | – | – | 3·43 | 1·46 | 30·08 | 7 | 20 | 18 | 24 | 26 | 16 |
| 12 | 0·54 | 3·38 | 9·49 | 36·38 | 2·53 | – | – | – | – | 5·31 | 2·54 | 18·05 | 8 | 38 | 19 | 32 | 43 | 36 |
| 13 | 0·73 | 3·22 | 9·65 | 30·72 | 1·59 | – | 14 | 8 | – | 5·21 | 2·16 | 37·59 | 9 | 19 | 23 | 32 | 41 | 25 |
| 14 | 2·64 | 2·57 | 19·26 | 8·98 | 2·87 | $2·549 \times 10^{-4}$ | – | – | – | 5·09 | 2·67 | 29·32 | 10 | 28 | 24 | 34 | 42 | 119 |
| 15 | 2·82 | 2·40 | 20·38 | 7·19 | 2·78 | $1·605 \times 10^{-5}$ | – | – | – | 6·08 | 3·03 | 23·85 | 10 | 33 | 23 | 36 | 48 | 48 |
| 16 | 1·47 | 2·77 | 8·79 | 23·61 | 2·26 | $1·476 \times 10^{-3}$ | 36 | 26 | – | 4·64 | 2·60 | 29·32 | 9 | 26 | 20 | 30 | 37 | 48 |
| 17 | −2·68 | 2·02 | 0·57 | 79·72 | 1·40 | – | – | – | – | – | – | – | – | – | – | – | – | – |
| 18 | −2·24 | 2·93 | 1·50 | 70·99 | 1·57 | – | – | – | – | 6·27 | 3·54 | 34·59 | 11 | 34 | 22 | 48 | 54 | 46 |
| 19 | 2·34 | 1·20 | 2·84 | 1·46 | 1·06 | $1·530 \times 10^{-3}$ | 8 | 5 | – | 3·84 | 1·88 | 21·05 | 7 | 23 | 18 | 27 | 34 | 32 |
| 20 | 2·61 | 0·92 | 2·29 | 0·53 | 1·16 | $2·961 \times 10^{-3}$ | – | – | – | 3·42 | 1·83 | 20·30 | 7 | 22 | 16 | 22 | 31 | 27 |
| 21 | −1·17 | 3·38 | 2·01 | 54·02 | 1·36 | – | – | – | – | 4·33 | 2·30 | 30·08 | 9 | 21 | 19 | 28 | 34 | 49 |
| 22 | 1·67 | 2·60 | 10·93 | 18·14 | 2·38 | – | – | – | – | 4·13 | 2·00 | 29·32 | 8 | 17 | 20 | 28 | 32 | 37 |
| 23 | 3·71 | 1·48 | 28·85 | 1·46 | 1·93 | $4·378 \times 10^{-3}$ | 70 | 58 | – | 5·52 | 2·23 | 12·78 | 7 | 36 | 19 | 30 | 39 | 32 |
| 24 | 0·67 | 3·91 | 24·26 | 42·04 | 4·68 | – | – | – | 70 | 7·53 | 3·32 | 19·55 | 10 | 42 | 27 | 43 | 59 | 56 |
| 25 | 3·04 | 2·21 | 21·75 | 8·04 | 3·59 | – | – | – | 30 | 5·21 | 2·63 | 19·55 | 10 | 33 | 22 | 36 | 44 | 41 |
| 26 | 2·71 | 2·97 | 24·23 | 14·42 | 3·36 | – | 134 | 74 | 60 | 6·54 | 3·71 | 24·06 | 11 | 35 | 33 | 44 | 53 | 64 |
| 27 | 0·20 | 3·19 | 7·27 | 43·66 | 2·64 | – | – | – | – | 4·88 | 2·82 | 30·83 | 10 | 25 | 22 | 35 | 41 | 38 |
| 29 | 0·48 | 2·75 | 3·10 | 30·06 | 1·33 | – | 1 | 1 | 290 | 4·86 | 3·27 | 21·05 | 9 | 28 | 19 | 30 | 49 | 45 |
| 30 | 0·70 | 2·73 | 6·10 | 33·82 | 1·15 | – | – | – | 220 | 4·91 | 2·58 | 17·29 | 8 | 29 | 21 | 30 | 41 | 34 |

| | | | | | | | | | | | | | | | | | | |
|---|---|---|---|---|---|---|---|---|---|---|---|---|---|---|---|---|---|---|
| 31 | −0·05 | 3·11 | 9·54 | 49·22 | 2·86 | 1·746 × $10^{-2}$ | – | – | 205 | 4·55 | 2·47 | 39·85 | 10 | 17 | 23 | 38 | 40 | 31 |
| 32 | 1·18 | 1·79 | 1·96 | 10·34 | 2·14 | 1·114 × $10^{-3}$ | – | – | 345 | 3·41 | 2·02 | 39·10 | 9 | 9 | 22 | 29 | 39 | 15 |
| 35 | 0·34 | 3·04 | 1·57 | 30·39 | 2·15 | – | – | – | 340 | 3·96 | 2·45 | 25·56 | 9 | 25 | 19 | 28 | 41 | 25 |
| 36 | 2·95 | 2·45 | 28·31 | 8·69 | 2·93 | 8·805 × $10^{-5}$ | – | – | 110 | 5·62 | 2·54 | 22·56 | 9 | 33 | 24 | 32 | 42 | 36 |
| 37 | 0·11 | 2·48 | 2·29 | 20·58 | 1·43 | – | 10 | 10 | – | 5·42 | 3·63 | 19·55 | 9 | 27 | 18 | 31 | 61 | 40 |
| 38 | 1·70 | 1·75 | 2·35 | 8·34 | 2·47 | 1·195 × $10^{-2}$ | 37 | 21 | 180 | 4·08 | 2·62 | 21·05 | 8 | 24 | 19 | 27 | 41 | 51 |
| 39 | 4·90 | 3·00 | 62·11 | 4·61 | 6·05 | – | – | – | 60 | 9·05 | 3·75 | 28·57 | 11 | 45 | 39 | 45 | 76 | 72 |
| 40 | −1·75 | 2·75 | 2·16 | 61·76 | 3·39 | – | – | – | – | 4·58 | 2·23 | 39·84 | 10 | 22 | 22 | 34 | 40 | 54 |
| 41 | 3·62 | 1·79 | 2·24 | 0·92 | 2·94 | 9·588 × $10^{-3}$ | – | – | 60 | 5·28 | 2·27 | 20·30 | 8 | 31 | 21 | 30 | 42 | 41 |
| 42 | 2·63 | 1·12 | 2·09 | 2·70 | 0·61 | 1·745 × $10^{-3}$ | 69 | 66 | 350 | 3·68 | 2·61 | 13·53 | 7 | 31 | 16 | 23 | 46 | 27 |
| 43 | 0.34 | 1·35 | 0·97 | 12·70 | 1·54 | – | – | – | 400 | 5·33 | 4·32 | 22·56 | 11 | 30 | 18 | 32 | 76 | 57 |
| 44 | 1·52 | 2·12 | 3·01 | 16·14 | 1·75 | – | – | – | 180 | 3·46 | 1·81 | 27·82 | 8 | 20 | 19 | 27 | 30 | 45 |
| 45 | −2·96 | 2·50 | 0·93 | 81·59 | 2·20 | – | – | – | – | 3·94 | 2·14 | 35·84 | 9 | 20 | 20 | 30 | 38 | 27 |
| 46 | −0·68 | 3·77 | 9·25 | 50·37 | 1·87 | – | – | – | – | 5·73 | 2·98 | 30·83 | 10 | 31 | 24 | 35 | 49 | 50 |
| 47 | 3·35 | 0·79 | 7·39 | 0·12 | 0.92 | 1·270 × $10^{-3}$ | 33 | 22 | – | 5·27 | 2·42 | 14·29 | 8 | 33 | 16 | 27 | 42 | 34 |
| 48 | 2·70 | 1·03 | 3·52 | 0·52 | 1·34 | 7·478 × $10^{-3}$ | 42 | 26 | – | 4·11 | 2·40 | 13·53 | 8 | 29 | 17 | 25 | 45 | 31 |
| 49 | 2·96 | 1·04 | 3·01 | 1·18 | 1·04 | – | – | – | – | 4·86 | 2·31 | 16·54 | 8 | 28 | 17 | 25 | 39 | 29 |
| 50 | 3·61 | 0·76 | 15·01 | 0·04 | 1·47 | – | – | – | – | 5·25 | 2·22 | 17·29 | 8 | 31 | 17 | 27 | 38 | 35 |
| 51 | 1·88 | 1·91 | 3·01 | 11·31 | 1·29 | – | 52 | 33 | 110 | 3·52 | 1·52 | 31·58 | 8 | 16 | 18 | 25 | 29 | 25 |
| 52 | 2·63 | 2·06 | 8·87 | 9·58 | 3·09 | – | – | – | 80 | 4·21 | 1·67 | 24·81 | 7 | 26 | 18 | 25 | 31 | 32 |
| 53 | 1·03 | 2·74 | 11·11 | 27·59 | 4·37 | 2·356 × $10^{-3}$ | 40 | 34 | 280 | 2·84 | 2·62 | 36·84 | 10 | 8 | 26 | 32 | 37 | 81 |
| 54 | 2·73 | 2·11 | 12·57 | 6·70 | 2·57 | – | – | – | 70 | 4·05 | 1·72 | 29·32 | 8 | 22 | 20 | 29 | 32 | 30 |
| 55 | 2·96 | 2·74 | 24·17 | 10·26 | 3·45 | – | – | – | 60 | 4·74 | 2·05 | 30·83 | 8 | 22 | 22 | 29 | 38 | 32 |
| 56 | 5·90 | 1·87 | 79·96 | 0·00 | 5·93 | 6·225 × $10^{-6}$ | 305 | 295 | −10 | 10·36 | 4·34 | 22·89 | 12 | 55 | 33 | 47 | 84 | 91 |
| 57 | 5·60 | 1·90 | 73·10 | 0·12 | 5·20 | – | – | – | −60 | 9·39 | 3·93 | 21·81 | 11 | 50 | 31 | 43 | 75 | 80 |

| | | | | | | | | | | | | | | | | | | |
|---|---|---|---|---|---|---|---|---|---|---|---|---|---|---|---|---|---|---|
| 58 | 2·07 | 2·39 | 10·27 | 12·33 | 4·19 | – | – | – | 200 | 5·11 | 3·80 | 19·55 | 8 | 31 | 71 | 31 | 44 | 50 |
| 59 | 3·36 | 2·33 | 28·66 | 4·84 | 4·09 | $3{\cdot}777 \times 10^{-3}$ | 95 | 66 | 155 | 5·99 | 2·71 | 24·06 | 9 | 31 | 34 | 34 | 46 | 49 |
| 60 | −0·18 | 4·22 | 15·83 | 48·35 | 1·18 | – | – | – | 130 | 8·30 | 3·46 | 19·55 | 10 | 43 | 26 | 40 | 60 | 57 |
| 61 | 3·94 | 2·56 | 41·29 | 3·31 | 5·62 | $8{\cdot}871 \times 10^{-6}$ | 134 | 102 | 100 | 7·85 | 3·62 | 24·81 | 11 | 40 | 30 | 43 | 65 | 72 |
| 62 | 3·04 | 3·56 | 38·86 | 19·71 | 6·32 | $1{\cdot}354 \times 10^{-6}$ | 176 | 135 | 125 | 8·14 | 4·22 | 32·33 | 12 | 40 | 44 | 47 | 70 | 81 |
| 63 | 1·69 | 4·25 | 20·01 | 21·49 | 0·52 | – | – | – | 100 | 5·23 | 2·03 | 28·57 | 8 | 26 | 24 | 32 | 38 | 55 |
| 64 | 4·75 | 1·88 | 49·48 | 0·00 | 4·50 | $1{\cdot}826 \times 10^{-6}$ | – | – | 75 | 7·70 | 3·07 | 17·74 | 9 | 42 | 26 | 41 | 58 | 59 |
| 65 | 5·85 | 1·82 | 81·49 | 0·00 | 7·53 | – | 310 | 190 | 55 | 10·37 | 4·30 | 19·05 | 11 | 56 | 33 | 49 | 86 | 86 |
| 66 | 5·99 | 2·13 | 74·88 | 0·32 | 7·14 | $1{\cdot}335 \times 10^{-5}$ | 300 | 180 | 60 | 10·32 | 4·59 | 23·44 | 11 | 55 | 36 | 48 | 88 | 97 |
| 67 | 5·00 | 2·41 | 59·13 | 1·97 | 5·40 | – | – | – | 55 | 9·14 | 3·91 | 21·62 | 11 | 51 | 32 | 47 | 73 | 78 |
| 68 | 3·00 | 1·85 | 12·25 | 4·45 | 1·73 | – | 60 | 52 | 230 | 4·39 | 1·76 | 25·56 | 8 | 21 | 19 | 28 | 31 | 27 |
| 69 | 2·07 | 2·30 | 6·65 | 11·22 | 1·11 | $8{\cdot}561 \times 10^{-4}$ | 29 | 16 | 270 | 4·57 | 0·89 | 24·81 | 8 | 21 | 19 | 27 | 32 | 24 |
| 70 | 2·95 | 1·96 | 14·84 | 4·09 | 1·00 | – | – | – | – | 4·26 | 1·70 | 30·83 | 8 | 18 | 19 | 27 | 30 | 19 |
| 71 | 2·89 | 2·79 | 22·10 | 10·13 | 0·85 | – | – | – | – | 4·50 | 1·87 | 24·06 | 8 | 22 | 18 | 27 | 32 | 32 |
| 72 | 2·90 | 1·60 | 10·64 | 1·81 | 2·77 | – | – | – | – | 5·44 | 2·41 | 32·84 | 9 | 27 | 22 | 32 | 41 | 37 |
| 73 | 3·63 | 1·64 | 19·72 | 0·10 | 2·67 | $1{\cdot}799 \times 10^{-5}$ | 47 | 33 | – | 5·21 | 2·08 | 19·55 | 8 | 26 | 20 | 29 | 37 | 40 |
| 74 | 3·47 | 1·30 | 14·04 | 0·28 | 1·58 | – | – | – | – | 4·82 | 1·88 | 16·54 | 7 | 26 | 16 | 25 | 32 | 30 |
| 75 | 3·93 | 1·64 | 23·78 | 0·10 | 1·79 | – | – | – | – | 5·74 | 2·08 | 12·78 | 7 | 29 | 18 | 27 | 37 | 37 |
| 76 | 3·07 | 1·59 | 14·26 | 0·00 | 0·59 | – | – | – | – | 4·62 | 1·94 | 5·26 | 5 | 25 | 14 | 24 | 35 | 37 |
| 77 | 3·25 | 1·18 | 9·42 | 0·12 | 1·84 | – | 44 | 26 | – | 4·52 | 1·73 | 15·79 | 7 | 26 | 15 | 23 | 29 | 29 |
| 78 | 4·89 | 1·81 | 54·77 | 0·00 | 5·73 | $6{\cdot}482 \times 10^{-6}$ | 245 | 150 | – | 8·51 | 3·34 | 15·79 | 9 | 47 | 26 | 41 | 66 | 66 |
| 79 | 3·70 | 2·29 | 33·57 | 0·35 | 3·41 | – | – | – | – | 7·42 | 3·44 | 7·52 | 8 | 43 | 24 | 35 | 62 | 61 |
| 80 | 2·86 | 1·56 | 13·70 | 0·18 | 1·16 | – | 24 | 21 | – | 3·24 | 1·24 | 5·26 | 4 | 21 | 11 | 16 | 23 | 22 |
| 81 | 2·75 | 1·44 | 10·04 | 0·00 | 1·04 | – | – | – | – | 3·41 | 1·40 | 3·76 | 4 | 20 | 11 | 18 | 26 | 24 |
| 82 | 2·27 | 1·15 | 4·30 | 0·14 | 0·40 | – | 21 | 16 | – | 1·97 | 1·14 | 9·77 | 4 | 17 | 11 | 14 | 20 | 10 |

| | | | | | | | | | | | | | | | | | | |
|---|---|---|---|---|---|---|---|---|---|---|---|---|---|---|---|---|---|---|
| 83 | 3·83 | 2·04 | 34·76 | 0·80 | 2·09 | – | – | – | – | 6·30 | 2·51 | 14·16 | 8 | 34 | 19 | 49 | 46 | 34 |
| 84 | 6·63 | 1·59 | 95·37 | 0·00 | 9·65 | $2·594 \times 10^{-6}$ | – | – | 50 | 12·58 | 5·02 | 16·96 | 12 | 70 | 39 | 56 | 110 | 109 |
| 85 | 4·42 | 2·44 | 53·39 | 4·00 | 3·84 | – | 56 | 37 | 115 | 9·83 | 3·91 | 17·29 | 10 | 48 | 22 | 42 | 77 | 45 |
| 86 | 2·92 | 2·06 | 14·79 | 5·60 | 1·84 | – | – | – | 280 | 4·59 | 1·77 | 27·07 | 8 | 24 | 20 | 34 | 31 | 27 |
| 87 | 4·99 | 1·81 | 63·18 | 0·00 | 4·01 | – | 215 | 135 | 380 | 8·86 | 3·60 | 16·54 | 10 | 46 | 28 | 47 | 69 | 63 |
| 88 | 3·54 | 1·41 | 13·53 | 0·00 | 2·56 | – | – | – | 160 | 5·53 | 2·16 | 11·28 | 6 | 30 | 21 | 25 | 40 | 35 |
| 89 | 3·06 | 1·89 | 13·37 | 4·78 | 1·71 | – | – | – | 90 | 6·78 | 2·55 | 12·03 | 8 | 37 | 22 | 36 | 47 | 41 |
| 90 | 5·63 | 2·00 | 70·93 | 0·00 | 5·50 | $1·973 \times 10^{-5}$ | 275 | 215 | 100 | 9·97 | 4·02 | 19·55 | 10 | 52 | 33 | 50 | 80 | 89 |
| 91 | 6·89 | 1·49 | 95·88 | 0·00 | 11·70 | – | – | – | −5 | 12·93 | 5·37 | 23·30 | 13 | 69 | 40 | 58 | 108 | 113 |
| 92 | 6·06 | 1·81 | 85·70 | 0·00 | 6·26 | $6·897 \times 10^{-5}$ | – | – | 100 | 11·14 | 4·55 | 17·81 | 11 | 58 | 34 | 49 | 90 | 90 |
| 93 | 3·70 | 2·60 | 37·48 | 4·76 | 6·40 | $7·853 \times 10^{-6}$ | 110 | 71 | 0 | 7·11 | 3·80 | 30·08 | 11 | 38 | 31 | 41 | 62 | 83 |
| 94 | 4·48 | 2·29 | 45·43 | 1·55 | 6·43 | $5·096 \times 10^{-5}$ | – | – | −25 | 8·22 | 3·63 | 17·29 | 10 | 44 | 31 | 40 | 64 | 83 |
| 95 | 2·91 | 3·63 | 38·26 | 16·31 | 4·19 | $5·709 \times 10^{-4}$ | 164 | 90 | 50 | 7·08 | 3·50 | 36·84 | 11 | 32 | 28 | 39 | 57 | 56 |
| 96 | 2·63 | 3·51 | 35·56 | 20·53 | 5·34 | – | – | – | 130 | 8·82 | 4·00 | 31·58 | 11 | 43 | 31 | 51 | 69 | 65 |
| 97 | 2·08 | 3·67 | 28·27 | 26·74 | 8·73 | $8·062 \times 10^{-5}$ | – | – | 70 | 6·88 | 3·34 | 31·58 | 11 | 30 | 30 | 42 | 56 | 58 |
| 98 | 2·30 | 2·93 | 22·36 | 15·35 | 3·77 | – | – | – | 85 | 5·64 | 2·79 | 40·60 | 10 | 24 | 71 | 44 | 45 | 40 |
| 99 | 2·71 | 2·76 | 24·77 | 7·61 | 3·85 | – | 160 | 130 | 145 | 4·28 | 2·46 | 40·60 | 10 | 13 | 26 | 34 | 38 | 47 |
| 100 | 5·59 | 1·99 | 70·08 | 0·00 | 7·36 | – | – | – | 120 | 8·96 | 3·61 | 20·30 | 11 | 46 | 30 | 49 | 70 | 70 |
| 102 | 3·77 | 2·41 | 42·49 | 6·55 | 4·04 | $5·293 \times 10^{-6}$ | 102 | 72 | 20 | 8·22 | 3·32 | 23·31 | 10 | 42 | 22 | 39 | 61 | 36 |
| 103 | 3·46 | 2·87 | 43·54 | 8·53 | 5·95 | $1·271 \times 10^{-5}$ | 171 | 91 | 75 | 7·46 | 3·22 | 37·14 | 11 | 30 | 28 | 51 | 62 | 43 |
| 104 | 4·41 | 2·50 | 51·44 | 4·38 | 8·26 | – | – | – | 35 | 8·57 | 3·75 | 22·56 | 11 | 44 | 37 | 46 | 68 | 64 |
| 105 | 4·17 | 2·63 | 42·99 | 3·93 | 6·62 | – | – | – | 60 | 7·65 | 3·28 | 18·75 | 10 | 40 | 25 | 51 | 60 | 65 |
| 106 | 5·99 | 1·91 | 83·62 | 0·00 | 9·67 | $4·757 \times 10^{-6}$ | 365 | 290 | 145 | 1·89 | 4·90 | 20·95 | 12 | 64 | 36 | 53 | 101 | 99 |
| 107 | 5·87 | 1·89 | 86·45 | 0·00 | 4·63 | $5·401 \times 10^{-6}$ | 79 | 64 | −55 | 1·84 | 5·26 | 11·49 | 11 | 61 | 31 | 74 | 100 | 76 |
| 108 | 5·14 | 3·00 | 73·13 | 8·06 | 7·18 | – | – | – | 140 | 0·47 | 4·91 | 12·90 | 11 | 65 | 30 | 54 | 106 | 94 |

| | | | | | | | | | | | | | | | | | | |
|---|---|---|---|---|---|---|---|---|---|---|---|---|---|---|---|---|---|---|
| 109 | 6·46 | 1·51 | 96·44 | 0·00 | 9·65 | – | 615 | 410 | 120 | 12·15 | 5·37 | 15·66 | 12 | 69 | 40 | 60 | 110 | 123 |
| 110 | 6·15 | 1·69 | 90·30 | 0·00 | 8·40 | – | 315 | 260 | 150 | 11·04 | 4·79 | 15·05 | 11 | 69 | 35 | 49 | 105 | 109 |
| 111 | 5·77 | 1·78 | 87·47 | 0·11 | 7·78 | – | – | – | 130 | 11·00 | 4·91 | 17·29 | 12 | 41 | 33 | 47 | 67 | 95 |
| 113 | 6·80 | 1·50 | 96·54 | 0·00 | 9·52 | – | – | – | 15 | 12·47 | 5·21 | 20·00 | 12 | 69 | 38 | 59 | 111 | 118 |
| 114 | 6·05 | 1·74 | 82·48 | 0·00 | 8·48 | 1·391 $\times 10^{-2}$ | 355 | 235 | 70 | 11·00 | 4·62 | 15·80 | 12 | 60 | 35 | 55 | 91 | 97 |
| 115 | 4·87 | 1·96 | 65·84 | 1·55 | 5·43 | – | 217 | 180 | 140 | 8·63 | 3·57 | 15·79 | 10 | 48 | 28 | 43 | 66 | 65 |
| 116 | 5·47 | 1·94 | 71·91 | 0·00 | 6·66 | – | – | – | 85 | 10·53 | 4·65 | 13·14 | 11 | 59 | 23 | 53 | 92 | 58 |
| 117 | 3·82 | 2·86 | 39·48 | 8·17 | 3·89 | – | – | – | 105 | 6·98 | 2·72 | 20·09 | 9 | 37 | 26 | 39 | 58 | 51 |
| 118 | 1·67 | 3·56 | 18·54 | 20·45 | 2·95 | – | 38 | 32 | 270 | 3·88 | 1·50 | 41·35 | 9 | 14 | 22 | 29 | 31 | 18 |
| 119 | 2·30 | 3·75 | 26·74 | 19·39 | 4·58 | – | – | – | 75 | 6·10 | 2·48 | 31·58 | 10 | 28 | 27 | 35 | 51 | 42 |
| 120 | 3.66 | 2·92 | 42·12 | 4·89 | 5·28 | – | | – | 190 | 7·07 | 3·07 | 29·92 | 11 | 30 | 32 | 40 | 62 | 55 |
| 121 | 3·58 | 3·65 | 50·26 | 18·38 | 6·58 | – | 250 | 180 | 140 | 11·56 | 4·94 | 20·22 | 13 | 61 | 35 | 55 | 97 | 111 |
| 122 | −0·92 | 3·61 | 12·99 | 62·90 | 2·32 | – | – | – | 70 | 8·13 | 4·61 | 33·83 | 13 | 41 | 44 | 49 | 72 | 133 |
| 123 | 4·30 | 2·14 | 41·90 | 2·28 | 4·50 | 7·333 $\times 10^{-6}$ | – | – | 30 | 8·22 | 3·32 | 18·05 | 10 | 44 | 28 | 44 | 63 | 69 |
| 124 | 5·14 | 2·44 | 65·35 | 0·00 | 6·86 | 6·525 $\times 10^{-6}$ | 211 | 200 | 70 | 11·25 | 4·66 | 18·86 | 12 | 61 | 35 | 57 | 93 | 98 |
| 125 | −0.04 | 3·92 | 11·44 | 40·74 | 1·63 | – | – | – | 80 | 4·72 | 2·21 | 42·11 | 10 | 19 | 31 | 34 | 30 | 41 |
| 126 | 1·23 | 4·09 | 16·82 | 26·34 | 3·46 | – | – | – | 85 | 4·87 | 1·96 | 36·84 | 9 | 22 | 24 | 30 | 39 | 36 |
| 127 | −0·03 | 3·72 | 9·76 | 43·88 | 4·01 | – | – | – | 130 | 6·02 | 2·81 | 25·56 | 10 | 30 | 25 | 35 | 45 | 53 |
| 128 | 3·43 | 3·34 | 41·47 | 13·77 | 7·58 | 6·283 $\times 10^{-2}$ | 455 | 355 | 75 | 8·70 | 5·14 | 31·37 | 14 | 45 | 51 | 52 | 82 | 107 |
| 129 | 2·64 | 3·62 | 34·83 | 23·10 | 4·98 | 2·262 $\times 10^{-2}$ | – | – | 100 | 9·74 | 5·10 | 24·81 | 12 | 48 | 34 | 46 | 79 | 98 |
| 130 | −0.09 | 4·06 | 17·70 | 51·39 | 5·26 | – | 150 | 135 | 125 | 10·82 | 4·57 | 21·05 | 12 | 54 | 31 | 54 | 80 | 77 |
| 131 | 2·34 | 2·22 | 15·77 | 4·15 | 3·70 | 3·789 $\times 10^{-2}$ | 148 | 94 | 135 | 6·48 | 3·67 | 33·08 | 11 | 32 | 25 | 35 | 53 | 59 |
| 132 | 4·86 | 2·57 | 56·19 | 2·44 | 5·96 | 1·248 $\times 10^{-1}$ | – | – | 30 | 9·37 | 3·93 | 23·08 | 12 | 49 | 33 | 58 | 82 | 82 |
| 133 | 5·19 | 1·87 | 67·46 | 0·00 | 7·18 | 6·104 $\times 10^{-6}$ | 375 | 310 | 50 | 9·94 | 4·18 | 19·55 | 11 | 53 | 32 | 50 | 79 | 83 |
| 135 | −0·18 | 4·38 | 18·00 | 49·72 | 5·50 | – | – | – | 75 | 7·89 | 4·01 | 28·57 | 12 | 41 | 33 | 47 | 67 | 95 |

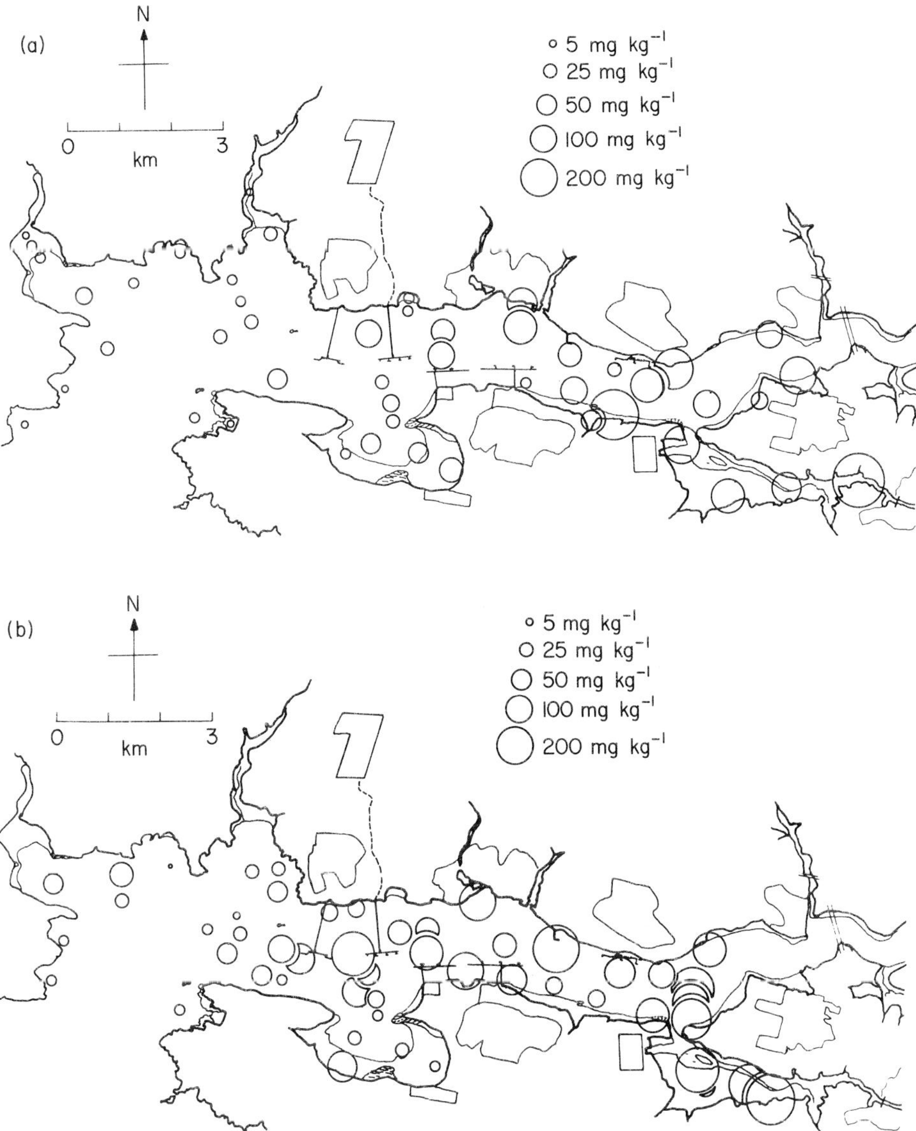


FIG. 4. Aliphatic hydrocarbon concentrations in Milford Haven sediments (a) 1978–79, (b) 1982 surveys.

### *Major and trace element concentrations*

The major elements Al, Fe, K and Ti were comparatively resistant to the HCl pretreatment and generally reflected the distribution of fine sediment in the estuary. Calcium, Mg, Na, P and Mn were partially leached by the HCl pretreatment. Among the trace elements

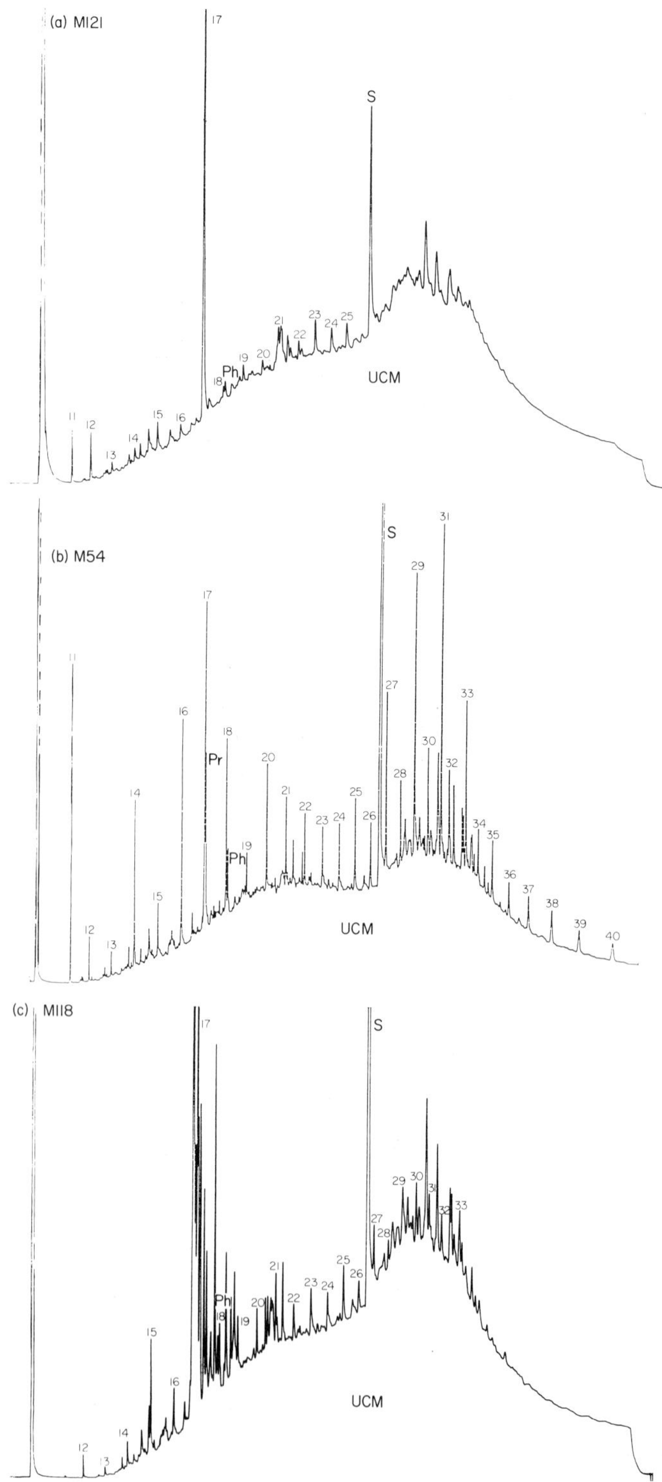
(a) M121
17
S
11
12
13
14
15
16
18
Ph
19
20
21
22
23
24
25
UCM
(b) M54
S
31
29
17
11
27
33
16
18
Pr
30
32
20
28
14
21
25
22
23
24
26
34
35
Ph
19
15
36
37
38
12
39
40
13
UCM
(c) M118
17
S
29
30
31
32
33
27
28
25
21
26
23
24
Ph
22
20
18
19
15
16
14
12
13
UCM

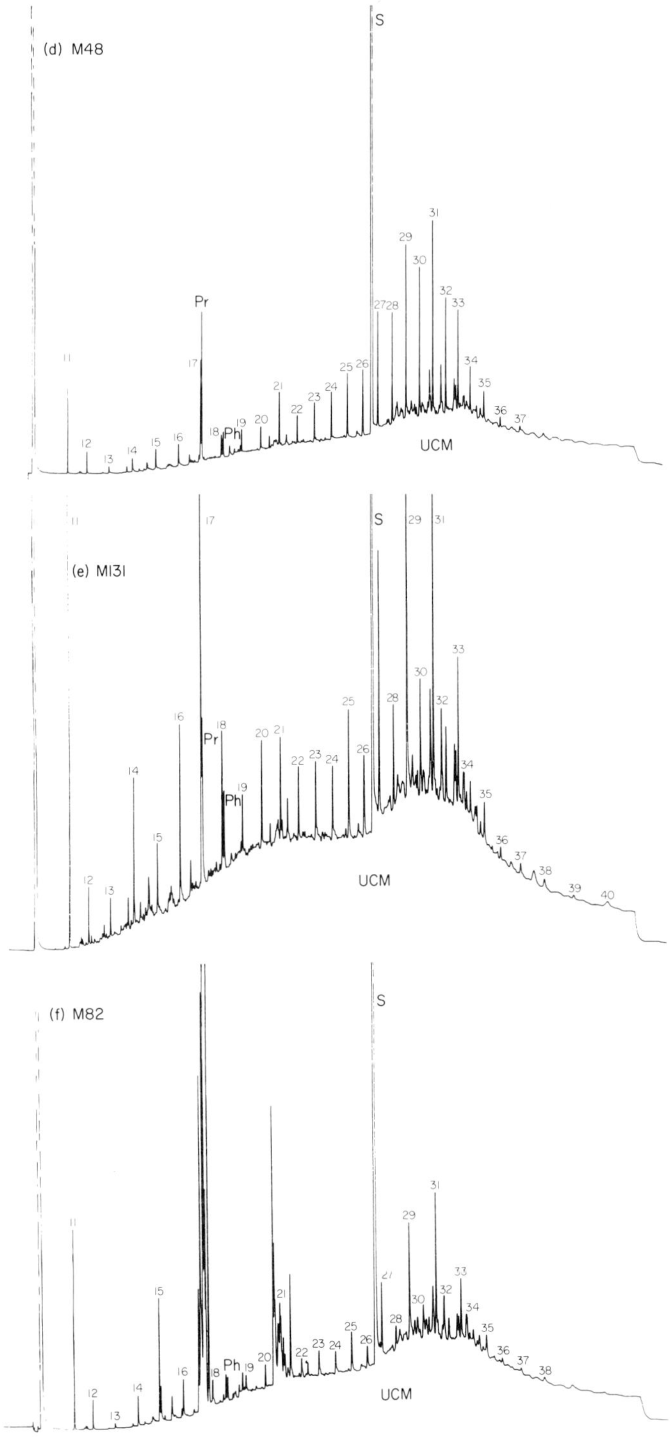

FIG. 5. Representative GLC traces of aliphatic hydrocarbon fractions from Milford Haven sediments (1982). Numbers = n-alkanes (n-$C_{11}$, n-$C_{12}$ etc.); Pr = pristane; Ph = phytane; UCM = unresolved complex mixture; S = squalane (internal standard).

which were usually leached from the sediment by dilute HCl, (Ba, Co, Cr, Ni, Sr, V, Y and Zn), all but Ba, Sr and Y are contoured in Fig. 6 a–f. The total metal concentrations from which these extractable fractions were derived are presented in Table 5 with total Al, Fe and the weight lost after HCl treatment.

The overall trace element concentrations were comparatively low for an estuary with substantial areas of fine sediment. The spatial distribution of the extractable fraction of all the trace elements mapped in Fig. 6 a–f, however, shows a clear enrichment in the central jetty area and in some of the finer sediments upstream.

Cobalt, Cr and Zn also showed slightly enriched concentrations in parts of the outer estuary around the Dale Peninsula. Among these, Co is unusual in that it was enriched in coarse, poorly-sorted, permeable sediments (Table 6). Unlike all the other elements mapped, Co was also positively correlated with the HCl soluble fraction, suggesting that carbonate sediments may be a sink for this element and that its fate may be biologically-mediated. In contrast, Cr and, to a lesser extent, V showed negative relationships with the coarser carbonate sediments (Table 6). This was especially noticeable in the maerl-rich sediments between the Esso and Amoco jetties, the bioclastic sands east of the Dale Peninsula and the shelly muds of Pwllcrochan Flats (Fig. 6b). In all these locations no Cr was removed by the HCl treatment in spite of the total Cr concentrations of from 20 to 38 mg $kg^{-1}$. It is possible that the HCl was neutralized by the carbonate at these stations, or that the source rock for the clastic fraction of these sediments is Cr-rich.

The extractable Cu was generally >50% of the total Cu in the sediments, whereas extractable Ni, V and Zn were usually <50% of their respective total concentrations. All of these elements, however, showed a strong degree of spatial correspondence to areas of industrial activity. Copper was unusual, however, in that the concentrations were not so strongly positively-correlated with % mud, % Al and sediment organic content, nor so negatively-correlated with Redox potential as was the case with the other elements in this group (Table 6). Copper may be one of the commoner elements of refinery effluents (Table 2), in addition to being a contaminant found in sewage effluents (Santschi *et al.* 1984).

## *Sedimentology*

### *Grain size distributions*

Some of the 1982 sediment parameters are presented in Table 5. General trends were similar in both surveys, so the following account is based mainly on the 1982 results. The Ø mean ranged from −2·96 Ø (Station 45) to 6·89 Ø (Station 91). Sediments varied widely from coarse pebbles to very fine sand, silt and clay. Those in the mouth were generally higher in percentage gravel and lower in percentage mud. Values of Ø standard deviation were mainly from 1·0 to 3·0 but ranged from 0·76 (Stations 2 and 50) to 4·38 (Station 135). Most sediments were thus very poorly-sorted, although well-sorted sands were found in some bays close to the mouth.

Values of Ø skewness ranged from −0·92 (Station 40) to 3·54 (Station 82). Negative values, indicating that the size distribution had a poorly-sorted coarse 'tail', were found at wave-exposed stations with a pebble or shell granule fraction, and also at many intertidal and shallow subtidal stations upstream. In these latter areas the 'tail' consisted of sand in otherwise muddy sediments. The predominantly positive Ø skewness values emphasize that, even in areas of coarser sediments, deposition of small quantities of fine material can occur. The sands found in Angle Bay, Sandyhaven Bay, Watwick and offshore from Gelliswick all illustrate this point. The Ø kurtosis ranged from 1·72 (Station 65) to 22·95 (Station 47). Lower kurtosis values were found at stations with multi-modal size distributions and these predominated except in zones of wave or current-sorted sands.

Areas of maximum wave activity were usually close to the mouth or on west-facing upper intertidal areas upstream. The highest concentration of mud in the estuarine mouth was at Station 26 (24·2%). This area of Dale Roads is sheltered by Dale Point and separated by a zone of higher wave action from the similar areas north of the Esso and Amoco jetties. Both areas are characterized by moderate wave action and relatively low current speeds.

Increasing percentage mud with distance eastwards from the mouth is illustrated in Fig. 2. The main sedimentological factor upstream appears to be current speed. Close to the Gulf jetty, current velocity in the main channel is sufficient to winnow much of the fine mud, leaving a coarse shell gravel lag deposit (Stations 122, 125–127 and 135). Lateral gradients from the main channel were also apparent. Stations 86, 118 and 119, for example, had a distinctive very fine sand mode, but this was not found in muddier sediments higher in the intertidal zone on Pwllcrochan Flats.

### *Organic matter content*

In general, the percentage organic matter (Table 5) was high in muddy sediments remote from the mouth of the estuary. A tongue of high organic content sediments extended westwards to encompass the Amoco and Esso jetties. The lowest values were found in well-sorted sandy sediments of Sandyhaven, Angle Bay and West Angle Bay. Some low values were also found at Stations 86, 122 and 125 in the main channel where sediments were coarse and poorly-sorted. The highest value of 11·7% was recorded at Station 91 close to the town of Milford Haven.

### *Hydraulic conductivity*

The hydraulic conductivity of the sediments (Table 5) ranged through six orders of magnitude from $1{\cdot}25 \times 10^{-1}$ cm s$^{-1}$ (Station 132) to $1{\cdot}35 \times 10^{-6}$ cm s$^{-1}$ (Station 62). The

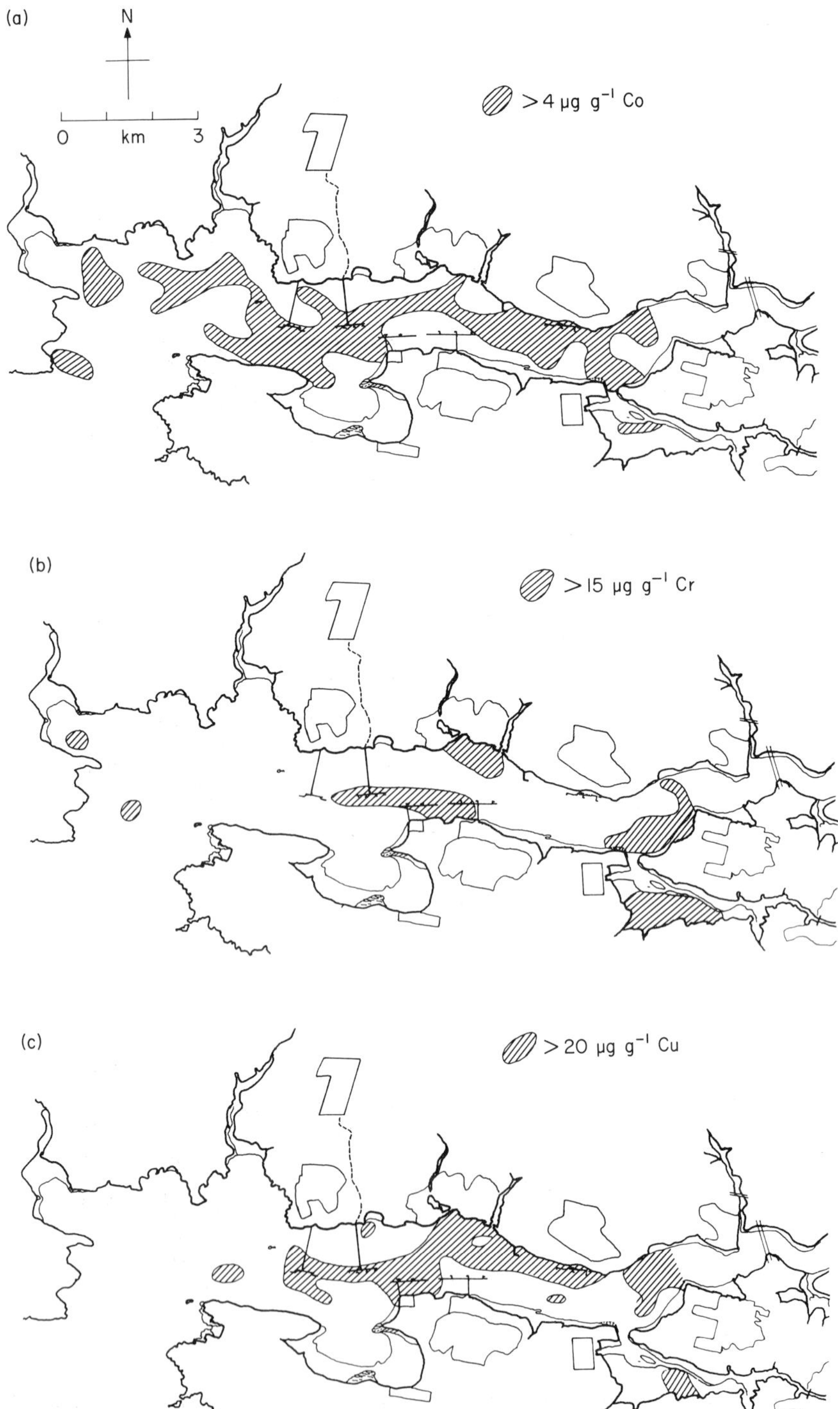


FIG. 6.

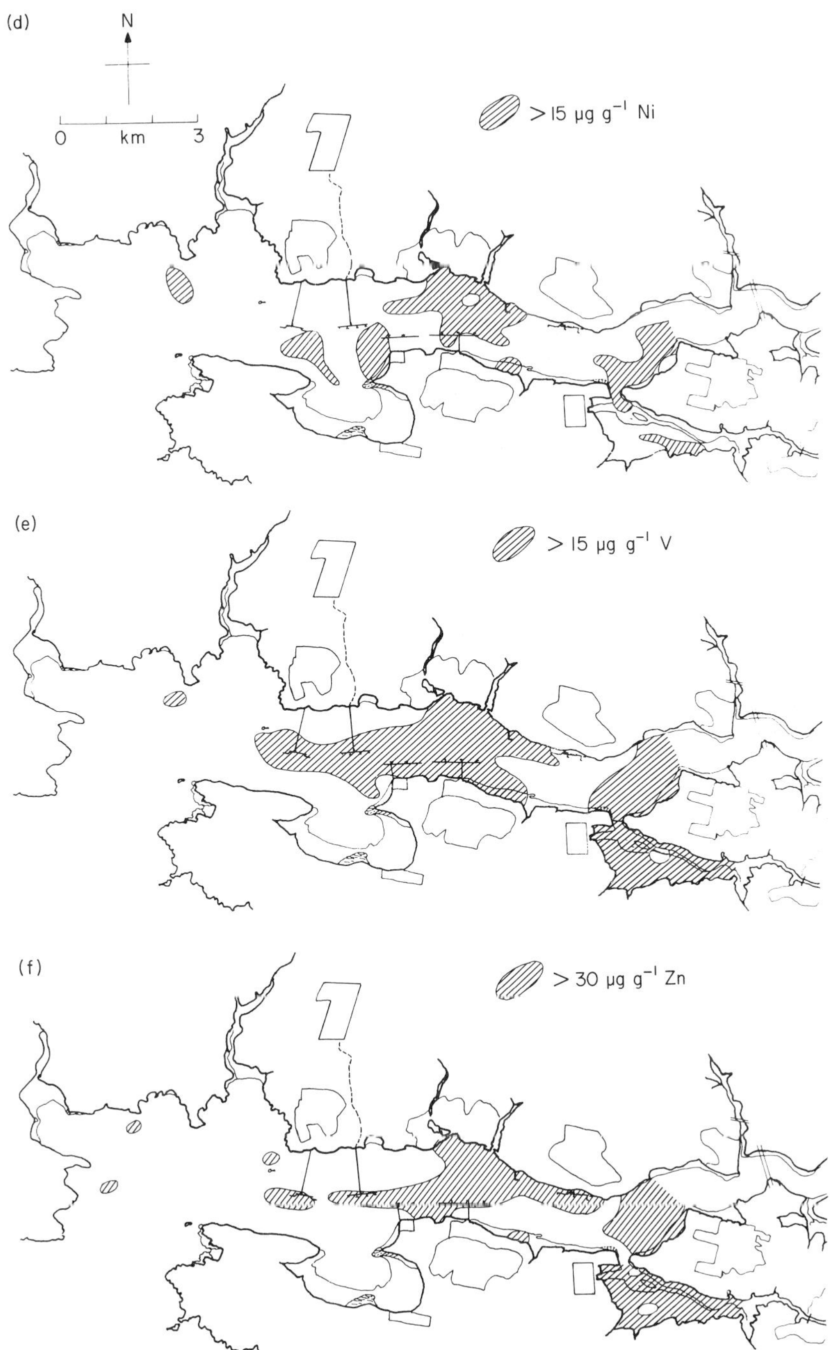


FIG. 6. Extractable trace elements in sediments for the 1982 Milford Haven survey: (a) Co; (b) Cr; (c) Cu; (d) Ni; (e) V; (f) Zn.

general trend was that more permeable sediments were found in the estuarine mouth ($10^{-2}$ and $10^{-3}$ cm $s^{-1}$). Muddier sands both in the mouth (Stations 14, 15 and 36) and in the central jetty area ranged from $10^{-4}$ to $10^{-6}$ cm $s^{-1}$. The much finer muds of the Pembroke River, Pwllcrochan and Pennar Flats were all at the lower end of this range, but the channel sediments from the inner area showed much more variability. Stations 114, 128, 129, 131 and 132 were all highly permeable and examination of the cores showed that the range in permeability from $10^{-1}$ to $10^{-2}$ cm $s^{-1}$ in the inner reaches was always associated with densely-burrowed muddy sediments. The lower hydraulic conductivity readings were recorded at Stations 62, 64 and 84. The latter two stations were composed of compacted, possibly eroding, muddy sand and mud. It is possible that Station 62 although coarser, was over-consolidated, having been sampled in one of the dredged berths at the Esso jetty. Both these factors would tend to reduce hydraulic conductivity.

## DISCUSSION

Some of the relatively large changes in hydrocarbon concentrations between 1978 and 1982 may be attributed to sedimentological factors such as the erosion of fine sediment, as perhaps at Stations 80 and 82 in Angle Bay and Station 85 on Pwllcrochan Flats where both mud and oil content of the sediments decreased. Comparison between the two surveys is hampered, however, by the precision of station relocation, the patchiness and other differences in sediment type at the sites selected for each survey, and changes in analytical techniques.

Nevertheless, the mean value of aliphatic hydrocarbon concentrations increased by 33% between the two surveys, with only slight changes in overall sediment characteristics. The drop in the mean % gravel from 1978 to 1982 (17·3% to 7·32%) might have been expected to increase the relative proportion of the generally more contaminated finer material in the samples used for hydrocarbon analysis, and this is reflected in the shift in Ø mean diameter from fine to very fine sand (2·87 Ø to 3·44 Ø). However, the comparability of the stations used is borne out by examination of the average % mud values for the two surveys (35·6% in 1978–79 and 34·4% in 1982), which were extremely close. Given the strong correlation between % mud and oil concentrations (Table 6 and Fig. 7) and the fact that % mud averages have, if anything, decreased, the increase in mean oil content was tested (using the *t*-test) but because of the very large variances was found to be insignificant. An increase in 1982 in the density of stations around the Esso and Amoco jetties was probably a contributory cause of the apparent increase. However, if only the 1982 stations in this area which were close to the 1978 survey positions are considered, the local mean aliphatic hydrocarbon concentration had increased by 1982 from 53 to 105 mg $kg^{-1}$.

The GLC traces of the aliphatic fractions may be interpreted as follows. The relatively large concentration of UCM in all samples indicates that the bulk of the hydrocarbons are

TABLE 6. Pearson correlation coefficients for the dependent variables against all other variables. For variable abbreviations see Table 3. Critical values (two-tailed test): ** $P < 0·001$; * $P < 0·01$; all other coefficients shown were significant at $P < 0·05$.

| | TCO | TCR | TCU | TNI | TV | TZN | THC | ALI | CO | CR | CU | NI | V | ZN |
|---|---|---|---|---|---|---|---|---|---|---|---|---|---|---|
| MEAN | 0·23* | 0·67** | 0·30** | 0·48** | 0·59** | 0·46** | 0·71** | 0·69 | – | 0·61** | – | 0·25* | 0·60** | 0·38** |
| SD | 0·41** | – | 0·31** | 0·26* | – | 0·20 | – | – | 0·33** | – | 0·20 | 0·26* | – | 0·20 |
| MUD | 0·61** | 0·87 | 0·53** | 0·81** | 0·87** | 0·75** | 0·82** | 0·81** | – | 0·73** | 0·32** | 0·45** | 0·80** | 0·61** |
| COA | – | −0·25* | – | – | – | – | – | – | 0·25* | −0·27* | – | – | −0·19 | – |
| COM | 0·75** | 0·79** | 0·66** | 0·82** | 0·85** | 0·80** | 0·90** | 0·88** | – | 0·61** | 0·42** | 0·46** | 0·77** | 0·67** |
| AL | 0·76** | 0·96** | 0·60** | 0·91** | 0·96** | 0·83* | 0·79** | 0·79** | – | 0·77** | 0·36** | 0·5[illegible]** | 0·86** | 0·73** |
| | | | | | | | | | | | | | | |
| FE | 0·85** | 0·88** | 0·67** | 0·89** | 0·95** | 0·88** | 0·78** | 0·77** | – | 0·68** | 0·44** | 0·42** | 0·80** | 0·77** |
| SOL | 0·32** | −0·43** | 0·19 | – | −0·21 | – | – | – | 0·48** | −0·45** | – | – | −0·20 | – |
| RED | −0·39** | −0·47** | −0·31* | −0·50** | −0·38** | −0·43** | −0·34 | −0·32 | – | −0·32* | – | −0·50** | −0·43 | −0·42** |
| PER | 0·33 | – | – | – | – | – | 0·36 | – | 0·32 | – | – | – | – | – |
| TCO | 1·0 | 0·65** | 0·67** | 0·85** | 0·79** | 0·79** | 0·72** | 0·72** | 0·44** | 0·44** | 0·40** | 0·49** | 0·63** | 0·66** |
| TCR | | 1·0 | 0·52** | 0·84** | 0·94** | 0·80** | 0·78** | 0·79** | – | 0·85** | 0·34** | 0·44** | 0·86** | 0·72** |
| | | | | | | | | | | | | | | |
| TCU | | | 1·0 | 0·66** | 0·60** | 0·65** | 0·82** | 0·81** | 0·18 | 0·33** | 0·82** | 0·43** | 0·53** | 0·57** |
| TNI | | | | 1·0 | 0·90** | 0·80** | 0·73** | 0·73** | 0·19 | 0·65** | 0·39** | 0·70** | 0·80** | 0·68** |
| TV | | | | | 1·0 | 0·85** | 0·81** | 0·81** | – | 0·76** | 0·38** | 0·44** | 0·89** | 0·73** |
| TZN | | | | | | 1·0 | 0·87** | 0·87** | – | 0·63** | 0·41** | 0·37** | 0·75** | 0·91** |
| THC | | | | | | | 1·0 | 0·98** | – | 0·63** | 0·53** | 0·45** | 0·73** | 0·80** |
| ALI | | | | | | | | 1·0 | – | 0·67** | 0·54 | 0·45** | 0·73** | 0·79** |
| | | | | | | | | | | | | | | |
| CO | | | | | | | | | 1·0 | – | 0·20 | 0·39** | – | 0·25* |
| CR | | | | | | | | | | 1·0 | 0·24* | 0·39** | 0·77** | 0·60** |
| CU | | | | | | | | | | | 1·0 | 0·31** | 0·40** | 0·40** |
| NI | | | | | | | | | | | | 1·0 | 0·53** | 0·36** |
| V | | | | | | | | | | | | | 1·0 | 0·69** |
| ZN | | | | | | | | | | | | | | 1·0 |

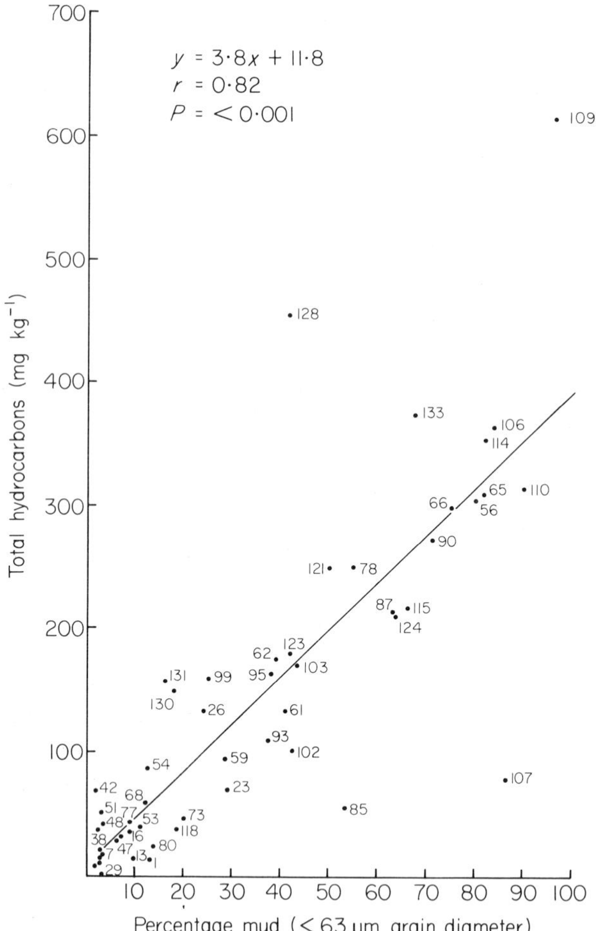


FIG. 7. Total hydrocarbons and percentage mud (1982) scatterplot and linear regression.

from degraded petrogenic sources. Although the boiling point distribution of the UCM suggests that the major source may be crude oil and/or heavy fuel oil, it may equally well represent the residue of degradation of a wide variety of oil types. In certain traces the presence of an homologous series of n-alkanes from n-$C_{11}$ to n-$C_{35}$ indicates a minor component derived from relatively recent inputs of crude oil. Enhanced n-alkane concentrations in the range in n-$C_{28}$ to n-$C_{36}$ and n-$C_{33}$ to n-$C_{40}$ observed, e.g. at Station 66, are similar to distributions associated with waxy sludges from the tanks of crude oil tankers (Wong, Green & Cretney 1976); Geyer & Giammona 1980).

In most cases, however, the resolved peaks represent mainly biogenic hydrocarbons. n-Alkanes in the range n-$C_{15}$ to n-$C_{34}$ with a predominance of odd carbon numbered n-alkanes are characteristic of leaf waxes of higher plants (Eglinton & Thompson 1962). The group of peaks including n-$C_{17}$ and pristane are thought to be of algal origin (Shaw & Wiggs 1979) and the group around n-$C_{22}$ are of algal or microbial origin (Farrington *et al.* 1977). n-Alkanes from n-$C_{13}$ to n-$C_{20}$ with a predominance of even carbon numbers are thought to be derived by microbial modification of algal lipids (Tibbetts *et al.* 1982).

THC concentrations showed a significant correlation ($r = 0{\cdot}82$, $P < 0{\cdot}001$) with % mud in the sediments (Fig. 7). A similar correlation was found in the 1978-79 data, when the correlation coefficient was found to increase slightly when aliphatic hydrocarbon concentration was plotted against % sediment particles < 15 $\mu$m in diameter. This may be explained by their higher specific surface area. The correlation coefficient did not increase significantly when fractions finer than 15 $\mu$m were tested, perhaps because all available oil was already adsorbed or because of the tendency of fine silt and clay to flocculate into aggregates of about this diameter. There was also a better fit for the 1978 subtidal data ($r = 0{\cdot}80$) than for the intertidal data ($r = 0{\cdot}69$) which may be perhaps explained by the stranding of floating oil, which may penetrate the sediment in intertidal areas. As in the 1978 survey there were a few anomalous values of hydrocarbon concentration in 1982 (Fig. 7). For example, Station 109 in the Pembroke River had a higher residual (+240 mg $kg^{-1}$) THC value than would be predicted by the general relationship. This station is near a sewage outfall, which may contain hydrocarbons from road run-off. In contrast, Station 107 nearby had an anomalously low residual (−260 mg $kg^{-1}$) THC value. Intertidal mudflats in this area may be stable with little or no sedimentation during the period since industrialization, although the high extractable metal concentrations, particularly Ni, may call this explanation into question. Station 85 on Pwllcrochan Flats (negative residual −160 mg $kg^{-1}$) has already been mentioned in connection with possible erosion of recent sediments to expose material deposited before the arrival of the oil industry. Station 128 had a high positive residual (+290 mg $kg^{-1}$), perhaps implicating as a point source the activities at Newton Noyes jetty.

The multiple regression analysis highlighted some of these anomalies, but in calculating a multivariate and better fit to the overall data, other anomalies were found to be 'explained'. This suggests that several substrates may in fact compete for the contaminant load in the water column. However, Stations 42, 56, 109 and 133 were extreme positive outliers. All but Station 109 are comparatively close to oil industry activity and all have evidently received pollutant hydrocarbon inputs which cannot be very accurately predicted from known sediment characteristics. Station 42 in particular was unusual in that the GLC trace showed an exceptionally high boiling range UCM. This station is adjacent to the main anchorage for crude oil and product tankers awaiting berths or prior to departure from the port. More rigorous analysis of such residuals may suggest zones of the estuary suited to further, more detailed field studies.

Table 6 illustrates the very high degree of intercorrelation of many of the contaminants with % mud, % organic matter, % Al, % Fe and redox state of the sediments. The multiple regression analyses (Table 7) confirmed and refined these general relationships. It was noticeable that in spite of the high bivariate correlation between % mud and % organic matter, only the latter variable of the two was included in the hydrocarbon equations: % mud may have been 'repeating' statistical information already supplied by % organic matter, so among the grain size variables Ø mean was selected as explaining the highest proportion of residual variance unexplained by % organic matter.

The very high levels of explanation of THC and aliphatic hydrocarbons provided by % organic matter (Table 7) suggest the possibility that the central zone of the estuary represents a 'biological' turbidity maximum zone. The area near the jetties and refinery effluent inputs is characterized by very high plankton productivity at low energy periods during the year, when clastic suspended sediments concentrations are very low (Williams & Jolly 1975). This means that biological scavenging of contaminants (passive sedimentation upon death and bioconcentration of contaminants in faecal pellets of zooplankton) may be the dominant transfer process in the clearer waters of the outer and central estuary (Skei 1981). Depending on the rate of vertical mixing and biodegradation in the sediment, this organic-rich, superficial layer will be repeatedly resuspended to act as a substrate for hydrocarbon and metal adsorption. Organic coatings of clastic particles elsewhere in the estuary will perform the same function. Santschi *et al.* (1984) proposed a similar model for the outer reaches of Rhode Island Sound where little or no net sedimentation is taking place.

In addition to the influence of refinery effluent, the strength of the multivariate relationship probably lies in the as yet unquantitified inputs of organic matter, oil and metals in road run-off and sewage. Copper was a powerful explanatory variable in conjunction with % organic matter and Ø mean in equations developed with both THC and aliphatic hydrocarbons as dependent variables (Table 7). It is significant that Cu is often a dominant trace element in both refinery and sewage effluents. It may well be that these equations, although estuary-specific, could predict contaminant distributions elsewhere. Armansson *et al.* (1985) have indeed found non-residual Cu to be an excellent indicator of refinery effluent areas of effect in Southampton Water. Caution must be advised with organic matter as a variable, however, because in the case of the often very high sediment hydrocarbon concentrations near the Fawley Refinery (up to *c.* 20 000 mg $kg^{-1}$) the % organic matter is partly made up by the THC, leading to potential problems of colinearity in the statistical procedures.

Bivariate and multivariate regression results using the extractable trace elements as dependent variables (Tables 6 and 7), show a general dependence of extractable metal concentrations upon their respective total element concentrations ($r^2$ values ranging from 0·476 for Ni to 0·822 for Zn). This suggests that a similar picture would have emerged from consideration of spatial distribution of either total or extractable metals (Luoma & Bryan 1981). However, the slightly weaker bivariate correlations of the extractable fractions with sediment parameters when compared to the equivalent correlation coefficient using total metal concentrations, implied that point sources and their potential area of biological effect may be more readily detected using extractable metal data. This was confirmed by multiple regression analysis, where in contrast to the extractable elements mapped in Fig. 6, the overriding feature of the total concentrations was a reflection of the principle sediment gradients (Table 7). Aluminium, Fe and Ø mean were found to account for up to 91% of the percentage variation of the total metal concentrations, although important and mainly negative contributions to the equations

TABLE 7. Summary of results of the multiple regression analyses. The values are the regression coefficients in the final equations. The values in parentheses indicate the percentage variability explained by the inclusion of that variable ($r^2 \times 100$). For variable names, see Table 3

| Independent variables | Dependent variables: TCO | TCR | TCU | TNI | TV | TZN | THC | ALI | CO | CR | CU | NI | V | ZN |
|---|---|---|---|---|---|---|---|---|---|---|---|---|---|---|
| MEAN | — | −3·03 (0·9) | −3·10(14·9) | - | - | −11·28 (2·4) | 20·74 (6·1) | 15·35 (5·8) | - | - | - | - | - | - |
| SD | - | −3·46 (1·1) | - | - | - | - | - | - | - | - | - | - | - | - |
| COM | - | −1·88 (1·7) | - | - | - | - | 21·86(80·4) | 15·50(77·3) | - | - | - | - | - | - |
| AL | - | 5·07(90·9) | - | 4·06(82·4) | 4·41(91·3) | - | - | - | - | - | - | - | - | |
| FE | 1·51 (71·2) | - | - | - | 9·90 (3·6) | 3·69(75·8) | - | - | - | - | - | −4·12(19·6) | - | - |
| SOL | 0·10 (20·3) | −0·45 (3·4) | - | 0·29 (4·3) | - | −1·31 (7·5) | - | - | 0·07(20·5) | - | - | - | - | - |
| RED | −0·003(1·2) | - | - | - | - | - | - | - | - | - | - | - | - | - |
| PER | — | −100·24 (0·7) | −112·3 (4·5) | - | - | −454·10 (4·4) | - | - | - | - | - | - | - | - |
| TCO | - | - | - | - | - | - | - | - | - | - | - | - | - | - |
| TCR | - | - | - | - | - | - | - | - | - | 0·432(71·0) | - | - | - | - |
| TCU | - | - | - | - | - | - | 6·65(8·8) | 5·08(9·2) | - | - | 0·80(66·0) | - | - | - |
| TNI | - | - | - | - | - | - | - | - | - | - | 0·19 (3·3) | 0·66(47·6) | - | - |
| TV | - | - | - | - | - | - | - | - | - | - | - | - | 0·31(78·5) | - |
| TZN | - | - | - | - | - | - | - | - | - | - | - | - | - | 0·52(82·2) |
| THC | - | - | 0·10(65·6) | - | - | 0·30 (9·0) | - | - | - | - | - | - | - | - |
| ALI | - | 0·10 (1·1) | - | - | - | - | - | - | - | - | - | - | - | - |
| Constant | 2·72 | 26·02 | 22·15 | [illegible]·68 | −4·72 | 67·68 | −173·24 | −130·58 | 1·84 | −6·43 | 3·72 | −0·14 | −4·83 | −7·21 |
| Overall *F* | 111·70 | 1176 | 50·10 | 8[illegible]·72 | 240·9 | 545·9 | 174·59 | 104·76 | 9·24 | 79·27 | 37·16 | 33·77 | 117·98 | 148·93 |
| $\Sigma(r^2 \times 100)$ | 92·7 | 99·7 | 85·0 | 86·7 | 94·9 | 99·1 | 95·3 | 92·3 | 20·5 | 71·0 | 69·3 | 67·2 | 78·5 | 82·2 |
| *P* | <0·001 | <0·001 | <0·001 | <0·001 | <0·001 | <0·001 | <0·001 | <0·001 | <0·01 | <0·001 | <0·001 | <0·001 | <0·001 | <0·001 |

were also provided by the percentage HCl soluble fraction, redox potential, and hydraulic conductivity.

It has been noted in many other studies that there are correlations between contaminants and sediments (e.g. Olsen *et al.* 1978; Marchand 1980). This is to some extent a reflection of the surface characteristics of the particles but is also due to the energy level at the site of deposition. Under high-energy conditions only coarser particles are deposited, and may be frequently disturbed by waves and current action. They also consist mainly of carbonates, quartz and detrital minerals which are relatively unreactive to aqueous contaminants. Any contaminated finer material is thus 'diluted' by this coarse material. Conversely, under low-energy conditions fine sediment is deposited and may remain undisturbed for considerable periods. These particles tend to accumulate in deeper subtidal reaches, harbours and shoals where flocculation may increase their settling velocity (Prentice *et al.* 1968; Olsen *et al.* 1978). They also move to the innermost and highest intertidal mudbanks (Van Straaten & Kuenen 1958). The differential of 40 cm $s^{-1}$ current speed in favour of the flood tide in Milford Haven and the localized existence of density-related currents might both augment the latter process, because water near the bed floods for longer than it ebbs and is also higher in suspended sediment concentration (Williams & Jolly 1975).

Fine sediments contain a large proportion of clay minerals and organic matter which have a high capacity of adsorption and complexation of contaminants (Ho & Karim 1978; Boehm & Quinn 1978; Wade & Quinn 1980). They may also be anoxic, which reduces the rate of hydrocarbon degradation, and densely-burrowed, which increases hydraulic conductivity and surface area of the seabed. In an estuary where contaminants are distributed evenly through the water body the highest concentrations, therefore, will be found associated with low-energy areas, characterized by sediments with high mud, organic, Al and Fe concentrations.

The cycling of fine particles from dense sediment suspensions near the estuarine bed into mobile, dilute suspensions means that in the period since industrialization a 'pool' of increasingly contaminated particles may have been building up. This, coupled with the mixing energy of the tides perhaps explains the general similarity of the GLC traces, in which there are no obvious, discrete pollutant types which can be attributed to known inputs. It remains to be seen from analyses of contaminant fluxes, radionuclide-dated sediment cores and sediments sampled from higher reaches of the estuary in the 'geological' turbidity maximum zone, to what extent contaminants are transferred permanently into the sediments or are retained in the pool of mobile and stationary sediment suspensions (Kirby & Parker 1982). The simultaneous analysis of organic and inorganic fractions in the sediments has, in the meantime, provided a useful means of discriminating between regional and local components in the transport and fate of estuarine contaminants.

## ACKNOWLEDGMENTS

We would like to thank the Institute of Petroleum and the Welsh Office for financial support. Some of the work was carried out during one of the authors' tenure of NERC

Industrial Studentship (GT6/82/AAPS/2). The contributions of many colleagues at the Oil Pollution Research Unit and the Geochemistry Laboratory at the University of London King's College are gratefully acknowledged.

## REFERENCES

**Ackroyd, T.N.W. (1969).** *Laboratory Testing in Soil Engineering.* Chapter 8. Permeability. Soil Mechanics Ltd, London.

**Addy, J.M. (1976).** Preliminary investigations of the sublittoral macrofauna of Milford Haven. *Marine Ecology and Oil Pollution* (Ed. by J.M. Baker), pp. 91–130. Applied Science Publishers/Institute of Petroleum, London.

**Agemian, H. & Chau, A.S.Y. (1976).** Evaluation of extraction techniques for the determination of metals in aquatic sediments. *The Analyst,* **101**, No. 1207, 761–767.

**Armannson, H., Burton, J.D., Jones, G.B. & Knap, A.H. (1985).** Trace metals and hydrocarbons in sediments from the Southampton Water region, with particular reference to the influence of oil refinery effluent. *Marine Environmental Research,* **15**, 31–34.

**Baker, J.M.** (Ed.) **(1976).** *Marine Ecology and Oil Pollution.* Applied Science Publishers/Institute of Petroleum, London.

**Baker, J.M., Crothers, J.H., Little, D.I., Oldham, J.H. & Wilson, C.M. (1984).** Comparison of the fate and ecological effects of dispersed and non-dispersed oil in a variety of intertidal habitats. *Oil Spill Chemical Dispersants: Research, Experience and Recommendations, STP 840* (Ed. by T.E. Allen), pp. 239–279. American Society for Testing and Materials, Philadelphia.

**Blackman, R.A.A., Baker, J.M., Jelly, J. & Reynard, S. (1973).** The Dona Marika oil spill. *Marine Pollution Bulletin,* **4**, 181–182.

**Boehm, P.D. & Quinn, J.G. (1978).** Benthic hydrocarbons of Rhode Island Sound. *Estuarine and Coastal Marine Science,* **6**, 471–494.

**Byers, S.C., Mills, E.L. & Stewart, P.L. (1978).** A comparison of methods of determining organic carbon in marine sediments, with suggestions for a standard method. *Hydrobiologia,* **58**, 1, 43–47.

**Chester, R. & Voutsinou, F.G. (1981).** The initial assessment of trace metal pollution in coastal sediments. *Marine Pollution Bulletin,* **12**, 84–91.

**Clarke, R.C., Jr & Brown, D.W. (1977).** Methods of analysis for petroleum hydrocarbons. *Effects of Petroleum on Arctic and Sub-arctic Marine Environments and Organisms.* Vol. 1 (Ed. by D.C. Malins), pp. 31–62. Academic Press, New York.

**Crapp, G.B. (1971).** Monitoring the rocky shore. *The Ecological Effects of Oil Pollution on Littoral Communities* (Ed. by E.B. Cowell), pp. 102–113. Institute of Petroleum, London.

**Dias, N.S. (1960).** *The plankton of Milford Haven.* M.Sc. thesis, University College of Wales, Swansea.

**Dicks, B. & Hartley, J.P. (1982).** The effects of repeated small oil spillages and chronic discharges. *Philosophical Transactions of the Royal Society of London,* **B.297**, 285–307.

**Eglinton, G. & Thompson, S. (1962).** Hydrocarbon constituents of the wax coating of plant leaves: a taxonomic survey. *Nature (Lond.)* **193**, 739.

**Farrington, J.N., Frew, N.H., Gschwend, P.M. & Tripp, P.W. (1977).** Hydrocarbons in the cores of northwestern Atlantic continental margin sediments. *Estuarine and Coastal Marine Science,* **5**, 793–808.

**Geyer, R.A. & Giammona, C.P. (1980).** Naturally occurring hydrocarbons in the Gulf of Mexico and Caribbean Sea. *Marine Environmental Pollution.* Vol. 1. Hydrocarbons (Ed. by R.A. Geyer), pp. 73–105. Elsevier, Amsterdam.

**Ho, C.L. & Karim, H. (1978).** Impact of adsorbed petroleum hydrocarbons on marine organisms. *Marine Pollution Bulletin,* **9**, 156–162.

**Hull, C.H. & Nie, N.H. (1981).** New procedures and facilities for releases 7–9. *Statistical Package for the Social Sciences.* McGraw-Hill, New York.

**Kirby, R. & Parker, W.R. (1982).** The distribution and behaviour of fine sediment in the Severn estuary and inner Bristol Channel, UK. *Dynamics of Turbid Coastal Environments. Special Volume. Canadian Journal of Fisheries and Aquatic Sciences,* **40**, 83–95.

**Krumbein, W.C. (1936).** The application of logarithmic moments to size-frequency distributions of sediments. *Journal of Sedimentary Petrology,* **6**, 35–47.

**Krumbein, W.C. (1959).** The 'sorting-out' of geological variables illustrated by regression analysis of beach firmness. *Journal of Sedimentary Petrology,* **29**, 575–87.

**Little, D.I., Staggs, M.F. & Woodman, S.S.C. (1984).** Sample pretreatment and size analysis of poorly-sorted cohesive sediments by sieve and electronic particle counter. *Transfer Processes in Cohesive Sediment Systems* (Ed. by W.R. Parker & D.J.J. Kinsman), pp. 47–74. Plenum Press, New York.

**Luoma, S.N. & Bryan, G.W. (1981).** A statistical assessment of the form of trace metals in oxidised estuarine sediments employing chemical extractants. *The Science of the Total Environment,* **17**, 165–196.

**Marchand, M.H. (1980).** The Amoco Cadiz oil spill, distribution and evaluation of hydrocarbon concentrations in seawater and marine sediments. *Environment International,* **4**, 421–29.

**Mook, D.H. & Hoskin, C.M. (1982).** Organic determinations by ignition: caution advised. *Estuarine, Coastal and Shelf Science,* **15**, 697–699.

**Moyse, J. & Nelson-Smith, A. (1963).** Zonation of animals and plants on rocky shores around Dale, Pembrokeshire. *Field Studies,* **1**, 1–31.

**Nelson-Smith, A. (1964).** *Some aspects of the marine ecology of Milford Haven, Pembrokeshire.* Ph.D. thesis, University College of Wales, Swansea.

**Nelson-Smith, A. (1965).** Marine biology of Milford Haven: the physical environment. *Field Studies,* **2**, 155–188.

**Nelson-Smith, A. (1967).** Marine biology of Milford Haven: the distribution of littoral plants and animals. *Field Studies,* **2**, 435–477.

**Nelson-Smith, A. & Case, R.M. (1984).** Aspects of the shore and sublittoral ecology of the Daucleddau Estuary (Milford Haven). *Proceedings of the Zoological Journal of the Linnaean Society,* **80**, 177–190.

**Nie, N.H., Hull, C.H., Jenkins, J.G., Steinbrenner, K. & Brent, D.H. (1975).** *Statistical Package for the Social Sciences, 2nd edn.* McGraw-Hill, New York.

**Olsen, C.R., Simpson, H.J., Bopp, R.F., Williams, S.C., Peng, T.H. & Deck, B.L. (1978).** A geochemical analysis of the sediments and sedimentation in the Hudson estuary. *Journal of Sedimentary Petrology,* **48**, 2, 401–418.

**Prentice, J.E., Beg, I.R., Colleypriest, C., Kirby, R., Sutcliffe, P.J.C., Dobson, M.R., D'Olier, B., Elvines. M.F., Kilenyi, T.I., Maddrell, R.J. & Phinn, T.R. (1968).** Sediment transport in estuarine areas. *Nature (Lond.),* **218**, 1207–1210.

**Santschi, P.H., Nixon,S., Pilson, M. & Hunt, C. (1984).** Accumulation of sediments, trace metals (Pb, Cu) and total hydrocarbons in Narragansett Bay, Rhode Island. *Estuarine, Coastal and Shelf Science,* **19**, 427–449.

**Shaw, D.G. & Wiggs, J.N. (1979).** Hydrocarbons in Alaskan marine intertidal algae. *Phytochemistry,* **18**, 2025–2027.

**Skei, J. (1981).** The entrapment of pollutants in Norwegian fjord sediments — a beneficial situation for the North Sea. *Holocene Marine Sedimentation in the North Sea Basin: Special Publications of the International Association of Sedimentologists,* 5 (Ed. by S.-D. Nio, R.T.E. Shüttenhelm & Tj.C.E. van Weering), pp. 461–481.

**Tibbetts, P.J.C., Rowland, S.J., Tovey, L.L. & Large, R. (1982).** Investigation of the sources of aliphatic hydrocarbons in the mussel *Mytilus edulis* from North Sea oil production platforms by capillary glc and cgc-ms. *Toxicology and Environmental Chemistry,* **5**, 177–193.

**Van Straaten, L.M.J.U. & Kuenen, Ph.H. (1958).** Tidal action as a cause of clay accumulation. *Journal of Sedimentary Petrology,* **28**, 406–413.

**Wade, T.L. & Quinn, J.G. (1980).** Incorporation, distribution and fate of saturated petroleum hydrocarbons in sediments from a controlled marine ecosystem. *Marine Environmental Research,* **3**, 15–33.

**Walsh, J.N. (1980).** The simultaneous determination of the major, minor and trace constituents of silicate rocks using inductively-coupled plasma spectrometry. *Spectrochimica Acta,* **35B**, 107–111.

**Whitfield, M. (1969).** Eh as an operational parameter in estuarine studies. *Limnology and Oceanography,* **14**, 546–548.

**Williams, B.R.H. (1971).** The Milford Haven: an appraisal of the suitability of existing hydrographical, climatological and biological data for constructing a mathematical model for use in predicting the effects of effluent discharged in the Haven. *Report No. BL/C/1357.* ICI Ltd, Brixham.

**Williams, D.J.A. & Jolly, R. (1975).** *Milford Haven: survey of pollution and mathematical modelling of pollutant distribution and transport.* Report from Department of Chemical Engineering, University College of Wales, Swansea.

**Wong, C.S., Green D.R. & Cretney, W.J. (1976).** Distribution and source of tar on the Pacific Ocean. *Marine Pollution Bulletin,* **7**, 102–106.

**Woodman, S.S.C. & Little, A.E. (1985).** Rocky shore monitoring in Milford Haven. *Oil and Petrochemical Pollution,* **2**, 79–91.

# Trace and major elements in the atmosphere at rural locations in Great Britain, 1972–81

P.A. CAWSE

*Environmental & Medical Sciences Division,*
*AERE Harwell, Didcot, Oxon, OX11 0RA*

## SUMMARY

**1** Concentrations of trace and major elements have been measured in air particulate, sampled continuously at four non-urban sites in Great Britain from 1972 to 1981. At three of the stations, rainwater and dry deposition were also analysed.

**2** The measurements of some thirty-six elements have provided information on the rural atmosphere for comparison with separate studies in urban areas where increases in heavy metals occur. A seasonal variation has been recorded at the rural locations, highest concentrations of elements in air occurring in winter months. A general decrease in concentrations was found over the 10-year period; measurements are continuing.

**3** The deposition inventory of elements has been obtained, including sulphate and nitrate, from separate analysis of soluble and insoluble fractions of rainwater samples.

## INTRODUCTION

### *Trace and major elements in the atmosphere*

Trace and major elements are introduced to the atmosphere by both natural and artificial (anthropogenic) sources. The primary sources of particulate matter in air are the resuspension by wind of soil, deposited dust and sea-spray aerosol, the combustion of fossil fuels, forest fires and volcanic activity. In addition, transformation of gaseous species such as volatile metals, $SO_2$ and nitrogen oxides to particulate forms by photochemical and gas phase reactions will also contribute to the aerosol, defined as a system of fine solid or liquid particles in gaseous suspension that is collectively referred to as 'particulates'.

The total suspended particulate matter (TSP) in the atmosphere generally contains 80–90% of inorganic material, the remainder consisting of organic compounds and biological debris. This aerosol and associated elements are mainly distributed in the lower troposphere, extending to 11–17 km above the Earth's surface depending on latitude, and becomes transported in the zonal circulation before deposition to land and water surfaces. However, particles that are injected through the troposphere to the stratosphere are subject to a global distribution, as observed after testing of nuclear weapons (Peirson & Cambray 1965) and after eruption of the Fuego volcano in Guatemala in 1974 (Volz 1975).

Aerosols are transferred to the Earth's surface by dry deposition through impaction and diffusion for smaller particles (diameter $< 10$ $\mu$m) and by sedimentation for larger particles. In addition, scavenging of aerosols by precipitation occurs by in-cloud (rainout) and below-cloud (washout) processes. Information on trace and major elements in the

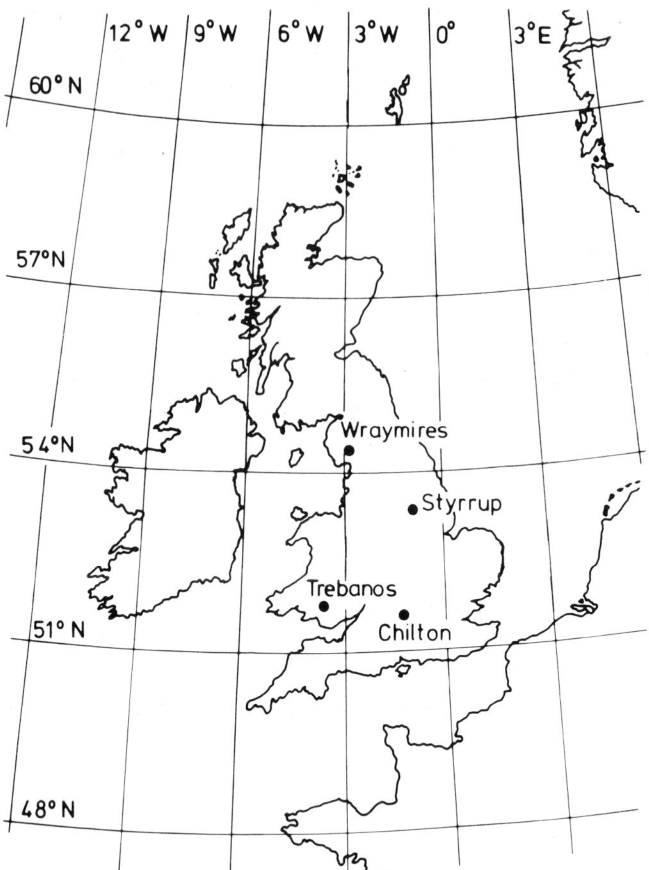


FIG. 1. The location of sampling stations.

atmosphere is important in relation to plant and animal health, and knowledge is required of biogeochemical cycling to assess the disturbance to natural cycles caused by artificial inputs of elements that may function as nutrients or pollutants. The diverse nature of current studies on inorganic particulate matter in the atmosphere is described in a recent review (Cawse 1982).

## *The present programme of measurements*

Concentrations of elements in air particulate and deposition have been measured continuously at four non-urban locations (Fig. 1) over the 10-year period 1972–81. The research project was initiated and maintained by the Natural Environment Research Council for the first 2 years; subsequently it was made on behalf of the Department of the Environment and the Welsh Office and further measurements are in progress. Originally, there was a network of seven non-urban stations that included sampling at Collafirth

(Shetland Islands), Plynlimon (Powys) and Leiston (Suffolk) but detailed multi-element analysis of samples from these sites was terminated in 1978. At Chilton (Oxon) and Wraymires (Cumbria), preliminary measurements of element concentrations in air are also available for 1971.

The objectives of the present study are to provide information on: (i) element concentrations in rural areas to compare with measurements in urban atmospheres; (ii) the seasonal and long-term variation in concentrations of elements in air particulate and in deposition; (iii) particle-size associations of elements; (iv) the annual and long-term (cumulative) deposition inventory of elements to the ground. During the period of the study, parallel measurements have been made under separate contracts concerned with the input of trace elements into the North Sea (Cambray, Jefferies & Topping 1979), and with urban measurements in Swansea (Pattenden 1974) and in Walsall (Cawse & Turner 1981; Pattenden 1977). In these separate projects, comparisons of atmospheric composition with non-urban data from the rural network are instructive.

## METHODS

Air particulate, rainwater and the dry deposit is sampled continuously, with monthly sample changes. All samples are collected at 1·5 m above ground, according to procedures developed for the initial pilot studies at Wraymires (Cawse & Peirson 1972; Peirson *et al.* 1973).

Cellulose filters (Whatman 40) are used to retain air particulate, and are contained in a polypropylene filter-holder. Air is drawn through the filter using an electrically-driven pump; the volume of air filtered is about 250 $m^3$ per month and is measured by meter. Rainwater is collected in a polythene funnel and bottle, and dry deposition collected on a horizontal sheet of Whatman 541 filter paper mounted 12 cm beneath a plastic cover to shield the surface from rain.

Samples of rainwater are filtered (Whatman 42 paper) to separate the soluble and insoluble fractions for analysis. The rainwater is stored at − 15 °C, both before and after the bulking procedure which is carried out each quarter year. The monthly rainfall at each site is recorded by a standard Meteorological Office raingauge at 0·3 m above ground.

The monthly air filters and dry deposition samples are now bulked for quarterly analysis, but in earlier studies they were analysed each month. Further, the analysis of dry deposition is restricted to two of the sites, Chilton and Styrrup , whereas rainwater and air particulate is analysed at all four locations (Fig. 1). During the bulking procedure of air filters and rainwater for quarterly analysis, the integrity of part of the monthly samples is retained should they be required for analysis of specific elements at a later date.

Elements routinely analysed are: Ag, Al, As, Au, Br, Ca, Cd, Ce, Cl, Co, Cr, Cs, Cu, Eu, Fe, Hg, I, In, K, La, Mg, Mn, Mo, Na, Ni, Pb, Rb, Sb, Sc, Se, Sm, Th, Ti, V, W, and Zn. Instrumental neutron activation analysis (INAA) is the preferred technique for the majority of analyses, but Cd, Cu, Ni and Pb are more conveniently measured by X-ray fluorescence and atomic absorption. Nitrate is analysed in rainwater by U.V.-spectrophotometry, and sulphate by the barium-thoronol procedure.

The accuracy of INAA, X-ray fluorescence, atomic absorption and colorimetry was validated by analysis of a standard reference material (National Bureau of Standard orchard leaves) and by an analytical intercomparison of air filters, also involving gamma-photon activation analysis (Cawse 1976).

Details of the four sites shown in Fig. 1, including their exact location and altitude are as follows.

*Chilton* (Harwell) Oxon. Grid Ref: SU 468861, altitude 130 m.
In central southern England, about 90 km from the English Channel. The sampling apparatus is situated on the westerly perimeter of the Atomic Energy Research Establishment, where a meteorological station and monitoring for radioactive fallout has been established for some years. The surroundings are predominantly rural, with large areas devoted to cereal cultivation.

*Styrrup* Nottinghamshire. Grid Ref: SK 606898, altitude 15 m.
Situated in the East Midlands, 2 km to the south-west of Bircotes colliery and 9 km to the north-east of Dinnington colliery. The sampling station is placed in a rural area, at the Institute of Geological Sciences local groundwater research compound, but industrial influences are very strong and 25 km to the west lies Sheffield industrial complex with steelworks and coal mining.

*Trebanos* near Swansea, Glamorganshire. Grid Ref: SN 712023, altitude 23 m.
In the River Tawe valley, 11 km inland from Swansea Bay. The site is just on the outer limit of the industrial zone, which has major activities of oil refining at Llandarcy and steelworks at Port Talbot. Smelting of nickel takes place at Clydach, only 2 km to the south-west of Trebanos sampling station.

*Wraymires* near Windermere, Lancashire. Grid Ref: SD 362974, altitude 84 m.
A rural site near Lake Windermere, about 25 km from the Irish Sea and 32 km to the north-east of Barrow-in-Furness, where there is shipbuilding industry. At 40 km to the south of Wraymires is the Heysham oil refinery. The station is considered to be fairly clean with moderate maritime influence and a high annual rainfall; there is hardly any arable farming in the area which might give rise to local soil dust.

## RESULTS AND DISCUSSION

The analytical results have been summarized to show the long-term annual average concentration of elements in air, dry deposition and total (dry + wet) deposition. The detailed results for each station are available separately as a computer archive of data which is held at AERE Harwell, Environmental and Medical Sciences Division.

The following parameters can be derived from the measurements: (i) percentage elemental composition on a dry weight basis, of the total suspended particulate; (ii) the dry deposition velocity ($V_g$) that relates to particle size; (iii) the ratio of dry to total deposition ($D/T$); (iv) total and soluble deposition to the ground; (v) washout factors ($W$), or concentration of element in rain relative to that in air; (vi) enrichment factors of elements in air particulate normalized to Sc, i.e. the ratio of an element to Sc in air divided by the ratio of the same elements in average soil.

*Air particulate*

Element concentrations found in air at the four stations are listed in Table 1 as averages for the 10-year period 1972–81 at Chilton, Styrrup and Trebanos, and from mid 1970 to 1981 at Wraymires. The lowest concentrations of TSP and elements were recorded at Wraymires, which is situated in a region of relatively high rainfall; highest concentrations occurred at Styrrup with the other sites intermediate, and this relative order was observed in each year of the study. As a general observation, elements analysed in the non-urban aerosol may, with few exceptions, be broadly classified into the following four main groups according to the concentrations recorded.

(a) High, with levels usually $>1000$ ng kg$^{-1}$ air: Cl, and also S (as $SO_4$) and N (as $NH_4$ and $NO_3$). The measurements of S and N were only made in 1972–73 (Cawse 1974).

(b) Medium, with levels usually from 100 to 1000 ng kg$^{-1}$ air: Na, Mg, Al, K, Ca, Fe, Zn and Pb.

(c) Low, with levels usually from 1 to 100 ng kg$^{-1}$ air: V, Cr, Mn, Ni, Cu, As, Se, Br and Sb.

(d) Very low, with levels usually $<1$ ng kg$^{-1}$ air: Sc, Co, Mo, Ag, In, Cs, La, Ce, Sm, Eu, W, Au, Hg and Th.

Exceptions to this grouping are given by relatively high concentrations of Co and Ni at Trebanos, which is attributed to the local smelting of nickel.

A comparison of the air quality at Wraymires with the city atmosphere of Swansea (Table 1) indicates that urbanization and industry has resulted in order of magnitude increases in air concentrations of Co, Ni, Zn, As, Br and Pb. Considering the Central European 'clean air background' concentrations measured 3752 m above sea level at the Jungfraujoch, Switzerland by Dams & De Jonge (1976), it is apparent that the non-urban concentrations in Great Britain show order of magnitude increases in Cl, Na, K and Mg due to strong maritime influence, with similar increases in concentrations of the heavy metals V, Cr, Mn, Fe, Co, Cu, Zn, As, Sb and Pb.

The average annual percentages of elements in air particulate at Styrrup were, on a dry weight basis: Cl 9·7%, Ca 3·1%, Na 2·2%, K 2·1%, Fe 1·5%, Mg 1·0%, Pb 0·5%, Zn 0·5%, Br 0·2%, Mn 0·1% and Cu 0·07%. Other trace elements listed (Table 1) were present at 0·03%. It was established previously (Cawse 1974) that some 50% of the dry weight of TSP consists of $SO_4^{--}$, $NO_3^-$ and $NH_4^+$, the remainder being carbon and organic material. It may be noted that the World Health Organization has proposed an interim TSP guideline of 60–90 $\mu$g m$^{-3}$ as an annual mean (WHO 1979) beyond which it is considered that there is a risk to health of long-term exposure.

Industrial influences at Styrrup are indicated by the average annual ratio of Cl/Na 4·3 compared with corresponding ratios of 2·7, 2·2 and 2·2 at Chilton, Trebanos and Wraymires respectively, exceeding the ratio 1·8 in bulk seawater; combustion of coal is the most likely source of the excess non-maritime Cl. The ratio Pb/Br at the four stations was

TABLE 1. Average annual elemental concentrations in air, 1972–81 (ng kg$^{-1}$‡)

| | Non-urban | | | | Urban |
|---|---|---|---|---|---|
| Element | Chilton, Oxon | Styrrup, Notts | Trebanos, W.Glam. | Wraymires, Cumbria* | Central Swansea† |
| Na | 660 | 830 | 960 | 800 | 1960 |
| Mg[s] | 270 | 810 | 280 | 270 | |
| Al | 195 | 410 | 210 | 155 | 370 |
| Cl | 1750 | 3600 | 2100 | 1750 | 4600 |
| K[s] | 590 | 840 | 800 | 580 | |
| Ca[s] | 740 | 1350 | 850 | 580 | |
| Sc | 0·057 | 0·11 | 0·066 | 0·038 | 0·16 |
| Ti[s] | <18 | <33 | <23 | <16 | |
| V | 9·0 | 12 | 9·4 | 6·2 | 21 |
| Cr | 2·4 | 9·9 | 8·6 | 1·8 | 6·1 |
| Mn | 12 | 35 | 12 | 8·8 | 25 |
| Fe | 230 | 550 | 300 | 195 | 940 |
| Co | 0·28 | 0·48 | 7·8 | 0·20 | 4·5 |
| Ni | 5·3 | 7·6 | 74 | 3·7 | 66 |
| Cu | 15 | 26 | 24 | 14 | 57 |
| Zn | 80 | 185 | 105 | 48 | 310 |
| As | 2·9 | 12 | 5·3 | 2·2 | 15 |
| Se | 1·3 | 3·1 | 1·5 | 0·95 | 2·7 |
| Br | 36 | 87 | 39 | 23 | 320 |
| Rb | <2 | <5 | <5 | <1 | |
| Mo[s] | <0·7 | <1 | <0·8 | <0·5 | |
| Ag[s] | 0·24 | 0·40 | 0·23 | 0·13 | |
| Cd[s] | <2 | <4 | 3·4 | <2 | |
| In[s] | 0·059 | 0·097 | 0·049 | 0·071 | |
| Sb | 1·8 | 4·2 | 2·2 | 1·4 | 4·0 |
| I | <3 | <4 | <3 | <1 | |
| Cs | 0·15 | 0·28 | 0·15 | 0·10 | 0·27 |
| La | 0·48 | 0·69 | 0·68 | 0·41 | |
| Ce | 0·35 | 0·55 | 0·41 | 0.22 | 0·80 |
| Sm[s] | 0·021 | 0·048 | 0·030 | 0·021 | |
| Eu | 0·008 | 0·009 | 0·007 | 0·004 | |
| W | <0·6 | <1 | <0·6 | <0·5 | |
| Au[s] | 0·009 | 0·005 | 0·005 | 0·004 | |
| Hg | <0·1 | <0·5 | <0·2 | <0·1 | |
| Pb | 100 | 195 | 115 | 55 | 500 |
| Th | 0·045 | 0·080 | <0·07 | 0·025 | 0·073 |
| TSP ¶ | 22 | 37 | 26 | 18 | |

*Mid 1970–81 at Wraymires.
†Central Swansea data from Pattenden (1974).
‡1 m$^3$ air at 15 °C, 760 mmHg (Standard Cubic Metre) = 1·226 kg.
[s]Data for Mg,K,Ca,Ti,Mo,Ag,Cd,In,Sm, and Au are available over the period 1975–81 and are averaged accordingly.
¶ $\mu g\ kg^{-1}$

in the range 2·2–2·9, near to the ratio of 2·6 in ethyl fluid added to petrol (Lininger *et al.* 1966), and implicates automobile exhaust as the major source of these elements. Ratios of rare elements derived from the average data in air particulate (Table 1) are in close agreement with ratios in average soil (Table 2). This suggests suspension of dust by agricultural activities and mining operations.

TABLE 2. Ratios of rare earth elements in air particulate compared with average soil.

| | Ce/Eu | Ce/Sm | Eu/Sm |
|---|---|---|---|
| Air particulate at non-urban sites | 55 | 14 | 0·25 |
| Median soil concentrations (Bowen 1979) | 50 | 11 | 0·22 |

The ratio of Mg/K in soil is 0·36 (Bowen 1966) and 3·4 in bulk seawater; at the four sites this ratio was from 0·35 to 0·96 (Table 1), which implies that their source was mainly the suspension of soil dust rather than maritime-derived aerosol. Further, the ratios of K/Na range from 0·7 to 1·0 and exceed the ratio in seawater of 0·036, the ratio in average soil being 2·2 (Bowen 1966).

The quarterly concentrations of some elements recorded in air are shown in Figs 2, 3, 4 and 5 for each site; this covers the period 1971–81 at Wraymires, mid 1971–81 at Chilton and 1972–81 at the two other stations. The seasonal fluctuations in absolute concentrations of elements in air were evident for all sites: these differences were examined by comparison of levels in the 'winter' (quarters 1 and 4) and 'summer' (quarters 2 and 3) periods (Table 3). Although total suspended particulate showed little elevation in winter, increases of $>50\%$ occurred in concentrations of many elements, namely Na, Cl, V, Cr, As, Sb and Pb. A greater increase, up to three times the summer level, was shown by the element Br. This feature is attributed to increased combustion of fossil fuels and the persistence of inversion layers in winter. Other elements such as Al, Sc, Fe and Ce showed little seasonal difference in concentration which suggests that they are introduced to the atmosphere mainly from natural sources, e.g. soil dust, whereas the heavy metals and Br are derived from industry.

Relatively high concentrations of particulate Br were measured at all stations during the winter of 1972–73, including the three initially in the network (Cawse 1975); subsequently the levels returned to previous values (Figs 2–5), and it was noted that the 'peak' period coincided with intense volcanic activity at Heimaey in Iceland causing outgassing beneath the sea, which would result in the release of halogens.

It is evident from Figs 2–5 that downward trends have occurred in concentrations of many elements over the last 10 years. The magnitude of these trends was examined by fitting simple regression lines to all quarterly results for each station, to the end of 1981. The significant changes are shown in Table 4; downward trends were recorded at all sites for Al, Sc, V, Cr, Mn, Co, Zn, Br, Cs and Pb, and at three sites for Fe and As. Evidence for corresponding decreases in Na and Cl was slight, indicating that changes in the

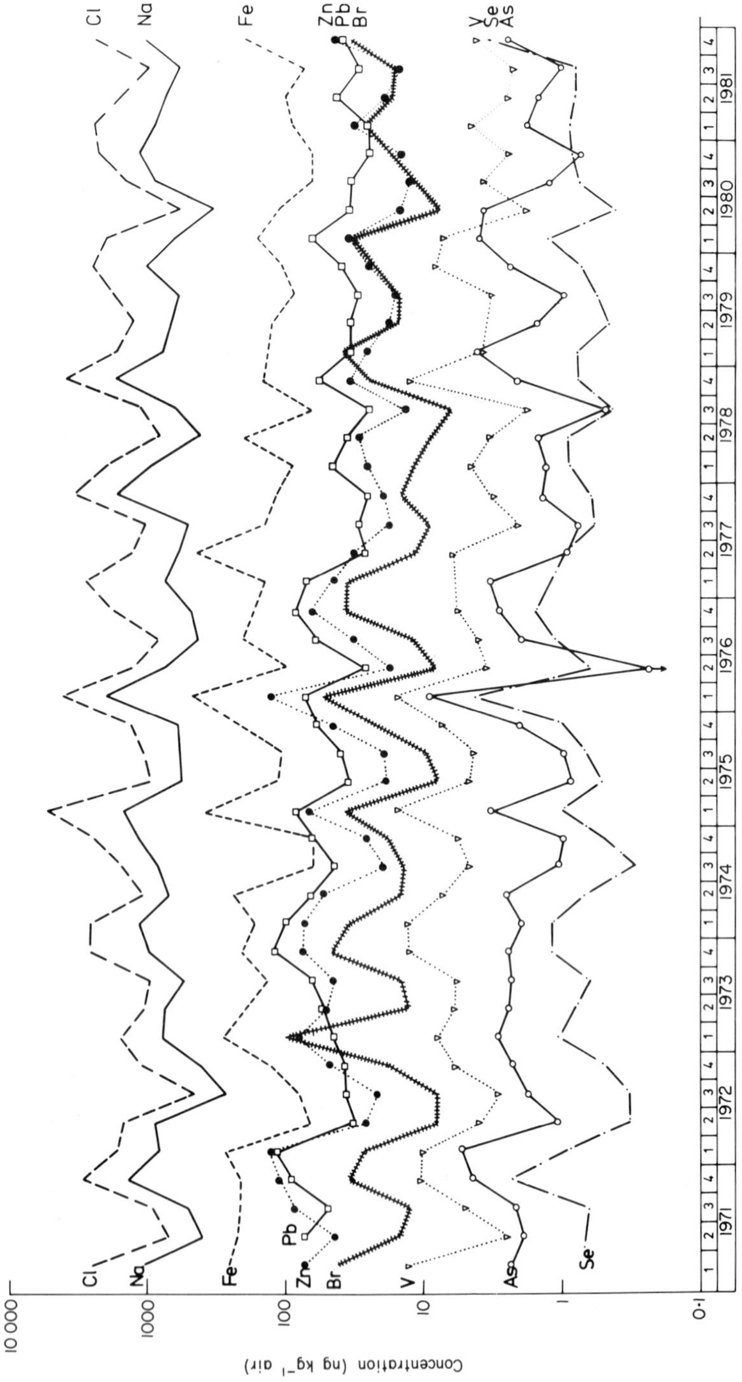

FIG. 2. Quarterly elemental concentration in air near ground level at Wraymires (1971–81).

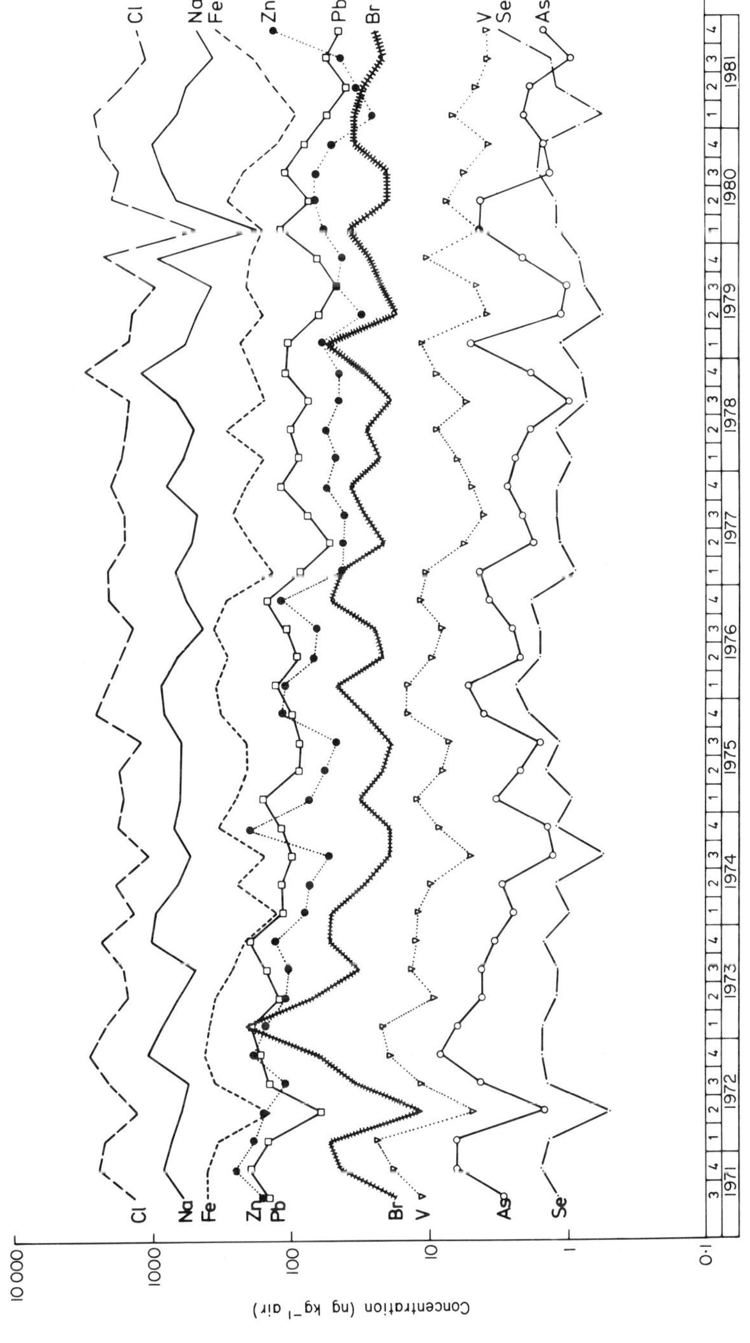


FIG. 3. Quarterly elemental concentration in air near ground level at Chilton (1971–81).

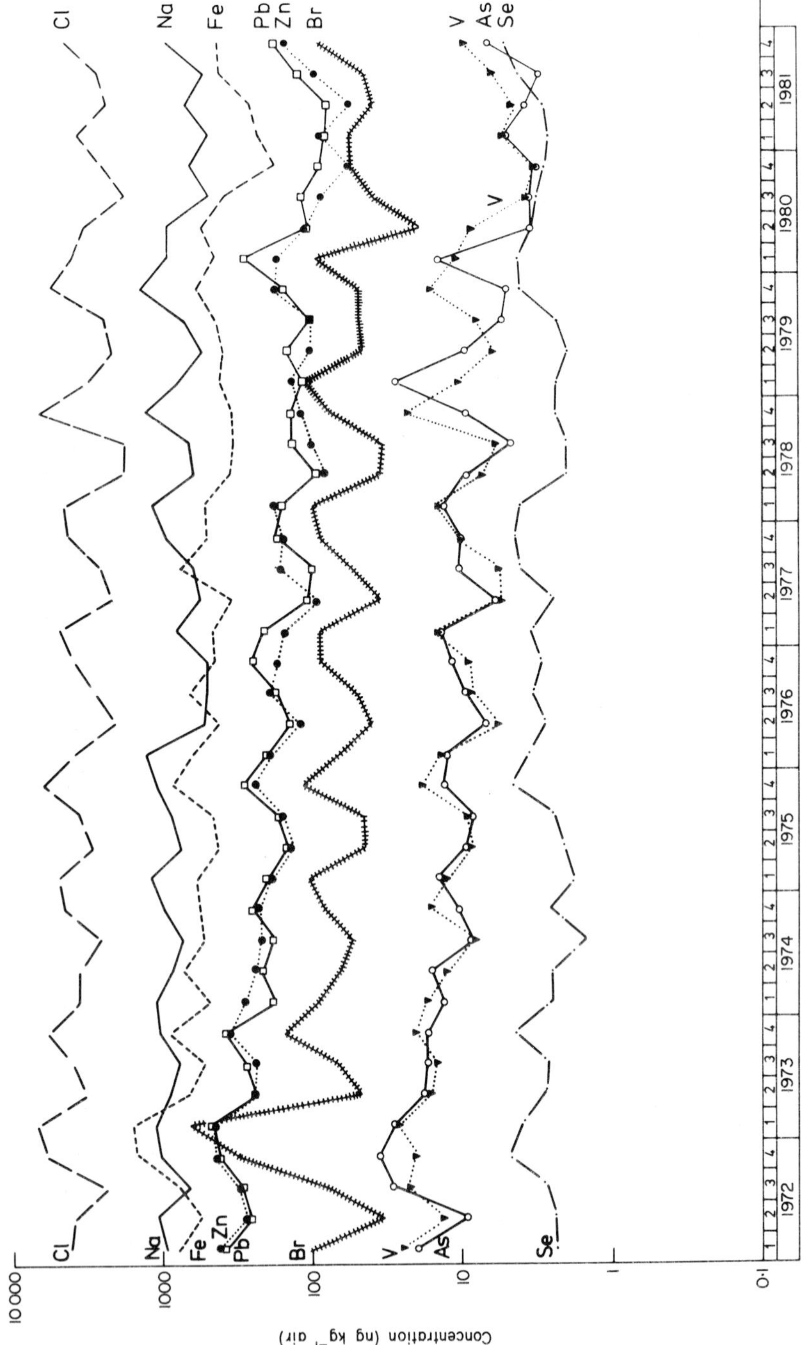


FIG. 4. Quarterly elemental concentration in air near ground level at Styrrup (1972–81).

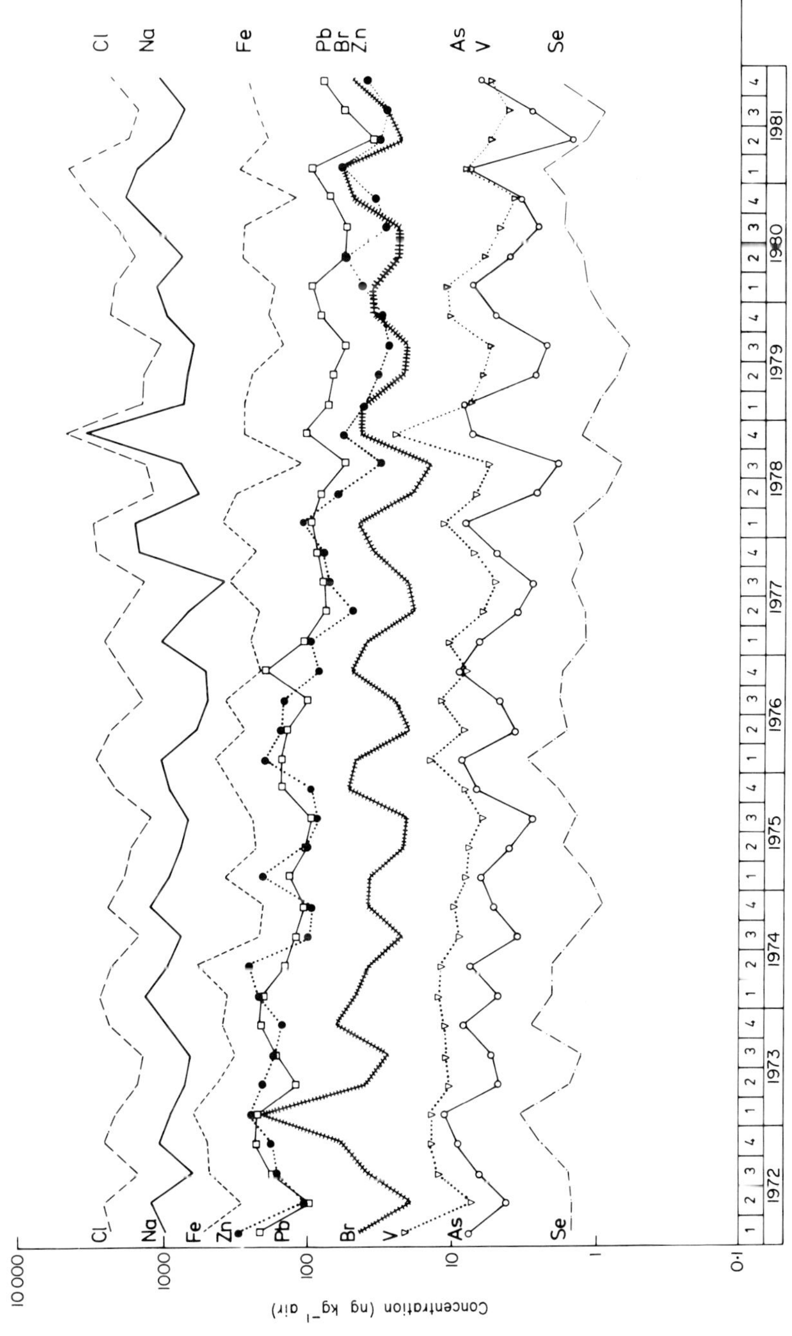


FIG. 5. Quarterly elemental concentration in air near ground level at Trebanos (1972–81).

concentrations of maritime aerosols were unimportant and were not associated with trends in other elements. It is also established that the downward trends for heavy metals generally exceed those shown by elements that are associated with material of direct terrestrial origin, such as Al, Sc, Fe and Ce.

A decrease in industrial emissions owing to the economic recession and/or improvements in emissions control and dispersion by taller stacks are considered to be the likely causes of downward trends in element concentrations. At Chilton, for example, the concentrations of Zn, As and Pb in 1981 were 39, 29 and 31%, respectively, of their levels in 1972–73, compared to the decline in black smoke which was 37% (DOE 1983). However, this decrease in Pb of 61% in 10 years exceeds the 20% reduction in Pb emissions from petrol-engined vehicles over the same period (DOE 1983) suggesting that emissions of Pb from industrial sources, e.g. coal burning were also reduced significantly.

Enrichment factors derived from the ratios of elements to Sc in air relative to the ratios of the same elements in average soil (Bowen 1966) for twenty-two elements in air particulate are shown in Table 5. At Trebanos the enrichments for Co and Ni were an order of magnitude higher than recorded at other stations (Table 5) and demonstrate the influence of a local source, i.e. smelting of nickel; the highest enrichments were found in 1975, reaching 140 for Co and 300 for Ni.

TABLE 3. Ratios of mean elemental concentrations in air for winter/summer for 1972–81.

| Element | Ratio of winter to summer concentrations at | | | |
|---|---|---|---|---|
| | Chilton, Oxon | Styrrup, Notts | Trebanos, W. Glam | Wraymires Cumbria* |
| Na | 1·4 | 1·4 | 1·6 | 1·7 |
| Mg | 1·2 | 1·0 | 0·99 | 1·5 |
| Al | 0·9 | 1·0 | 0·95 | 1·4 |
| Cl | 1·4 | 1·7 | 1·6 | 2·2 |
| K | 0·94 | 1·4 | 2·7 | 1·9 |
| Ca | 0·91 | 1·1 | 2·2 | 1·5 |
| Sc | 1·0 | 1·2 | 0·96 | 1·2 |
| V | 1·6 | 1·6 | 1·5 | 2·1 |
| Cr | 1·4 | 1·5 | 1·9 | 1·7 |
| Mn | 1·4 | 1·5 | 1·3 | 1·6 |
| Fe | 1·0 | 1·3 | 1·1 | 1·0 |
| Co | 1·3 | 1·5 | 1·1 | 1·8 |
| Ni | 1·3 | 1·7 | 1·2 | 1·2 |
| Cu | 1·1 | 2·2 | 1·4 | 1·3 |
| Zn | 1·4 | 1·5 | 1·3 | 1·9 |
| As | 1·6 | 1·5 | 2·0 | 1·8 |
| Se | 1·3 | 1·3 | 1·4 | 2·0 |
| Br | 1·9 | 2·9 | 2·3 | 2·8 |
| Sb | 1·8 | 1·6 | 1·4 | 1·9 |
| Cs | 1·3 | 1·2 | 0·93 | 1·7 |
| Ce | 0·8 | 0·93 | 0·93 | 1·1 |
| Pb | 1·3 | 1·5 | 1·5 | 1·6 |
| TSP | 0·9 | 1·0 | 0·96 | 1·3 |

*Mid 1970–81.

TABLE 4. Trends in elemental concentrations in air, 1972–81.

| Element | Slope†, per cent year $^{-1}$ at: Chilton, Oxon | Styrrup, Notts | Trebanos, W. Glam. | Wraymires, Cumbria* |
|---|---|---|---|---|
| Na | − 3·3 | − 3·5 | nst | nst |
| Mg | nst | nst | nst | −14·4 |
| Al | − 7·3 | − 8·9 | − 5·4 | − 8·0 |
| Cl | nst | − 4·4 | nst | nst |
| K | nst | −12·6 | nst | nst |
| Ca | nst | − 8·6 | nst | − 9·8 |
| Sc | − 3·2 | − 7·6 | − 7·3 | − 5·1 |
| V | −12·0 | −12·0 | − 8·0 | − 7·5 |
| Cr | − 7·6 | −11·9 | − 6·8 | − 8·6 |
| Mn | −11·0 | −10·5 | − 9·0 | − 9·4 |
| Fe | − 6·0 | −11·2 | − 9·1 | nst |
| Co | − 7·0 | − 9·5 | − 5·2 | −10·1 |
| Ni | nst | − 6·8 | nst | nst |
| Cu | nst | nst | − 7·0 | nst |
| Zn | −13·0 | −15·0 | −19·7 | −16·1 |
| As | −10·0 | −14·3 | − 6·0 | nst |
| Se | nst | nst | − 6·2 | nst |
| Br | −10·0 | −14·4 | − 7·4 | − 5·1 |
| Sb | nst | − 4·7 | nst | − 7·6 |
| Cs | −12·4 | −18·8 | −19·7 | −13·7 |
| Ce | nst | − 4·4 | nst | nst |
| Pb | − 9·6 | −12·4 | −12·9 | −10·8 |
| TSP | − 5·1 | − 5·1 | nst | nst |

†Slope is expressed as per cent year$^{-1}$ of the mean annual air concentration from 1972–81, or mid 1970–81 at Wraymires.
*mid 1970–81.
nst, No significant trends.

The mechanism of enrichment has been related to the high volatility of chalcophilic elements such as Zn, As, Se, Sb and Pb compared to the lithophilic group that includes Al, Sc, Fe and Ce, with the result that volatile metals condense on the smaller size fraction of air particulate which possesses relatively high surface area (Heindryckx 1976). Accordingly, there is separation of elements into two classes of low and high enrichment as indicated by the non-urban results (Table 5). Further, there is a general uniformity of enrichment factors at the stations which suggests similarity of origin or good atmospheric mixing (Peirson, Cawse & Cambray 1974; Peirson & Cawse 1979). High enrichments of metals have also been reported by other workers at rural locations in the northern Hemisphere, in Japan (Mamuro *et al.* 1972) and Norway (Rahn 1972) and reflect the influence of industrial activities. The incineration of urban refuse can result in exceptionally high enrichments of metals in particulate emissions (Greenberg, Zoller & Gordon 1978). In contrast, measurements away from the industrial latitudes of Europe have shown that in the very rural location of Bagauda, northern Nigeria, enrichment factors for the volatile metals are two orders of magnitude lower than in non-urban Britain (Beavington & Cawse 1978).

Quarterly enrichment factors recorded at Wraymires from 1971 to 81 for V, Zn, As, Br, Sb and Pb are shown in Fig. 6. The seasonal increase in enrichment of these elements in winter is pronounced, in accordance with their increases in air concentrations relative to Sc (Table 3). The winter and summer average enrichments of elements in air particulate are listed in Table 6. It is generally observed that elements of low enrichment (less than 5) show the least seasonal difference at all stations, whereas for those of higher enrichment increases of 1·5–2-fold in winter are usual. Vanadium has been selected as a 'marker' for emissions from oil burning in studies on source contributions (Kowalkczyk, Choquette & Gordon 1978) and from the present work this element shows a winter increase in enrichment averaging 55% at the four stations.

The increase in enrichments of Na and Cl in winter occurs at all sites (Table 6) and is attributed to an increase in rough weather causing generation and transfer of sea-spray aerosol to inland regions. At Styrrup however, the increase in enrichment of Cl in winter (51%) exceeds that for Na (24%) and coal-burning in this mining region is considered responsible for the excess Cl.

### *Deposition velocity of elements and enrichment*

From the continuous measurements of dry deposition at Chilton, Styrrup and Wraymires an inverse relationship has been demonstrated between $V_g$ of the elements, which is related

TABLE 5. Mean annual air particulate enrichment factors, 1972–81 (normalized to Sc)

| Element | Chilton, Oxon | Styrrup, Notts | Trebanos, W. Glam. | Wraymires, Cumbria* |
|---|---|---|---|---|
| Na | 13 | 8·3 | 16 | 23 |
| Mg | 6·7 | 10 | 6·0 | 10 |
| Al | 0·35 | 0·37 | 0·31 | 0.40 |
| Cl | 2200 | 2300 | 2300 | 3300 |
| K | 5·2 | 3·8 | 6·1 | 7·6 |
| Ca | 6·7 | 6·3 | 6·6 | 7·8 |
| Sc | 1·0 | 1·0 | 1·0 | 1·0 |
| V | 11 | 7·9 | 10 | 12 |
| Cr | 2·9 | 6·5 | 9·3 | 3·4 |
| Mn | 1·7 | 2·6 | 1·5 | 1·9 |
| Fe | 0·76 | 0·93 | 0·84 | 0·94 |
| Co | 4·4 | 4·0 | 110 | 4·7 |
| Ni | 16 | 12·1 | 195 | 17 |
| Cu | 94 | 81 | 125 | 130 |
| Zn | 195 | 240 | 230 | 180 |
| As | 59 | 125 | 94 | 69 |
| Se | 760 | 970 | 810 | 860 |
| Br | 890 | 1100 | 830 | 850 |
| Sb | 32 | 38 | 33 | 36 |
| Cs | 3·1 | 3·0 | 2·6 | 3·1 |
| Ce | 0·86 | 0·70 | 0·86 | 0·83 |
| Pb | 1250 | 1250 | 1250 | 1250 |

*Mid 1970–81.

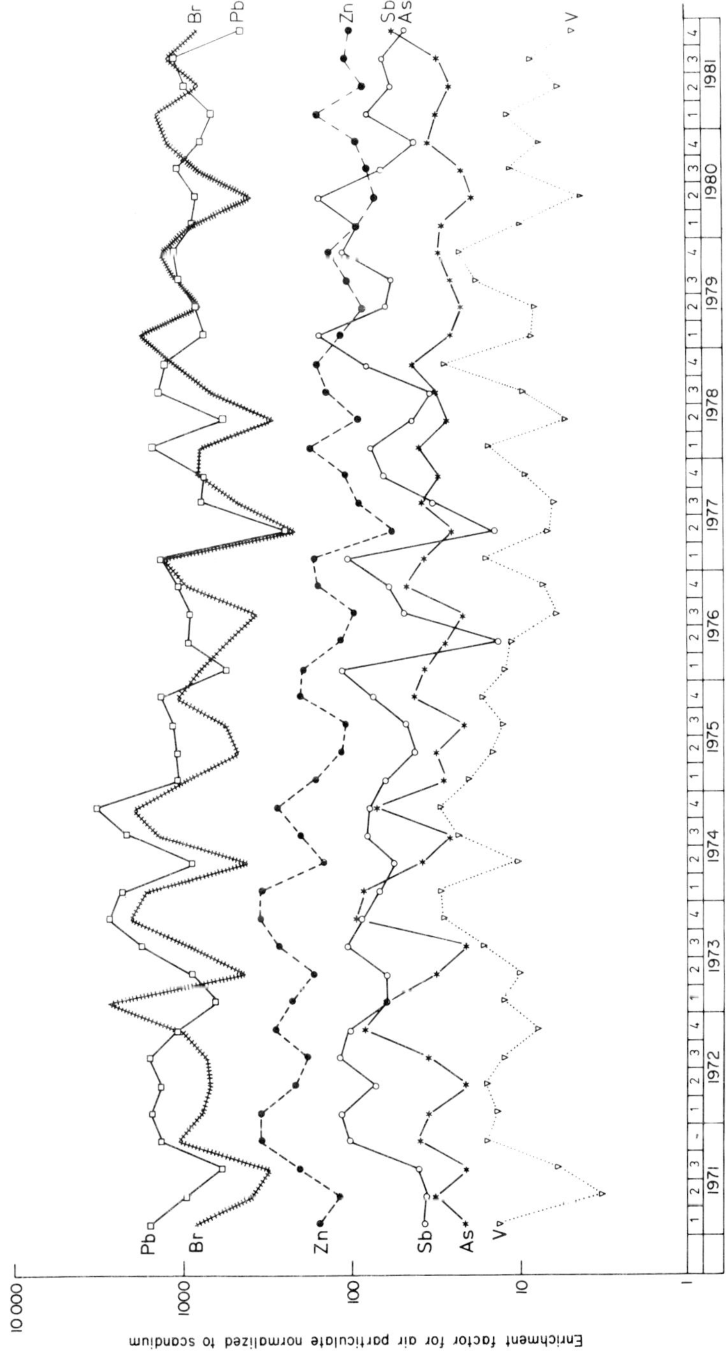


FIG. 6. Enrichment factors for air particulate at Wraymires (1970–81).

to particle size, and their enrichment factors as shown in Fig. 7 which is based on 10 years of data at Styrrup, Notts. The deposition velocity is derived from:

$$V_g\ (\text{cm s}^{-1}) = \frac{\text{rate of dry deposition, in } \mu\text{g cm}^{-2}\ \text{s}^{-1}}{\text{concentration in air, in } \mu\text{g cm}^{-3}} \quad \text{(Chamberlain 1960)}$$

By operation of an Andersen size-selective sampler (impactor) in parallel with the standard sampling method it has been established (Cawse 1974) that a $V_g$ of 0·5 cm $s^{-1}$ corresponds to a mass median diameter of ~ 1 μm (relative density).

Relatively low deposition velocity below 0·5 cm $s^{-1}$ (and hence small particle size), is typical of combustion-derived aerosols of high enrichment factor, e.g. As, Se, Sb and Pb, produced by industry or motor vehicles. Higher $V_g$ (larger particle size) is typical of elements with low enrichment that originate from soil disturbance, mining or quarrying, such as Al, Ce, Eu and Fe (Fig. 7).

### *Total (wet and dry) deposition*

The average annual total deposition of elements at three of the non-urban stations is listed in Table 7, together with measurements made in 1979 in the industrial West Midlands. At Trebanos, detailed analysis of rainwater was discontinued after 1976. Apart from the

TABLE 6. Mean air particulate enrichment factors (normalized to Sc) for winters (quarters 1 & 4) and summers (quarters 2 & 3), 1972–81.

| Element | Chilton, Oxon | | Styrrup, Notts | | Trebanos, W. Glam. | | Wraymires, Cumbria* | |
|---|---|---|---|---|---|---|---|---|
| | Winter | Summer | Winter | Summer | Winter | Summer | Winter | Summer |
| Na | 15 | 11 | 9·2 | 7·4 | 21 | 12 | 27 | 19 |
| Mg | 7·3 | 6·2 | 9·9 | 11 | 6·1 | 5·9 | 11 | 8·8 |
| Al | 0·33 | 0·37 | 0·34 | 0·39 | 0·31 | 0·32 | 0·43 | 0·37 |
| Cl | 2500 | 1900 | 2800 | 1850 | 2900 | 1700 | 4000 | 2300 |
| K | 5·0 | 5·5 | 4·2 | 3·3 | 9·3 | 3·4 | 9·0 | 6·0 |
| Ca | 6·3 | 7·1 | 6·1 | 6·3 | 9·1 | 3·9 | 8·3 | 7·1 |
| Sc | 1·0 | 1·0 | 1·0 | 1·0 | 1·0 | 1·0 | 1·0 | 1·0 |
| V | 14 | 8·7 | 9·3 | 6·4 | 12 | 8·1 | 14 | 8·4 |
| Cr | 3·4 | 2·5 | 7·3 | 5·5 | 13 | 6·1 | 3·8 | 2·8 |
| Mn | 2·0 | 1·5 | 3·0 | 2·2 | 1·7 | 1·3 | 2·1 | 1·7 |
| Fe | 0·78 | 0·76 | 0·98 | 0·87 | 0·91 | 0·79 | 0·85 | 1·0 |
| Co | 5·1 | 3·9 | 4·4 | 3·4 | 115 | 105 | 5·3 | 3·9 |
| Ni | 19 | 14 | 14 | 9·5 | 220 | 180 | 17 | 17 |
| Cu | 95 | 94 | 105 | 55 | 145 | 105 | 135 | 125 |
| Zn | 230 | 165 | 260 | 210 | 260 | 200 | 210 | 135 |
| As | 74 | 46 | 145 | 105 | 130 | 63 | 79 | 53 |
| Se | 820 | 700 | 1050 | 900 | 950 | 680 | 1050 | 650 |
| Br | 1150 | 620 | 1550 | 620 | 1200 | 510 | 1150 | 510 |
| Sb | 40 | 24 | 44 | 32 | 41 | 26 | 43 | 28 |
| Cs | 3·5 | 2·8 | 3·1 | 2·8 | 2·7 | 2·8 | 3·6 | 2·6 |
| Ce | 0·78 | 0·97 | 0·64 | 0·78 | 0·85 | 0·89 | 0·81 | 0·86 |
| Pb | 1450 | 1100 | 1400 | 1100 | 1500 | 1000 | 1200 | 890 |

*Mid 1970–81.

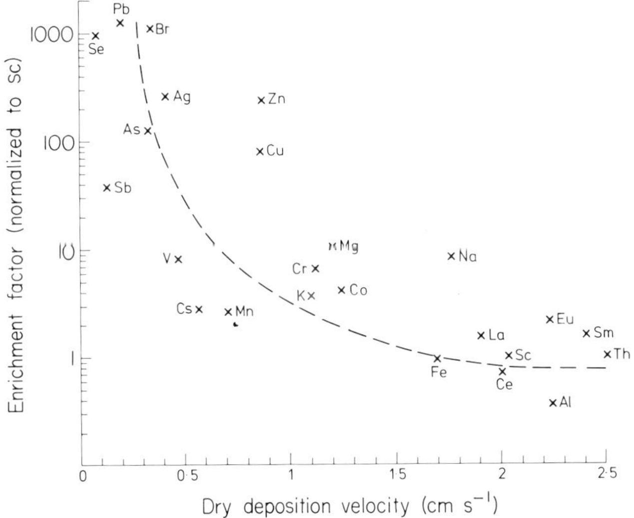


FIG. 7. Average annual enrichment factors and dry deposition velocities of elements in air particulate at Styrrup, Notts (1972–81).

maritime-derived elements which showed the highest inputs at Wraymires, greatest deposition of elements occurred at Styrrup. Nevertheless, in the urban situation at Walsall, order of magnitude increases occurred in the annual deposition of Cr, Cu and Zn compared with rural sites, together with some five times the deposition of Fe, As, Se, Sb and Pb.

In agreement with previous results obtained from operation of the sampling network of stations from 1971–75, a direct relationship was established between element solubility in rainwater and their enrichment factors (Cawse 1980). Many plant nutrients and potentially toxic pollutants of high solubility had high enrichments in the range 10–700 (normalized to Sc), which is important with respect to mobility and cycling in the terrestrial environment. The solubility of total deposition is generally >80% for Na, Cl, K, Ca, Se and Br, and in the range 50–80% for Mg, V, Cr, Mn, As, Co, Ni, Cu, Cd, Sb and Pb. Low solubility (<25%) is found for Al, Sc and Fe.

Ratios of $SO_4$/Na in the soluble deposition at Chilton, Wraymires and Styrrup were 1·8, 1·4 and 2·6, respectively. Considering that the ratio is 0·23 in bulk seawater, the large excess of non-maritime sulphate is attributed to industrial sources. Thus, the industrial contribution to sulphate input was 91·2% at Styrrup in the East Midlands, but decreased to 87% at Chilton and 84% at Wraymires. The average annual deposition of $SO_4$ and $NO_3$, and the pH of rainwater is in general agreement with data in recent reviews of acid deposition in Europe (WSL 1983; Buckley-Golder 1984) although long-term measurements are clearly inadequate.

Measurements of the atmospheric deposition in 1980 of some trace metals at rural locations in Europe (FAO European Co-Operative Network) have been summarized by Mosbaek, Tjell & Hovmand (1982) and are comparable (Table 8) with average annual inputs at the non-urban locations in Britain (Table 7).

TABLE 7. Average annual total deposition of elements ($\mu g\ cm^{-2}\ y^{-1}$), 1972–81[†]

| Element | Non-urban<br>Chilton, Oxon | <br>Styrrup, Notts | <br>Wraymires, Cumbria | Urban<br>Walsall, W. Midlands (1979)[‡] |
|---|---|---|---|---|
| Na | 240 | 210 | 410 | |
| Mg | 54* | 58* | 105* | |
| Al | 39 | 150 | 24 | |
| Cl | 420 | 460 | 740 | |
| K | 160* | 92* | 200* | |
| Ca | 150 | 175 | 190 | |
| Sc | 0·004 | 0·012 | 0·003 | 0·017 |
| Ti | <4 | <5 | <7 | |
| V | 0·50 | 0·73 | 0·60 | |
| Cr | 0·19 | 0·61 | 0·28 | 1·2 |
| Mn | 0·94 | 2·0 | 1·1 | |
| Fe | 22 | 57 | 14 | 100 |
| Co | 0·024 | 0·060 | 0·016 | 0·05 |
| Ni | 0·74 | 0·58 | 1·1 | 1·1 |
| Cu | 1·7 | 2·4 | 2·5 | 53 |
| Zn | 4·9 | 10 | 4·8 | 41 |
| As | 0·13 | 0·40 | 0·19 | 0·7 |
| Se | 0·028 | 0·052 | 0·040 | 0·15 |
| Br | 2·0 | 2·9 | 4·6 | |
| Rb | <0·1 | <0·4 | <0·2 | |
| Mo | <0·09 | <0·1 | <0·1 | |
| Ag | <0·007 | <0·01 | <0·01 | <0·07 |
| Cd | <0·2 | <0·3 | <0·4 | <0·4 |
| | (0.052)* | (0·068)* | (<0·1)* | (0·3)* |
| In | <0·01 | <0·01 | 0·025 | |
| Sb | 0·036 | 0·080 | 0·040 | 0·2 |
| I | <0·3 | <0·4 | <0·7 | |
| Cs | 0·006 | 0·012 | 0·007 | |
| La | <0·05 | <0·1 | <0·08 | |
| Ce | 0·040 | 0·072 | 0·024 | |
| Sm | 0·004 | 0·005 | <0·003 | |
| Eu | 0·0007 | 0·001 | <0·0005 | |
| W | <0·07 | <0·08 | <0·1 | |
| Au | <0·001 | <0·0007 | <0·002 | |
| Hg | <0·01 | <0·01 | <0·02 | |
| Pb | 2·5 | 3·5 | 2·9 | 13 |
| Th | <0·005 | 0·011 | <0·005 | |
| $NO_3$ | 390* | 360* | 440* | |
| $SO_4$ | 430* | 550* | 580* | |
| pH | 4·0 | 4·0 | 4·0 | |
| Rainfall ($mm\ y^{-1}$) | 670 | 580 | 1800 | 740 |

*Measurement on soluble fraction.

[†] $\mu g\ cm^{-2} \times 0{\cdot}1 = kg\ ha^{-1}$.

‡Walsall data from Cawse & Turner (1981).

The average total deposition of Pb (Table 7) corresponds with the calculated fallout of Pb from automobile exhaust and industrial sources (Chamberlain 1983), and with measurements of Pb in cores of peat whereby the recent deposition rate is estimated as 2·3 $\mu g\ cm^{-2}\ y^{-1}$ (Livett, Lee & Tallis 1979).

The majority of elements showed increases in total deposition during the winter period, particularly at Wraymires (Table 9) in agreement with the increase in average annual rainfall in winter at that site compared to the other stations; the most noticeable increases of 2–3-fold were found for Na, Mg, Al, Cl, K, Ca, V, Mn, Ni, As, Sb, Cs and Pb. The deposition of sulphate and nitrate at Chilton and Styrrup showed little change in winter, but increased deposition of 2–3-fold was evident for Na, Cl and Br at both sites and increases in winter deposition were frequently observed for several heavy metals, notably V, Cr, Ni and Pb.

### *Washout factors*

The washout factor ($W$) relates the concentration of an element in rain to that in air (Chamberlain 1960). It is known that a relatively large scavenging efficiency (SCE, i.e. collision efficiency × retention efficiency) of precipitation occurs for small particles in the range 0·01–0·05 μm diameter, which is attributed to their increased hygroscopic component (Radke, Eltgroth & Hobbs 1978). Washout factors derived from the total (wet + dry) deposition may be excessive for elements that are predominant in dusty locations, e.g. Al, Sc and Fe, where the dry deposition contributes excessively to the true rainwater concentrations derived from the wet deposition. Therefore, it is preferable to use soluble deposition data to derive washout factors ($W_{sol}$), and average values found at Chilton from 1972 to 1981 can be classified as follows.

Low $W_{sol}$ ($<500$): Al 340, Sc 100, Cr 430, Fe 260, Se 310, Sb 190, Ce 230, Pb 300.

Intermediate $W_{sol}$ (500–1000): V 590, Mn 930, Co 790, As 660.

High $W_{sol}$ (1000–3000): Na 5500, Cl 3700, Ni 1500, Cu 1040, Zn 1010.

Low $W_{sol}$ is associated with a local, near to ground source of emission, e.g. Pb from automobile exhaust or Al from soil dust. Elements with relatively high $W_{sol}$ values are considered to be present mainly at the rain-forming altitude (~3 km) and are more easily scavenged by rainfall, particularly if associated with small particle-size, e.g. Mn, Zn, or if highly hygroscopic (Na, Cl).

TABLE 8. Atmospheric deposition at rural locations in Europe in 1980.

| Location | Deposition rate ($\mu g\ cm^{-2}\ y^{-1}$) | | | Rainfall ($mm\ y^{-1}$) |
|---|---|---|---|---|
| | Pb | Cd | Zn | |
| Faroe Islands | 5·5 | 0·08 | 5.6 | 1617 |
| Karvatn, Norway | 1·7 | 0·06 | 5.2 | 1200 |
| Birkenes, Norway | 8·9 | 0·35 | 18 | 1001 |
| Sweden | 4·6 | 0·07 | 6·5 | 744 |
| Jutland, Denmark | 9·6 | 0·12 | 9·5 | 635 |
| Zealand, Denmark | 12·8 | 0·18 | 19 | 916 |

### *Dry deposition*

Dry deposition of elements generally contributed less to the total deposition at Wraymires than at other sites; for example, the ratio of dry to total deposition (*D/T*) was in the range 0·2–0·4 at Chilton and Styrrup for the elements Na, Mg, Cl, K, Ca, V, Mn, Co, Sb, Cs and Ce but showed *D/T* from 0·06 to 0·2 for these elements at Wraymires where the average annual rainfall is some 3-fold greater than the other two sites. *D/T* ratios at Wraymires can be grouped as follows.

Low *D/T* (0·04–0·1): Na 0·08, Mg 0·09, Cl 0·08, K 0·06, V 0·07, Co 0·07, As 0·04, Se 0·05, Br 0·05, Sb 0·1.

Medium *D/T* (0·1–0·2): Ca 0·11, Cr 0·11, Mn 0·13, Cu 0·12, Ce 0·17.

High *D/T* (0·2–0·7): Al 0·21, Sc 0·68, Fe 0·39, Ni 0·40, Zn 0·21, Cs 0·23, Pb 0·38.

TABLE 9. Ratios of mean concentrations in total deposition for winter/summer 1972–81*

| Element | Chilton, Oxon | Styrrup, Notts | Wraymires, Cumbria |
|---|---|---|---|
| Na | 2·7 | 2·3 | 3·1 |
| Mg | 1·8 | 1·7 | 2·9 |
| Al | 1·3 | 3·3 | 2·5 |
| Cl | 2·8 | 2·7 | 3·2 |
| K | 1·3 | 1·1 | 2·3 |
| Ca | 1·4 | 1·4 | 2·8 |
| Sc | 1·0 | 2·0 | 1·8 |
| V | 1·6 | 1·7 | 2·8 |
| Cr | 1·2 | 1·8 | 0·6 |
| Mn | 1·1 | 1·2 | 2·1 |
| Fe | 1·2 | 1·7 | 1·7 |
| Co | 1·4 | 1·0 | 1·7 |
| Ni | 1·8 | 1·3 | 3·4 |
| Cu | 0·8 | 1·4 | 1·5 |
| Zn | 1·2 | 0·8 | 1·4 |
| As | 1·3 | 1·0 | 2·0 |
| Se | 1·3 | 0·8 | 1·9 |
| Br | 2·4 | 2·4 | 1·3 |
| Sb | 1·1 | 0·9 | 2·0 |
| Cs | 1·0 | 1·0 | 2·0 |
| Ce | 1·1 | 1·3 | 1·0 |
| Pb | 1·3 | 1·3 | 2·3 |
| $NO_3$ | 1·1 | 1·0 | 1·9 |
| $SO_4$ | 1·3 | 1·0 | 1·6 |
| Rainfall | 1·2 | 1·1 | 1·6 |

*Winter = quarters 1 and 4; Summer = quarters 2 and 3.

TABLE 10. Trends in elemental concentrations in total deposition, 1972–81

| Element | Slope*, per cent year$^{-1}$ at: Chilton, Oxon | Styrrup, Notts | Wraymires, Cumbria |
|---|---|---|---|
| Na | nst | nst | nst |
| Mg | nst | − 8·0 | nst |
| Al | −14 | −30 | nst |
| Cl | nst | nst | nst |
| K | nst | nst | +26 |
| Ca | + 7·8 | nst | +25 |
| Sc | − 7·0 | − 8·9 | nst |
| V | − 9·0 | −17 | nst |
| Cr | −10 | −15 | nst |
| Mn | − 5·8 | − 6·5 | nst |
| Fe | − 9·5 | −12 | nst |
| Co | − 8·1 | −10 | nst |
| Ni | nst | nst | nst |
| Cu | + 6·6 | nst | nst |
| Zn | nst | nst | − 6·3 |
| As | nst | − 11 | nst |
| Se | nst | nst | nst |
| Br | nst | nst | nst |
| Sb | nst | − 9·1 | nst |
| Ce | nst | − 6·2 | nst |
| Cs | −16 | −20 | − 8·3 |
| Pb | − 5·6 | nst | nst |
| $NO_3$ | nst | nst | nst |
| $SO_4$ | nst | −9·5 | nst |
| pH | nst | nst | nst |
| Rainfall | nst | + 3·8 | nst |

*Slope is expressed as per cent year$^{-1}$of the mean annual deposition 1972–81.
nst, no significant trends.

### *Trends in total deposition*

The quarterly total deposition of elements, rainfall and pH was examined for long-term trends over the years 1972–81 (Table 10). At Wraymires, no significant downward trends were found for the twenty-two elements listed except for decreases in Zn and Cs inputs of 6–8% y$^{-1}$ of the mean annual depositions. However, at Chilton and Styrrup there were significant downward trends for more elements, particularly at Styrrup for Al, V, Cr, Fe, As, Co and Cs with decreases in total deposition from 10 to 30% y$^{-1}$. No significant trends were evident for deposition of $NO_3$ and $SO_4$ or rainwater pH at any of the three stations. Increasing trends were only observed for Ca and Cu at Chilton, and both K and Ca at Wraymires; unusually high deposition of K and Ca was recorded in months with high rainfall as confirmed by their low *D/T* ratio, and a probable source of these elements may be washout of dust from fertilizers and limestone used in agriculture.

### *Long-term effects of deposition*

The effect of atmospheric deposition of elements on soil concentrations in the long-term, assuming that the average deposition rates (1972–81) are maintained over 30 years, may be made by reference to analysis of total element concentrations in a rendzina soil at Chilton, Oxon in 1973 (Cawse 1974). Lead deposition would have a marked effect, since atmospheric inputs would raise the soil concentration in the surface 5 cm from 3 mg Pb $kg^{-1}$ in 1973, to 18 mg Pb $kg^{-1}$ by the year 2003. On a similar basis, increases in soil concentrations of other elements relative to their levels in 1973 would be: V +3·8%, Cr +1·0%, Co +0·6%, Ni +13%, Zn +11% and As +8% given mixing to 5 cm depth.

## CONCLUDING COMMENTS

Continuous measurements of the chemical composition of air particulate and deposition in rural Britain are necessary to establish the long-term trends as influenced by industrial activity, changes in methods of energy production and emissions control policy, e.g. lowering of lead levels in petrol. It has been shown that the concentrations in air of many elements of potential toxicity have decreased over the period 1972–81, but this may not continue.

The interactions between gaseous pollutants and atmospheric trace metals is an important consideration; it is known that the aqueous phase oxidation of $SO_2$ to $SO_4^{--}$ is catalysed by Mn in particular (Penkett, Jones & Eggleton 1979) and also that the formation of $HNO_3$ aerosol from $NO_2$ is increased by Cu, Mo and Zn (ten Brink *et al.* 1978).

Further information is needed on the biogeochemical cycling of trace and major elements to assess disturbances caused to terrestrial and aquatic ecosystems, bearing in mind the possibility of synergistic effects.

## ACKNOWLEDGMENTS

The effort of staff from the Environmental and Medical Sciences Division who were associated with the analysis of samples is gratefully acknowledged. The survey was initially sponsored by the Natural Environment Research Council for 3 years, and is now sponsored by the Department of the Environment. Routine operation of the sampling stations was made possible by co-operation of staff from the Freshwater Biological Association, the Institute of Geological Sciences and Borough of Lliw Valley Environmental Health Department.

## REFERENCES

**Beavington, F. & Cawse, P.A. (1978).** Comparative studies of trace elements in air particulate. *Science of the Total Environment,* **10**, 239–244.

**Bowen, H.J.M. (1966).** *Trace Elements in Biochemistry.* Academic Press, London.

**Bowen, H.J.M. (1979).** *Environmental Chemistry of the Elements.* Academic Press, London.

**ten Brink, H.M., Bontje, J.A., Poelstra, H.S. & van de Vate, J.F. (1978).** The interaction between $NO_x$, $O_3$ and airborne particles. *Atmospheric Pollution 1978* (Ed. by M.M. Benarie), pp. 239–244. Elsevier, Amsterdam.

**Buckley-Golder, D.H. (1984).** *Acidity in the Environment.* Energy Technology Support Unit, Harwell, Report R23. HMSO, London.

**Cambray, R.S., Jefferies, D.F. & Topping, G. (1979).** The atmospheric input of trace elements to the North Sea. *Marine Science Communications,* **5**, 175–194.

**Cawse. P.A. (1974).** *A survey of atmospheric trace elements in the UK (1972–73).* AERE Harwell Report R-7669. HMSO, London.

**Cawse, P.A. (1975).** Anomalous high concentrations of particulate bromine in the atmosphere of the UK. *Chemosphere,* **1**, 107–112.

**Cawse, P.A. (1976).** *Intercomparison of air filter deposits for trace elements.* AERE Harwell Report R-8191. HMSO, London.

**Cawse, P.A. (1980).** Deposition of trace elements from the atmosphere in the UK. *Inorganic Pollution and Agriculture,* pp. 22–46. Reference Book 326, Ministry of Agriculture, Fisheries & Food. HMSO, London.

**Cawse, P.A. (1982).** Inorganic particulate matter in the atmosphere. *Environmental Chemistry,* Vol. 2, pp. 1–69. Royal Society of Chemistry, London.

**Cawse, P.A. & Peirson, D.H. (1972).** *An analytical study of trace elements in the atmospheric environment.* AERE Harwell Report R-7134. HMSO, London.

**Cawse, P.A. & Turner, G.S. (1981).** *Trace elements in vegetables from allotments in Walsall in 1979.* AERE Harwell Report G-1653. UKAEA, Harwell, Oxon.

**Chamberlain, A.C. (1960).** Aspects of the deposition of radioactive and other gases and particles. *International Journal of Air Pollution,* **3**, 63–88.

**Chamberlain, A.C. (1983).** Fallout of lead and uptake by crops. *Atmospheric Environment,* **17**, 693–706.

**Dams, R. & de Jonge, J. (1976).** Chemical composition of Swiss aerosols from the Jungfraujoch. *Atmospheric Environment,* **10**, 1079–1084.

**DOE (1983).** *Department of the Environment Digest of Environmental Pollution and Water Statistics, No. 5, 1982. Air Pollution: Additional Tables.* HMSO, London.

**Greenberg, R.R., Zoller, W.H. & Gordon, G.E. (1978).** The contribution of refuse incineration to urban aerosols. *Proc. 4th Joint Conf. Sensing Environmental Pollutants,* pp. 817–820. Am. Chem. Soc., Washington, D.C.

**Heindryckx, R. (1976).** Comparison of the mass-size functions of the elements in the aerosols of Gent industrial district with data from other areas. Some physico-chemical implications. *Atmospheric Environment,* **10**, 65–71.

**Kowalczyk, G.S., Choquette, C.E. & Gordon, G.E. (1978).** Chemical element balances and identification of air pollution sources in Washington D.C. *Atmospheric Environment,* **12**, 1143–1153.

**Lininger, R.L., Duce, R.A., Winchester, J.W. & Matson, W.R. (1966).** Chlorine, bromine, iodine and lead in aerosols from Cambridge, Massachusetts. *Journal of Geophysics Research,* **71**, 2457–2463.

**Livett, E.A., Lee, J.A. & Tallis, J.H. (1979).** Lead, zinc and copper analysis of British blanket peats. *Journal of Ecology,* **67**, 865–891.

**Mamuro, T., Matsuda, Y., Mizohatu, A. & Matsunami, T. (1972).** Relationships among various elements in the atmosphere over Osaka district. *Annual Report Radiation Center Osaka Prefecture,* **13**, 1–15.

**Mosbaek, H., Tjell, J.C. & Hovmand, M.F. (1982).** Atmospheric deposition of trace elements to agricultural soils and plants. *Newsletter from the FAO European Co-Operative Network on Trace Elements, 1st Issue,* pp. 27–30. State University Gent, Belgium.

**Pattenden, N.J. (1974).** *Atmospheric concentrations and deposition rates of some trace elements measured in the Swansea–Neath–Port Talbot area.* AERE Harwell Report R-7729. HMSO, London.

**Pattenden, N.J. (1977).** *Multi-element airborne dust measurements for Walsall Metropolitan Borough.* AERE Harwell Report EMS (77)09. UKAEA, Harwell, Oxon.

**Peirson, D.H. & Cambray, R.S. (1965).** Fission product fallout from the nuclear explosions of 1961 and 1962. *Nature (Lond.),* **205**, 433–440.

**Peirson, D.H. & Cawse, P.A. (1979),** Trace elements in the atmosphere. *Philosophical Transactions Royal Society Lond.,* **B.288**, 41–49.

**Peirson, D.H., Cawse, P.A. & Cambray, R.S. (1974).** Chemical uniformity of airborne particulate material, and a maritime effect. *Nature (Lond.),* **251**, 675–679.

**Peirson, D.H., Cawse, P.A., Salmon, L. & Cambray, R.S. (1973).** Trace elements in the atmospheric environment. *Nature (Lond.),* **241**, 252–256.

**Penkett, S.A., Jones, B.M.R. & Eggleton, A.E.J. (1979).** A study of $SO_2$ oxidation in stored rainwater samples. *Atmospheric Environment,* **13**, 139–147.

**Radke, L.F., Eltgroth, M.W. & Hobbs, P.V. (1978).** Precipitation scavenging of aerosol particles. *Proc. Am. Met. Soc. Conf. Cloud Physics & Atmospheric Electricity,* pp. 44–48. American Met. Soc., Boston, Mass.

**Rahn, K.A. (1972).** *Study of national air pollution by combustion, Progress Report INW.* University Liege, Gent, Belgium.

**Volz, F.E. (1975).** Volcanic twilights from the Fuego eruption. *Science,* **189**, 48–50.

**WHO (1979).** *Environmental Health Criteria, No.8.* World Health Organisation, Geneva.

**WSL (1983).** *Acid Deposition In the United Kingdom.* UK Review Group on Acid Rain, Chairman C.F. Barrett. Warren Spring Laboratory, Stevenage.

# Facts and fallacies in wet deposition modelling

H. M. ApSIMON, A. J. H. GODDARD,
P. M. MANNING* AND K. SIMMS

*Nuclear Power Section, Department of Mechanical Engineering, Imperial College, Exhibition Road, London SW7 2BX*

## SUMMARY

**1** Following a reactor accident, relatively high contamination at ground level can occur, even at quite long distances from the source, if the pollutant cloud encounters intense precipitation. To estimate such contamination and its extent properly, it is necessary to take into account the spatial and temporal structure of rain patterns and their motion.
**2** Currently, models of wet deposition are rather crude. Source meteorology is usually used and is clearly inadequate. Furthermore, no allowance is made for the dynamic nature of rainfall, which occurs as a result of vertical air motions and convergence; nor for the different scavenging mechanism operating in and below cloud.
**3** Meteorological information available on these aspects of wet deposition is reviewed, and their importance and inclusion in modelling and prediction of resulting ground contamination is indicated. Some of the pitfalls of simple modelling procedures are illustrated.

## INTRODUCTION

The scenario in which an accidental release of radioactivity from a nuclear power station is envisaged to travel a considerable distance from the source in dry weather, and is then scavenged by heavy rain, has caused particular concern in reactor risk assessments because it suggests a possible way in which relatively severe contamination could be caused over a major population centre. It should be stressed that such a scenario has a very low probability of occurrence. Nevertheless, the modelling of wet deposition when radioactive releases encounter spatially and temporally varying precipitation areas is currently very crude.

This paper points out some of the problems and the areas of uncertainty in trying to predict consequences of accidental releases of radionuclides. These potential consequences may be divided into two types. If there are areas where contamination and resulting dose to individuals might exceed certain threshold levels than there may be immediate early effects and possible deaths. This can only occur over localized areas and in the event of a very major accident with a very low likelihood of occurrence. But at lower exposures there are also latent effects such as induced cancers for which there is no recognized threshold of dose, and which are statistically related to the collective dose

*Present address: British Gas, Watson House, Fulham.

summed over the whole exposed population. In this case the prime requirement is to assess the total exposure and not necessarily to have a detailed breakdown of the spatial distribution of exposure among the population.

## CURRENT MODELLING

Current consequence assessment techniques for nuclear accidents, e.g. those used in CRAC (VSNRC 1975) or MARC (Clarke & Kelly 1981), are based on the use of a Gaussian plume model in which the release is represented as a plume travelling downwind from the source. To represent variable meteorological conditions this plume is treated as a sequence of contiguous segments each representing a different hour of travel of the release. In each segment the meteorological conditions correspond to those at the source in the corresponding hour after the release. Thus, if the windspeed increases during a certain hour at the source the corresponding plume segment extends further in the alongwind direction. Material is only washed out in segments corresponding to periods of rain at the source. Fig. 1 gives schematic illustration of a situation in which the plume travels for 5 hours while it is dry at the source and it then rains.

When it rains on the plume it is assumed that the plume is depleted uniformly throughout its depth at a prescribed rate, $\Lambda\ s^{-1}$. Thus, over each square metre of ground a fraction of the material $\Lambda$ in the column of air above is assumed to be instantly deposited on the ground. $\Lambda$ is the washout coefficient and may be varied with rainfall rate. This is a very simple washout model.

It is argued that the segmented Gaussian plume treatment is equivalent to assuming that meteorological conditions are changing everywhere synchronously with those at the source. This is fallacious. What the segmented plume model really depicts is material dispersing and travelling downwind from the source and encountering stationary bands of

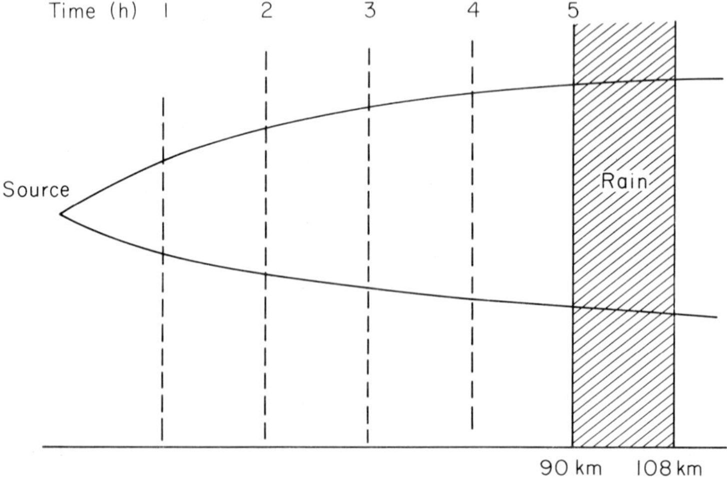


FIG. 1. Segmented Gaussian plume model.

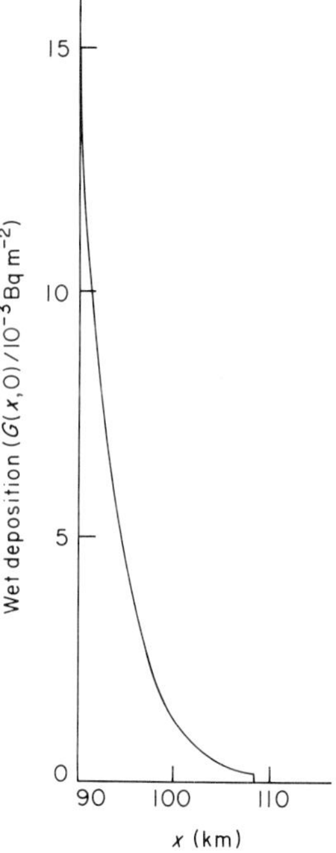


Fig. 2. Predicted pattern of wet deposition along plume axis using segmented Gaussian plume approach.

precipitation lying across the entire width of the plume for however long the material may take to pass beneath it. The resulting predicted pattern of wet deposition has a sharp peak at the leading edge of the cloud band where all the material released first encounters precipitation, and the flux of material thereafter falls off exponentially with alongwind distance during precipitation. It is often assumed that the effective washout rate increases linearly with rainfall rate $J$, in which case the peak of deposition will be higher and sharper for high rainfall rates and low windspeeds.

Fig. 2 provides an example of the predicted variation of wet deposition along the plume axis as a function of distance from the source using this segmented plume approach. In this instance the wind-speed transporting the pollutant is uniform at 5 m $s^{-1}$ with 5 hours of travel in dry conditions followed by an hour of heavy rain at a rate $J$ of 25 mm $h^{-1}$. Uniform washout at a rate of $\Lambda = 5 \times 10^{-5} J s^{-1}$ has been assumed and lateral spreading of the release corresponds to Pasquill category D conditions. The amount released was taken as a nominal value of $10^6$ Bq.

## SIMPLE ALTERNATIVE APPROACH

A simple alternative model may be used to demonstrate the hidden assumptions in the Gaussian plume approach. To start with, intense rain is likely to be highly localized in space and time and hence it may not be realistic to assume that it could occur over the whole width of the plume. Ritchie, Brown & Wayland (1980) attempted to allow for this in a statistical way based on the analysis by Austin & Houze (1972) of rainfall structure within nine storms over the New England area. However, a Gaussian plume model was used.

The difficulty with this is that it does not allow for the fact that dispersion of the release in the alongwind direction will be at least as great as the cross-wind spread, even for a release of very short duration. Spread out in this way, it may take a long time to pass beneath a rain cloud; longer in fact than the duration of precipitation. Also, the rainclouds themselves may not be stationary but may move with quite a different windspeed aloft from the lower-level pollutant release.

Disregarding the limitations of the finite duration of a rainstorm, as a simple illustration we can consider a uniform rectangular raincloud moving downwind with speed $v$, and a pollutant cloud moving at a speed $u$. Suppose the raincloud has alongwind extent $R$ and the pollutant release is spread over a distance $s$. The situation is depicted in Fig. 3. In this particular case $v$ is greater than $u$ and the raincloud is overtaking the polluted air. There are three separate phases: (a) as the raincloud gradually overlaps the pollutant cloud; (b) during complete overlap; (c) as the raincloud passes above the leading edge of the release. The resulting pattern of deposition is illustrated in the lower part of Fig. 3.

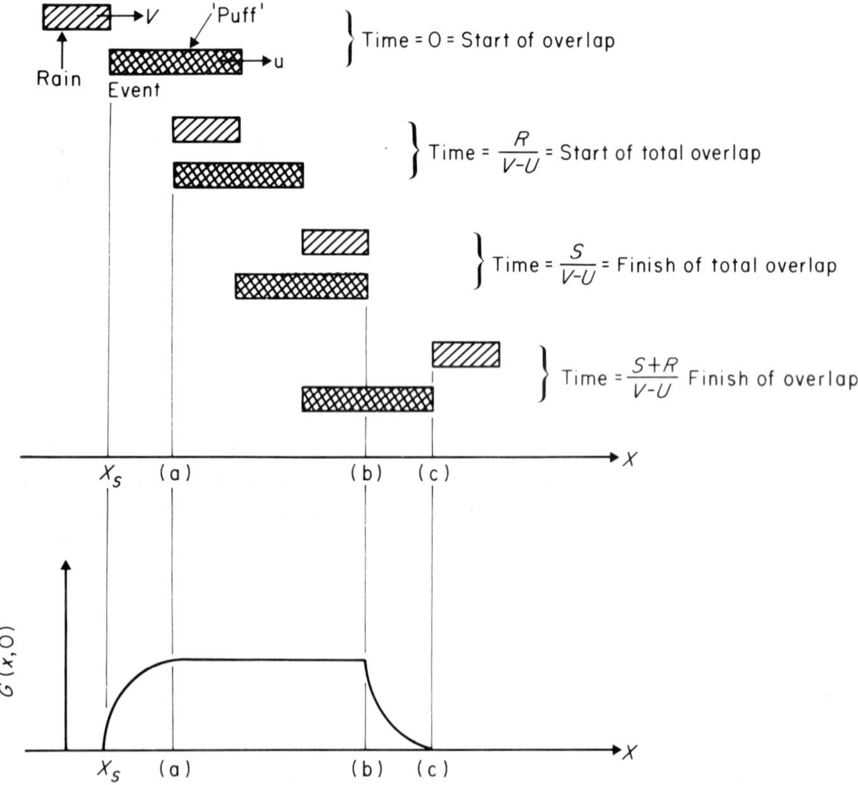


FIG. 3. Simple scheme with rain clouds and polluted air in motion.

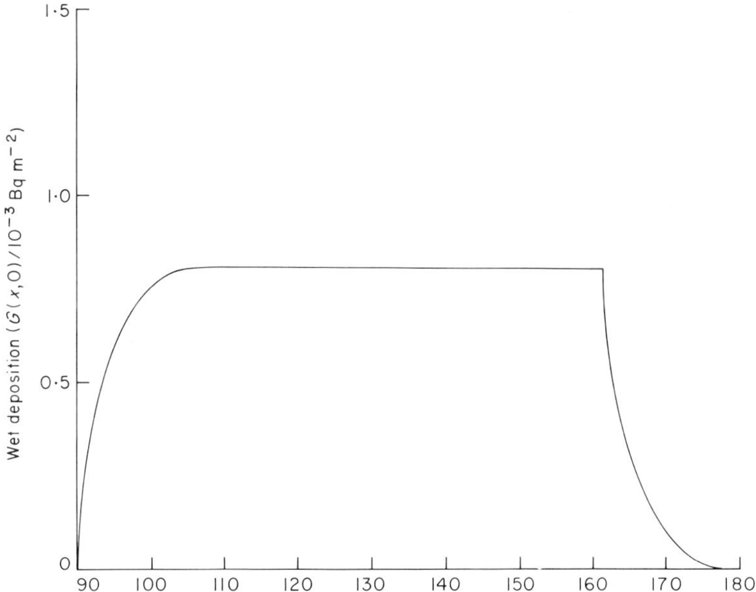


Fig. 4. Downwind pattern of wet deposition predicted with simple alternative model. $V = 10$ ms$^{-1}$, $U = 5$ m s$^{-1}$, $R = 18$ km, $S = 1$ hour release.

There is no sharp peak as the Gaussian plume model would imply, but the deposition is spread over a distance range somewhat greater than the downwind length of the pollutant cloud.

Fig.4 gives a more quantitative illustration of this difference. The situation in Fig. 2 corresponds to a rain band between 90 and 108 km downwind of the source. Thus, for Fig. 4 the raincloud was also assumed to be of length $R = 18$ km, but moving with a speed of 10 m s$^{-1}$ to overtake the polluted air beneath moving at 5 m s$^{-1}$. The alongwind spread of the polluted air depends on the duration of release, and in Fig. 4 was taken as appropriate to a $10^6$ Bq release over 1 hour. The maximum deposition is $8 \times 10^{-4}$ Bq m$^{-2}$ for the same release, a factor of 20 difference. This difference could determine whether or not a threshold level was exceeded. Even an instantaneous release treated with this alternative approach still leads to a maximum deposition an order of magnitude less than the Gaussian plume prediction. The two models agree if, and only if, the raincloud is stationary and permanent, which is an unrealistic assumption!

The interpretation of Eulerian observations at the source should also be considered. If the raincloud was moving at 10 m s$^{-1}$ and took 1 hour to pass over the source this would imply an alongwind extent of 36 km instead of 18 km. The extent of the cloud cannot be deduced from the surface windspeed. In the example given above a more elongated raincloud would lead to a slightly extended range of wet deposition but little change in the maximum, since the pollutant is almost completely removed by the shorter cloud anyway. This would not necessarily be true for other examples. A more comprehensive description with further examples may be found in ApSimon *et al.* (1985). However, Fig. 4 is sufficient to demonstrate the potential effect of the assumptions implicit in the segmented Gaussian plume model.

## PRECIPITATION DATA

A difficulty in introducing a better treatment of the spatial and temporal variation of precipitation arises in the availability of suitable data. Rain-gauges are sparsely spaced and give poor time-resolution, so that rainfall cannot sensibly be interpolated. Intense local showers can be localized on the scale of a few kilometres. Moreover, because the pollutant release is moving, it is difficult to prove that the mean frequency and duration of rainfall encountered corresponds to those observed at a stationary point: hence the controversy about the difference between Lagrangian and Eulerian statistics. However, modern radar techniques now provide a much better spatial and temporal resolution of rainfall according to a graded scale of intensities on a spatial grid of 2–5 km every 5 minutes. A new radar was commissioned by the UK Meteorological office in January 1985 at Chenies. It is intended to use data from this new station in conjunction with a puff model for the pollutant release to investigate the problems described above more realistically.

## PROCESSES OF PRECIPITATION

The modelling described so far has all been based on a simple washout model. However, the processes involved in precipitation can imply quite a different behaviour. These processes are summarized in Fig. 5. Imagining a parcel of ascending moist air: when this becomes saturated cloud droplets will form, including those formed by nucleation on soluble particulate pollutants. The cloud droplets may then gradually coalesce or accrete

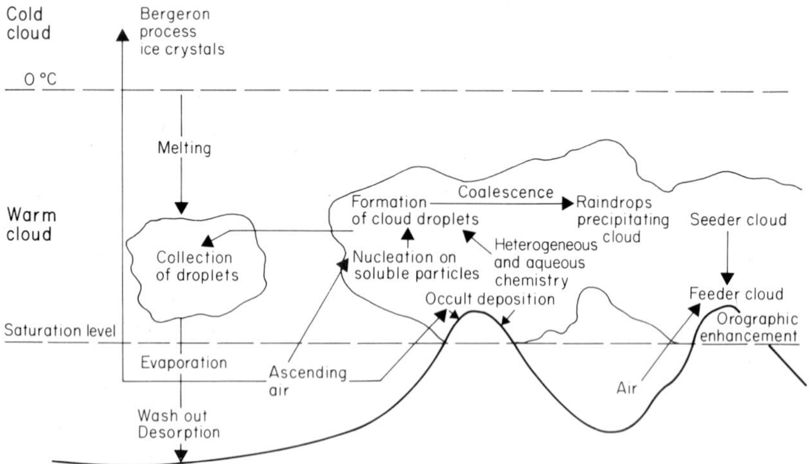


FIG. 5. Wet deposition processes.

with larger raindrops, and will efficiently absorb soluble gases. These are the rainout processes by which pollutants are transferred to cloud water in warm clouds. But if the cloud rises higher, extending above the freezing level, then the presence of ice crystals becomes significant. Since the equilibrium vapour pressure with respect to ice is less than that with respect to water at sub-zero temperatures, any small ice crystals at that height will grow at the expense of evaporating cloud water. Hence, droplets may evaporate leaving dry pollutant exported from the boundary layer. The growing ice crystals will fall and may melt to form raindrops at lower levels, and will accumulate cloud water with or without pollutants from lower level clouds. The different physical processes within cold clouds imply different efficiencies in transferring pollutants to precipitation. As the precipitation falls from the cloud it will evaporate, and perhaps as little as 20% will reach the ground (Rogers 1979). Complete evaporation of smaller droplets may resuspend pollutants scavenged at higher levels. The surviving drops may take up further pollutants beneath cloud level by washout. Thus, precipitation is very different from a constant spectrum of raindrops falling through the atmosphere, and the processes of transfer from air to precipitation vary markedly with height. In some cases precipitation systems will merely lift pollutant out of the lowest layers of the atmosphere without depositing them.

Attempts have been made to derive simple effective removal parameters applicable to pollutants at different heights. Thus, for example, Scott (1982) suggested two different formulae to apply to cold cloud above the freezing level and warm cloud beneath for scavenging of soluble particulate.

$$\Lambda = 0{\cdot}88\, J\, (\mathrm{h}^{-1}) \text{ in cold cloud}$$
$$\Lambda = 1{\cdot}26\, J^{0{\cdot}78}\, (\mathrm{h}^{-1}) \text{ in warm cloud}$$

where $J$ is the rainfall rate in mm $\mathrm{h}^{-1}$ and varies with height. However, these equations are based on the assumption that the particulate has acted to nucleate small droplets which are then scavenged by snowflakes or raindrops in cold and warm clouds respectively. Effects of evaporation are ignored and vertical velocities assumed small. It is also difficult to generalize about cold clouds because of the critical dependence on the presence or absence of ice crystal nuclei.

Topography also plays a role. Over mountains air may be forced to rise above the saturation-level, forming cloud. This may enhance wet deposition in two ways. Raindrops falling from higher levels will grow by accretion with droplets leading to heavier rainfall: the so-called 'feeder cloud–seeder cloud' mechanism of orographic rainfall enhancement. Whether or not the pollutant concentration in rainwater is higher or lower on the hilltops will depend on the relative pollutant burdens in the higher-level seeder cloud and the feeder cloud on the the hill top. On the other hand, if there is no precipitation from an overlying seeder cloud, the cloud droplets in the mountain-top cloud may still take up pollutants efficiently and settle on the ground or be intercepted by trees, vegetation, etc. as the cloud sweeps across. This is the mechanism called occult wet deposition whose possible importance has been pointed out in relation to localized high acid deposition at high altitudes and is discussed elsewhere in these proceedings by Unsworth & Crossley (p. 125). Occult wet deposition could also cause enhanced deposition of radionuclides on high land in the event of a nuclear accident. There was some evidence of localized deposition of this

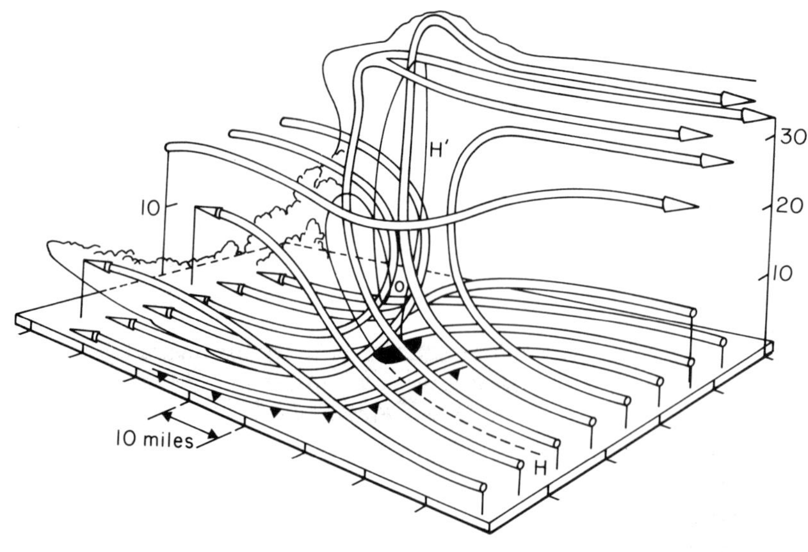


FIG. 6. Three-dimensional structure of convective storm. (From Browning & Ludlam 1962.)

type following the Windscale accident in 1957, which took place in dry weather.

The uptake of pollutant may be further complicated by chemistry. For example, $SO_2$ is taken up by cloud droplets and oxidized by hydrogen peroxide and ozone; however, the oxidation of $SO_2$ by ozone is reduced with increasing acidity so that the process is self-limiting. On the other hand, releases of ammonia over agricultural land can neutralize the acidity of droplets allowing greater uptake of $SO_2$. Similar processes probably apply to important radionuclides such as $^{131}I$. However, there is a considerable uncertainty about the chemical form of radionuclides released in accident conditions, with resulting uncertainty about the mechanisms of deposition. The importance attached to better definition of source terms for accidental releases of radioactivity is discussed by Cooper & Underwood (1985).

With such complexity to consider, it is difficult to justify simple relationships between an effective washout coefficient and rainfall intensity at ground level.

## DYNAMICAL EFFECTS OF PRECIPITATION

It is also evident from the above discussion that precipitation involves organized dynamical systems which spatially redistribute pollutant material. The picture of pollutants moving passively beneath the clouds is not very realistic. This is illustrated below by systematic motion in different types of rain system.

### *Convective rain storms*

These are more prevalent over land in summer with surface heating from insolation. A typical dynamical structure for a long-lived severe storm is illustrated in Fig. 6. There is a complex convolution of ascending air (the updraught) and descending air (the downdraught). Moist air ascending in the updraught follows a path distorted by wind shear; hence the forward overhanging 'anvil' shape of convective storm clouds. As precipitation forms it falls out of the updraught into the unsaturated air beneath. Evaporative cooling and to a lesser extent viscous drag in this lower level air give rise to the downdraught. This three-dimensional circulation within the storm system of updraught and compensating downdraught is necessary to sustain precipitation. For modelling purposes it is useful to resolve the motion into two components along the direction of maximum windshear, the 'wind lattice circulation', and a 'tropical circulation' along the perpendicular direction (Tam 1979). From the point of view of pollutant deposition, this kind of system is capable of sucking in a large volume of polluted air from the boundary layer in a sustained updraught, and depositing it over a relatively small area beneath governed by the motion of the core of the storm. As an approximate estimate the total volume $V$ of air drawn into the updraught may be deduced from the humidity mixing ratio (g of water per kg of air) $x$, the total rainwater reaching the ground $w$ (kg) and the efficiency of removal of water from that air $E$. This efficiency includes allowance for evaporation beneath the precipitating updraught into the downdraught and is typically of the order of 50%.

$$\frac{V}{\mathrm{e}} = \frac{W.10^3}{xE\,\rho_{air}}\ \mathrm{m}^3$$

where $\rho_{air}$ is the density of air ($1{\cdot}293$ kg m$^{-3}$). $W$ can be estimated from radar data integrating the rainfall rate over the area covered by the storm. $x$ is typically of the order of a few g kg$^{-1}$ of air. The volume of air $V$ would be drawn from the boundary layer of the atmosphere and might or might not encompass the whole of an accidental release of radionuclides.

### *Frontal rain systems*

These are much larger in scale and often incorporate convective cells, but again imply systematic movement resulting from the intruding wedge of warm air. Fig. 7a, based on Browning (1985), shows the dynamical pattern with the warm conveyor belt of air typically a few hundred km across and 1–2 km deep ascending slowly ahead of the trailing edge constituting the cold front. This warm conveyor belt carries humid air from warmer southerly latitudes which is the source of moisture for precipitation.

Cold overlaying air is forced up and over the warmer conveyor belt. This leads to instability and mesoscale areas of rain. The regions of precipitation are indicated by dots. Fig. 7b shows the vertical structure through the system of fronts along the line AB in Fig. 7a. As precipitation falls out of the conveyor belt system it scavenges air of completely different origin ahead of the leading edge of the warm front. Thus, rainout processes and washout processes take place within different air masses. It is clear that air parcels may traverse quite complex trajectories following the conveyor belt system.

Both the above examples of the small-scale convective system and the larger-scale frontal system illustrate the difficulties of trying to predict transport of pollutants and where wet deposition would result purely from observations of surface winds. Yet this is frequently the basis even of sophisticated air-pollution dispersion models. In the case of the frontal situation, forecasting models give a fair representation of the three-dimensional windfield within a frontal system and could be used to plot air-parcel trajectories and forecast the regions where wet deposition could occur. This could be useful for accidental releases caught up in the conveyor belt system: in particular for *post-facto* analysis of measurements of lower levels of radioactivity at long distances from the source. Convective systems are smaller and are not resolved on the coarse grids used by forecasting models. Their presence could be deduced from radar data but the passage of such a system overhead during an accidental release could lead to exceedingly complicated patterns of contamination.

## CONCLUSIONS

Simple modelling techniques currently used to assess wet deposition in studies of potential consequences of nuclear accidents may be very innaccurate. Some hidden assumptions in

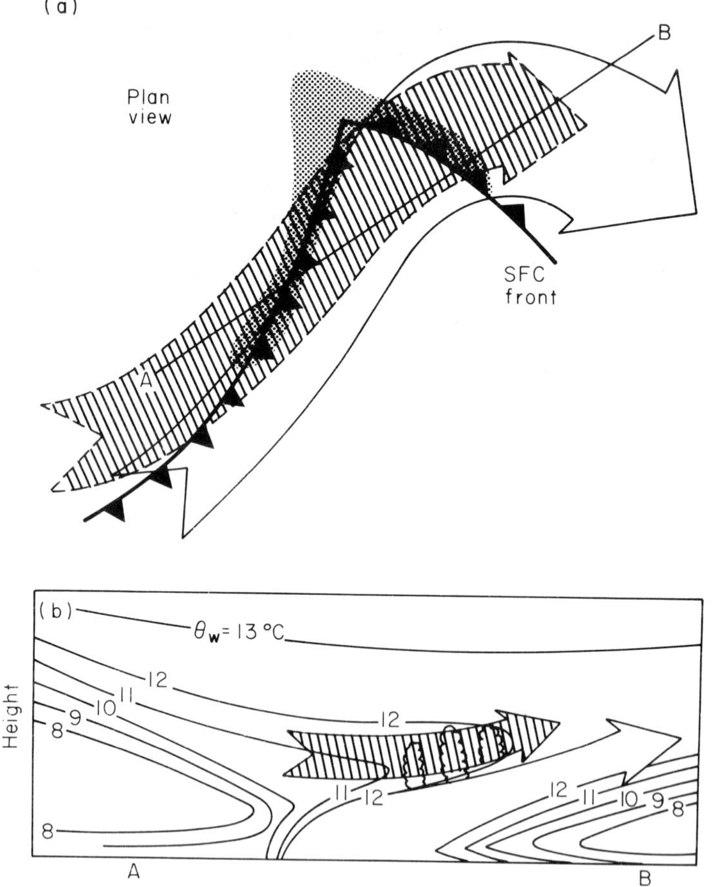


FIG. 7. Dynamical structure of a frontal rain system. (From Browning 1985.)

the segmented Gaussian plume have been demonstrated using a simple alternative model. Physical and chemical processes involved in precipitation have been contrasted against the simple idealized model of washout, and the problems of simple relationships with rainfall rate discussed. Finally it has been emphasized that precipitating systems of different types involve highly organized circulation patterns which will affect the path followed by a pollutant release and the location of subsequent deposition. All these factors influence the potential pattern of contamination from wet deposition of radionuclides following an accidental release from a nuclear installation.

## REFERENCES


**ApSimon, H.M., Goddard, A.J.H., Manning, P. & Simms, K. (1985).** Estimating patterns of contamination from wet deposition for probabilistic consequence assessment. *Proc. CEC Workshop on Methods of Assessing the Off-site Radiological Consequences of Nuclear Accidents.* Commission of the European Communities, Luxembourg.

**Austin & Houze (1972).** Analysis of the structure of precipitation patterns in New England. *Journal of Applied Meteorology,* **II**, 926–935.

**Browning, K.A. (1985).** Conceptual models of precipitation systems. *Meteorological Office Radar Research Laboratory, Research Report No 43.* RSRE, Malvern.

**Browning, K.A. & Ludlam, F.H. (1962).** Airflow in convective storms. *Quarterly Journal Royal Meteorological Society,* **88**, 117–135.

**Clarke, R.H. & Kelly, G.N. (1981).** *MARC — The NRPB Methodology for assessing radiological consequences of accidental releases of activity NRPB-R127.* HMSO, London.

**Cooper, P. & Underwood, B. (1985).** The impact of source term characteristics and the processes that modify them post release on dry and wet deposition rates. *Final report on task 1 of extension to contract SR011 80UK(B).* Safety & Reliability Directorate, UKAEA.

**Ritchie, L.T., Brown, W.D. & Wayland, J.R. (1980).** *Effects of rainstorms and runoff on consequences of nuclear reactor accidents.* SAND 76-0429. Sandia National Laboratories, Alberquerque.

**Rogers, R. (1979).** *A Short Course in Cloud Physics.* Pergamon Press, Oxford.

**Scott, B. (1982).** Theoretical estimates of the scavenging coefficient for soluble aerosol particles as a function precipitation type rate and altitude. *Atmospheric Environment,* **16**, 753–1762.

**Tam, H.Y. (1979)** *Cummulonimbus convection and parameterisation.* PhD thesis. University of London.

**USNRC (1975).** *Reactor Safety Study.* WASH1400, AppVII (NUREG-75/014). United States Nuclear Regulatory Commission.

# Capture of wind-driven cloud by vegetation

M. H. UNSWORTH AND A. CROSSLEY
*Institute of Terrestrial Ecology, Bush Estate,*
*Penicuik, Midlothian, Scotland EH26 0QB*

## SUMMARY

1 The capture of wind-driven cloud water by vegetation has long been known to be important hydrologically in some high, arid regions, but only recently has received attention as a pathway for chemical deposition.

2 The main physical processes in cloud water deposition are turbulent impaction and sedimentation. Impaction is most efficient at high windspeeds and onto small targets such as conifer needles. Sedimentation rates depend on drop sizes, and are independent of underlying vegetation.

3 Knowledge of aerodynamic properties of foliage, isolated trees and canopies allows deposition rates to be estimated. Such estimates agree well with the few available measurements.

4 In some meteorological conditions, cloud water on foliage may partially evaporate at the same time as it is depositing. In these circumstances chemical concentrations on foliage surfaces could be larger than in the original cloud drops. Methods of estimating this concentration mechanism are reviewed.

5 There are few published measurements of cloud water chemistry, and separation of rainfall from cloud water is difficult. Observations in the UK and USA suggest that highly polluted cloud water is associated with long and short-range transport from industrial regions. Insoluble particles are also contained in the cloud water and so are efficiently deposited on vegetation, whereas dry particles are usually very inefficiently deposited.

## INTRODUCTION

The ecological significance of water captured from clouds by vegetation has fascinated biologists for many years and has lead to an extensive literature on 'mist precipitation', 'fog drift' or 'occult precipitation' (thoroughly reviewed by Kerfoot 1968). The literature reveals that there is much confusion over the physical processes involved in water capture, and that there have been very few quantitative studies from which generalizations can be drawn.

One of the earliest recorders of this water input was Gilbert White, who wrote 'In heavy fogs at elevated situations particularly, trees are perfect alembics*. An amazing amount of water may be distilled by one tree in a night's time, by condensing the vapour which trickles down the twigs and boughs so as to make the ground below quite in a float' (White 1789). Although White did not interpret the main physical process correctly, he clearly

*An apparatus used for distilling.

recognized the importance of trees in capturing moisture and redistributing it on the ground beneath them.

It is well established that cloud water captured in this way may have great hydrological significance in some otherwise arid zones, for example in California (Byers 1953; Azevedo & Morgan 1974), Hawaii (Mordy & Hurdis 1955; Ekern 1964), South Africa (Nagel 1956) and Tenerife (Ceballos & Ortuño 1952; Garcia-Prieto, Ludlam & Saunders 1960). As an inverse of this reasoning, trees can be planted for absorbing moisture from wind-driven clouds and fog, with the aim of improving the visibility downwind (Hori 1953).

More recently, it has been appreciated that input of pollutant chemicals dissolved in captured cloud water may be of much more importance in some regions than the hydrological inputs of the water itself (Chamberlain 1975; Unsworth 1980). Cloud water inputs are not measured by normal rain gauges, and so 'occult deposition' may be an important, but currently hidden, source of sulphur and nitrogen to vegetation in areas commonly enveloped in wind-driven cloud (Unsworth 1980).

In this paper the processes involved in the transfer of cloud water to vegetation are discussed, the implications for chemical inputs and their effects on leaves are considered, and preliminary observations of cloud water and its chemical content are reported.

## PRINCIPLES AND PROCESSES

Clouds consist of fine droplets of water, typically in the size range 1–60 $\mu$m diameter (Fig. 1). The drop size depends partly on the type of cloud and its lifetime, and partly on the

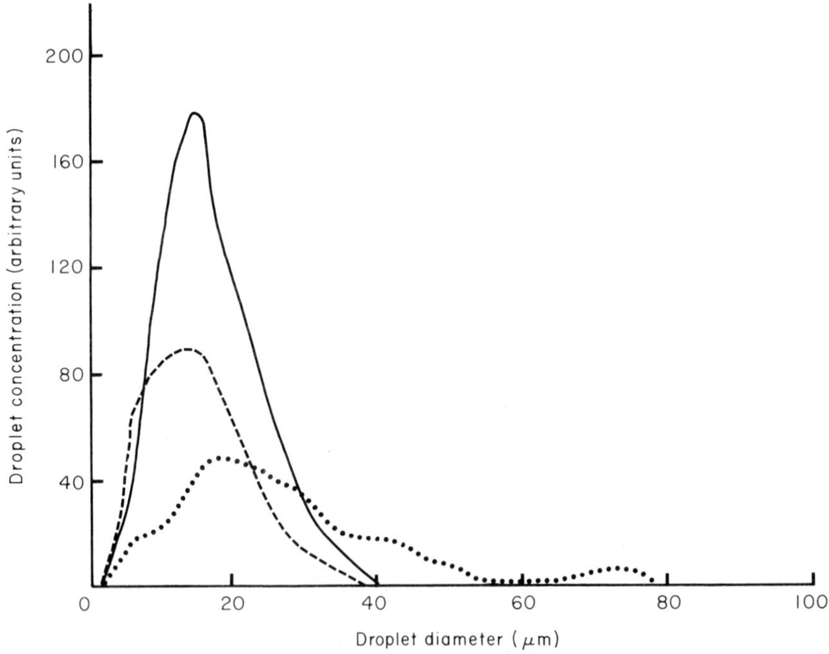


FIG. 1. Typical size distribution of droplets in various types of cloud. Fair weather cumulus (———), stratocumulus (----) and stratus (· · · ·).

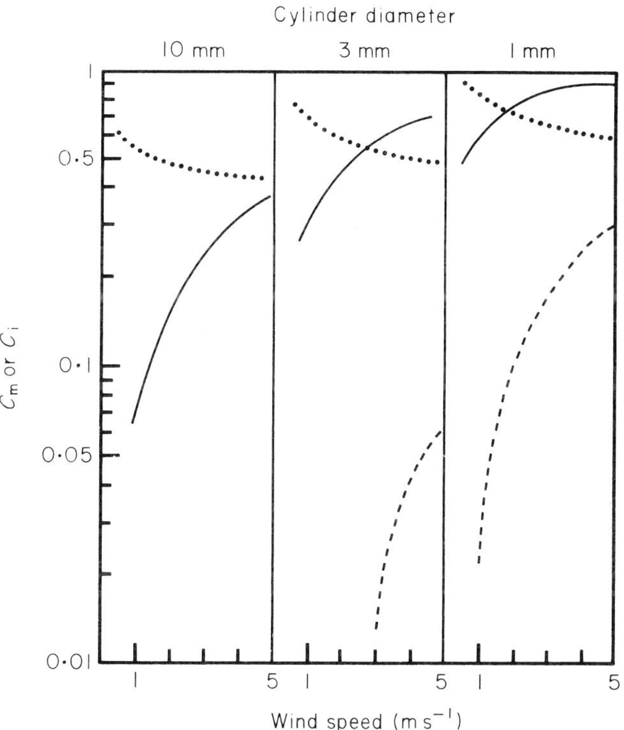


FIG. 2. Efficiencies of impaction $C_i$, for particles, diameter 5 μm (----) and 20 μm (———), and the efficiency of momentum transfer $C_m$ (· · · ·) for transfer to cylinders of three diameters (From Chamberlain 1975).

chemical nucleii around which the drops have formed. In a developing cloud, drops grow by condensation and coagulation until they may become large enough to fall from the cloud as rain. Cloud which envelops hill tops is usually either advected stratus, or is formed orographically, i.e. it is produced when moist air rises up a hillside, is cooled, and water vapour condenses into droplets. Orographic cloud can be a particularly important site for chemical deposition, because a cloud acts rather like a filter, trapping material from the air moving up the slope. Within clouds the air is usually very close to saturation (Mason 1971); in the interior of clouds that are not in contact with the ground the air is often super-saturated with a relative humidity between 0·1% and 1% above saturation; when clouds are in contact with the ground, especially when a cloud is thin, the small amount of solar energy penetrating the cloud can cause local heating at the ground and consequently the relative humidity near the surface may fall below 100%.

The exchange of water between a wind-driven cloud and the surface can take place either through transfer of droplets, or as water vapour exchange (Shuttleworth 1977). For droplets, two mechanisms operate: impaction is the transfer of a drop driven by the wind against an object; sedimentation is the vertical movement of drops, an inevitable consequence of the force of gravity on them. Water vapour from the cloud may condense at the surface if the surface temperature is below the dewpoint of the air, but there are also circumstances (discussed later) in which evaporation of water from the surface to the atmosphere may take place at the same time as droplets are depositing. The transfer of droplets and small particles to the surface by impaction has been reviewed thoroughly by Chamberlain (1975) and Chamberlain & Little (1981). The efficiency of the mechanism depends on drop size, target size, and wind speed. Figure 2 illustrates some principles by showing the impaction efficiency of two drop sizes onto cylindrical targets of three diameters over a range of wind speeds. Also shown in the figure are the efficiencies with which the momentum is transferred from the moving air to the targets. The figure illustrates that small targets and large drops are especially effective in deposition, particularly at the higher wind speeds. For typical cloud drops with a diameter of 20 $\mu$m, Fig. 2 demonstrates that the efficiencies of impaction to targets between 1 mm and 5 mm diameter, typical of blades of grass or pine needles, are close to those for momentum. Consequently, one may make use of the extensive literature describing wind forces on vegetation to estimate the deposition rates of drops to individual leaves.

When leaves or needles are assembled on a shoot, branch, or as part of a whole tree, some of the elements are sheltered from the force of the wind and from wind-driven cloud by other leaves. Consequently, a branch or a whole tree has a lower efficiency for

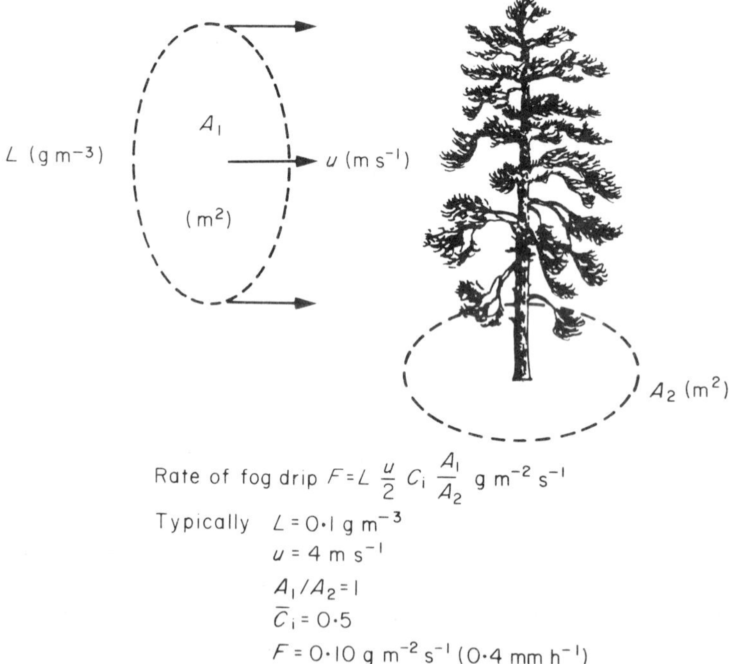


FIG. 3. Estimate of the rate of horizontal interception of wind-driven cloud water by an isolated tree, assuming that (i) mutual sheltering of branches leads to a bulk impaction efficiency $C_i$ for the tree of 0·50 (an average figure, smaller than the impaction efficiency to isolated cylinders, see Fig. 2), (ii) liquid water content $L$ = 0·1 g m$^{-3}$, (iii) wind speed $u$ = 4 m s$^{-1}$, and (iv) aspect ratio $A_1/A_2$ = 1.

capturing momentum and droplets than would be estimated by scaling up from the individual leaves or needles. Landsberg & Thom (1971) estimated that the 'shelter effect' reduced the wind force by a factor of between 2 and 3 for shoots and full trees. On the basis that the factor (2) also applies to the efficiency of impaction of wind-driven cloud, Fig. 3 illustrates a rough calculation for the rate of fog drip that might be expected beneath an isolated tree in cloud.

The calculation in Fig. 3 does not apply when there are extensive areas of vegetation, because the transfer is no longer by horizontal motion but relies on the downward turbulent transport of drops from air moving over a uniform canopy. The micrometeorology of such transfer to vegetation canopies has been studied extensively for momentum, heat, carbon dioxide and water vapour, but there are very few observations of droplet transfer. A pioneer of such studies was Hori (1953) who, with his colleagues in Japan, studied the diffusion of fog water to forests, using a number of physical techniques. Dollard & Unsworth (1983) used an aerodynamic method to measure the turbulent flux of wind-driven fog drops to rough grassland. Table 1 summarizes some of their results and illustrates that the deposition rates were two orders of magnitude smaller than estimates for an isolated tree. Table 1 also shows the rates of deposition by sedimentation of cloud drops to grass and demonstrates that, for this surface, sedimentation supplied about one-third of the total water drop input.

TABLE 1. Deposition rates of wind-driven cloud to grass by turbulent impaction ($D_t$) and sedimentation ($D_s$)

| | Wind speed (m $s^{-1}$) | Liquid water content (mg $m^{-3}$) | $D_t$ | $D_s$ (mg $m^{-2}$ $s^{-1}$) | $D_{total}$ |
|---|---|---|---|---|---|
| 29 Jan 1981 | 3·2 | 103 | 5·0 | 1·6 | 6·6 |
| | 3·8 | 79 | 2·7 | 1·3 | 4·0 |
| | 3·8 | 57 | 2·0 | 0·9 | 2·9 |
| 31 Jan 1981 | 2·8 | 228 | 2·1 | 4·3 | 6·4 |
| | 3·3 | 101 | 2·7 | 2·4 | 5·1 |
| | 4·2 | 72 | 3·4 | 1·7 | 5·1 |
| 31 March 1981 | 2·7 | 90 | 2·9 | 1·5 | 4·4 |
| | 3·9 | 72 | 1·5 | 1·3 | 2·8 |
| 3 April 1981 | 1·6 | 200 | 1·3 | 3·4 | 4·7 |
| | 1·1 | 65 | 0·7 | 1·1 | 1·8 |

From Dollard & Unsworth (1983). For conversions 1 mg $m^{-2}$ $s^{-1}$ = 3·6 $\mu$m $h^{-1}$.

Analysis of the results of Hori (1953) and Dollard & Unsworth (1983) suggests that the turbulent flux of cloud water to extensive areas of vegetation can be estimated from knowledge of the vertical variation of windspeed and temperature over the surface. Table 2 (after Unsworth 1984) illustrates some calculations of deposition rates to a forest, using windspeed observations analysed by Oliver (1971) and Stewart & Thom (1973) over a pine forest at Thetford, Norfolk. At a wind speed of 4 m $s^{-1}$ the turbulent deposition rate is about three times smaller than that estimated in Fig. 3 for an isolated tree. Deposition rates to the edges of forests and to dominant trees projecting above a forest canopy are likely to lie between the two extremes of these calculations.

TABLE 2. Calculated rates of deposition of wind-driven cloud to a forest by turbulent impaction $D_t$ and sedimentation $D_s$

| Windspeed $u$ (m s$^{-1}$) | 2 | 4 | 6 | 8 |
|---|---|---|---|---|
| $D_t$ (mg m$^{-2}$ s$^{-1}$) | 17·4 | 35·4 | 52·8 | 70·8 |
| $D_s$ (mg m$^{-2}$ s$^{-1}$) | 2·0 | 2·0 | 2·0 | 2·0 |
| $D_{total}$ (mg m$^{-2}$ s$^{-1}$) | 19·4 | 37·4 | 54·8 | 72·8 |

Values of $D_t$, taken from Unsworth (1984), are based on aerodynamic roughness factors measured over Thetford forest, a mixed forest, about 15·5 m high, of *Pinus sylvestris* and *Pinus nigra* var. *maritima* (Oliver 1971). The estimated value of $D_s$ is based on values calculated by Dollard & Unsworth (1983) from measurements of the drop-size spectrum in cloud. A cloud liquid-water concentration of 0·1 g m$^{-3}$ is assumed. For conversions, 1 mg m$^{-2}$ s$^{-1}$ = 3·6 $\mu$m h$^{-1}$.

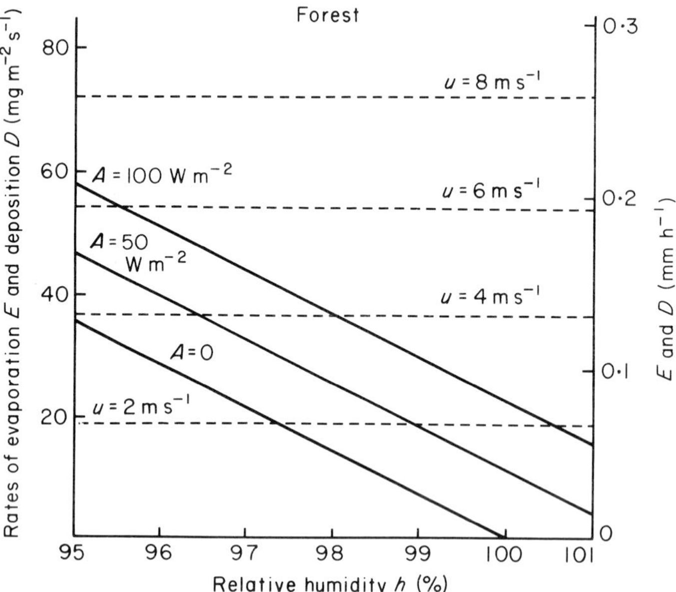


FIG. 4. The calculated dependence of evaporation rate $E$ (———) from a wet forest canopy on relative humidity for three values of available energy A (from Unsworth 1984); also shown are total deposition rates $D$ of cloud water (----), at four wind speeds $u$, taken from Table 2. The values of $E$ are calculated using the Monteith–Penman equation,

$$E = \frac{1}{\lambda}\frac{\Delta A + \rho c_p e_s (1-h)/r}{\Delta + \gamma}$$

where $\lambda$ is the latent heat of vaporization of water, $\Delta$ is the rate of change of saturation vapour pressure with temperature at air temperature, $\rho c_p$ is the volumetric specific heat of dry air $e_s$ is the saturation vapour pressure at air temperature, $h$ is the relative humidity, $A$ is the available energy, $\gamma$ is the psychrometric constant and $r$ is the aerodynamic resistance to transfer of water vapour from the canopy. It was assumed that air temperature was 10 °C and the aerodynamic resistance was taken as 6 s m$^{-1}$, the average value derived from measurements at Thetford forest.

## EVAPORATION AND CHEMICAL CONCENTRATION

Evaporation of water intercepted by a forest canopy is recognized to be an important component of the water balance of forested systems (Stewart 1977). Hydrologists have been concerned mainly with the evaporation of intercepted water after a period of rain, but evaporation rates may also be of some significance for different reasons during periods when vegetation is enveloped in cloud or mist. Evaporation from a wet forest canopy can be well estimated by the Monteith Penman equation (Monteith 1973; Stewart 1977). Over a limited range of relative humidity close to 100% the evaporation rate is approximately a linear function of humidity for a given amount of available energy (e.g. solar radiation). Figure 4 illustrates evaporation rates calculated over Thetford Forest for three values of available energy (Unsworth 1984); estimates of deposition from Table 2 are also shown. Because there is normally very little energy within a cloud, typical evaporation rates are very low, and are of much less importance hydrologically than the rates that would occur when the cloud cleared and radiation increased. However, Fig. 4 shows that, for example, at a typical wind speed of 4 m $s^{-1}$ the evaporation rate from the canopy may approach the deposition rate of cloud water if the relative humidity falls to 97%. If this was the case, then water would evaporate from the canopy at the same rate as it was depositing, but any non-volatile chemicals would remain on the leaf surface and would be in solution at concentrations higher than they occurred in the original cloud drops. Such a concentration mechanism would arise continuously as long as the available energy and humidity remained fixed. It is easy to demonstrate in such cicrcumstances that calculated concentrations of chemicals could be an order of magnitude larger than in the cloud drops alone. For example, Unsworth (1984) showed that the pH on surfaces of vegetation exposed to wind-driven cloud with pH 3·5, could fall as low as pH 2·5 if the humidity was a few per cent below saturation. Experimental investigations to test these predictions will need to take into account the possibility of chemical exchange across leaf cuticle which would alter the concentration within any liquid film on a leaf surface (Adams & Hutchinson 1984).

Figure 5 shows comparable calculations to Fig. 4 for deposition and evaporation rates to an extensive grass sward. The deposition rates, based on the work of Dollard & Unsworth (1983), are much lower than to the aerodynamically rougher forest canopy, and the evaporation rate changes less with humidity and more with available energy. Thus, for grass surfaces, the balance between deposition and evaporation is likely to be most sensitive to available energy.

## CHEMISTRY OF CLOUD WATER

There have been no long-term observations of cloud water chemistry at mountain sites in Britain, and so our knowledge of the chemistry is based on only a few days of observations at present. Evidence from Britain (Dollard, Unsworth & Harvey 1983; Gervat 1985) and from North America (Falconer & Falconer 1980) suggests that cloud water is generally of lower pH than rainfall at a given site and this is consistent with the origin of cloud droplets and raindrops.

Table 3 shows some observations on individual occasions (from Daum, Schwartz & Newman 1984), and illustrates the pronounced variability in chemical concentration that can be found between samples. This variability probably depends partly on the trajectory of the air mass in which the cloud is formed, and partly on evaporation and condensation processes taking place within a cloud. The table shows that the highest concentrations were associated with cloud of very low liquid water content; when the product of concentration and liquid water content was calculated there was much more uniformity between different samples in the total amount of material per unit volume of air. This observation that chemicals in cloud water are most concentrated in thin clouds may be important for inducing biological responses; the analysis in the previous section indicates that the surface concentration mechanism will operate effectively when there is adequate radiation present, and this would be most likely if the cloud was thin.

Within Britain the published observations of cloud water (Dollard, Unsworth & Harvey 1983; Gervat 1985) indicate that highly polluted cloud water probably occurs rather frequently. Currently we are developing techniques for collecting cloud water on a routine basis and for separating it from inputs of rain. Table 4 shows some preliminary observations from our field measurements in which we find that cloud water often has a large content of insoluble material and a low pH. Similar values of pH were reported by the other observers in Britain noted above. Other measurements of cloud water at mountain

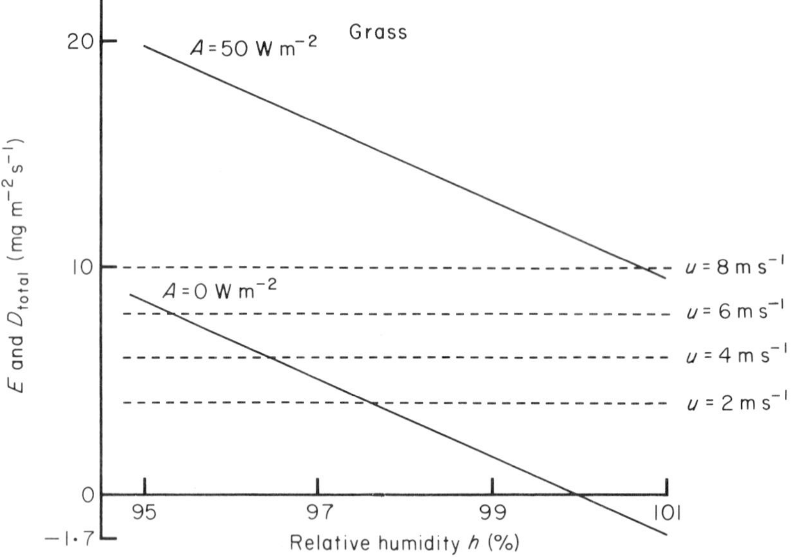


FIG. 5. The calculated dependence of evaporation rate $E$ (———) from a wet canopy of rough grassland on relative humidity for two values of available energy $A$. The calculations use the same method as in Fig. 4, but the aerodynamic reistance $r$ is taken as 25 s m$^{-1}$, assuming a wind speed dependence $r = 100/u$ and typical windspeed $u = 4$ m s$^{-1}$. Also shown are calculations of total deposition rates of cloud water $D_{total}$ (----) at four wind speeds. Turbulent deposition $D_t$ is calculated as $D_t = L/r$ with $L$, the liquid water content $= 0{\cdot}1$ g m$^{-3}$; deposition by sedimentation $D_s$ is assumed the same as in Table 2; and $D_{total} = D_t + D_s$.

TABLE 3. Variation of cloud water chemical concentration $C$( eq $l^{-1}$) and volumetric chemical burden $LC$(neq $m^{-3}$) in clouds with liquid water content $L$(g $m^{-3}$)

| Sample | Liquid water content, $L$ (g $m^{-3}$) | Chemical concentration, $C$ ( eq $l^{-1}$) $H^+$ | $NH_4^+$ | $SO_4^{2-}$ | $NO_3^-$ | Volumetric burden, $LC$ (neq $m^{-3}$) $H^+$ | $NH_4^+$ | $SO_4^{2-}$ | $NO_3^-$ |
|---|---|---|---|---|---|---|---|---|---|
| 1 | 0·37 | 162 | 54 | 86 | 89 | 60 | 20 | 32 | 33 |
| 2 | 0·12 | 151 | 62 | 96 | 81 | 63 | 26 | 40 | 34 |
| 3 | 0·45 | 166 | 76 | 138 | 122 | 75 | 34 | 62 | 55 |
| 4 | 0·25 | 145 | 55 | 102 | 83 | 36 | 14 | 26 | 21 |
| 5 | 0·15 | 631 | 328 | 556 | 177 | 95 | 49 | 83 | 27 |
| 6 | 0·10 | 617 | 213 | 552 | 184 | 62 | 21 | 55 | 18 |

Measurements recalculated from Daum, Schwartz & Newman (1984), who made observations during aircraft flights near Charleston, South Carolina.

TABLE 4. Examples of the particulate content and pH of rain and cloud water collected at sites near Edinburgh, Scotland

| Date | Wind direction | Collection period (days) | Sample type | Collected volume (ml) | pH | Particulate content (mg $l^{-1}$) |
|---|---|---|---|---|---|---|
| Dunslair Heights | | | | | | |
| 6.2.85 | SE | 1 | Cloud water | 169 | 2·79 | 25 |
| Castlelaw | | | | | | |
| 24.4.85 | NW | 5 | Cloud water | 139 | 3·96 | 50 |
| 24.4.85 | NW | 5 | Cloud water + rain | 372 | 4·09 | 43 |
| 24.4.85 | NW | 5 | Rain | 71 | 4·40 | 3 |
| 15.5.85 | S | 9 | Cloud water | 2500 | 3·38 | 102 |
| 15.5.85 | S | 9 | Cloud water + Rain | 2500 | 3·44 | 89 |
| 15.5.85 | S | 9 | Rain | 513 | 3·87 | 10 |

Dunslair Heights (elevation 602 m) is about 30 km south of Edinburgh and Castlelaw (elevation 408 m) is about 10 km south-west of Edinburgh.

sites in southern Germany (Georgii & Schmitt 1985), north-eastern North America (Falconer & Falconer 1980), California (Waldman *et al.* 1982) and Japan (Okita 1968) show comparable values of pH, suggesting that acidic cloud water is common in regions where there are large emissions of the necessary gaseous precursors, primarily sulphur dioxide and nitrogen oxides.

Although the occurrence of large concentrations of dissolved substances in cloud water are increasingly well documented, little attention seems to have been paid to the composition and possible effects of the large amounts of solid material deposited in cloud water. Figure 6 shows a scanning electron micrograph of material filtered from a sample of cloud water and illustrates that much of the solid component appears to be pulverized fuel

FIG. 6. Scanning electron micrograph of particulate material collected in a sample of cloud water, showing spherical particles (arrowed), probably of pulverized fuel-ash from industrial combustion, and coagulated particles, probably arising from domestic and low-temperature combustion of carbonaceous material.

ash deriving from power station combustion, and carbon particles arising from domestic and industrial fuel use. Gervat (1985) calculated trajectories for a number of his cloud water samples. Figure 7 illustrates some of the trajectories and the associated pH and sulphate concentrations; the most polluted cloud water arose when wind directions brought material from industrial and urban areas to the east of his site.

## CONCLUSIONS

On the basis of our understanding of the processes controlling cloud water deposition, it is possible to devise models for predicting deposition rates to extensive, uniform vegetation. Estimates of deposition to isolated trees or shrubs, to the edges of forests or hedgerows, and to individual tall trees projecting above a forest canopy are much more uncertain, but may be of more importance from the viewpoints of biological effects.

Far more needs to be known about the causes and sources of variation in the chemical content of cloud water. There is a need to establish a chemical climatology for cloud water in the same way that a climatology is being developed for understanding temporal and spatial variations in acid rain deposition.

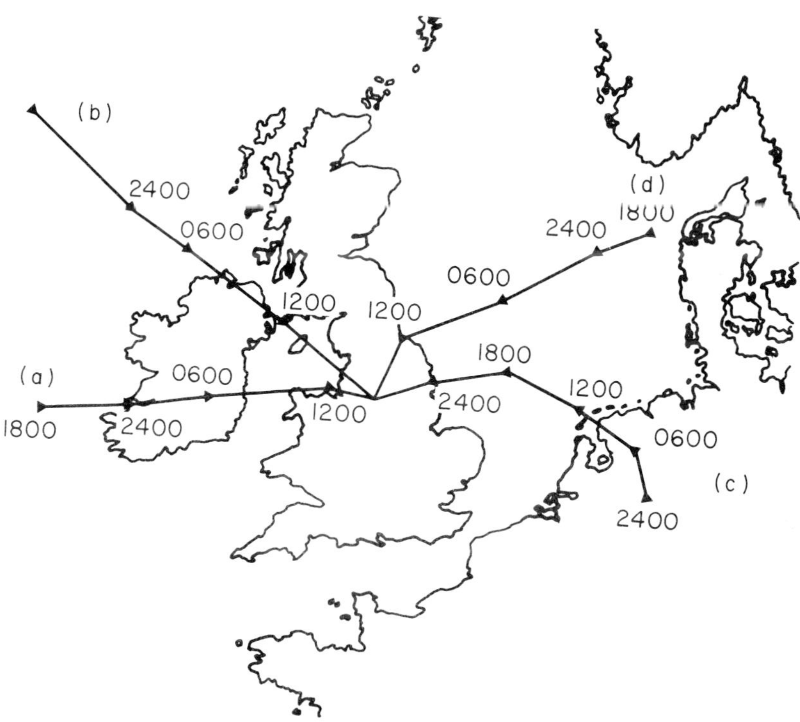


FIG. 7. Surface air mass trajectories (geostrophic) for air arriving at a cloud chemistry monitoring site in the southern Pennines (from Gervat 1985).

| | | Daily range | |
|---|---|---|---|
| | | pH | $SO_4$ (eq $l^{-1}$) |
| (a) | Arriving approx. 15.00 h, 13.4.83 | | |
| (b) | Arriving approx. 18.00 h, 13.4.83 | 3·5–5·0 | 200–800 |
| (c) | Arriving approx. 09.00 h, 27.4.83 | | |
| (d) | Arriving approx. 18.00 h, 27.4.83 | 2·5–2·9 | 600–2500 |

The observation that cloud water sometimes contains large amounts of insoluble particles indicates another previously 'hidden' deposition pathway. Particles of less than one micrometre diameter are deposited very inefficiently in dry conditions, and so their incorporation in cloud drops must result in larger amounts of particle deposition on vegetation at high elevations than occurs at lower levels where cloud and mist are less frequent.

The suggestion that evaporation allows surface concentrations of chemicals on vegetation in cloud to become large requires experimental verification. In principle, the mechanism could have important implications concerning biological responses to chemical inputs, but there are many uncertainties in the micrometeorology, surface chemistry and biological interactions which need to be investigated before we can understand clearly the relation between vegetation and the cloud water deposited on it.

## ACKNOWLEDGMENTS

Some of the experimental work described here was supported by the Natural Environment Research Council, the UK Department of the Environment and the Commission of the European Communities.

## REFERENCES

**Adams, C.M. & Hutchinson, T.C. (1984).** A comparison of the ability of leaf surfaces of three species to neutralize acidic rain drops. New *Phytologist*, **95**, 463–478.

**Azevedo, J. & Morgan, D.L. (1974).** Fog precipitation in coastal California forests. *Ecology,* **55**, 1135–1141.

**Byers, H.R. (1953).** Coast redwoods and fog drip. *Ecology,* **34**, 192–193.

**Ceballos, L. & Ortuño, F. (1952).** *Forests and moisture in the Canary Islands.* Montes, Madrid.

**Chamberlain, A.C. (1975).** The movement of particles in plant communities, *Vegetation and the Atmosphere, I. Principles* (Ed. by J.L. Monteith), pp. 155–203. Academic Press, London.

**Chamberlain, A.C. & Little, P. (1981).** Transport and capture of particles by vegetation. *Plants and their Atmospheric Environment* (Ed. by J. Grace, E.D. Ford & P.G. Jarvis), pp. 147–173. Blackwell Scientific Publications, Oxford.

**Daum, P.H., Schwartz, S.E. & Newman, L. (1984).** Acidic and related constituents in liquid water stratiform clouds. *Journal of Geophysical Research,* **89, (D1)**, 1447–1458.

**Dollard, G.J. &Unsworth, M.H. (1983).** Field measurements of turbulent fluxes of wind-driven fog drops to a grass surface. *Atmospheric Environment,* **17**, 775–780.

**Dollard, G.J., Unsworth, M.H. & Harvey, M.J. (1983).** Pollutant transfer in upland regions by occult precipitation. *Nature (London)*, **302**, 241–242.

**Ekern, P.C. (1964).** Direct interception of cloud water on Lanaihale, Hawaii. *Proceedings of the Soil Science Society of America,* **28**, 419–421.

**Falconer, R.E. & Falconer, P.D. (1980).** Determination of cloud water acidity at a mountain observatory in the Adirondak Mountains of New York State. *J. Geophysical Research,* **85**, 7465–7470.

**Garcia-Prieto, P.R., Ludlam, F.H. & Saunders, P.M. (1960).** The possibility of artificially increasing rainfall at Tenerife in the Canary Islands. *Weather (London),* **15**, 39–51.

**Georgii, H.W. & Schmitt, G. (1985).** Methoden und ergebnisse der nebelanalyse. *Staub Reinhaltung der luft,* **45** (6), 260–264.

**Gervat, G.P. (1985).** Clouds at ground level: samples from the southern Pennines. *Central Electricity Research Laboratories Report No. TPRD/L/270/N84.,* CERL, Leatherhead.

**Hori, T.** (Ed.) **(1953).** *Studies of fog in relation to fog-preventing forest.* Tanne Trading Co. Ltd., Sapporo, Hokkaido, Japan.

**Kerfoot, O. (1968).** Mist precipitation on vegetation. *Forestry Abstracts,* **29**, 8–20.

**Landsberg, J.J. & Thom, A.S. (1971).** Aerodynamic properties of a plant of complex structure. *Quarterly Journal of the Royal Meteorological Society,* **97**, 565–570.

**Mason, B.J. (1971).** *The Physics of Clouds* (2nd ed). Clarendon Press, Oxford.

**Monteith, J.L. (1973).** *Principles of Environmental Physics.* Edward Arnold, London.

**Mordy, W.A. & Hurdis, J. (1955).** The possible importance of fog drip in Hawaiian watersheds. *Bulletin of the American Meteorological Society, Easton, Pennsylvania,* **36**, 495 (Abstract only).

**Nagel, J.F. (1956).** Fog precipitation on Table Mountain. *Quarterly Journal of the Royal Meteorological Society,* **82**, 452–460.

**Okita, T. (1968).** Concentration of sulphate and other inorganic materials in fog and cloud water and in aerosol. *Journal of the Meterological Society of Japan,* **46**, 120–127.

**Oliver, H.R. (1971).** Wind profiles in and above a forest canopy. *Quarterly Journal of the Royal Meterological Society,* **97**, 548–553.

**Shuttleworth, W.J. (1977).** The exchange of wind-driven fog and mist between vegetation and the atmosphere. *Boundary Layer Meterology,* **12**, 463–489.

**Stewart, J.B. (1977).** Evaporation from the wet canopy of a pine forest. *Water Resources Research,* **13**, 915–921.

**Stewart, J.B. & Thom, A.S. (1973).** Energy budgets in pine forest. *Quarterly Journal of the Royal Meteorological Society,* **99**, 154–170.

**Unsworth, M.H. (1980).** Dry deposition of gases and particles onto vegetation. *Methods for studying acid precipitation in forest ecosystems: definitions and research requirements* (Ed. by I.A. Nicholson, I.S. Paterson & F.T. Last), pp. 9–15. Institute of Terrestrial Ecology, Cambridge.

**Unsworth, M.H. (1984).** Evaporation from forests in cloud enhances the effects of acid deposition. *Nature (London),* **312**, 262–264.

**Waldman, J.D., Munger, J.W., Daniel, J.J., Flagan, R.C., Morgan, J.J. & Hoffman, M.R. (1982).** Chemical composition of acid fog. *Science,* **218**, 677–679.

**White, G. (1789).** *The Natural History of Selbourne* (facsimile edition published by Gresham Books, 1979).

# Transport of acidity through ecosystems

R. A. SKEFFINGTON
*CEGB, Central Electricity Research Laboratories,*
*Kelvin Avenue, Leatherhead, Surrey KT22 7SE, UK*

## SUMMARY

**1** This paper is intended to clarify the concept of acidity and to introduce a method of calculating the flux of acidity through ecosystems.

**2** The Bronsted-Lowry definition describes acids as proton donors; bases as proton acceptors. Acidity is thus a relative property and a substance can be simultaneously both an acid and a base. The ionized alkali metals ($Ca^{2+}$, $Mg^{2+}$) are not bases, nor are the anions of strong acids ($SO_4^{2-}$, $NO_3^-$, $Cl^-$) themselves acids. Some metal cations (e.g. $Al^{3+}$) possess acidity by co-ordinating water molecules.

**3** The strength of acids is measured by the ease with which they will donate a proton to water. Acids only partly dissociated at the pHs found in the natural environment are called weak acids. They are important because of their ability to resist pH change (buffering).

**4** A neutral solution is one where the acidic properties balance the basic ones. For water this occurs at pH 7 at 25 °C, but at higher pH when the temperature is lower.

**5** There are two common uses for the term 'acidity': hydrogen ion concentration (an intensity factor); and base neutralizing capacity or BNC — a capacity factor. These two uses are commonly confused.

**6** BNC is usually measured by titrating to an equivalence point, where the concentrations of acidic species are equal to those of basic species. The equivalence points of carbonic acid are used in natural waters for historical reasons. The effect of this is to define acidity (and its converse, alkalinity) with respect to carbonic acid. Equivalence points do not occur at fixed pH levels but vary with concentration. It is argued that organisms are affected by pH rather than the BNC with respect to carbonic acid, and thus BNC to a fixed pH would be a more appropriate quantity to calculate. A method for calculating BNC to a fixed pH level is then described.

**7** It is shown that it is impossible to calculate a proton budget, as free protons are not conservative species. It is, however, possible to calculate a BNC budget. A practical example of a BNC budget is given. It is suggested that this technique could yield useful information about the sources and transport of acidity in ecosystems.

## INTRODUCTION

The problems and possible problems associated with the acidity of precipitation are currently arousing a lot of interest (e.g. Drabløs & Tollan 1980; Watt Committee 1984). Among other effects, it is claimed that acid rain is the principal cause of the acidification of soils and freshwaters observed in certain parts of the world (e.g. Evans *et al.* 1981). This has led to the idea that unless the acid load in precipitation (both wet and dry deposition) can be reduced, 'acidification of the environment' (Swedish Ministry of Agriculture 1982) will continue to worsen. Clearly the input of acid from the atmosphere imposes an acid load on ecosystems which receive it. On the other hand it has long been recognized that internal production of acid in ecosystems is also significant. The comparison of internal acid generation with external inputs is an unsolved problem. Authors who have attempted to tackle it have come to radically different conclusions, from those who believe that internal acid production is more important (Rosenqvist 1977, 1978; Krug & Frink 1983) to those who believe that external inputs have the greatest influence at least in sensitive systems (Van Breemen, Mulder & Driscoll 1983, 1984; Havas, Hutchinson & Likens 1984). In spite of all the interest in acid rain, the concepts of "acid" and "acidity" are not generally well understood. This is perhaps not surprising as our understanding has developed in an *ad hoc* manner, and confusing terminology abounds. Nevertheless, some ecologists appear to be using concepts of acidity which were abandoned by chemists in the 1920s. In this paper some of the concepts relating to acidity are clarified, some common misconceptions are pointed out, and a more precise definition of acidity is developed. These ideas are then applied to the transport of acidity through ecosystems.

## FUNDAMENTAL CONCEPTS

### *What is an acid?*

The modern concept of acids and bases originated with J. N. Brönsted and M. Lowry in 1922–23. They defined an acid as a proton donor and a base as a proton acceptor. Thus, sulphuric acid is an acid because it can donate protons:

$$H_2SO_4 \rightarrow 2H^+ + SO_4^{2-} \quad (1)$$

Ammonia is a base because it can accept them:

$$NH_3 + H^+ \rightarrow NH_4^+ \quad (2)$$

Free protons probably do not occur in aqueous solution, as water itself accepts protons: in other words it can act as a base. Reactions (1) and (2) thus should be written:

$$\underset{\text{acid}}{H_2SO_4} + \underset{\text{base}}{2H_2O} = \underset{\text{acid}}{2H_3O^+} + \underset{\text{base}}{SO_4^{2-}} \quad (3)$$

$$\underset{\text{base}}{NH_3} + \underset{\text{acid}}{H_2O} = \underset{\text{acid}}{NH_4^+} + \underset{\text{base}}{OH^-} \quad (4)$$

Note the symmetrical character of equations (3) and (4). Acids and bases react in pairs so that an acid reaction with a base always generates another acid and base. Note also that in reaction (3) water is acting as a base and in reaction (4) as an acid. One misconception is that there is one set of compounds called "acids" and another set called "bases" and these are mutually exclusive. Reactions (3) and (4) illustrate that acidity is a *relative* property, and that the same compound can behave both as an acid and a base. There are many other compounds in the natural environment which do so (e.g. $HCO_3^-$, $HSO_4^-$, $AlOH^{2+}$).

Another common misconception is that the ionized alkaline earth metals ($Ca^{2+}$, $Mg^{2+}$) are bases. Consideration of the stable structure of the $Ca^{2+}$ ion shows that the likelihood of it accepting a proton is remote. These ions are referred to as "basic cations" not because they are bases but because they are signs that there has in the past been a reaction with a base such as Ca metal, or more likely $CaCO_3$ or a calcium alumino-silicate:

$$CaCO_3 + H_3O^+ = Ca^{2+} + HCO_3^- + H_2O \quad (5)$$

$$\underset{\text{anorthite}}{CaAl_2Si_2O_8} + 2H_3O^+ = Ca^{2+} + Al_2Si_2O_5(OH)_4 + H_2O \quad (6)$$

Here a relatively strong acid ($H_3O^+$) is being replaced by a very weak one ($Ca^{2+}$). The presence of large amounts of $Ca^{2+}$ in a water body thus tends to be associated with less acidic conditions, hence it has become known as a "basic cation".

Similarly the anions of the strong mineral acids ($SO_4^{2-}$, $NO_3^-$, $Cl^-$) are not in themselves acid, though often referred to as such. This should be obvious if acids are proton donors: these ions have no protons to give. Their presence in water *may* be a sign that some acid-forming reaction has occurred in the ecosystem in the past, such as oxidation of a sulphide:

$$H_2S + 2O_2 = 2H^+ + SO_4^{2-} \quad (7)$$

or it may not as in the solution of gypsum:

$$CaSO_4 \cdot 2H_2O_{(S)} \rightarrow Ca^{2+} + SO_4^{2-} + 2H_2O \quad (8)$$

The gypsum may have originated from an acid-forming reaction earlier in geological time, but this cannot count as an acid-forming reaction *in the ecosystem*. A corollary of these considerations is that it is not possible to deduce the contribution of sulphuric and nitric acids to the acidity of a sample if there are ions present other than $H^+$, $SO_4^{2-}$ and $NO_3^-$. In rain, there are always other ions present: it is therefore not possible to attribute acidity in a single rain sample, though Fowler *et al.* (1982) introduced, without explanation, a method for attributing acidity given a number of different samples distributed in space.

A further class of acid needs to be mentioned. Solutions of aluminium salts have pronounced acidic properties, but if the Bronsted concept is adhered to it is not obvious what is the source of the protons that supply this acidity. Some metal ions in solution are able because of their electronic structure to co-ordinate water molecules. The acidity of the co-ordinated water molecules is much greater than that of free water, because of the electron-attracting power of the central metal cation. Al can co-ordinate six water molecules, so $Al^{3+}$ should thus be written $[Al(H_2O)_6]^{3+}$ (Fig. 1), and its acidic properties expressed as in reaction (9)

$$[Al(H_2O)_6]^{3+} + H_2O = [Al(H_2O)_5OH]^{2+} + H_3O^+ \quad (9)$$

As there are six water molecules associated with $Al^{3+}$ there is the potential to transfer six protons, though at the pH range normally found in natural waters only four transfers occur, the final Al species being $[Al(H_2O)_2(OH)_4]^-$. The strongest acids of this type are $Al^{3+}$, $Fe^{3+}$ and $Mn^{2+}$, which are consequently sometimes called "acidic cations". Older textbooks attribute the acidity of salt solutions of these metals to "hydrolysis", but these reactions are in principle no different from other acid–base reactions.

### *Strength of acids and bases: strong and weak acids*

Logically, the strength of an acid should be measured by its readiness to donate a proton. This needs to be defined with respect to the base which is accepting the proton. For comparative purposes water is usually used as the standard base, which is useful because all

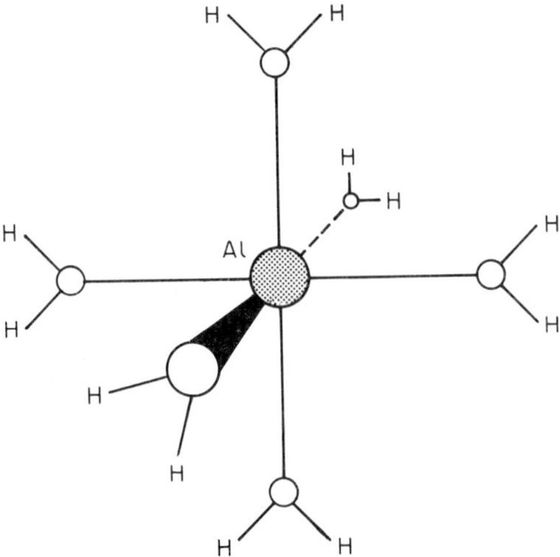


FIG. 1. Hydrated aluminium ion. Each water molecule has the capacity to lose a proton, making $Al^{3+}$ a potentially hexavalent acid.

biochemical reactions occur in aqueous solution. The reaction is thus:

$$HA + H_2O \rightarrow H_3O^+ + A^- \qquad (10)$$

where Ha is any acid and $A^-$ the conjugate base. The equilibrium constant $K_a$ for this reaction

$$K_a = \frac{(H_3O^+)(A^-)}{(HA)(H_2O)} \qquad (11)$$

is the natural measure of acid strength. (Parentheses denote activities, which approximate to concentrations at low ionic strength.) Thermodynamic convention allows us to write reaction (11) as

$$K_a = \frac{(H^+)(A^-)}{(HA)} \qquad (12)$$

since the activity of water is nearly constant. The negative logarithm of $K_a$ ($pK_a$, by analogy with pH) defines the acid strength. The $pK_a$ of acids varies widely: some values are shown in Table 1. From eqn (12) it can be calculated that an acid is 99% dissociated if the pH is 2 units higher than the $pK_a$, 50% dissociated at $pK_a$, and 99% associated if the pH is 2 units lower. At pH values commonly found in the natural environment, (4–10), it can thus be seen from Table 1 that the strong mineral acids (HCl, $HNO_3$, $H_2SO_4$) are virtually completely dissociated. Acids with a $pK_a$ above about 3–4 tend to be called weak acids, but as Table 1 shows, this distinction between strong and weak acids is not absolute. Note that some organic acids are quite strong and are capable of affecting the pH of even acid natural waters. Weak acids are important in the natural environment because they provide buffering.

### *Buffering*

Buffering is best illustrated with an example. If 1 $cm^3$ of 1M HCl is added to 1 litre of $10^{-3}$M NaCl at pH 7, the pH will drop to 3. To depress the same volume of a solution

TABLE 1. $pK_a$ of some acids at 25 °C. From Stumm & Morgan (1981) Table 3.2, Albert & Serjeant (1971).

| Acid | Reaction | $pK_a$ |
|---|---|---|
| Hydrochloric | $HCl + H_2O \rightarrow H_3O^+ + Cl^-$ | -3 |
| Sulphuric | $H_2SO_4 + H_2O \rightarrow H_3O^+ + HSO_4^-$ | -3 |
| Nitric | $HNO_3 + H_2O \rightarrow H_3O^+ + NO_3^-$ | -1 |
| Oxalic | $C_2O_4H_2 + H_2O \rightarrow H_3O^+ + C_2O_4H^-$ | 1·27 |
| Bisulphate | $HSO_4^- + H_2O \rightarrow H_3O^+ + SO_4^{2-}$ | 1·96 |
| Phosphoric | $H_3PO_4 + H_2O \rightarrow H_3O^+ + H_2PO_4^-$ | 2·15 |
| Formic | $HCOOH + H_2O \rightarrow H_3O^+ + HCOO^-$ | 3·75 |
| Acetic | $CH_3COOH + H_2 \rightarrow H_3O^+ + CH_3COO^-$ | 4·7 |
| Aluminium (aq) | $Al(H_2O)_6^{3+} + H_2O \rightarrow H_3O^+ + Al(H_2O)_5OH^{2+}$ | 4·9 |
| Carbonic | $H_2CO_3 + H_2O \rightarrow H_3O^+ + HCO_3^-$ | 6·35 |
| Dihydrogen phosphate | $H_2PO_4^- + H_2O \rightarrow H_3O^+ + HPO_4^{2-}$ | 7·21 |
| Ammonium | $NH_4^+ + H_2O \rightarrow H_3O^+ + NH_3$ | 9·3 |
| Silicic acid | $Si(OH)_4 + H_2O \rightarrow H_3O^+ + SiO(OH)_3^-$ | 9·5 |
| Bicarbonate | $HCO_3^- + H_2O \rightarrow H_3O^+ + CO_3^{2-}$ | 10·3 |
| Hydrogen phosphate | $HPO_4^{2-} + H_2O \rightarrow H_3O^+ + PO_4^{3-}$ | 12·3 |
| Silicate | $SiO(OH)_3^- + H^2O \rightarrow H_3O^+ + SiO_2(OH)_2^{2-}$ | 12·6 |

containing $10^{-3}$M sodium acetate from pH 7 to pH 3 would take 1·98 $cm^3$ of 1 M HCl. The sodium acetate solution is said to be buffered because it resists the pH change, as extra protons have to be added to re-protonate the acetate ion. Conversely, in a base titration, protons from the dissociating acetic acid have to be neutralized. The buffer intensity, $\beta$, can be defined as

$$\beta = \frac{dC_B}{dpH} = \frac{dC_A}{-dpH} \quad (13)$$

where $C_B$ is the amount of base and $C_A$ the amount of acid added (Stumm & Morgan 1981, p.159). The buffer intensity is a maximum at the $pK_a$ of the weak acid concerned, important information when the likely effects of acids on water pH are being considered.

### *Neutralization*

A neutral solution ought to be one where the acidic properties are exactly balanced by the basic properties. As already noted, 'acid' and 'base' are relative terms, and for the above statement to be meaningful a reference state needs to be defined. One obvious reference is pure water. Water has both acidic and basic properties and can express them by donating a proton to itself.

$$H_2O + H_2O = H_3O^+ + OH^- \quad (14)$$

By analogy with eqn (12), the equilibrium constant for this reaction is:

$$K_w = [H_3O^+][OH^-] = [H^+][OH^-] \quad (15)$$

$K_w$ is usually called the ion product of water and at 25 °C has the value $10^{-14}$. At the neutral point, $[H^+] = [OH^-]$ and so at 25 °C $[H^+] = [OH^-] = 10^{-7}$M, and the solution therefore has pH 7·00. Note that pH 7 is not neutral by definition. $K_w$ increases with temperature and so at the water temperatures more often found in temperate climates, neutral pH is somewhat higher than 7: at 0 °C, $K_w$ is $10^{-14.93}$ and hence the neutral point is at pH 7·47.

But at 25 °C, we can speak of any pH $<7$ as 'acid' and any pH $>7$ as 'alkaline'.

## ACIDITY AND ALKALINITY

Clearly it would be useful to have a quantitative measure of the degree to which a solution departs from neutrality on either the acid or the base side, and the terms 'acidity' and 'alkalinity' naturally suggest themselves. Unfortunately, 'acidity' is used in the literature in two quite distinct ways:

(a) to mean hydrogen ion concentration — an intensity factor;
(b) to mean the amount of base which needs to be added to reach some reference point — a capacity factor.

'Alkalinity' is used only in the second sense, so it would be logical to abandon the use of 'acidity' to mean $[H^+]$, but it is probably too well entrenched. The term 'base neutralizing capacity' or BNC (Stumm & Morgan 1981, p.163) is thus coming into use to replace the second sense of 'acidity'. The corresponding term for alkalinity is 'acid neutralizing capacity' or ANC.

### *Measuring ANC and BNC: titration*

ANC measures the amount of acid which must be added to a solution to reach some reference point. The difficulty lies in defining the reference point and determining whether it has been attained. The traditional reference points in acid–base titrations are *equivalence points,* i.e. points where the equivalent sum of the acidic species equals the equivalent sum of the basic species. Remembering always that acidity is a relative property, a reference state for the terms 'acidic' and 'basic' in the previous sentence must be defined. This reference state is a pure solution of the weak acid species in question, e.g. for the $CO_2$/$H_2O$ system there are three equivalence points corresponding to pure solutions of $H_2CO_3$, $NaHCO_3$ or $Na_2CO_3$. A formal definition of an equivalence point might be: the point where there is no net excess of either acidic or basic species over those resulting from the dissolution and ionization of the pure substance which defines the reference point.

To make this clearer, dissolving $CO_2$ in water gives a solution of carbonic acid, where:

$$[H^+] = [HCO_3^-] + 2[CO_3^{2-}] + [OH^-] \qquad (16)$$

This is the carbonic acid equivalence point. Titrating this solution with NaOH, the ion balance is:

$$[Na^+] + [H^+] = [HCO_3^-] + 2[CO_3^{2-}] + [OH^-] \qquad (17)$$

At the $NaHCO_3$ equivalence point, the sodium concentration equals by definition the sum of the carbon species derived from $NaHCO_3$, i.e.

$$[Na^+] = [HCO_3^-] + [CO_3^{2-}] + [H_2CO_3] \qquad (18)$$

substituting this into eqn (17) gives:

$$[H_2CO_3] + [H^+] = [CO_3^{2-}] + [OH^-] \qquad (19)$$

which thus defines the $NaHCO_3$ equivalence point. Continuing the titration to the $Na_2CO_3$ equivalence point, as the $Na^+$/C balance in $Na_2CO_3$ solutions is:

$$[Na^+] = 2[HCO_3^-] + 2[CO_3^{2-}] + 2[H_2CO_3] \qquad (20)$$

Substituting for $Na^+$ in eqn (17) gives:

$$2[H_2CO_3] + [HCO_3^-] + [H^+] = [OH^-] \quad (21)$$

which thus defines the $Na_2CO_3$ equivalence point. In the case of a strong acid or base the only acid or basic species present are $H^+$ and $OH^-$, so at the equivalence point $[H^+] = [OH^-]$ and the pH of the equivalence point is 7·0 at 25 °C (eqn (15)).

Equivalence points are useful as reference levels for two reasons, one practical and one theoretical. The practical reason is that it can be shown that the rate of change of pH with added acid (or base) is a maximum at an equivalence point (e.g. Stumm & Morgan 1981, p.160). The end point of a potentiometric titration can thus be determined by plotting dpH against $C_A$ where $C_A$ is the amount of acid added (see Fig. 2). More traditionally, a pH indicator which changes colour in the pH range close to the equivalence point can be used

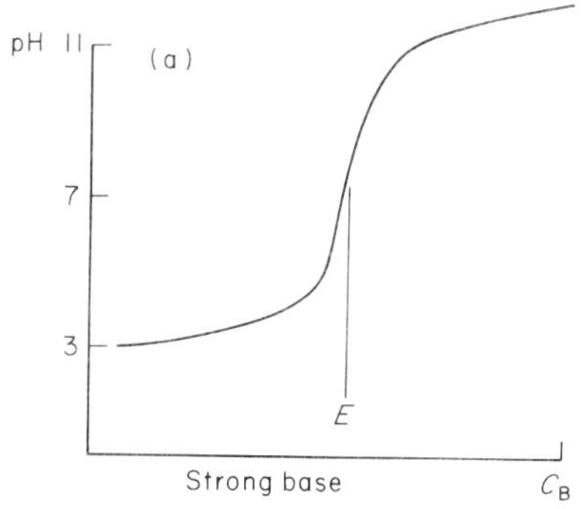


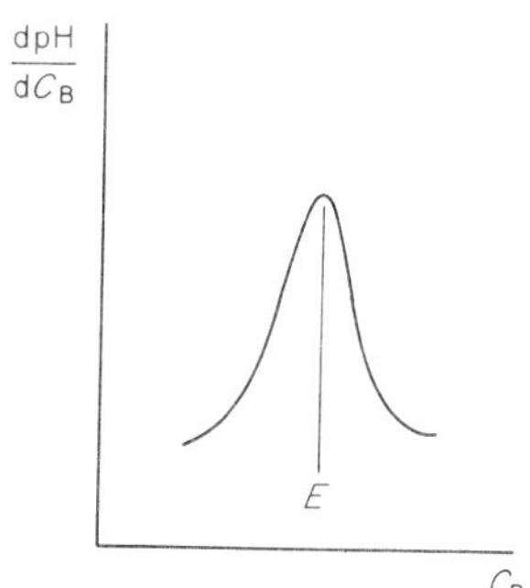


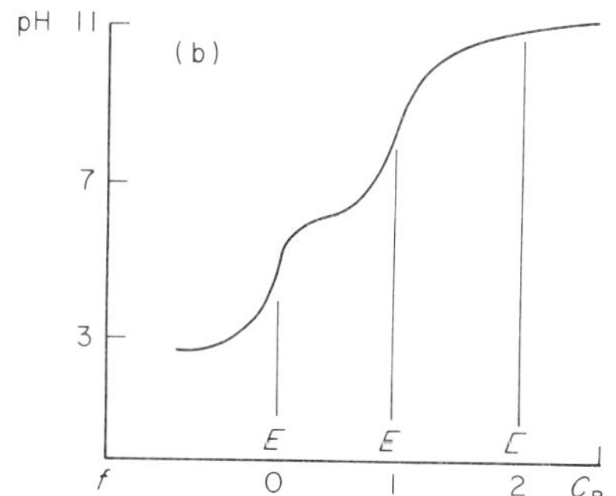


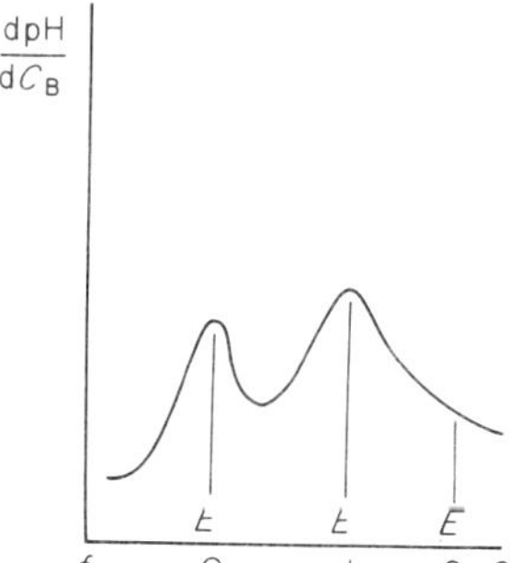


FIG. 2. (a) Titration of a strong acid with strong base. (b) Titration of a weak acid (carbonic acid) with strong base. *E* marks the positions of equivalence points in each case.

with reasonable accuracy because of the large pH shift caused by a small acid addition (Fig. 2). The theoretical advantage of equivalence points is that they allow ANC and BNC to be calculated easily in the presence of a weak acid. To understand this, consider a natural water sample containing various cations derived from strong bases $[SB^+]$ equivalents $l^{-1}$, and anions derived from strong acids $[SA^-]$ equivalents $l^{-1}$, with dissolved $CO_2$. The ion balance is:

$$[SB^+] - [SA^-] + [H^+] = [HCO_3^-] + 2[CO_3^{2-}] + [OH^-] \qquad (22)$$

If $[SB^+] > [SA^-]$ the pH will be above that of the $H_2CO_3$ equivalence point, and conversely if $[SB^+] < [SA^-]$ the pH will be below. If we titrate the former example with acid to the $H_2CO_3$ equivalence point, eqn (22) becomes:

$$[SB^+] - [SA^-] + [H^+] = [HCO_3^-] + 2[CO_3^{2-}] + [OH^-] + C_A \qquad (23)$$

where $C_A$ is the concentration of acid added. Since at the equivalence point eqn (16) applies, the weak acid species can be eliminated from eqn (23), giving:

$$[SB^+] - [SA^-] = C_A = \text{ANC} \qquad (24)$$

In other words the amount of acid required to reach the equivalence point (the ANC with respect to that point) can be calculated by subtracting the concentrations of anions derived from strong acids from the concentrations of cations derived from strong bases in the original sample. Similarly, the base neutralizing capacity is given by

$$\text{BNC} = C_B = [SA^-] - [SB^+] \qquad (25)$$

for the case where $[SA^-] > [SB^+]$. Equations (24) and (25) cannot be used unmodified if there is more than one weak acid present.

From the ecological point of view, there is a very substantial disadvantage with using equivalence points as references for ANC and BNC, which is that they do not represent fixed pH levels. For instance, the pH at which eqn (16) holds varies from around 7 where $[H_2CO_3] < 10^{-7}$M to about 4·2 at $[H_2CO_3]$ of $10^{-2}$M. However, the equivalence points of carbonic acid are used in the system for ANC and BNC measurement in natural waters for historical reasons.

### *ANC and BNC in natural waters: alkalinity*

Like 'acidity', the use of the term 'alkalinity' in natural waters is also confusing, though logical, and for this we probably have to thank the pioneer limnologists. They discovered that most natural waters, if titrated with acid, had substantial buffering capacity, due largely to the presence of carbonic acid. They expressed this buffering capacity by calling it 'alkalinity', thereby implicitly defining alkalinity with respect to carbonic acid as a reference point. Stumm & Morgan (1981) have clarified the situation by providing rigorous conceptual definitions for alkalinity and acidity, as discussed above. The total inorganic carbon in high conductivity waters is such that the pH of the equivalence point (eqn (16)) is around 4·5. Hence, such a water at pH 5·5 can be said to have 'alkalinity' (in relation to carbonic acid) but still be 'acid' (in relation to water). To illustrate some of these points, Table 2 shows the composition of six theoretical waters differing in alkalinity and total inorganic carbon, which is varied by equilibrating them with different concentrations of $CO_2$. Note that alkalinity is $C_B - C_A$, that two waters can have the same alkalinity but differ in pH, that alkalinity is not the same thing as $HCO_3^-$ though it may approximate to $HCO_3^-$

TABLE 2. Components of alkalinity in various waters at 10 °C at 1 atmosphere pressure. All units in $\mu$eq $l^{-1}$ except $pCO_2$ (atm), $H_2CO_3$* ($\mu$M) and pH. The asterix indicates that the $H_2CO_3$ is really dissolved $CO_2$ rather than carbonic acid.

| | Water 1 | Water 2 | Water 3 | Water 4 | Water 5 | Water 6 |
|---|---|---|---|---|---|---|
| $pCO_2$ | $10^{-3\cdot5}$ | $10^{-1\cdot5}$ | $10^{-3\cdot5}$ | $10^{-1\cdot5}$ | $10^{-3\cdot5}$ | $10^{-1\cdot5}$ |
| $H^+$ | 2·42 | 24·2 | 0·12 | 9·82 | 50·12 | 59·82 |
| pH | 5·62 | 4·62 | 6·93 | 5·01 | 4·30 | 4·22 |
| $OH^-$ | 0·00 | 0·00 | 0·02 | 0·00 | 0·00 | 0·00 |
| $H_2CO_3$* | 17·11 | 1711 | 17·11 | 1711 | 17·11 | 1711 |
| $HCO_3^-$ | 2·42 | 24·2 | 50·07 | 59·82 | 0·12 | 9·82 |
| $CO_3^{2-}$ | 0·00 | 0·00 | 0·03 | 0·00 | 0·00 | 0·00 |
| $SO_4^{2-}$ | 0·00 | 0·00 | 0·00 | 0·00 | 50·00 | 50·00 |
| $Ca^{2+}$ | 0·00 | 0·00 | 50·00 | 50·00 | 0·00 | 0·00 |
| Alkalinity | 0·00 | 0·00 | 50·00 | 50·00 | −50·00 | −50·00 |

at high pH, and that altering the amount of $H_2CO_3$ does not change the alkalinity (because $H_2CO_3$ is the species which defines the reference point).

*Other alkalinity parameters:*

Proton reference levels other than that in eqn (16) can be used. The other two equivalence points of the carbonic acid system, representing solutions of sodium bicarbonate or sodium carbonate, can be used as reference levels. There are again a variety of names for the alkalinity and acidity parameters at these equivalence points. Titration to the $NaHCO_3$ end-point, where eqn (19) holds, gives the '$CO_2$-acidity' from the acid side and 'p-alkalinity' or 'phenolphthalein alkalinity' from the alkaline side (so called from the indicator phenolphthalein, which changes colour at about the pH represented by eqn (19)). At the $Na_2CO_3$ end-point, where eqn (21) holds, titration from the acid side gives 'acidity' and from the base side 'caustic alkalinity' or 'hydroxide alkalinity'. The algebraic sum of the acidity and alkalinity terms at any equivalence point is zero (e.g. alkalinity + mineral acidity = 0) so they can be plotted on the same scale, and addition of the substance whose solution the equivalence point represents does not affect the alkalinity parameters at that point (e.g. $CO_2$-acidity is unaffected by addition of $NaHCO_3$).

*Acid rain*

The discussion above should make it clear that it is quite logical to define acid rain as rain with pH $< 5\cdot6$. This has nothing to do with rain pH in the absence of man-made emissions, but follows from the definition of acidity in natural waters. Pure water in equilibrium with atmospheric $CO_2$ has a pH of 5·6 at 10 °C and zero alkalinity by definition: rain with a pH less than this is therefore acid rain. This theoretical concept implies nothing about the pre-industrial pH of rain, which was probably less than 5·6 owing to natural S and N emissions (Charlson & Rodhe 1982).

### *Measuring alkalinity*

Alkalinity should be measured as noted above by titrating to the equivalence pH as defined by eqn (16). Classical limnology considered waters of high pH which were thus high in inorganic C: for these waters the equivalence pH is not far from 4·5, which is the pH at which methyl orange indicator changes colour. Classical alkalinity titrations thus use methyl orange to define the end-point. For waters with low concentrations of inorganic carbon, such as rainwater or very soft lake waters the methyl orange end-point is completely inappropriate because the equivalence pH is much higher (5·6 for rainwater) and because the relative error involved if the equivalence pH is not attained exactly is much larger than for high alkalinity waters. For instance, if Water 3 in Table 2 is titrated to the methyl orange end-point (pH 4·5) the measured [alk] is 81 $\mu$eq $l^{-1}$ instead of the correct 50. For low alkalinity waters, the total fixed end-point method given in most textbooks is inappropriate, and Gran titration should be used (see Mackereth, Heron & Talling 1978, for a simple version). Alternatively, Henriksen (1982) suggested a method for estimating the true alkalinity from a total fixed end-point determination.

Units of alkalinity are another rich source of confusion. The former practice was to express it as 'mg $l^{-1}$ $CaCO_3$', giving the false impression that an alkalinity titration was a determination of $CaCO_3$, and that alkalinity was something to do with Ca. As $HCO_3^-$ is often the weak acid present in highest concentration, the result of an alkalinity titration is sometimes expressed as [$HCO_3^-$], which should be avoided because it gives the impression that the titration is an assay which measures only $HCO_3^-$. The rational units are units of equivalence, e.g. milli-equivalents per litre.

### *ANC and BNC at fixed pH values*

Of all the species of living organism, only a few individuals of *Homo sapiens* are interested in the results of titrations. Species living in water are affected by the pH of the solution, not the alkalinity with respect to carbonic acid. An organism in Water 2 of Table 2 is experiencing ten times the [$H^+$] of one in Water 1, and is quite indifferent to the fact that the alkalinity is the same. It therefore makes sense to define ANC and BNC with respect to a fixed pH level, rather than an equivalence point. This is perfectly possible, though the calculations are harder to do and the determination possibly less precise.

### *Calculation of ANC and BNC*

To illustrate the calculation, consider a solution of 100 $\mu$Mole $l^{-1}$ of an acid HA of $pK_a$ 4·7 in a solution of pH 4·0 whose composition is shown in the top line of Table 3. To calculate

TABLE 3. To illustrate the calculation of $BNC_{pH6}$

| | $H^+$ | $OH^-$ | HA | $A^-$ | $Na^+$ | $Cl^-$ |
|---|---|---|---|---|---|---|
| Solution pH 4·0 | 100·0 | 0·0001 | 83·4 | 16·6 | 100·0 | 183·4 |
| Solution pH 6·0 | 1·0 | 0·01 | 4·8 | 95·2 | 277·6 | 183·4 |

the BNC to pH 6 ($BNC_6$), the base required to titrate $H^+$ and the base required to titrate HA between pHs 4 and 6 must be added. This is the algebraic sum of the titratable species at pH 4 less the sum of the same species at pH 6, all in equivalents. Titratable species are defined as species which change in concentration during the titration. Therefore in this case:

$$[BNC_6] = ([H^+]_{pH4} - [OH^-]_{pH4} - [A^-]_{pH4} - ([H^+]_{pH6} - [OH^-]_{pH6} - [A^-]_{pH6}) \quad (26)$$

But as the electroneutrality condition gives for the original solution (top line of Table 3):

$$[H^+]_4 + [Na^+]_4 = [A^-]_4 + [Cl^-]_4 + [OH^-]_4 \quad (27)$$

and thus

$$[H^+]_4 - [OH^-]_4 - [A^-]_4 = [Cl^-]_4 - [Na^+]_4 = SA^- - SB^+ \quad (28)$$

So eqn (26) becomes

$$[BNC_6] = SA^- - SB^+ - [H^+]_{pH6} + [OH^-]_{pH6} + [A^-]_{pH6} \quad (29)$$

As $[H^+]$ and $[OH^-]$ at pH 6 are known, all that is required to calculate the BNC is $SB^+$ and $SA^-$ at pH 4 and $[A^-]$ at pH 6. If the total concentration of HA + $A^-$ ($A_T$) can be determined analytically, then $[A^-]$ at pH 6 can be calculated if the $pK_a$ of the weak acid HA is known. Since the degree of dissociation or ionization fraction $\alpha = [A^-] \div A_T$ (Stumm & Morgan 1981, p.153) varies only with pH at constant temperature (and to a very minor extent with ionic strength), $[A^-]$ can be obtained from a knowledge of pH and $A_T$. Values of $\alpha$ for most acid-base systems of interest in freshwater are tabulated in Table 4.

TABLE 4. Ionization fractions of weak acids common in fresh water

| | | pH | | | | | |
|---|---|---|---|---|---|---|---|
| Acid | Species | 4·0 | 4·5 | 5·0 | 6·0 | 7·0 | 8·0 |
| Carbonic | $H_2CO_3$ | 0·996 | 0·987 | 0·960 | 0·706 | 0·194 | 0·023 |
| | $HCO_3^-$ | 0·004 | 0·013 | 0·040 | 0·294 | 0·806 | 0·972 |
| | $CO_3^{2-}$ | 0·000 | 0·000 | 0·000 | 0·000 | 0·000 | 0·005 |
| Aluminium* | $Al^{3+}$ | 0·909 | 0·735 | 0·387 | 0·008 | 0·000 | 0·000 |
| | $Al(OH)^{2+}$ | 0·085 | 0·218 | 0·364 | 0·070 | 0·000 | 0·000 |
| | $Al(OH)_2^+$ | 0·006 | 0·047 | 0·246 | 0·475 | 0·011 | 0·000 |
| | $Al(OH)_4^-$ | 0·000 | 0·000 | 0·003 | 0·447 | 0·989 | 1·000 |
| Ammonium | $NH_4^+$ | 1·000 | 1·000 | 1·000 | 1·000 | 0·995 | 0·952 |
| | $NH_3$ | 0·000 | 0·000 | 0·000 | 0·000 | 0·005 | 0·048 |
| Silicic | $Si(OH)_4$ | 1·000 | 1·000 | 1·000 | 1·000 | 0·997 | 0·969 |
| | $SiO(OH)_3^-$ | 0·000 | 0·000 | 0·000 | 0·000 | 0·003 | 0·031 |
| Organic | HA | 0·613 | 0·334 | 0·137 | 0·016 | 0·002 | 0·000 |
| $pK_a = 4·2$ | $A^-$ | 0·387 | 0·666 | 0·863 | 0·984 | 0·998 | 1·000 |
| Organic | HA | 0·934 | 0·613 | 0·334 | 0·048 | 0·005 | 0·000 |
| $pk_a = 4·7$ | $A^-$ | 0·166 | 0·387 | 0·666 | 0·952 | 0·995 | 1·000 |

*Aluminium calculated using equations in Driscoll (1980) assuming ionic strength of 200 $\mu$M and no significant $F^-$ or $SO_4^{2-}$ complexing. All calculations assume 25 °C: three-place accuracy is for illustrative purposes only.

These values may be of interest to toxicologists trying to determine which chemical species is toxic to freshwater organisms. Returning to the example in Table 3,

$$[BNC_6] = [Cl^-] - [Na^+] - 1 + 0·01 + 95·2 = 177·6\ \mu eq\ l^{-1}$$

Inspection of the second line of Table 3 shows that the addition of 177·6 $\mu$eq $l^{-1}$ of NaOH to the solution in the first line results in an electrically-neutral solution at pH 6. More generally,

$$[BNC_{pH=x}] = SA^- - SB^+ - 10^{-x} + 10^{-(14-x)} - (\sum WA - \sum WB)_x \quad (30)$$

where $SA^-$ is the concentration of anions of strong acids in the solution, $SB^+$ is the concentration of cations of strong bases in the solution, $\sum WA$ and $\sum WB$ are the equivalent sums of the concentrations of weak acids and bases respectively at pH$x$, excluding $H^+$ and $OH^-$. Similarly,

$$[ANC_{pH=x}] = SB^+ - SA^- + 10^{-x} - 10^{-(14-x)} + (\sum WA - \sum WB)_x \quad (31)$$

To take another example, what is the $BNC_{pH6}$ of a solution containing 100 $\mu$M $Na^+$ and 609 $\mu$M $Cl^-$? From eqn (30),

$$[BNC_6] = [Cl^-] - [Na^+] - 1 + 0{\cdot}01 - [3 \times [Al^{3+}] + 2 \times [Al(OH)^{2+}] + [Al(OH)_2^+] - [Al(OH)_4^-]) \quad (32)$$

Using Table 4 to determine the Al species at pH 6,

$$[BNC_6] = 609 - 100 - 1 + 0{\cdot}01 - (3 \times 0{\cdot}85) + (2 \times 7{\cdot}0) + 47{\cdot}5 - 44{\cdot}7)$$
$$= 489\ \mu eq\ l^{-1}$$

This procedure works equally well for mixtures of weak acids. Stumm & Morgan (1981, pp. 165, 187), suggested a method for dealing with mixtures of weak acids in an equivalence point titration, but this is valid only if the other acids are fully protonated or deprotonated at the equivalence point. One major assumption of eqns (30) and (31) is that the acid species remain in solution, e.g. that Al does not precipitate or $CO_2$ degas as the pH rises. Note that unless the pH is 7, when $10^{-x} = 10^{-(14-x)}$ in eqns (30) and (31), water itself has some ANC or BNC.

## TRANSPORT OF ACIDITY THROUGH ECOSYSTEMS

The above discussion should have made it clear that it is necessary to define very carefully what is meant by acidity before discussing its movement through ecosystems. The two definitions (capacity factor or intensity factor) have tended to be employed interchangeably with resulting confusion. This section examines what can be said about the passage of both senses of 'acidity' through ecosystems, starting with the intensity factor $[H^+]$.

### *Passage of $H^+$]*

It is possible to measure the flux of free protons from one ecosystem compartment to another, where there is a flow of water which can be collected and analysed. This is simply

$$F_{H^+}(t) = V(t) \times 10^{-pH} \quad (33)$$

where $F_H{}^+$ is the flux of $H^+$ in moles over time $t$, $V(t)$ is the volume of water in litres passing in time $t$, and pH is the pH of $V(t)$. This ability to measure proton fluxes makes it tempting to construct 'proton budgets', but this is impossible because free protons are not conservative species. This is illustrated in Fig. 3. The word 'budget' is derived from the Old French for 'small bag', the implication being that what was put into the bag could ultimately be taken out again. Thus in Fig. 3, if the box represents an ecosystem and two

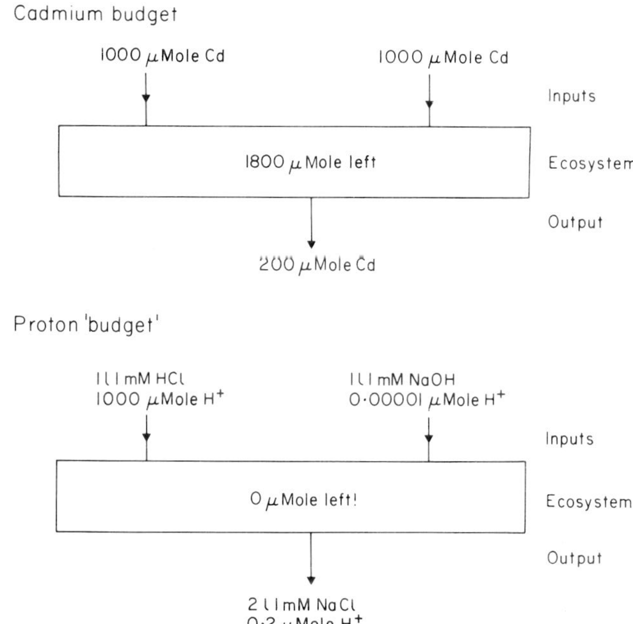


FIG. 3. A cadmium budget can be constructed, but a proton budget can not.

fluxes each of 1000 μmoles of Cd are measured entering the ecosystem and a single flux of 200 μmoles is measured leaving, then the ecosystem must have gained 1800 μmoles Cd provided all fluxes have been measured. Protons do not behave like this. The lower half of Fig. 3 shows a simplified example in which a flux of HCl reacts with a flux of NaOH and the reaction products (NaCl and $H_2O$) are allowed to leave the ecosystem. None of the 'missing' 999·8 μmoles of $H^+$ are left behind. Hydrogen *atoms* are of course conserved, and a budget could be constructed for them, but not protons. If proton budgets cannot be constructed, why measure the flux of $H^+$ at all? The reason is that the $H^+$ *concentration* is a master variable, controlling a variety of biological and chemical reactions. The $H^+$ *flux* is part of the flux of BNC, for which budgets *can* be constructed, as is discussed below.

## *Passage of BNC*

Defining BNC as in eqn (30) relative to pH 7 then the relationship in Fig. 3 looks different. The acid solution has a $BNC_7$ of 1 meq $l^{-1}$, and the alkaline solution of $-1$ meq $l^{-1}$. The resultant salt solution has a $BNC_7$ of 0. The amount of $BNC_7$ is thus conserved (not the concentration). With BNC at other pHs the calculation has to take into account that water itself possesses some BNC (see above). Thus in the above example, the acid solution has $BNC_5$ of 990 μeq $l^{-1}$, the alkaline solution of $-1010$ μeq $l^{-1}$ and the salt of $-10$ μeq $l^{-1}$. The amount of $BNC_5$ is conserved, taking into account that there are 2 l of salt solution. If, however, 1·5 l of water were to evaporate reducing the salt volume to 0·5 l, as pure water has a $BNC_5$ of $-10$ $\mu$eq $l^{-1}$, there is a loss of $-15$ $\mu$eq from the system and thus $-5$ $\mu$eq remain in the output. This is consistent with a $BNC_5$ value of $-10$ $\mu$eq $l^{-1}$ in the 0·5 l of output NaCl.

It may be possible to calculate the BNC flux even if no convenient flow of water can be collected. The most important situation is the calculation of the acidifying effect of plants on soil. From eqn (30), if a plant takes up $SB^+$ in excess of $SA^-$ there will tend to be a flux of BNC out of the plant, i.e. the soil gains BNC (becomes more acid). The magnitude of the acidification can be calculated by subtracting the change in $SA^-$ from the change in $SB^+$ in the plant compartment over a period of time. In practice this approximates to:

$$\Delta BNC_{(soil)} = \Delta Na + \Delta K + 2\Delta Ca + 2\Delta Mg - 2\Delta S - \Delta Cl \pm \Delta N \qquad (34)$$

where $\Delta X$ is in moles per unit time.

The assumption is made that the speciation of the ions taken up is known. For example, if Ca was taken up as a neutral complex rather than as $Ca^{2+}$, the calculation would be in error. The major problem of this sort is with N, which is quantitatively the most important element taken up from soil. Uptake of $NH_4^+$ leads to an outward BNC flux: uptake of $NO_3^-$ to an inward BNC flux. This effect provides the best experimental evidence for the reality of these calculations as it has long been known that plants grown in solution culture with $NO_3^-$ as sole N source raise the solution pH, whereas plants grown with $NH_4^+$ depress it (Hewitt 1966, p.205). The effect of plants on the BNC of solution cultures could be used to test these calculations in a more quantitative way. To avoid these problems with N it is necessary to make an arbitrary assumption about N speciation, unless there is independent evidence which can be brought to bear. Another assumption to eqn (34) is that weak acids are not taken up or given out to any significant extent. If the only weak acid which is taken up is $HCO_3^-$, then a positive $\Delta BNC_{soil}$ in eqn (33) approximates to a proton flux out of the plant and a negative $\Delta$BNC to a proton flux into the plant. An example of this approach is Nilsson, Miller & Miller (1982).

## A PRACTICAL EXAMPLE

To illustrate how these ideas might be applied, data from a study of a Douglas fir ecosystem in Western Oregon, USA will be used (Sollins *et al.* 1980). The study measured fluxes between various ecosystem compartments, including (unusually) titratable alkalinity to pH 4·5. It is therefore possible to construct a measured $ANC_{4 \cdot 5}$ budget and also to attempt a calculated one using eqn (31), though all the relevant chemical species were not measured. Fig. 4 shows both calculated and measured $ANC_{4 \cdot 5}$ budgets, abstracted from data in Sollins *et al.* (1980) Table 1, and making the assumptions given in the legend to the Figure. As water has an $ANC_{4 \cdot 5}$ of 31·6 $\mu$eq $l^{-1}$, evaporation has to be taken into account in the budget. The measured budget shows an intuitively realistic picture, with most of the ANC being added to the water flow by the mineral soil, and the output water being a mixture of water from mineral soil, litter, throughfall and rain and therefore of intermediate composition. Sollins *et al.* (1980) have assumed that all water leaving the catchment flows through the mineral soil, whereas this is clearly not the case. The calculated budget reflects uncertainty about the weak acid concentrations, especially (as the authors state) the inorganic carbon anions and the organic carbon anions.

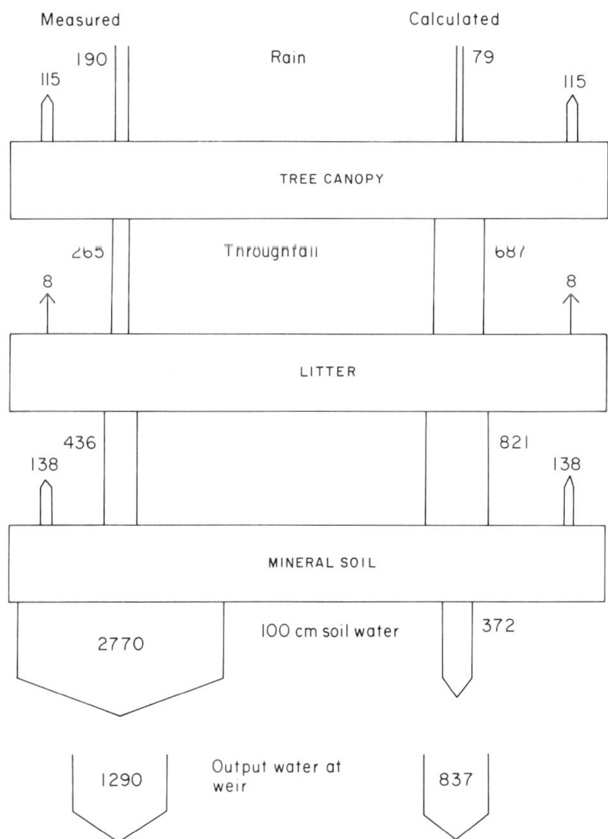


FIG. 4. Calculated and measured $ANC_{4.5}$ budgets, from Sollins *et al.* (1980). The calculated budget assumed that all the Kjeldahl N is $NH_4^+$ and that all solutions are in equilibrium with atmospheric $CO_2$ ($pCO_2 = 10^{-3.5}$). Some strong acid anion has been added to the 'weir' calculated flux to make up the anion deficit.

One note of caution should be sounded: as ANC and BNC are measured on a relative and not an absolute scale, it is not very meaningful to calculate proportions. Thus in Fig 4, though the tree canopy adds 75 equiv $ha^{-1} y^{-1}$ of $ANC_{4.5}$ and the litter 171 equiv $ha^{-1} y^{-1}$, it is not useful to say that the litter adds 2·3 times as much as the trees. If $ANC_7$ had been calculated a different proportion would have been obtained. To calculate proportions in that way is analogous to calculating that a body at 10 °C is twice as hot as one at 5 °C. Nevertheless, the above ANC budget does show that the litter is more important than the tree canopy in adding $ANC_{4.5}$. Perhaps the standard pH for ANC/BNC calculation should be that of the output water, if that is what is of most interest, though this would require a more sophisticated approach to calculation than that adopted in this paper.

## CONCLUSION

A proper understanding of the concept of acidity is essential if ecosystem studies are to be designed to answer questions about acidity inputs and internal acidity generation. The study of BNC and ANC budgets should provide a useful tool for such studies.

## ACKNOWLEDGMENTS

I would like to thank Mr A.H. Webb for constructive criticism. This work was carried out at the Central Electricity Research Laboratories and is published by permission of the Central Electricity Generating Board.

## REFERENCES

**Albert A. & Serjeant, E.P. (1971)** *The Determination of Ionization Constants.* Chapman and Hall, London.

**Charlson, R.J. & Rodhe, H. (1982)** Factors controlling the acidity of natural rainwater. *Nature (Lond.),* **295**, 683–685.

**Drabløs, D. & Tollan, A., (Eds.), (1980)** *Ecological Impact of Acid Precipitation.* SNSF Project, Oslo–Ås, Norway.

**Driscoll, C.T. (1980)** *Chemical characterisation of some dilute acidified lakes and streams in the Adirondak region of New York State.* PhD thesis, Cornell University.

**Evans, L.S., Hendrey, G.R., Stensland, G.J., Johnson, D.W. & Francis, A.J.** (1981) Acid precipitation: considerations for an air quality standard. *Water, Air and Soil Pollution,* **16**, 469–509.

**Fowler, D., Cape, J.N., Leith, I.D., Paterson, I.S. Kinnaird, J.W. & Nicholson, I.A. (1982)** Rainfall acidity in northern Britain, *Nature (Lond.),* **297**, 383–386.

**Havas, M., Hutchinson, T. C. & Likens, G. E. (1984)** Red herrings in acid rain research. *Environmental Science and Technology,* **18**, 176–186.

**Henriksen, A. (1982)** Alkalinity and acid precipitation research. *Vatten,* **38**, 83–85).

**Hewitt, E.J. (1966)** *Sand and Water Culture Methods Used in the Study of Plant Nutrition.* Commonwealth Agricultural Bureaux Technical Communication No. 22, Farnham Royal, UK.

**Krug, E.C. & Frink, C.R. (1983)** Acid rain on acid soil: a new perspective. *Science,* **221**, 520–525.

**Mackereth, F.J.H., Heron, J. & Talling, J.F. (1978)** *Water analysis: some revised methods for limnologists.* Freshwater Biological Association (UK), Scientific Publication No. 36, Ambleside UK.

**Nilsson, S.I., Miller, H.G. & Miller, J.D. (1982)** Forest growth as a possible cause of soil and water acidification – an examination of the concepts. *Oikos,* **39**, 40–49.

**Rosenqvist, I.T. (1977)** *Sur jord-Surt vann.,* Ingeniorforlaget A/S, Oslo, Norway.

**Rosenqvist, I.T. (1978)** Alternative sources for acidification of river water in Norway. *Science of the Total Environment,* **10**, 39–49.

**Sollins, P., Grier, C.C., McCorison, F.M., Cromack, K., Fogel, R. & Fredriksen, R.L. (1980)** The internal element cycles of an old-growth Douglas fir ecosystem in Western Oregon. *Ecological Monographs,* **50**, 261–285.

**Stumm, W. & Morgan, J.J. (1981)** *Aquatic Chemistry,* 2nd edn. John Wiley, Chichester.

**Swedish Ministry of Agriculture (1982)** *Acidification Today and Tomorrow.* Stockholm.

**Van Breemen, N., Mulder, J. & Driscoll, C.T. (1983)** Acidification and alkalinization of soils. *Plant and Soil,* **75**, 283–308.

**Van Breemen, N., Mulder, J. &. Driscoll, C.T. (1984)** Acid deposition and internal proton sources in acidification of soils and waters. *Nature (Lond.),* 307, 599–604.

**Watt Committee on Energy (1984)** *Acid Rain.* Report No. 14, London.

# Modification of rainfall chemistry by a forest canopy

J. N. CAPE*, D. FOWLER*, J. W. KINNAIRD,
I. A. NICHOLSON AND I. S. PATERSON
*Institute of Terrestrial Ecology, Hill of Brathens, Glassel, Banchory, Kincardineshire AB3 4BY*
*and *Institute of Terrestrial Ecology, Bush Estate, Penicuik, Midlothian EH26 0QB*

## SUMMARY

**1** Results are reported for an experiment to measure the chemical composition of rainfall above a forest and of throughfall (drip) and stemflow below the canopy. Samples were taken every 2 weeks for 1 year from four replicate plots each with eight throughfall and four stemflow collectors. The forest, in central Scotland, of Scots pine was *c.* 30 years old.

**2** On average, throughfall made up 53% of above-canopy rainfall, stemflow 11% and the balance (36%) was lost by evaporation. Annual sulphate deposition in rainfall, throughfall and stemflow was 1·4, 3·1 and 1·5 g S $m^{-2}$, respectively.

**3** Continuous measurements of $SO_2$ concentration and meteorological variables, and measurements (by eddy correlation) of dry deposition of $SO_2$ to the forest, allowed the input by dry deposition to be estimated for the same period. This could not account for the increased flux of sulphur below the canopy, and was only 8% of the flux in three winter months.

**4** A simple model is used to estimate the amount of sulphate available for leaching and the rate of leaching by rain. It is concluded that the increased flux of sulphur below the canopy results from leaching of sulphate from plant tissues. Air pollution may increase the 'pool' of available sulphate.

## INTRODUCTION

As rainfall passes through a forest canopy the water is spatially redistributed over the forest floor as throughfall, which includes drips and non-intercepted water, and stemflow which reaches the forest floor by running down tree trunks. Some water is retained by the canopy and evaporated, and this is known as interception loss (Rutter 1975). As well as being redistributed, there may be substantial changes in chemical composition, with losses or gains of material through interaction with the vegetation. Possible mechanisms for change include the wash-off of surface material (previously deposited from the atmosphere) and exchange of material from within the plant. In the case of a net gain this latter would be called leaching. There have been many experiments to investigate the dependence of throughfall and stemflow chemistry on tree species, particularly in relation to nutrient cycling studies, and these have been reviewed recently by Parker (1983).

In cases where the deposition of sulphate is greater in throughfall and stemflow than in rain, it has been suggested that air pollutants, sulphur dioxide and particulate sulphate, may be responsible for the observed increase in S deposition (Miller & Miller 1980). Larger additional inputs of sulphate have been observed at a site in S. Norway when compared with a more northerly site (Horntvedt, Dollard & Joranger 1980), at a site in Alberta, Canada, closer to a large point source of $SO_2$ (Baker, Hocking & Nyborg 1976), and at sites in the south-eastern USA closer to major power stations (Parker, Lindberg & Kelly 1980). Of particular interest is the way in which additional material may be incorporated into throughfall and stemflow. What are the relative contributions of wash-off and leaching? Parker (1983) has reviewed the possible mechanisms and approaches to answering this question, and showed great diversity in the published data. Estimates of the contribution of air pollutants to the increased sulphur deposition in throughfall and stemflow were between 13 and 100%, but he indicated a number of studies suggesting that leaching is more important than surface wash-off. In this type of discussion a distinction must be made between material (gaseous or particulate) which is deposited on the surface of vegetation and which is subsequently removable by washing, and material (e.g. $SO_2$ gas) which enters the plant at one time and may be leached out at some later time.

In order to investigate the partitioning between wash-off and leaching, it is necessary to make continuous measurements of $SO_2$ gas and sulphate particle concentrations and to have sufficient data on the meteorological variables (e.g. occurrence of rain, solar radiation) which control the uptake of sulphur dioxide by vegetation. Such integrated studies have been attempted before over a short period (Bache 1977, Horntvedt *et al.* 1980); but the problems to be overcome are not trivial (Galloway & Parker 1980).

The results presented here come from such an integrated study, which in addition to rainfall, throughfall and stemflow composition measurements includes continuous measurements of $SO_2$ gas, solar radiation and wind speed, daily records of rainfall occurrence, and the direct measurement of $SO_2$ uptake at the same site (by eddy correlation) during both day-time and night-time and by dry and wet canopy surfaces.

## METHODS

The study forest (Devilla Forest) of Scots pine *(Pinus sylvestris* L.*)* is situated in central Scotland, 30 km north-west of Edinburgh (Ordnance Survey grid ref. 958 894). The trees were planted in 1952 at a spacing of 1·6 m in rows 1·3 m apart, giving a simply-structured community, without shrub layer or significant field layer, with *c.* 3900 trees $ha^{-1}$. In 1979–80 the trees were 9-10 m in height.

Rainfall amount was measured using Nipher-shielded rain collectors 12·7 cm in diameter (Nipher 1878), and rainwater for chemical analysis was collected using 20 cm diameter polypropylene funnels. Three sets of these collectors were installed (above the forest) at a height of 12 m on towers which could be lowered to facilitate cleaning and replacement of funnels. Water was collected in darkened polypropylene bottles at ground level, having passed through a continuous length of pvc tubing, and sampled every 2 weeks. The towers were placed in a triangle, *c.* 500 m apart.

Within the forest study area 4 plots (15 × 12 m) were established for collection of throughfall and stemflow. In each plot sixteen collectors were placed at random; eight standard 5 inch (12·7 cm) Meteorological Office pattern rain-gauges, each adjacent to a 20 cm polypropylene funnel which drained into a darkened polypropylene bottle. An estimate of canopy cover vertically above each throughfall site was made visually to the nearest 10%. Individual sites ranged from 20% to 90% canopy cover.

For stemflow measurements in each plot, stem diameters (at 1·3m) were stratified into four classes (0–7·6, 7·6–9·8, 9·8–11·8, 11·8–17·0 cm) and one tree in each class was selected at random, giving sixteen trees in all. Stemflow was collected by a moulded latex gutter (width 3 cm) fixed to the tree in a spiral round the stem which led to a darkened polypropylene bottle. The gap between the bark and gutter was sealed using silicone rubber. Samples from the throughfall and stemflow collectors were made every two weeks.

Rainwater samples were analysed for pH, conductivity, 'total' acid (titration to pH 10), $Ca^{2+}$, $Mg^{2+}$, $K^+$, $SO_4^{2-}$, $NO_3^-$ and $NH_4^+$. Throughfall and stemflow samples were analysed for all the above with the exception of $NO_3^-$ and $NH_4^+$.

The data presented here are for twenty-six collections from 29 March 1979 to 3 April 1980, and consider only the hydrology and sulphate fluxes during that period.

At the same site continuous measurements at 10-minute intervals were made above the canopy of $SO_2$, NO, $NO_x$, $O_3$, wind direction, wind speed and solar radiation (Nicholson *et al.* 1980). Annual mean concentrations for 1978 were 11·6 ppbV of $SO_2$, 12.3 ppbV NO, 26·7 ppbV $NO_x$ and 19·4 ppbV $O_3$. Measurements were also made early in 1979 of total particulate sulphate trapped on a paper filter (Whatman 41) at the inlet to the gas sampling line. Filter papers were extracted in water and analysed for sulphate.

In a related study at the same forest, measurements were made of the rate of uptake of $SO_2$ by the forest canopy using a micrometeorological method (Fowler & Cape 1983). Individual measurements averaged the net exchange of $SO_2$ between the forest and the atmosphere over 20 – 30 minutes. Fluxes of sensible heat and net radiation were also measured, enabling estimates of water vapour loss and hence the bulk stomatal conductance of the canopy to be estimated. The relative rates of stomatal and external surface uptake were then separated and in this way a simple model of $SO_2$ deposition onto a pine forest was constructed.

## RESULTS AND DISCUSSION

### *Hydrology*

Measurements of rainfall amount above the canopy showed very good agreement between the three collectors (for twenty-six collections, $r > 0{\cdot}98$) but showed systematic differences of between 2 and 5%. These small, yet statistically significant (at 5% level) differences illustrate the small sampling errors over the study area. Total annual rainfall was 883 mm, fortnightly values ranging from 4 mm to 90 mm.

Throughfall amount, averaged over all thirty-two collectors, was correlated with rainfall amount (Fig. 1), the regression (Fig. 2) showing an intercept not significantly different from zero (Table 1). For the nineteen throughfall collectors which were under a canopy cover of 70% or more, however, the regression gave a significant intercept equivalent to 2·5 mm rain (per fortnight) (Table 1). The amount of throughfall does not vary greatly with canopy cover, but does show a systematic difference between sparse cover (20–60%, $n = 13$) and dense cover (70–90%, $n = 19$), (Fig. 3). The annual average throughfall amount was 467 mm or 53% of above-canopy rainfall.

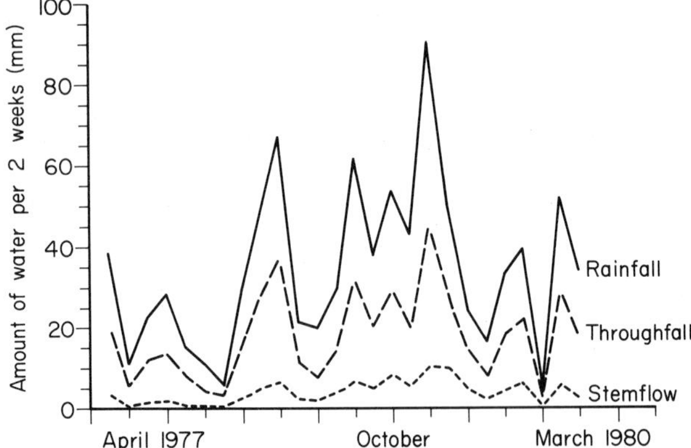


FIG. 1. The distribution of rainfall, throughfall and stemflow during the period March 1979 – April 1980 at Devilla Forest.

The amount of water collected as stemflow also varied with rainfall (Fig. 1), but was less well correlated with amount than was throughfall (Fig. 4). The intercept, as with throughfall, was not significantly different from zero (Table 1). Stemflow increased with the size of tree as measured by the stem cross-sectional area at 1·3 m (Fig. 5), but the relationship was weak (Table 1).

The average annual stemflow amount was 98 mm or 11% of above canopy rainfall. Amounts for each of the plots are summarized in Table 2. The loss of water through evaporation (interception loss) was therefore 36% (± 10%).

## *Sulphate in rain, throughfall and stemflow*

In rainfall there was greater variability between collectors for sulphate deposition than for rainfall amount (Fig. 6). This was probably caused by analytical errors, but a random

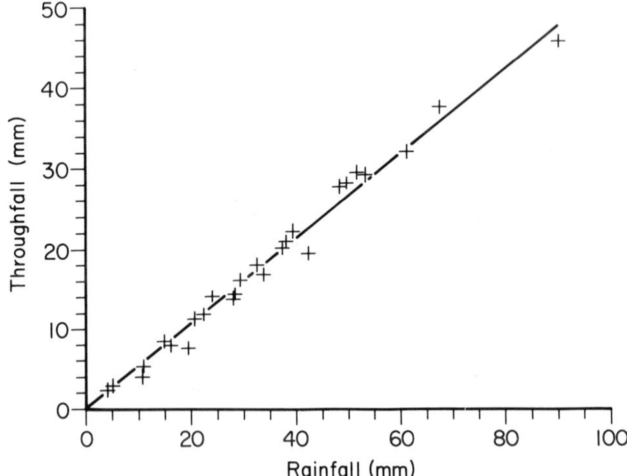


FIG. 2. The variation of throughfall with rainfall for twenty-six fortnightly collections at Devilla Forest, March 1979 – April 1980.

TABLE 1. Regresssion equations for lines shown in figures. Numbers in parentheses are standard errors of estimate

| Figure | | $r$ | d.f. |
|---|---|---|---|
| 2 | Throughfall (mm per 2 weeks) = 0·530 (±0·007) × rainfall (mm per 2 weeks) | 0·99 | 25 |
| | for canopy cover >70%, throughfall (mm per 2 weeks) = −1·25 (±0·48) + 0·498 (±0·012) × rainfall (mm per 2 weeks) | 0·99 | 17 |
| 4 | Stemflow (mm per 2 weeks) = 0·115 (±0.006) × rainfall (mm per 2 weeks) | 0·90 | 25 |
| 5 | Stemflow (mm $y^{-1}$) = 1·12 (±0·13) × stem area ($cm^2$) | 0·58 | 15 |
| 8 | Sulphate concentration ($\mu$M) = 3·2 (±0·2) × canopy cover (%) | 0·54 | 31 |
| 10 | Stemflow dep. sulphate (g S $m^{-2}$) = 14·4 (±0·6) × stemflow (mm $tree^{-1}$) | 0·95 | 15 |
| 9, 13 | ln (C, $\mu$M) = 5·76 (±0·10) − 0·020 (±0·005) × mm throughfall | 0·64 | 24 |
| 11 | ln (C, $\mu$M) = 6·67 (±0·09) − 0·10 (±0·02) × mm stemflow | 0·74 | 24 |

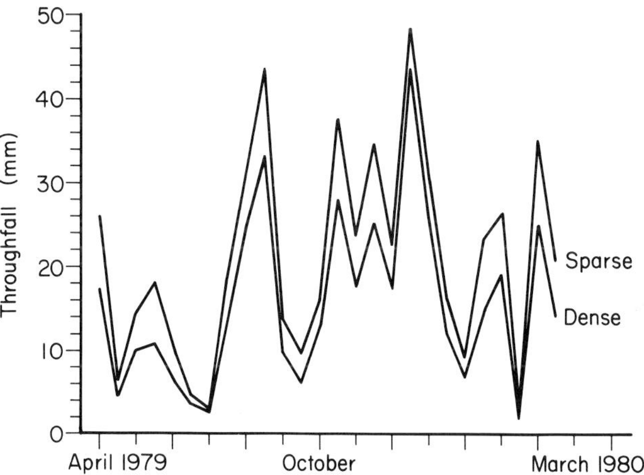


FIG. 3. The variation of throughfall with canopy cover; sparse cover (20–60%), dense cover (70–90%).

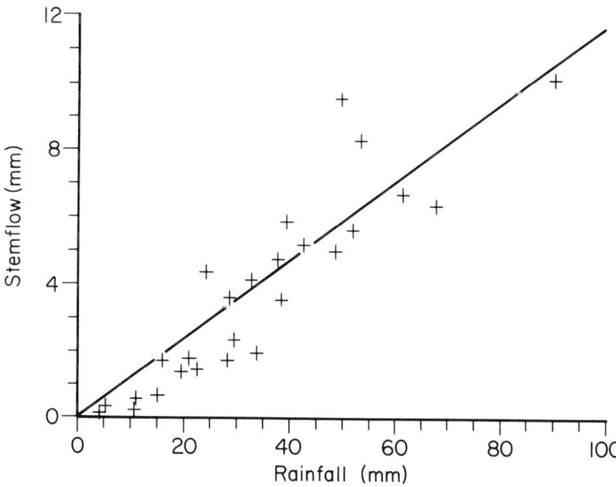


FIG. 4. The variation of stemflow with rainfall for twenty-six fortnightly collections at Devilla Forest, March 1979–April 1980.

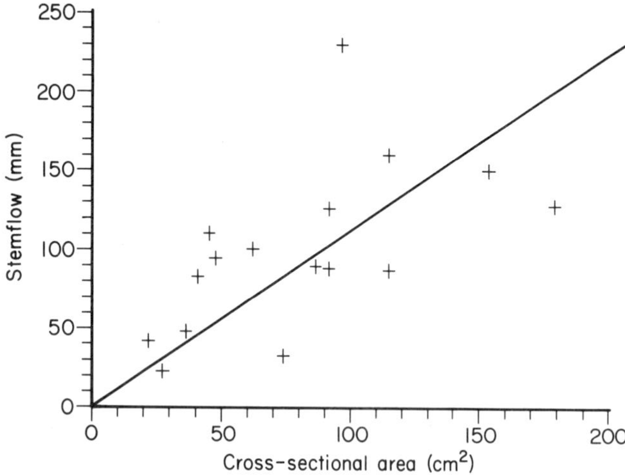


FIG. 5. The variation of stemflow with stem cross-sectional area (measured 1·3 m above ground), for sixteen trees.

TABLE 2. Amounts of water (mm) in rainfall, throughfall and stemflow at Devilla Forest, April 1979 – March 1980

| Plot | (i) | (ii) | (iii) | | Mean ± S.E. |
|---|---|---|---|---|---|
| Rain ($R$) | 857 | 888 | 904 | | 883 ± 17 |

| Plot | 1 | 2 | 3 | 4 | Mean ± S.E. |
|---|---|---|---|---|---|
| Throughfall ($T$) | 452 | 373 | 464 | 579 | 467 ± 49 |
| Stemflow ($S$) | 89 | 113 | 84 | 105 | 98 ± 8 |
| $T + S$ | 541 | 486 | 548 | 684 | 565 ± 48 |
| Interception ($R - T - S$) | 342 | 397 | 335 | 199 | 318 ± 48 |
| Est. canopy cover (%) | 61 | 71 | 70 | 65 | 67 ± 3 |

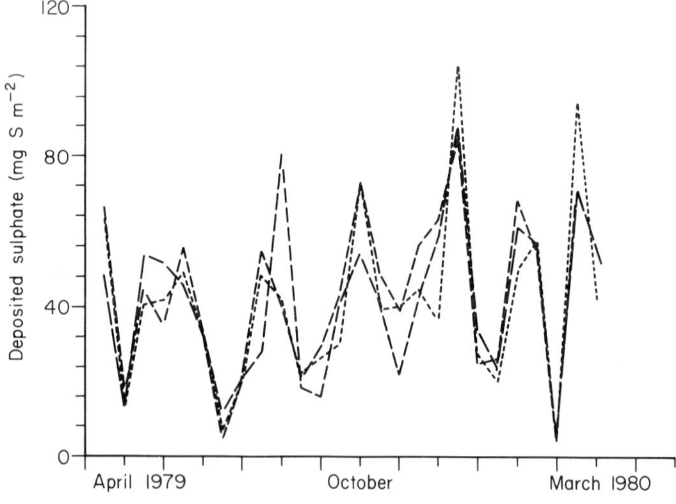


FIG. 6. The distribution of sulphate in above-canopy rain collected using three collectors, each separated by 500 m, during the period March 1979 – April 1980.

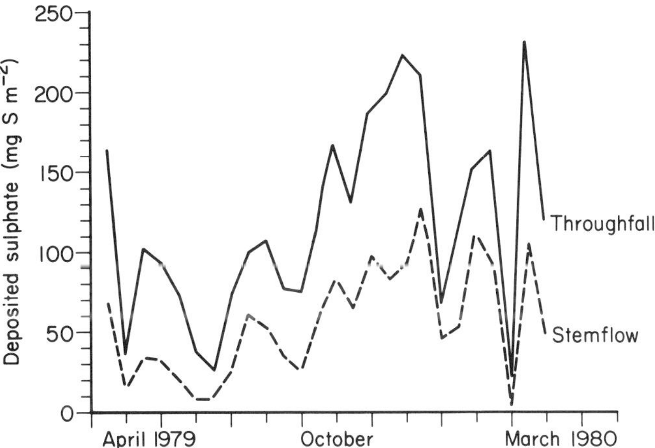


FIG. 7. Average deposition of sulphate in throughfall (———) and stemflow (----) for the period March 1979 – April 1980.

contamination of collectors would also explain the results. There were no significant differences between the amounts deposited on the three above-canopy collectors, the average annual input in rain being 1·37 g S m$^{-2}$, corresponding to a rainfall-weighted mean concentration of 48 $\mu$M $SO_4^{2-}$.

Sulphate deposited in throughfall and stemflow followed the same pattern throughout the year (Fig. 7). For throughfall, the amount deposited did not depend on canopy cover, as the smaller amounts of water at high canopy cover were offset by larger concentrations of sulphate in throughfall (Fig. 8). Again the relationship was not strong (Table 1) and there was considerable scatter. Sulphate deposition increased with increasing throughfall

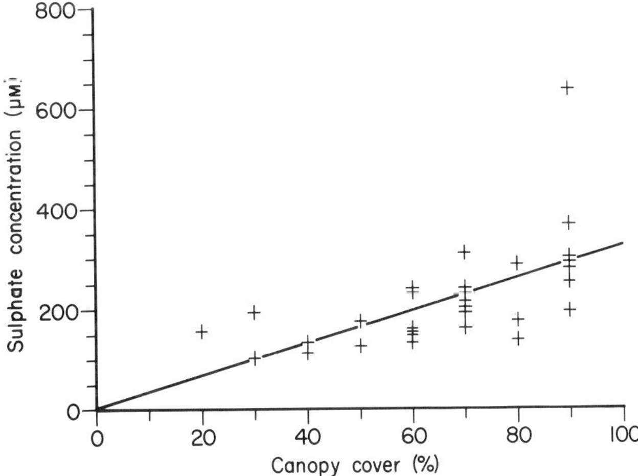


FIG. 8. The variation of average throughfall sulphate concentration with canopy cover for thirty-two collectors.

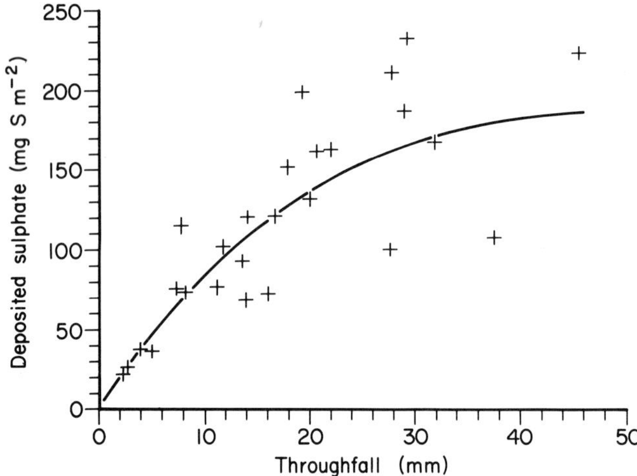


FIG. 9. The relationship between deposited sulphate in throughfall and throughfall quantity for twenty-six fortnightly collections. For explanation of the curve, see text.

amount (Fig. 9), and there was no evidence of a non-zero intercept, as would have been expected if 'wash-off' contributed greatly to the increased sulphate deposition. The average annual deposition of sulphate in throughfall was 3·08 g S m$^{-2}$, exceeding deposition in rainfall by a factor of 2·25.

The average amount of deposited sulphate in stemflow varied greatly from tree to tree (Fig. 10), being related linearly to the amount of stemflow (Table 1). This gives an average sulphate concentration of 450 ($\pm$ 20) $\mu$M, regardless of tree size. However, the average sulphate concentration did vary from collection to collection, decreasing with increasing volumes of stemflow. As with throughfall, a plot of of deposited sulphate against

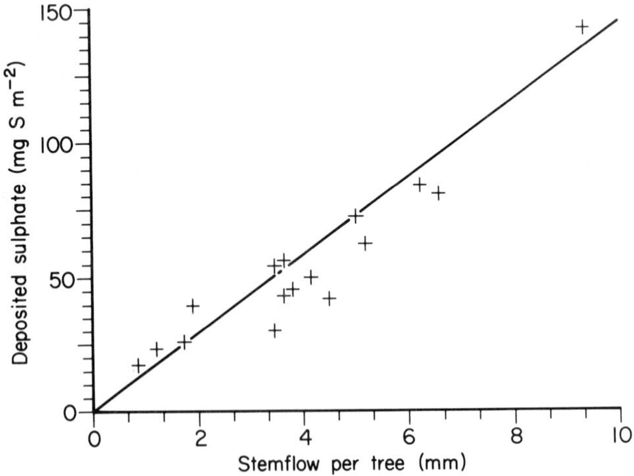


FIG. 10. The variation of average (2-weekly) deposited sulphate with stemflow amount for sixteen trees.

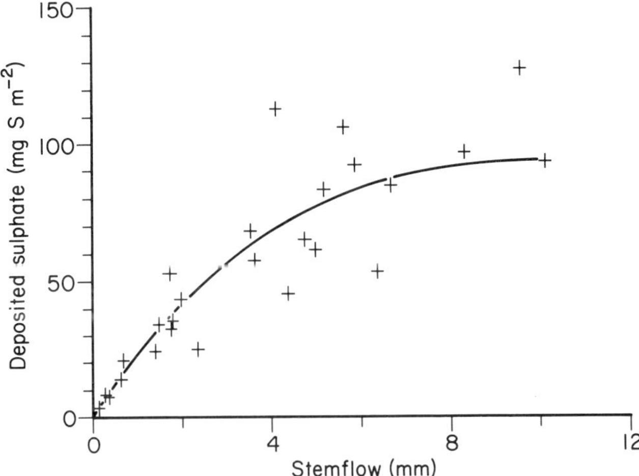


FIG. 11. The relationship between deposited sulphate in stemflow and stemflow quantity for twenty-six fortnightly collections. For explanation of the curve, see text.

stemflow amount (Fig. 11) showed no evidence of a non-zero intercept, again suggesting that 'wash-off' is not important. The annual average sulphate deposition in stemflow was 1·45 g S m$^{-2}$.

The annual values for sulphate deposition for each plot are tabulated separately in Table 3. The experimental design permits an estimate of the uncertainty due to spatial

TABLE 3. Sulphate deposition (g S m$^{-2}$) in rainfall, throughfall and stemflow, Devilla Forest, April 1979 – March 1980

| Plot | (i) | (ii) | (iii) | | Mean ± S.E. |
|---|---|---|---|---|---|
| Rain ($R$) | 1·34 | 1·43 | 1·34 | | 1·37 ± 0·04 |
| | | | | | |
| Plot | 1 | 2 | 3 | 4 | mean ± S.E. |
| Throughfall ($T$) | 3·60 | 2·57 | 2·93 | 3·20 | 3·08 ± 0·25 |
| Stemflow ($S$) | 1·30 | 1·93 | 1·16 | 1·42 | 1·45 ± 0·19 |
| Total = $T + S$ | 4·90 | 4·50 | 4·09 | 4·62 | 4·53 ± 0·20 |
| 'Excess' = $T + S - R$ | 3·53 | 3·14 | 2·72 | 3·26 | 3·16 ± 0·20 |

variability with a coefficient of variation for total sulphate deposition to the forest floor (in stemflow and throughfall) of around 8% of the average over four sites. Total annual sulphate deposition of 4·53 g S m$^{-2}$ exceeds deposition in rainfall by a factor of 3·3 and Fig. 12 shows how this 'excess' sulphate is distributed throughout the year. The pattern is closely related to rainfall distribution (Fig. 1).

### *Deposition of $SO_4^{2-}$ – containing particles and $SO_2$ gas*

#### *Particles*

Few data for sulphate particle concentrations are available for the period April 1979 – March 1980, but data are available for late 1978 and early 1979. 'Excess' deposited sulphate was found to be strongly correlated with air concentrations of particulate $SO_4^{2-}$ for this period.

The relationship obtained is at first sight surprising because if the excess deposited sulphate were due entirely to particle deposition then the rate of deposition required would be very large, equivalent to a deposition velocity ($v_g$) of 70 cm s$^{-1}$. Such deposition rates are only found for very large particles (>20 μm diameter) (Chamberlain 1975), and aerosol $SO_4^{2-}$ is found almost entirely in the particle size range 0·1–1·0 μm (Whitby 1978). Using the upper limits for deposition rates of 1 μm particles onto vegetation given by Chamberlain with measured air $SO_4^{2-}$ concentrations it is only possible to account for *c.* 5% of measured excess $SO_4^{2-}$ in throughfall and stemflow. We conclude that particle deposition cannot account for the observed 'excess' sulphate deposition. This example also serves to emphasize the spurious 'relationships which may be derived from correlation analysis. For the period studied (late 1978 – early 1979) particulate $SO_4^{2-}$ concentrations were also strongly correlated with rainfall amount which is the major factor contributing to sulphate deposition in throughfall and stemflow.

### *$SO_2$ gas*

Measurements of $SO_2$ deposition onto the forest in dry, early spring conditions showed typical daytime $v_g$ values of 0·3–0·7 cm s$^{-1}$. Stomatal uptake during the day accounted for > 70% of the net $SO_2$ flux in these springtime conditions. 'Winter' values (November–December) were similar except for the much shorter days, the 'night' values being observed soon after 15.00 GMT.

In order to estimate dry deposition of $SO_2$ for specific periods, it was assumed that at 'night' (defined as periods when short wavelength solar energy fluxes were < 100 W m$^{-2}$) all uptake was by external surfaces only. With large solar radiation values ('day') uptake

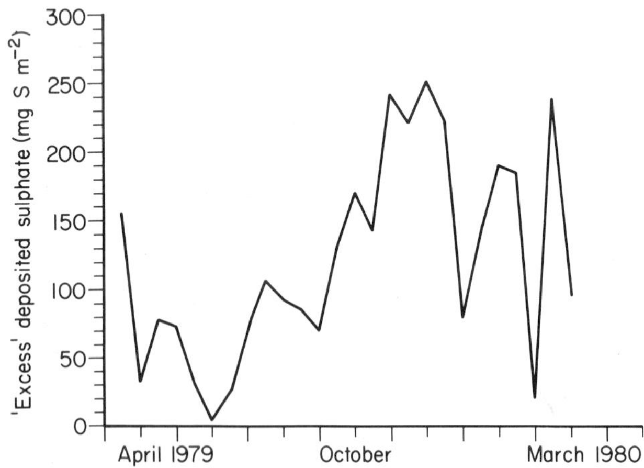


FIG. 12. Variation of 'excess' deposited sulphate (throughfall + stemflow – rainfall) for the period March 1979 – April 1980.

may be described by a stomatal component and a leaf-surface component acting in parallel, with the flux partitioned in the ratio 4:1 respectively (Fowler & Cape 1983).

Deposition of $SO_2$ onto the forest was calculated for each day during two 3-month periods, April–June and October–December 1979. The results, summarized in Table 4,

TABLE 4. Dry deposition (mg $m^{-2}$) of $SO_2$ onto Scots pine in Devilla Forest

| Month | Cumulative external deposit | Cumulative stomatal uptake |
|---|---|---|
| (a) Spring | | |
| 20 - 30 April | 25·1 | 44·4 |
| 1 - 31 May | 55·7 | 111·6 |
| 1 - 30 June | 47·1 | 125·0 |
| Total for 71 d | 127·9 | 281.0 |
| Total S deposition for 71 d | 64 | 140 |
| (b) Autumn | | |
| 4 - 31 October | 42.2 | 29.3 |
| 1 - 30 November | 57·2 | 26·0 |
| 1 - 27 December | 32·6 | 5·1 |
| Total for 84 d | 132.0 | 60.4 |
| Total S deposition for 84 d | 66 | 30 |

show that a total of 0·2 g S $m^{-2}$ were deposited during the spring period, of which almost 70% was due to stomatal uptake. For the winter period the total deposit was 0·1 g S $m^{-2}$ of which almost 70% was due to external uptake, the much smaller stomatal uptake being the consequence of short, 'dull' days.

## *Sources of excess sulphate in throughfall and stemflow*

The large increase in concentrations of $SO_4^{2-}$ in throughfall and stemflow over those in rainfall (by factors of 4 and 10 respectively) are due in part to the evaporation of surface water from the forest canopy. This evaporation (or 'interception') loss represents 36% of above-canopy rainfall and would, in the absence of no loss or gain in $SO_4^{2-}$ through other mechanisms, lead to an increase in $SO_4^{2-}$ concentration in throughfall and stemflow (combined), amounting on average to a factor of × 1·6. The observed increase in combined throughfall and stemflow concentration by a factor of × 5 shows a large net gain in sulphur as water proceeds downwards through the canopy.

The net gain in $SO_4^{2-}$ deposition to the forest floor over that in rain is due either to the removal of $SO_4^{2-}$ from external surfaces, 'washoff', or to leaching of $SO_4^{2-}$ from inter- and intra-cellular fluids, 'crown leaching', or a combination of the two.

From measurements of rainfall, stemflow and throughfall composition alone it is not possible to identify the contributions of wash-off and crown leaching independently. This has not deterred a number of workers from attempting to quantify crown leaching and wash-off from such measurements. Mayer & Ulrich (1974) used arguments based on the differences between summer and winter in measurements of throughfall and stemflow beneath a beech forest. The need to measure more variables has been appreciated recently, and the use of collectors which capture more small rain and cloud droplets than rain collectors (as well as particulate $SO_4^{2-}$ and $SO_2$ by dry deposition) has provided some additional information. However, the assumption that dry deposition onto such collectors is directly proportional to dry deposition onto trees (Miller & Miller 1980) is not valid. It is therefore not possible to separate wash-off and crown leaching for sulphur because of the lack of information on the uptake of $SO_2$ and particle $SO_4^{2-}$ onto external surfaces and *via* stomata. Bache (1977), recognizing the problems, made independent measurements of $SO_2$ deposition onto a Scots pine/Corsican pine forest in addition to measurements of rainfall, throughfall and stemflow composition. The technique for $SO_2$ deposition measurements lacked the precision required for the study, but Bache concluded that for $SO_4^{2-}$, crown leaching was more important than wash-off. Horntvedt *et al.* (1980) also combined estimates of dry deposition with throughfall, stemflow and rainfall composition measurements and reached similar conclusions.

From the measurements reported here, there was just sufficient dry deposition of $SO_2$ gas (total stomatal plus external uptake) during the spring period to account for the extra $SO_4^{2-}$ reaching the forest floor in throughfall and stemflow. However, only a third of this was deposited onto external surfaces, and particle deposition to the external surfaces could not make up the difference. There must, therefore, have been a substantial transfer of $SO_4^{2-}$ from within needles to the water running over needle surfaces and collected as throughfall and stemflow. For the winter period the total estimated dry deposition of $SO_2$ gas could only account for 8% of the excess $SO_4^{2-}$ in throughfall and stemflow. It is clear, therefore, that crown leaching is a major component of the enrichment. Is the concept of wash-off useful to these studies? Horntvedt *et al.* (1980) and others have shown that when plants are exposed to $SO_2$ labelled with the radioactive tracer $^{35}S$, a substantial fraction of deposited sulphur may be removed by washing with deionized water (up to 40%), but that after 1–30 minutes this fraction falls to $< 5\%$ indicating that dry deposited $SO_2$ is rapidly 'fixed' in or on the leaf. It is also clear from the results presented here that the decrease in $SO_4^{2-}$ concentration with increasing throughfall amount is too small to be consistent with the simple concept of surface wash-off which would predict a rapid decrease in concentration as throughfall volume increased. It appears that exchange between 'the leaf' and surface water is the major source of $SO_4^{2-}$ in throughfall and stemflow.

The removal of $SO_4^{2-}$ in leaching may be treated as a first order process such that

$$\frac{dC}{dV} = -kC$$

where $C$ is the concentration of leachable $SO_4^{2-}$ in the canopy and $dC/dV$ is the rate of change of concentration as $V$ mm of water pass through the canopy. Then

$$\ln(C) = \ln(C_o) - k.V$$

and a plot of $\ln(C)$ against $V$ would be linear. Data in Fig. 13 suggest a value for $C_o$ of 320 ($\pm 30$) $\mu M$ $SO_4^{2-}$ which would be the average initial concentration of throughfall and the 'volume constant' (equivalent to a time constant in most decay processes, and equal to a time constant if rainfall rates were constant) $k = 0{\cdot}020$ ($\pm 0{\cdot}005$) per mm throughfall. This theoretical expression is shown in Fig. 9 and is seen to account for the gross features of

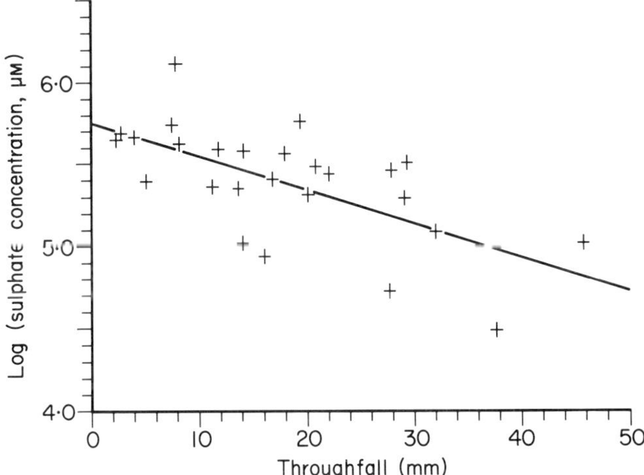

FIG. 13. Natural logarithm of throughfall $SO_4^{2-}$ concentration plotted against throughfall amount (see text and Fig. 9).

the throughfall measurements, in particular the zero intercept and curvature at large volumes of throughfall. The same approach may be applied to stemflow data (Table 1), yielding an initial concentration ($C_0$) of 790 (± 70) μM $SO_4^{2-}$ and a volume constant (k) of 0·099 (±0·019) per mm stemflow, for the data shown in Fig. 11. The volume constants for throughfall and stemflow may be expressed in terms of rainfall amount, (Table 2) giving 0·011 per mm rain in both cases. This indicates that the rate of depletion of the 'pool' of leachable $SO_4^{2-}$ is rather slow, and does not depend on the pathway of water through the forest.

While $SO_4^{2-}$ ions are removed from external surfaces as well as leached from within the vegetation the measurements presented here suggested that the entire free $SO_4^{2-}$ content of the foliage takes part in the exchange with water.

## CONCLUSIONS

The partitioning of rainwater inputs to forests into throughfall, stemflow and interception loss leads to changes in the chemical composition of the water reaching the forest floor. In the absence of exchange of soluble ions with the forest canopy the interception loss, which represents 36% of above-canopy rain, would lead to larger average concentrations (by a factor of 1·6) in water reaching the forest floor than in above-canopy rain, but no change in deposition. In practice, observed throughfall and stemflow concentrations of $SO_4^{2-}$ exceed those in rain by a factor of 5, deposition of $SO_4^{2-}$ to the forest floor being over three times the input in rain to the forest.

Direct measurements of dry deposition rates of $SO_2$ onto the forest canopy have enabled estimates to be made of the rate of deposition to external surfaces and separately to internal surfaces of the vegetation, the latter representing stomatal uptake. Air concentrations of $SO_2$ and particulate $SO_4^{2-}$ measured at the site then enabled estimates to be made of deposition of $SO_2$ over specific periods during which stemflow and throughfall measurements were made. A comparison of the measured and modelled fluxes show that leaching of $SO_4^{2-}$ from within the foliage is the major component of enrichment in throughfall and stemflow. Further, it appears from the measurements that the entire free $SO_4^{2-}$ content of the foliage takes part in the exchange on leaf surfaces, so that attempts to separate surface wash-off from crown leaching are unnecessary, and misleading. In areas polluted with $SO_2$ the additional in $SO_4^{2-}$ throughfall and stemflow may be interpreted as reflecting the larger available pool of $SO_4^{2-}$ for exchange. The results of exchange between leaf-surface fluids and the exchangeable $SO_4^{2-}$ pool may be treated as a first-order decay process, such that the concentration in throughfall and stemflow declines with rainfall amount. The 'decay constant' k = 0·01 per mm rainfall, is a volume constant, but for an average rainfall rate of 1 mm $h^{-1}$ is equivalent to a time constant of 1% $h^{-1}$. It would be of interest where data exist to apply similar analyses to those given above in order to investigate the effect of different atmospheric $SO_2$ concentrations on the initial concentrations ($C_0$), and volume constants (k) which characterize the throughfall and stemflow sulphate concentrations.

## ACKNOWLEDGMENTS

We wish to acknowledge the co-operation of the Forestry Commission in providing a suitable site for this study. Miss M. McPherson (ITE Banchory) was responsible for much of the sampling and site maintenance. All chemical analyses were performed by the Analytical Chemistry Service Section at ITE Merlewood.

## REFERENCES

**Bache, D.H. (1977).** Sulphur dioxide uptake and the leaching of sulphates from a pine forest. *Journal of Applied Ecology,* **14**, 881–895.

**Baker, J., Hocking, D. & Nyborg, M. (1976).** Acidity of open and intercepted precipitation in forests and effects on forest soils in Alberta, Canada. *Proceedings of the First International Symposium on Acid Precipitation and the Forest Ecosystem.* (Ed. by L.S. Dochinger & T.A. Seliga), pp. 779–790. USDA Forest Service Technical Report NE-23.

**Chamberlain, A.C. (1975).** The movement of particles in plant communities. *Vegetation and the Atmosphere,* Vol. I (Ed. by J.L. Monteith), pp. 115–201. Academic Press, London.

**Fowler, D. & Cape, J.N. (1983).** Dry deposition of $SO_2$ onto a Scots pine forest. *Precipitation Scavenging, Dry Deposition and Resuspension* (Ed. by H.R. Pruppacher, R.E. Semonin & W.G.N. Slinn), pp. 763–774. Elsevier, New York.

**Galloway, J.N. & Parker, G.G. (1980).** Difficulties in measuring wet and dry deposition in forest canopies and soil surfaces. *Effects of Acid Precipitation on Terrestrial Ecosystems* (Ed. by T.C. Hutchison & M. Havas), pp. 57–68. Plenum Press, New York.

**Horntvedt, R., Dollard, G.J. & Joranger, E. (1980).** Atmosphere–vegetation interactions. *Ecological Impact of Acid Precipitation* (Ed. by E. Drabløs & A. Tollan), pp. 192–193. SNSF, Oslo.

**Mayer, R. & Ulrich, B. (1974).** Conclusions on the filtering action of forests from ecosystem analysis. *Oecologia Plantarum,* **9**, 157–168.

**Miller, H.G. & Miller, J.D. (1980).** Collection and retention of atmospheric pollutants by vegetation. *Ecological Impact of Acid Precipitation* (Ed. by D. Drabløs & A. Tollan), pp. 33–40. SNSF, Oslo.

**Nicholson, I.A., Fowler, D., Paterson, I.S., Cape, J.N. & Kinnaird, J.W. (1980).** Continuous monitoring of airborne pollutants. *Ecological Impact of Acid Precipitation* (Ed. by D. Drabløs & A. Tollan). pp 144–145.

SNSF, Oslo.

**Nipher, F.E. (1878).** On the determination of the true rainfall in elevated gauges. *American Association for the Advancement of Science,* **27**, 105–108.

**Parker, G.G. (1983).** Throughfall and stemflow in the forest nutrient cycle. *Advances in Ecological Research,* **13**, 58–133.

**Parker, G.G., Lindberg, S.E. & Kelly, J.M. (1980).** Atmosphere–canopy interactions of sulfur in the southeastern United States. *Atmospheric Sulfur Deposition* (Ed. by D.S. Shriner, C.R. Richmond & S.E. Lindberg), pp. 477–494. Ann Arbor Science, Michigan.

**Rutter, A.J. (1975).** The hydrological cycle in vegetation. *Vegetation and the Atmosphere* (Ed. by J.L. Monteith), pp. 111–154. Academic Press, London.

**Whitby, K.T. (1978).** The physical characteristics of sulphur aerosols. *Atmospheric Environment,* **12**, 135–160.

# Transformations in rainwater chemistry on passing through forested ecosystems

H. G. MILLER, J. D. MILLER* AND JEAN M. COOPER
*Department of Forestry, University of Aberdeen, Aberdeen, AB9 2UU and *Macaulay Institute for Soil Research, Aberdeen, AB9 2QJ*

## SUMMARY

**1** Rainwater generally loses hydrogen ions at the leaf surface, but the extent of this change is found to vary with season, species and age of tree, there also being an interaction with acid input such that the seasonal effect is only found at the more polluted sites.
**2** The ground vegetation beneath the tree then further reduces rainwater acidity, but much of this is regained on passing through the humus layer.
**3** Movement of acidity to greater soil depths is in part related to the amount of sulphate introduced in the rain but, again, effects of both species and tree age have been detected. Nitrate, however, is retained in most ecosystems during the summer months but may pass through with the drainage water during winter.

## INTRODUCTION

Rainwater chemistry becomes much modified as the water passes, in turn, through the tree canopy, the ground vegetation layer, the soil humus and then the mineral soil. It can be surmised that the nature of these changes will depend not only on the soil and vegetation type but also the chemistry of the received rainfall and the concentrations of gases and salts, natural or anthropogenic, in the local atmosphere. Certain of the changes in the soil are well documented, although there is still uncertainty as to the effect of an increasing load of the pollutant-derived sulphate and nitrate anions in the received rainwater. Both these anions are potentially very mobile in soil unless retained by biological uptake, in the case of nitrate, or through anion absorption, in the case of sulphate (Johnson 1984). If not retained, sulphate and nitrate are lost in drainage water accompanied either by base cations or, more damagingly to stream life, by hydrogen and aluminium.

Prior to entering the soil, however, much of the rainwater will have passed over vegetation surfaces where a complex of reactions may occur leading to both the loss and gain of certain elements. Some elements — in particular nitrogen as either ammonium or nitrate — may be taken up by foliar absorption, an uptake that must be matched either by uptake of an accompanying ion of opposite charge or by a compensating efflux of ions from the leaf. Cations may also be exchanged at the leaf surface, usually, but not invariably, leading to a loss of hydrogen ions from rainwater and a gain of base cations, especially potassium. As a result, many vegetation canopies in areas of low atmospheric pollution will raise the pH of the rainwater passing through. This loss of elements from within the plant, either by direct efflux or cation-exchange, is crown leaching. Measurement of crown-leaching, however, is complimented by the fact that there is also wash-off by the rain of substances previously deposited from the atmosphere on leaf surfaces, a process that has been termed either the filter effect of the vegetation (Mayer & Ulrich 1974) or interception deposition (Ulrich 1983). In areas in which the atmosphere is heavily polluted with acid-forming gases or acidic particles the interception of these by vegetation could be expected to acidify subsequent throughfall. In addition there will be deposition of naturally-derived alkaline-dust and sea-spray that would tend to reduce acidity.

The chemistry of throughfall and stemflow collected in a forest is the net result of all these processes, the relative importance of each of which undoubtedly varies between elements. It is extremely difficult to interpret changes in terms of the relative contributions of the different processes involved. At one extreme some authorities would suggest that crown leaching is unimportant except for potassium and manganese (e.g. Ulrich 1983) whereas others maintain that the wide range of elements, including sulphate, are lost from the plant by this process (Tukey, Tukey & Whittwer 1958; Parker 1983; Miller 1984). Much more work is required before there can be any resolution of this difference of opinion. Recently, however, there has been a great increase in the amount of experimental information available and some of this is presented here to illustrate aspects of the problem.

## EFFECTS OF TREE CANOPY ON RAINWATER CHEMISTRY

### *Geographical effects*

The chemistry of rainwater above and below a canopy of Sitka spruce *(Picea sitchensis* (Bong.)Carr.) at Fetteresso forest (Grampian Region), in a moderately polluted region in eastern Scotland, is shown in Table 1. The presence of an intercepting surface, whether

TABLE 1. Weighted mean concentration of cations and anions ($\mu$eq $l^{-1}$) in bulk precipitation (i.e. rainwater collected above the canopy), throughfall, stemflow and in water collected beneath an artifical intercepting surface, comprised of a cylinder of polyethylene coated wire mesh (filter gauge). Samples were collected every 2 weeks over 5 years in a Sitka spruce (initially 25-years-old) stand at Fetteresso forest in eastern Scotland

| Ion | Bulk precipitation | Throughfall | Stemflow | Water in 'filter gauge' |
|---|---|---|---|---|
| $NH_4$ | 18 | 88 | 81 | 50 |
| K | 6 | 110 | 145 | 21 |
| Ca | 41 | 173 | 187 | 103 |
| Mg | 40 | 105 | 171 | 200 |
| Na | 190 | 406 | 684 | 937 |
| H | 72 | 77 | 195 | 119 |
| $NO_3$ | 33 | 62 | 78 | 100 |
| $PO_4$ | 0·7 | 9 | 7 | 0·8 |
| $SO_4$ | 169 | 489 | 713 | 392 |
| Cl | 166 | 397 | 718 | 942 |
| mm water per year | 879 | 517 | 86 | 2254 |

foliage or artificial, increases the concentration of all elements, including hydrogen ions. The fact that there is also an increase in the concentration of hydrogen ions in the water collected beneath an inert intercepting surface (filter gauge) suggests that at this site the decline in throughfall pH may be associated with the interception of acidic substance from the atmosphere. However, at none of the other five sites studied in this series (Table 2) did

TABLE 2. Weighted mean pH of bulk deposition, throughfall, stemflow and water from the filter gauge measured over 5 years beneath Sitka spruce at six sites around Scotland, all crops being aged 25 years at the start of the measurement period excepting Kilmichael which was 3 years older

| | Location | pH values | | | |
|---|---|---|---|---|---|
| | | Bulk precipitation | Throughfall | Stemflow | Filter gauge water |
| Leanachan | 56°52′N 4°56′W | 4·54 | 5·05 | 4·56 | 4·81 |
| Kilmichael | 56° 4′N 5°24′W | 4·30 | 4·93 | 3·68 | 4·50 |
| Strathyre | 56°20′N 4°20′W | 4·46 | 4·59 | 4·00 | 4·40 |
| Elibank | 55°37′n 3° 0′W | 4·20 | 4·35 | 3·67 | 4·21 |
| Kershope | 55° 8′N 2°42′W | 4·20 | 4·23 | 3·64 | 4·19 |
| Fetteresso | 56°58′N 2°20′W | 4·13 | 4·11 | 3·68 | 3·92 |

throughfall show a lower mean pH than did bulk precipitation, and indeed at only one other site, Strathyre, was there any suggestion that water in the filter gauge was more acid than bulk precipitation. At a further two sites, Elibank and Kershope, the filter gauge had little effect on water pH whereas at Kilmichael and Leanachan, both on the Scottish west coast, it appeared to markedly increase the pH. At these latter two sites the pH of throughfall water was considerably elevated and at Leanachan even the pH of stemflow water was marginally higher than that of bulk precipitation. As the filter gauges collected considerably more water than did the bulk precipitation gauges, changes in acidity (Table 2) or levels of other elements (Table 1) are not likely to be due to a concentration effect such as that which might result during evaporation of potential throughfall water.

Even at Fetteresso, where throughfall acidity was greater than that of bulk precipitation, the very much smaller volume of throughfall meant that the total transfer of acidity to the soil surface was less than that received in bulk precipitation (Table 3).

TABLE 3. Mean annual deposition of hydrogen ions (eq. $H^+$ $ha^{-1}$ $y^{-1}$) in bulk precipitation and in throughfall plus stemflow measured over 5 years beneath Sitka spruce at six sites around Scotland. Also shown in the proportion of the acidity reaching the forest floor that is in stemflow

| Forest | eq $H^+$ $ha^{-1}$ $y^{-1}$ in bulk precipitation | eq $H^+$ $ha^{-1}$ $y^{-1}$ in throughfall plus stemflow | Proportion of net acidity reaching forest floor in stemflow |
|---|---|---|---|
| Leanachan | 430 | 140 | 0·43 |
| Kilmichael | 970 | 350 | 0·63 |
| Strathyre | 480 | 390 | 0·44 |
| Elibank | 740 | 270 | 0·26 |
| Kershope | 750 | 590 | 0·36 |
| Fetteresso | 640 | 530 | 0·28 |

Indeed, at all the six sites studied there was a reduction, sometimes quite marked, in the flux of acidity reaching the surface due to the presence of the spruce canopy. However, it should be borne in mind that much of the acidity is transferred in stemflow which accounts for a quarter to two-thirds of the flux of hydrogen ions (Table 3) but concentrates this input over only about 0·5% of the area of the soil surface.

Tukey *et al.* (1958) demonstrated that very little chloride leaches from leaves. Any gain of this element beneath trees, therefore, would be due almost entirely to wash-off of interception deposition, the trees acting as inert filters in the same way as the filter gauge. In this case the ratio of amount of chloride beneath trees to the amount in the filter gauge should be a measure of the relative catching efficiency of the trees. Application of this ratio to the amounts of other elements collected by the filter gauge might then provide estimates of the amounts of these elements that enter the forest as a result of interception deposition. Such calculations for two sites are shown in Table 4. The assumptions involved cannot be

TABLE 4. Quantities of elements (kg $ha^{-1}y^{-1}$) predicted in throughfall (TF) plus stemflow (SF) at two sites by multiplying the amounts in the filter gauge by the ratio of quantity of chloride in throughfall plus stemflow to that in the filter gauge. Also shown are the measured amounts (kg $ha^{-1}y^{-1}$) in bulk precipitation and in throughfall plus stemflow. A large discrepancy between predicted and actual amounts in throughfall plus stemflow indicates a predominance of crown leaching whereas the lack of such a difference indicates a predominance of interception deposition

| | Leanachan | | | Fetteresso | | |
|---|---|---|---|---|---|---|
| Ion | TF + SF | | Measured in bulk | TF + SF | | Measured in bulk |
| | Pred. | Meas. | precipitation | Pred. | Meas. | precipitation |
| $NH_4$ | 0·45 | 2·00 | 0·50 | 2·3 | 7·7 | 2·3 |
| K | 2·60 | 22·90 | 4·40 | 2·8 | 27·1 | 2·1 |
| Ca | 8·60 | 23·30 | 9·60 | 6·7 | 22·0 | 7·3 |
| Mg | 7·20 | 8·40 | 5·30 | 8·0 | 9·2 | 4·2 |
| Na | 80·00 | 74·00 | 60·00 | 70·0 | 70·0 | 38·0 |
| $NO_3$ | 0·83 | 1·52 | 0·66 | 4·6 | 5·8 | 4·1 |

entirely correct, for on this basis the estimated amounts of calcium and potassium received by the trees through bulk deposition and interception are less than that actually measured in bulk deposition. Presumably the relative catching efficiency must vary between elements. However, the calculations do suggest that any additional ammonium, potassium and calcium beneath trees is largely derived from crown-leaching, the extra magnesium and sodium is primarily from interception deposition, with nitrate coming from both sources. This pattern of interception of magnesium, sodium and nitrate at the various sites is illustrated in Table 5. Interception of magnesium and sodium is greatest at Leanachan, Kilmichael and Fetteresso, all of which are within a few miles of the sea. Nitrate interception, however increases fairly uniformly on the west to east gradient, presumably reflecting the changing pollution environment across the country.

In Table 6 are shown similar measurements for four sites at Glen Tanar, in Grampian

TABLE 5. Weighted mean concentrations ($\mu$eq $l^{-1}$) of magnesium, sodium and nitrate in water collected over 5 years in replicated ($\times$ 4) filter gauges at each of six sites arranged on an approximately west (Leanachan and Kilmichael) to east transect across Scotland

| Forest | Mg | Na | $NO_3$ |
|---|---|---|---|
| Leanachan | 110 | 648 | 11 |
| Kilmichael | 125 | 521 | 23 |
| Strathyre | 42 | 197 | 15 |
| Elibank | 94 | 471 | 40 |
| Kershope | 99 | 385 | 60 |
| Fetteresso | 200 | 937 | 100 |

TABLE 6. Weighted mean concentration ($\mu$eq $l^{-1}$) of magnesium, sodium, nitrate and hydrogen, together with coefficients of variations, in water collected over 2 years in replicated ($\times$ 4) filter gauges and bulk precipitation gauges at four sites down an altitudinal transect at Glen Tanar. Altitude declines from sites 1 (420 m) to 4 (150 m); the gauges at sites 2 and 4 were somewhat protected by adjacent trees; site 1 is an exposed heather moor whereas site three is a large clear-felled area

| | | Mg | | H | | $NO_3$ | | $SO_4$ | |
|---|---|---|---|---|---|---|---|---|---|
| Gauge | Site | Mean | cv(%) | Mean | cv(%) | Mean | cv(%) | Mean | cv(%) |
| | 1 | 262 | 9 | 46 | 31 | 20 | 6 | 404 | 4 |
| Filter | 2 | 84 | 8 | 45 | 24 | 65 | 14 | 167 | 18 |
| | 3 | 151 | 21 | 22 | 31 | 75 | 19 | 212 | 14 |
| | 4 | 54 | 16 | 22 | 48 | NA | - | 118 | 9 |
| | 1 | 30 | 10 | 34 | 43 | 25 | 28 | 91 | 24 |
| Bulk precipitation | 2 | 30 | 24 | 22 | 58 | 12 | 32 | 66 | 28 |
| | 3 | 32 | 32 | 48 | 45 | NA | - | NA | - |
| | 4 | 22 | 9 | 41 | 67 | 10 | 22 | 107 | 55 |

NA, no result available.

Region, the Tanar being a tributary of the River Dee on its south side. Identical filter gauges were installed within a few kilometres of each other but at different altitudes, there being four gauges at each location. In this case the measurements are meaned over 2 years, rather than the 5 years for the study reported in Table 5. The Glen Tanar site (57°4′N, 2°46′W) is within the same general region as Fetteresso (cf. Table 5) but rather more distant from the sea. The results in Table 6 show a surprisingly large variation in element concentration in the filter gauge water despite the relative closeness of the four sample sites, a proximity that is reflected in fairly uniform concentrations in bulk precipitation collected at the different sites. Although the concentration of most elements is greatest at the upper slope sites, nitrate exhibits the reverse trend. Deposition of sulphate to the filter gauge parallels that of magnesium rather than that for nitrate, suggesting rather different atmospheric sources of the two anions. Thus, interception deposition can be expected to vary rather more markedly with geographical location than does the input in bulk

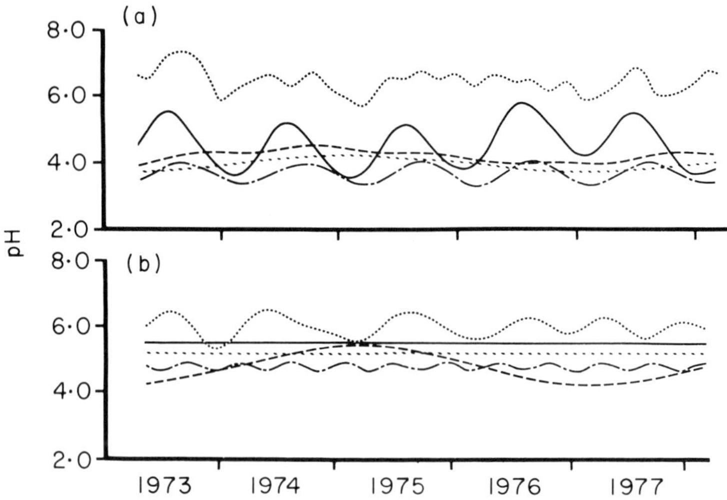


FIG. 1. Variations at Fetteresso forest on the Scottish east coast (a), and at Leanachan forest, on the relatively unpolluted west coast (b) in pH of bulk deposition (----), water collected in the 'filter gauge' (......), throughfall (———), stemflow (-·-·-·-) and streamwater (········) measured over 5 years in crops of Sitka spruce (*Picea sitchensis* (Bong.) Carr.) (Miller *et al.*, in press).

deposition. Furthermore, even within one pollution zone interception deposition may vary markedly with altitude.

### *Seasonal effects*

The prolonged period of collection at the six spruce sites already referred to in Tables 2, 3 and 5 enabled time-series analysis to search for seasonal effects (Miller *et al.*, in press). The pH results from two extreme sites, the relatively polluted east coast site at Fetteresso and the relatively unpolluted west coast site at Leanachan forest, are shown in Fig. 1. At none of the six sites was any seasonal trend detected in the acidity of either bulk precipitation or, rather surprisingly, the water collected in the filter gauges. Nor was there any annual oscillation in the pH of the throughfall or stemflow at the unpolluted sites (Fig. 1) but on moving east an oscillation developed with lowest pH values recurring during the second half of winter. Unexpectedly, the oscillation in stemflow pH appears to be slightly out of phase with that in throughfall pH. In some sites annual oscillations were also found in streamwater pH but this is thought to reflect the expected summer–winter variations in the proportion of base flow.

The complete absence of any annual variation in the acidity of filter-gauge water was somewhat surprising and accordingly the data from the Glen Tanar study were analysed separately by winter and summer months. In this case regular winter–summer differences were detected for almost all elements in most types of waters sampled, excepting acidity for which the pattern was not sufficiently consistent between sites to convincingly suggest a seasonal effect (Table 7).

TABLE 7. Variations between winter and summer in concentrations ($\mu$eq $1^{-1}$) of magnesium and hydrogen ions in bulk precipitation, filter-gauge water and throughfall beneath old Scots pine, Sitka spruce and birch at the Glen Tanar study site

| Element | Season | Bulk precipitation | | Filter-gauge water | | Throughfall beneath | | |
|---|---|---|---|---|---|---|---|---|
| | | Site 1 | Site 4 | Site 1 | Site 4 | Pine | Spruce | Birch |
| Mg | Winter | 66 | 49 | 414 | 144 | 239 | 187 | 77 |
| | Summer | 13 | 12 | 134 | 26 | 38 | 39 | 33 |
| | Winter | 51 | 25 | 343 | 79 | 142 | 103 | 53 |
| | Summer | 24 | 18 | 206 | 36 | 65 | 54 | 48 |
| H | Winter | 45 | 53 | 105 | 60 | 184 | 13 | 50 |
| | Summer | 26 | 34 | 36 | 41 | 128 | 14 | 1 |
| | Winter | 43 | 12 | 71 | 16 | 139 | 24 | 14 |
| | Summer | 34 | 24 | 8 | 5 | 17 | 32 | 9 |

## *Species effects*

The study at Glen Tanar, referred to earlier, compared rainwater chemistry beneath old (aged 110 years) and young (aged 45 years) Scots pine *(Pinus sylvestris L.),* Sitka spruce, larch *(Larix kaempferi* (Lamb) Carr.) and birch *(Betula pendula* Roth.) in three replicated plots of each vegetation type. The different tree species had rather a similar effect on the concentration of most elements other than hydrogen, for which some marked and rather puzzling differences were recorded (Table 8). Most species, including the young Scots

TABLE 8. Quantities of elements (kg $ha^{-1}$ $y^{-1}$) in bulk precipitation and tree throughfall plus stemflow (TF + SF) beneath various species at Glen Tanar. The values for bulk precipitation from site 3 are used as the input values not only for the young pine but also for the larch and spruce

| Element | | Site 1 Heather moor | Site 2 Old pine | Site 3 Young pine | Larch | Spruce | Site 4 Birch |
|---|---|---|---|---|---|---|---|
| Mg | Bulk precip. | 2·4 | 2·9 | 2·9 | | | 2·2 |
| | TF + SF | — | 6·0 | 4·0 | 5·9 | 4·7 | 3·9 |
| H | Bulk precip. | 0·23 | 0·18 | 0·37 | — | — | 0·33 |
| | TF + SF | — | 0·51 | 0·15 | 0·18 | 0·09 | 0·11 |
| $NO_3-N$ | Bulk precip. | 2·3 | 1·4 | 1·2 | — | — | 1·2 |
| | TF + SF | — | 2·7 | 2·6 | 2·9 | 3·4 | 4·7 |
| $SO_4-S$ | Bulk precip. | 10 | 9 | 10 | — | — | 14 |
| | TF + SF | — | 27 | 15 | 17 | 15 | 21 |

pine, reduced the flux of acidity beneath the trees. There was, however, a considerable difference between the two ages of pine; the older pine increased the flux of acidity whereas the younger reduced it.

Miller (1984) collated the available data on throughfall acidity from the literature and concluded that reductions in pH are largely limited to older (more than *c*. 60 years) conifers, with younger conifers and broadleaved species of all ages generally increasing pH unless the stand is situated in a very polluted environment. This hypothesis is broadly supported by research at the Institute of Terrestrial Ecology (M. Hornung, pers. comm.) which showed that Sitka spruce at a certain stage switches from reducing rainwater acidity (as has been observed in virtually all the studies reported here) to increasing it; although with this fast-growing species the change occurs between ages 25 and 30 rather than 60 years. Quite why there should be such a radical change in influence on throughfall chemistry is not clear. Suggestions based on increased interception area seem inadequate so it may relate to changes in the physiology of the tree as it ages.

### *Ground vegetation*

In the Glen Tanar study, water was sampled not only from beneath the tree but also beneath the humus layer (surface mineral soil in the case of the birch stand) using zero-tension lysimeters, and beneath the rotting zone (usually marked by the surface of the indurated soil horizon) using ceramic cup lysimeters to which a tension was applied. In addition, during the summer months, small troughs were inserted below the ground vegetation to collect ground vegetation throughfall. In order to include the latter values in any comparison, only data for the summer of 1981 are presented (Table 9), this being the first summer in the comparable input data in Table 7.

Ground vegetation appears to be remarkably effective at removing acidity from the received tree throughfall, although the difficulty of collecting this fraction of the rainwater uncontaminated by dead vegetation or insects should counsel caution over uncritical acceptance of the results (the coefficients of variation, however, between the ground vegetation throughfall results from three instrumental plots at each site where comparable to those for tree throughfall). As would be expected, considerable acidity was regenerated in the humus layer only to be neutralized again in the mineral soil. A puzzling aspect is the continuing high level of acidity at depth in the soil beneath the old pine, a feature that was found at all periods (mean hydrogen ion concentration over 2 years for the soil lysimeter for the old pine was 101 $\mu$eq $l^{-1}$ as against, for example 8 $\mu$eq $l^{-1}$ for the comparable samples from beneath the spruce).

Magnesium concentrations varied rather little, suggesting only limited leaching of this element. Nitrate, however, is clearly retained in all except the larch and spruce sites, although during the winter months no such retention was detected in either of the pine sites, the effect being limited to summer. Nitrate retention was consistently most marked beneath the birch despite the very much better soil at this site. There was little indication of any net retention of sulphate and this anion appears to be fairly mobile in all the soils covered by the experimental plots.

It should be borne in mind when considering the soil water values that although the

TABLE 9. Weighted mean concentrations ($\mu$eq $l^{-1}$) in water passing through ground vegetation and soil over the summer and autumn of 1981 (28 May–11 Nov) for the various sites at Glen Tanar. Tree throughfall data (bulk precipitation in the case of the heather moor) are given as input values

| Element | | Site 1 Heather moor | Site 2 Old pine | Site 3 Young pine | Larch | Spruce | Site 4 Birch |
|---|---|---|---|---|---|---|---|
| Mg | Tree throughfall | (13) | 38 | 25 | 48 | 39 | 33 |
| | Ground veg. TF* | 25 | NA | 52 | 47 | — | 52 |
| | Humus leachate | 34 | 72 | 43 | 70 | 73 | 93 |
| | Soil water† | 59 | 100 | 57 | 43 | 98 | 53 |
| H | Tree throughfall | (26) | 128 | 30 | 41 | 13 | 1 |
| | Ground veg. TF* | 13 | <1 | 1 | 9 | — | 1 |
| | Humus leachate | 85 | 150 | 149 | 10 | 49 | 3 |
| | Soil water† | <1 | 101 | 45 | 1 | 4 | <1 |
| $NO_3-N$ | Tree throughfall | (30) | 14 | 28 | 12 | 23 | 22 |
| | Ground veg. TF* | 20 | 54 | 9 | 36 | — | 133 |
| | Humus leachate | 2 | 4 | 9 | 30 | 24 | 58 |
| | Soil water† | 1 | 7 | 5 | 33 | 34 | 4 |
| $SO_4-S$ | Tree throughfall | (35) | 327 | 136 | 95 | 83 | 120 |
| | Ground veg. TF* | 331 | 110 | 98 | 274 | — | 120 |
| | Humus leachate | 100 | 118 | 134 | 44 | 113 | 135 |
| | Soil water† | 171 | 444 | 188 | 123 | 180 | 147 |

*Throughfall sampled beneath the ground vegetation, no ground vegetation present beneath the spruce.
†Soil water sampled with a suction lysimeter from beneath the rooting zone.
NA, No result available.

plots are identified by plant species the soils themselves do vary. The differences quoted, therefore, may reflect the soil as much or rather more than, the species carried.

## DISCUSSION

By being able to draw on results both from plots of the same species of tree at a wide range of locations and from plots of different tree species in close proximity, it becomes clear that great care should be taken in transferring results from a single-site study unless the underlying mechanisms are understood.

Trees can much reduce the acidity of water passing through their crowns, a process that is probably continued by the ground vegetation, where present. This reduction of acidity is at the expense of a loss of base cations, particularly potassium (Miller 1984), from the tree. These base cations have to be obtained by root uptake from the soil, an uptake process that has to be matched by efflux of hydrogen ions from root to soil. Accelerated crown leaching as a response to rainwater acidity, therefore, still ends up with the acid load being placed on the soil. Should the foliage-derived base cations in throughfall then come into contact with the relevant soil exchange surface, it is feasible that the root-derived hydrogen ions could be displaced by these base cations returning to the exchange surfaces from which they were originally derived. In this event there is a closed cycle of base cations while the sulphuric acid introduced from the atmosphere is reformed to pass on to deeper soil horizons. In practice not all the base cations will be exchanged. Thus, the mobile sulphate anions introduced from the atmosphere pass through the immediate surface soil accompanied by base cations, derived from the vegetation, and by hydrogen ions, either those that have avoided exchange reactions or those that have been displaced from soil surfaces.

Within the soil the hydrogen ions, from whatever source, may be subsequently exchanged for base cations, buffered by dissolution of aluminium or neutralized by reaction with soil minerals. However, it is clear from the Glen Tanar study that in certain soils, or soil-vegetation combinations, only a very limited reduction in water acidity may occur but the reasons for this are not clear.

Similarly, it still remains unclear why there should be a change with tree age in the nature of the reactions that occur at leaf surfaces.

## ACKNOWLEDGMENTS

The six Sitka spruce sites in the first study described were made available by the appropriate Conservators of the Forestry Commission and valuable assistance was provided by both local management and research staff on the Forestry Commission. The study at Glen Tanar was by kind permission of the owner and the sites were located with the help of I. Ross. Sample collection and analysis in this latter study were by W.R. Johnston and C.F. Flower.

## REFERENCES

**Johnson, D.W. (1984)** Sulphur cycling in forests. *Biogeochemistry,* **1**, 29–43.

**Mayer, R. & Ulrich, B. (1974).** Conclusions on the filtering actions of forests from ecosystems analysis. *Oecologia Plantarum.*, **9**, 157–168.

**Miller, H.G. (1984).** Deposition-plant-soil interactions. *Philosophical Transactions of the Royal Society of London.*, **B305**, 339–352.

**Miller, H.G., Miller, J.D. Cooper, J.M. & Flower, C.F. (in press).** Changes in acid flux through forest canopies. *Acid Rain and Forest Resources.* Proceedings of the Quebec Conference, 1983.

**Parker, G.G. (1983).** Throughfall and stemflow in the forest nutrient cycle. *Advances in Ecological Research,* **13**, 57–133.

**Turkey, H.G. Jr, Tukey, H.G. & Wittwer, S.H. (1958).** Loss of nutrients by foliar leaching determined by radioisotopes. *Proceedings of the American Horticultural Society,* **71**, 496–506.

**Ulrich, B. (1983).** Interaction of forest canopies with atmospheric constituents : $SO_2$, alkali and earth alkali cations and chloride. *Effects of Accumulation of Air Pollutants in Forest Ecosystems* (Ed. by B. Ulrich & J. Pankrath), pp. 33–45. D. Reidel, Holland.

# The microbial sulphur cycle in atmospheric-polluted soils

M. WAINWRIGHT AND WENDY NEVELL
*Department of Microbiology, University of Sheffield, Sheffield, S10 2TN*

## SUMMARY

**1** Micro-organisms play the major role in the cycling of sulphur in soils, being responsible for (a) the mineralization of organic sulphur to sulphate; (b) the oxidation of reduced forms of inorganic sulphur to sulphate; (c) the anaerobic reduction of sulphate to sulphides and (d) the immobilization of sulphate as organic sulphur.

**2** The importance of these reactions in atmospheric-polluted soils relates to the mobility of sulphate, since free sulphate affects the leaching of cations from soils, as well as influencing $Al^{3+}$ solubility. Sulphur oxidation reactions also produce sulphuric acid.

**3** Sulphur mineralization and oxidation reactions are likely to increase sulphate mobility in soils, while sulphate reduction and immobilization reactions tend to decrease the mobility of the ion.

**4** The microbial cycling of sulphur in atmospheric-polluted soils is discussed in this paper with particular attention to recent work from this laboratory.

## INTRODUCTION

Sulphur undergoes a number of transformations in soils, which together form the soil sulphur cycle (Fig. 1). These reactions are largely mediated by micro-organisms (Brown 1982), although physical-chemical reactions such as sulphate adsorption are also

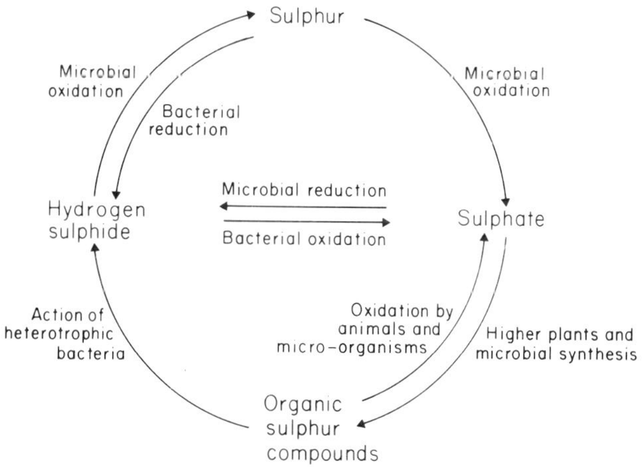


FIG. 1. The sulphur cycle in soils.

important. Since sulphate is the major form of the element available to plants it is the key ion in this process, and when mobile it can influence both cation leaching from soils and the transport of $Al^{3+}$ to surface waters (Johnson & Reuss 1984). The soil biological and chemical reactions which influence sulphate mobility will therefore have an important bearing upon soil acidity in atmospheric-polluted soils (Wainwright 1982; Lettl 1984; Wainwright & Nevell 1985).

Microbial mineralization and sulphur-oxidation reactions will tend to increase sulphate mobility as well as contributing $H^+$ ions (Abrahamsen & Stuanes 1983), while reduction, adsorption and immobilization will generally 'fix' the ion, thereby decreasing its concentration and reducing cation leaching. Such 'fixation' is likely only to be temporary however, as sulphate can be remobilized as the result of sulphur mineralization and oxidation reactions, as well as by physical-chemical desorption. It is therefore important that a full knowledge is gained of the reactions involved in the sulphur cycle in soils, and in particular of how these transformations operate in atmospheric-polluted soils.

Soils subjected to atmospheric pollution receive sulphur from the atmosphere as $SO_4^{2-}$ as well as $HSO_3^-$ and $SO_2$. As a result, vegetation and soils at polluted sites invariably have higher sulphur contents than similar, but unpolluted soils (Moss 1975). On litterfall, organic sulphur present in polluted vegetation will augment the sulphur content of these soils. Particulate pollution deposits, largely consisting of soot, may also provide a source of reduced sulphur which, on entering soils, is likely to be oxidized by micro-organisms to sulphate (Killham & Wainwright 1981).

The role of micro-organisms in the cycling of sulphur in atmospheric-polluted soils is reviewed here with emphasis on our own recent studies. Throughout, reference is made to two sites, the details of which are given in Table 1.

TABLE 1. Details of the two sites used.

| Characteristic | Site 1 (Coking works 500 m downwind) | Site 2 (Relatively unpolluted) |
|---|---|---|
| $SO_2$ | As high as 150–200 $\mu m^{-3}$ | 50–80 $\mu m^{-3}$ |
| Soot deposits | abundant | few |
| Soil: | | |
| Total C (%) | 14·6 | 9·8 |
| Total N (%) | 0·93 | 0·83 |
| pH | 3·3 | 4·9 |

Temperature rainfall, days on which snow lay, soil type, and altitude are essentially similar.

Two subcycles can be recognized within the sulphur cycle in soils: the sulphur mineralization–immobilization subcycle and the sulphur oxidation–reduction subcycle, summarized in Figs 2 and 5 respectively.

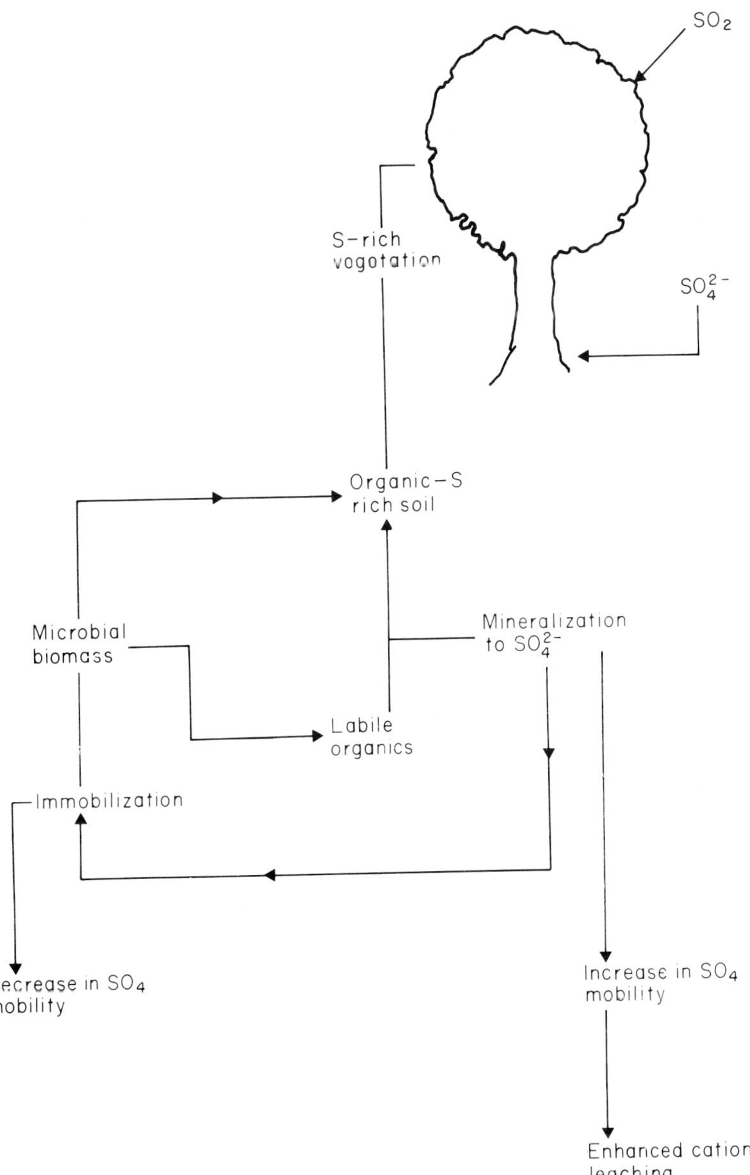


FIG. 2. The sulphur mineralization–immobilization subcycle.

## SULPHUR MINERALIZATION-IMMOBILIZATION SUBCYCLE

### *Sulphur mineralization*

The mineralization of sulphur from soil organic matter is a microbial process mediated by at least two processes (Strick *et al.* 1982): (i) the hydrolysis of sulphate esters, catalysed by plant and microbial sulphohydrolases and (ii) the oxidation of reduced carbon-bonded forms. The former reactions appear to be regulated by the need for sulphur as a nutrient, while the oxidations are believed to result from the requirement by micro-organisms of carbon or energy sources (Strick *et al.* 1982). Nyborg (1978) estimated that the net result of these reactions is the mineralization of 1–3% of the soil organic sulphur content each year. At sites exposed to atmospheric pollution, vegetation is enriched with organic sulphur and soils are enriched with both organic and inorganic sulphur. For example, leaves of *Acer pseudoplatanus* L. sampled downwind of a coking works site (Table 1) had a total sulphur content three times higher than leaves sampled from a relatively unpolluted site (Killham & Wainwright 1980). David, Mitchell & Nakas (1982) suggested that sulphate formation resulting from mineralization reactions may in fact be more important in forest soils than are exogenous inputs. Thus, if only 2% of the organic sulphur present in hardwood forest soils is mineralized, some 60 kg S $ha^{-1}$ would be released each year, which is typically five times more than the amount which enters in throughfall. The oxidation of organic S to sulphate yields two hydrogen ions (Reuss 1977).

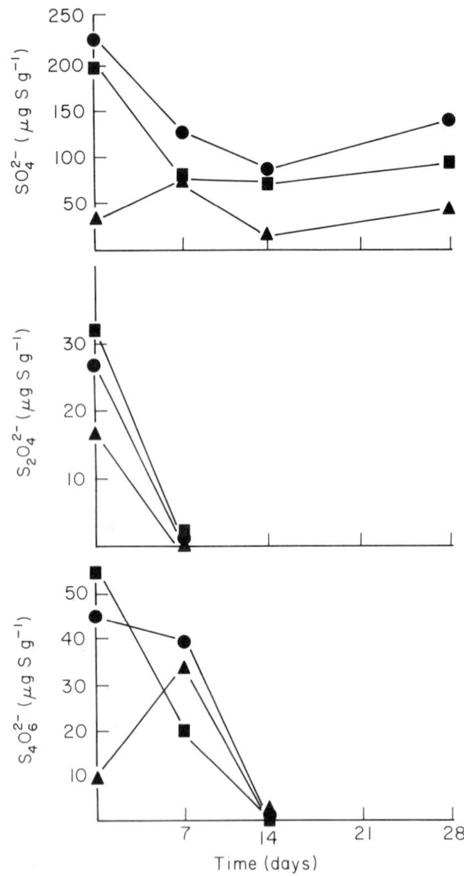


FIG. 3. Changes in S-oxyanion content following the addition of sycamore leaf litter to soils: (●) litter from a coking works site; (■) litter from a relatively unpolluted site; (▲) unamended soil.

The effect of adding sycamore litter from the two sites detailed in Table 1 on the sulphur content of soil is shown in Fig. 3. Addition of both leaf litter types (1% w/w) increased the concentration of thiosulphate, tetrathionate and sulphate in the soil. The concentration of sulphate, while showing an initial decline, eventually began to increase again after 28 days; presumably as the result of mineralization reactions. In contrast, the concentrations of the other two sulphur oxyanions decreased to zero, following sulphur immobilization or oxidation. In the past we have regarded these ions as originating solely from the microbial oxidation of reduced forms of inorganic sulphur; however, the fact that they can be released from polluted litter invalidates this view.

We have also recently investigated the ability of micro-organisms to liberate sulphate from leachates of sycamore leaf and litter samples obtained from the sites described in Table 1. Unsterilized leachates were used, allowing the indigenous microflora to grow with the leachates acting as sole nutrient source. Growth of the native microflora increased the sulphate content of the litter leachates (Fig. 4). Micro-organisms are therefore likely to liberate sulphate from litter leachates as well as from the litter itself.

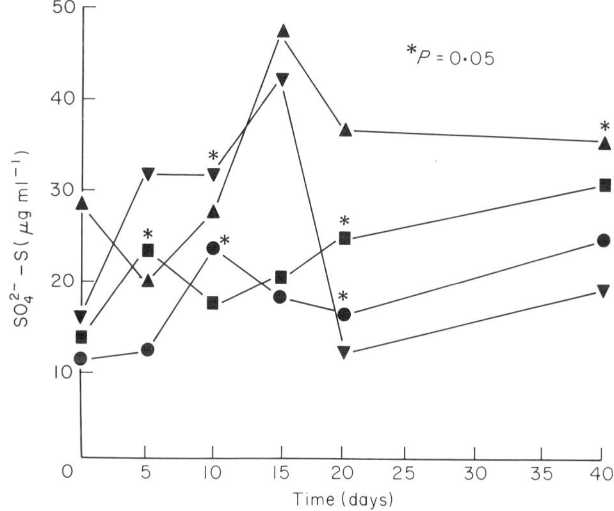


FIG. 4. Changes in sulphate concentration of sycamore litter leachates supporting the growth of indigenous micro-organisms. Coking works litter leached with water (●) and simulated acid rain (▲) ; relatively unpolluted litter leached with water (■) , and simulated acid rain (▲).

### *Sulphur immobilization*

The converse of sulphur mineralization is sulphur immobilization, resulting in the incorporation of sulphate into plant and microbial biomass. Fungi in particular appear to have a high requirement for sulphate (Raistrick & Vincent 1948). Although the soil microbial biomass contains only a small proportion of the total organic soil sulphur content, it is an extremely labile fraction. This process is of potential importance in atmospheric-polluted soils because it leads to a reduction in the mobility of the sulphate ion. Swank, Fitzgerald & Ash (1984) and Strickland & Fitzgerald (1984) suggested that the incorporation of sulphate into organic forms in soil could provide a buffer against the impact of acid rain on forest soils. At some point however, it is likely that the microbial sulphur pool would be again mineralized to release sulphur (Saggar, Bettany & Stewart 1981). Reuss & Johnson (1985) have suggested recently that biological and physical processes, particularly sulphate adsorption, are important processes in explaining how surface waters are acidified by atmospheric deposition through soil-mediated processes. They also suggested that these processes influence the recovery time of a soil once acid input has been discontinued.

## OXIDATION-REDUCTION SUBCYCLE

### *Sulphur oxidation*

Only a limited amount of non-biological sulphur oxidation occurs in soils, micro-organisms being by far the most important mediators of these reactions (Wainwright 1984). The role of the chemolithotrophic bacteria of the genus *Thiobacillus* in soil sulphur oxidation has generally been emphasized, although numerous heterotrophic micro-organisms also play a role. The relative importance of heterotrophs and chemolithotrophs in the process has yet to be determined (Wainwright 1984). Elemental sulphur is oxidized in soils to thiosulphate and tetrathionate (Nor & Tabatabai 1977; Skiba & Wainwright 1984), which can themselves be oxidized to sulphate:

$$Na_2S_2O_3 + 2O_2 + H_2O = Na_2SO_4 + H_2SO_4$$

$$2Na_2S_4O_6 + 7O_2 + 6H_2O = 2Na_2SO_4 + 6H_2SO_4$$

The end-product of these reactions is sulphate in the form of sulphuric acid. Thus, microbial sulphur oxidation tends to acidify soils, a fact which has been utilized in the reclamation of sodic soils (Rupela & Tauro 1973). As the main form of sulphur entering atmospheric-polluted soils is sulphate, there would appear to be little opportunity for sulphur oxidation to be of increased importance in atmospheric-polluted compared to non-polluted soils. Reduced forms of sulphur (including $SO_2$ and $HSO_3^-$) can however, reach soils from the atmosphere (Wainwright & Nevell 1985). These compounds appear to be largely chemically oxidized in soils (Ghiorse & Alexander 1976), although some microbial oxidation of sulphite also occurs (Wainwright & Johnson 1980). Certain types of atmospheric pollution deposits (APD) also contain reduced forms of sulphur, which are available for microbial oxidation. For example, the addition to soil of APD obtained from the surface of sycamore leaves polluted by a coking works can lead to an increase in the concentration of sulphur oxyanions in soil (Fig. 6; Wainwright & Killham 1982). Evidence for the occurrence of sulphur oxidation in soils receiving such particulate pollutants includes the following factors.

(a) Such contaminated soils contain greater populations of sulphur-oxidizing micro-organisms than similar, but relatively unpolluted soils (Wainwright 1979).
(b) Contaminated soils contain higher concentrations of the sulphur oxyanions $S_2O_3^{2-}$; $S_4O_6^{2-}$ and $SO_4^{2-}$ than uncontaminated soils, all of which are products of

microbial sulphur oxidation (Wainwright 1978, 1979).

(c) Polluted soils have higher activities of the enzyme thiosulphate cyanide sulphur transferase (rhodanese) which has been associated with the occurrence of sulphur oxidation in soils (Wainwright 1979).

(d) Finally, transfer of relatively unpolluted soil to sites exposed to heavy atmospheric pollution results in an increase in the numbers of sulphur-oxidizing micro-organisms and to an increase in the above-mentioned sulphur oxyanions (Killham & Wainwright 1984).

As these sulphur oxidations appear limited to those atmospheric-polluted soils which

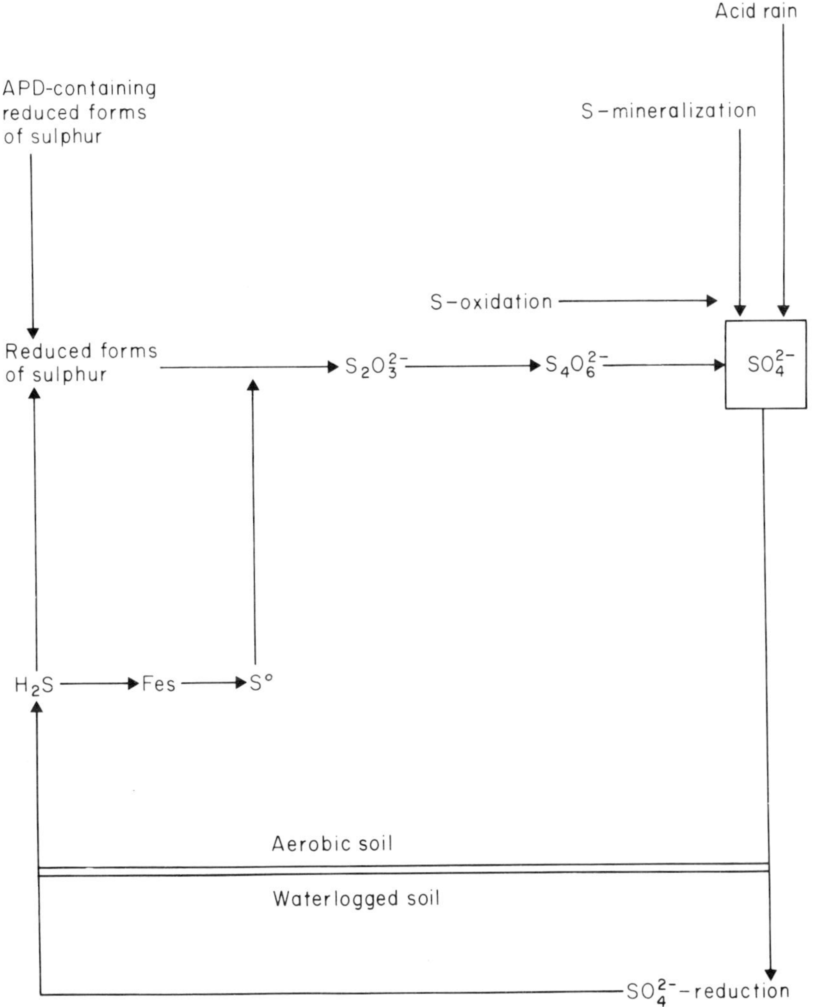


FIG. 5. The sulphur oxidation-reduction subcycle.

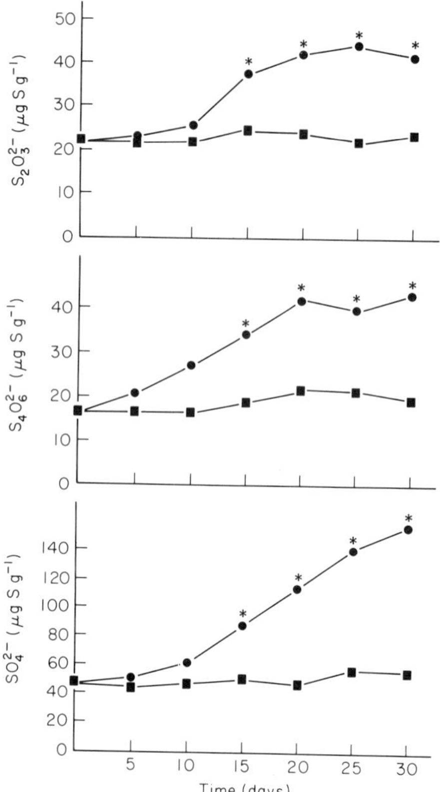


FIG. 6. Effect of adding atmospheric pollution deposits (from sycamore growing downwind of a coking works) on the S-oxyanion content of soil: (●) APD-amended soil; (■) control (*$P=0\cdot05$).

receive soot-rich particulates, the process might appear to be of only local significance. However, pollution particulates can be transported over long distances, resulting in so-called 'black episodes' (Brossett 1976), or to the appearance of black acidic snow (Davies *et al.* 1984). However, the ability of micro-organisms to release sulphate from these particulates has not yet been determined. A recent decision by the US Environmental Protection Agency to change its definition of particulate air pollution is likely to lead to an increase in the release of these particulates into their atmosphere (Sasanow 1985), thereby increasing their concentration in soils and availability for microbial oxidation. If other countries follow this retrograde step, the microbial oxidation of the sulphur in particulates is likely to become increasingly important in air-polluted environments.

## *Sulphate reduction*

This process occurs rapidly in soils which are anaerobic (i.e. waterlogged) and which contain large amounts of available organic matter (Brown 1982). Reduction reactions are carried out largely by obligate anaerobic dissimilatory sulphate-reducing bacteria. The product of sulphate reduction is $H_2S$, but although this gas is released from marine sediments and muds, it rarely occurs free in soils. This is because any free $H_2S$ rapidly combines with metals to form metal sulphides, particularly FeS. Sulphate reduction is generally thought to be pH limited, being optimal at alkaline pH. The process is therefore likely to be inhibited by the acid nature of most atmospheric-polluted soils. However, there is evidence to show that the process can occur in acid soils, including peats (K.A. Brown, pers. comm.).

Sulphate reduction should reduce the mobility of sulphate in atmospheric-polluted soils, by converting the ion to sparingly-soluble metal sulphides. In some circumstances however, sulphate reduction may prove detrimental since, if free $H_2S$ does occur, it can damage plant roots. On the re-imposition of aerobic conditions due to soil drainage any sulphides formed by reduction reactions will be rapidly oxidized to sulphuric acid.

Recent studies in this laboratory have shown that when brown earth soils are saturated with water or acid rain, there was an initial increase in soil sulphate concentration, rather than the expected decrease in the concentration of the ion due to sulphate reduction (Fig. 5). The redox potential (Eh) of these soils fell only slowly on saturation and never reached the negative values typical of anaerobic soils; soil pH on the other hand showed the typical increase (Fig. 7). These changes favour sulphur mineralization and sulphate desorption, which together could help explain the observed flush of sulphate (Swift 1985). However, sulphate loss did occur following prolonged waterlogging (Fig. 7), although no free $H_2S$ was formed. These results show that waterlogging need not lead to an immediate decrease in the mobility of sulphate in soils, but may in fact stimulate sulphur mineralization, thereby increasing the soil sulphate content. Saturating these soils with either acid rain (pH 3·8) or simulated acid rain (pH 4·0) had the same effect on soil sulphate concentration as did deionized water. The initial increases in sulphate concentration were higher however, when heavily atmospheric-polluted soils were saturated, so it is probable that the sulphate flush and subsequent sulphate reduction will be greater following the waterlogging of polluted soils compared with non-polluted soils.

As waterlogged soils dry out they become increasingly aerobic, so that any metal sulphides formed by sulphate reduction will be rapidly oxidized to elemental sulphur and sulphuric acid, e.g. for FeS (Harmsen 1954):

(1) $3H_2S + 2Fe(OH)_3 = 2FeS + S + 6H_2O$

(2) $H_2S + FeCO_3 = FeS + H_2O + CO_2$

(3) $4H_2S + 4Fe(OH)_3 + CH_2O = 4\,FeS + 11\,H_2O + CO_2$

The resultant FeS can then be chemically oxidized:

(4) $4FeS + 6H_2O + 3O_2 = 4\,Fe(OH)_3 + 4S$

Any elemental sulphur formed by reactions (1) and (4) can then in turn be oxidized by micro-organisms:

$$2S + 3O_2 + 2H_2O \rightarrow 2H_2SO_4$$

This oxidation reaction is best written as:

$$S + \frac{3}{2}O_2 + H_2O \rightarrow 2H^+ + SO_4^{2-}$$

showing that two hydrogen ions are formed.

Attempts to assess the importance of the microbial sulphur cycle in atmospheric-

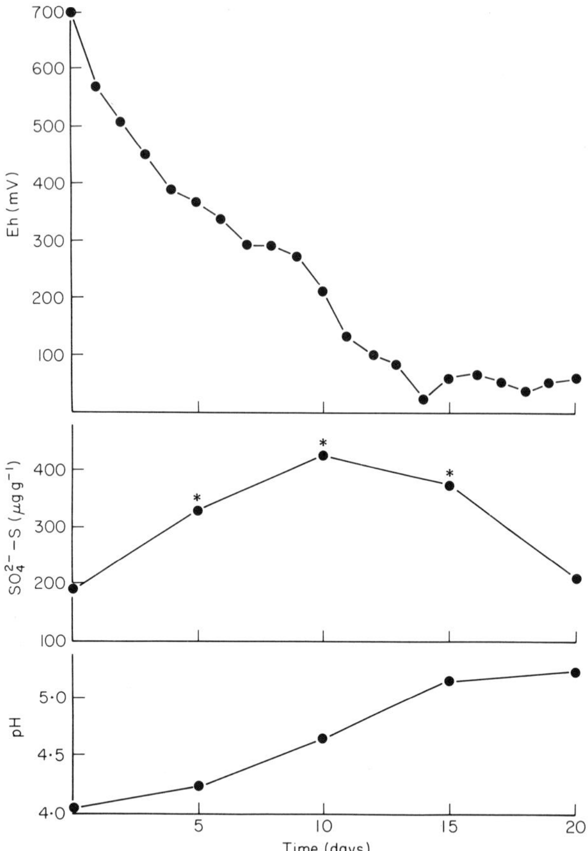


FIG. 7. Effect of saturating with simulated acid rain (pH 4·0) on the sulphate content, Eh and pH of a brown earth soil. (*Significant increase over initial sulphate concentration, $P=0\cdot05$.)

polluted soils are hindered by the fact that the individual reactions involved can occur simultaneously. Plants also assimilate sulphate, and further complications arise from the ability of soils to adsorb the ion. For example, if the sulphate released by sulphur oxidation is assimilated by plants the amount of acid which is formed will be balanced by the release of $2OH^-$ ions (or $HCO_3^-$) from the plant. The net change in $H^+$ ions for the overall process is therefore zero (Reuss 1977), and there is no resultant change in soil acidity. Although the sulphur cycle is balanced as far as acid production and consumption is concerned, the potential for base leaching is provided because lags may occur between the release of sulphate and its uptake by plants.

## CONCLUSIONS

Micro-organisms clearly play a major role in the cycling of sulphur in soils. As yet however, we have no quantitative measure of the overall importance of these transformations in atmospheric-polluted soils, nor of the relative importance of the individual processes involved. It is likely that sulphur mineralization reactions will be dominant, due to the increased amounts of organic sulphur occurring in these soils. The importance of the microbial sulphur cycle in soils exposed to atmospheric pollution relates mainly to its influence on the mobility of sulphate, and thereby on cation-leaching as well as to the internal production and consumption of $H^+$ ions. It should also be borne in mind that microbial transformation of nitrogen in soil, such as nitrification, will have a similar impact on soil acidity and cation leaching (Van Miegroet & Cole 1984). Clearly more research on the microbiology of atmospheric-polluted soils is needed before the impact of these changes on soil nutrient cycling in general can be determined.

## ACKNOWLEDGMENTS

We thank the Natural Environmental Research Council for provision of a postgraduate studentship for W.N., and for a previous studentship for work in this area.

## REFERENCES

**Abrahamsen, G. & Stuanes, A.O. (1983).** Effects of acid deposition on soil: An overview. *VDI-Berichte,* **500,** 279–287.

**Brossett, C. (1976).** Air-borne particles: black and white episodes. *Ambio,* **5,** 157–163.

**Brown, K.A. (1982).** Sulphur in the environment: a review. *Environmental Pollution (Series B),* **3,** 47-80.

**David, M.B., Mitchell, M.J. & Nakas, J.P. (1982).** Organic and inorganic sulfur constituents of a forest soil and their relationship to microbial activity. *Soil Science Society of America Journal,* **46,** 847–852.

**Davies, T.D., Abrahams, P.W., Tranter, M., Blackwood, I., Brimblecombe, P. & Vincent, C.E. (1984).** Black acidic snow in the remote Scottish Highlands. *Nature (London),* **312,** 58–61.

**Ghiorse, W.C. & Alexander, M. (1976).** Effect of micro-organisms on the sorption and fate of sulfur dioxide and nitrogen dioxide in soil. *Journal of Environmental Quality,* **5,** 227–230.

**Harmsen, G.W. (1954).** Observations on the formation and oxidation of pyrite in the soil. *Plant and Soil,* **4,** 324–348.

**Johnson, D.W. & Reuss, J.O. (1984).** Soil-mediated effects of atmospherically deposited sulphur and nitrogen. *Philosophical Transactions of the Royal Society, London B,* **305,** 383–392.

**Killham, K. & Wainwright, M. (1980).** A closed combustion method for the rapid determination of total S in atmospheric-polluted soils and vegetation. *Environmental Pollution (Series B),* **2,** 81–85.

**Killham, K. & Wainwright, M. (1981).** Microbial release of sulphur ions from atmospheric pollution deposits. *Journal of Applied Ecology,* **18,** 889–896.

**Killham, K. & Wainwright, M. (1984).** Chemical and microbiological changes in soils following exposure to heavy atmospheric pollution. *Environmental Pollution (Series A),* **33,** 121–131.

**Lettl, A. (1984).** The effect of atmospheric $SO_2$ pollution on the microflora of forest soils. *Folia Microbiologia,* **29,** 455–475.

**Moss, M.R. (1975).** Spatial patterns of sulphur accumulation by vegetation and soils around industrial centres. *Journal of Biogeography,* **2,** 205–222.

**Nor, Y.M. & Tabatabai, M.A. (1977)** Oxidation of elemental sulfur in soils. *Soil Science Society of America Journal,* **41,** 736–741.

**Nyborg, M. (1978).** Sulfur pollution and soils. *Sulfur in the Environment:* Vol. 2 *Ecological Impacts* (Ed. by J.O. Nriagu), pp. 359–390. John Wiley, New York.

**Raistrick, H. & Vincent, J.M. (1948).** A survey of fungal metabolism of inorganic sulphates. *Biochemical Journal,* **43,** 90–99.

**Reuss, J.O. (1977)** Chemical and biological relationships relevant to the effect of acid rainfall on the soil-plant system. *Water Air and Soil Pollution,* **7,** 461–478.

**Reuss, J.O. & Johnson, D.W. (1985).** Effects of soil processes on the acidification of water by acid deposition. *Journal of Environmental Quality,* **14**, 26–31.

**Rupela, O.P. & Tauro, P. (1973).** Utilization of *Thiobacillus* to reclaim alkali soils. *Soil Biology and Biochemistry,* **5**, 899–901.

**Saggar, S., Bettany, J.R. & Stewart, J.W.B. (1981).** Measurement of microbial sulfur in soil. *Soil Biology and Biochemistry,* **13**, 493–498.

**Sasanow, S. (1985).** Acid rain: a review of current controversy. *Soil Use and Management,* **1**, 34–36.

**Skiba, U. & Wainwright, M. (1984).** Oxidation of elemental S in coastal dune sands and soils. *Plant and Soil,* **77**, 87–95.

**Strick, J.E., Schindler, S.E., David, M.B., Mitchell, M.J. & Nakas, J.P. (1982).** Importance of organic sulfur constituents and microbial activity to sulfur transformation in an Adirondack forest soil. *North Eastern Environmental Science,* **1**, 161–169.

**Strickland, T.C. & Fitzgerald, J.W. (1984).** Formation and mineralization of organic sulfur in forest soils. *Biochemistry,* **1**, 79–95.

**Swank, W.T., Fitzgerald, J.W. & Ash, J.T. (1984).** Microbial transformations of sulfate in forest soils. *Science,* **223**, 182–184.

**Swift, R.S. (1985).** Mineralization and immobilization of sulphur in soil. *Sulphur in Agriculture,* **9**, 20–24.

**Van Miegroet, H. & Cole, D.W. (1984).** The impact of nitrification on soil acidification and cation leaching in a red alder ecosystem. *Journal of Environmental Quality,* **13**, 586–590.

**Wainwright, M. (1978).** Distribution of sulphur oxidation products in soils and on *Acer pseudoplatanus* L. growing close to sources of atmospheric pollution. *Environmental Pollution,* **17**, 153–160.

**Wainwright, M. (1979).** Microbial S-oxidation in soils exposed to heavy atmospheric pollution. *Soil Biology and Biochemistry,* **11**, 95–98.

**Wainwright, M. (1982).** Microbial oxidation of sulphur in soils subject to atmospheric sulphur deposition. *Proceedings of the International 1982 Sulphur Conference, London,* **1**, 427–437.

**Wainwright, M. (1984).** Sulfur oxidation in soils. *Advances in Agronomy,* **37**, 349–396.

**Wainwright, M. & Johnson, J. (1980).** Determination of sulphite in mineral soils. *Plant and Soil,* **54**, 299–305.

**Wainwright, M. & Killham, K. (1982).** Microbial transformations of some particulate pollution deposits in soils: a source of plant-available nitrogen and sulphur. *Plant and Soil,* **63**, 297–301.

**Wainwright, M. & Nevell, W. (1984).** Microbial transformations of sulphur in atmospheric polluted soils. *Reviews on Environmental Health,* **4**, 339–355.

# Pathways of fluoride transfer in terrestrial ecosystems

A. W. DAVISON
*Department of Plant Biology, University of Newcastle,*
*Newcastle Upon Tyne, NE1 7RU*

## SUMMARY

**1** The review summarizes published work on rates of deposition and transport of fluoride in ecosystems.
**2** Analytical problems, spatial and temporal variation, and the lack of a suitable isotope limit the scope and accuracy of measurements of deposition and transport.
**3** Rates of wet and dry deposition are summarized and the complex effects of rain in scrubbing, depositing and leaching fluorides are discussed. Reasons for differences in the rate of deposition on biological surfaces need further investigation.
**4** Data show that fluoride is lost from surfaces and is exported from ecosystems at a rapid rate. The known mechanisms are discussed and it is concluded that further work is needed.
**5** Soil is considered as both source and sink. Fixation by soil colloids is important in limiting transport and cycling.
**6** Pathways of transfer to animals are considered and it is concluded that too little is known about surface deposition, the effects of preening, differences in feeding strategy and the availability of different chemical species of fluoride.

## INTRODUCTION

In remote rural areas the concentration of fluorides in the air is at the limit of detection, i.e. around $0 \cdot 1\ \mu g\ m^{-3}$ or less. Entrained soil, marine aerosols, fires and volcanoes all increase the concentration locally. Industrial sources may increase the concentration above this background level by at least two orders of magnitude but the actual concentration and the time over which it occurs depend on a great many engineering, climatic and geographical factors. In British industrial cities, concentrations are commonly in the range $0 \cdot 1 - 1 \cdot 0\ \mu g\ m^{-3}$ for long periods of time. Accidental release of HF from chemical plants may lead to concentrations in excess of $50\ \mu g\ m^{-3}$ for periods of a few hours.

It has long been known that exposure to low concentrations of gaseous fluoride may cause visible injury and reduce the economic or aesthetic value of sensitive plant species. For example, the tissues of ripening peach fruit are visibly injured by exposure to as little as $0 \cdot 3\ \mu g\ m^{-3}$ for 12 h, 5 days a week (Weinstein & Alscher-Herman 1982). No other gaseous air pollutant is effective at such low concentrations (Weinstein & Alscher-Herman 1982); on the other hand resistant species, and they are in the majority, may withstand concentrations up to two orders of magnitude higher. Deposition on herbage increases fluoride intake by animals and if this is excessive it may lead to fluorosis, a condition that varies in severity from barely detectable tooth marking to severe crippling (NAS 1974;

EPA 1980). The first detectable signs of this condition are induced when herbage is exposed to concentrations of airborne fluoride greater than about 0·3 – 0·5 $\mu g\ m^{-3}$ and grazing is continuous for periods of several months. Higher concentrations induce progressively more severe symptoms over shorter time periods. Excessive fluoride may also affect human health but cases of fluorosis caused solely by exposure to airborne fluoride have rarely been substantiated even in old, poorly controlled industrial environments (WHO 1984).

The purpose of studying deposition is largely self-evident, i.e. to reveal the factors that control deposition and to elucidate pathways and rates of transfer so that groups at risk can be identified and undesirable effects prevented. With a suitable knowledge of deposition and transfer it is possible to optimize the efficiency of monitoring and to predict the impact of new or altered sources of emmission. The aim of this review is to summarize what is known about rates of deposition and transfer of fluorides, and to highlight those areas where more work is needed.

## *Methods and limitations in the study of fluoride deposition*

A study of the deposition and transfer of fluoride must involve measurement of concentrations in air as low as 0·1 $\mu g\ m^{-3}$ (perhaps for short time periods), in vegetation to less than 5 mg $kg^{-1}$ dry weight, and in animal tissues to much less than 1 mg $kg^{-1}$ on a wet weight basis. Unfortunately, fluoride presents some problems in this respect. Many other elements have isotopes that have been successfully used to study deposition over short time periods (Chamberlain 1970) and to measure rates of movement through trophic levels, but $^{18}F$ has too short a half-life (1·8 h) to be used for anything other than a few physiological studies. Analytically, fluorine has always presented problems and, although introduction of the specific ion electrode was a major advance, there are still frustrating problems in the analysis of biological materials. The result is that variation in results reported from different laboratories is large. Several studies (e.g. Jacobson & McCune 1969, 1972) involving collaborative analysis of a range of samples have shown the magnitude of this problem and how little progress there has been. Much of this variation is due to poor quality-control. Even with the best control, variation in repeated analysis of the same sample is commonly of the order of ± 10%. Accuracy cannot be readily measured because there is no absolute standard available.

Airborne fluoride concentrations are measured using a variety of techniques mostly involving trapping the gas or particles on filters or coated tubes with subsequent analysis in a laboratory. Deposition rate on alkali papers is often used as a practical and very useful alternative to monitoring by volumetric means, though there are limitations to the technique (Davison & Blackmore 1980). Automatic fast-response analysers have become commercially available in recent years but they are expensive and require both routine technical input and modification for regular field use. None appear to have been used in the study of deposition.

There are few published studies of spatial variation in airborne fluoride or of comparisons of sampling systems, especially under field conditions. Two comparative studies in Australia (Milne 1982; Horne 1982) demonstrated problems of sensitivity and relative accuracy with analytical techniques. The data in Table 1 show a comparison of six nominally identical air samplers operated on a 2 × 2 m grid at a field site for 29 days. In each case, air was drawn through an open-faced double filter paper protected from the rain (Davison & Blakemore 1980). All the daily readings were highly correlated but the regression co-efficients differed considerably, indicating systematic differences between samplers. The source of this variation is unknown but it demonstrates a restriction on accuracy in studies of deposition. As airborne fluoride concentrations are reduced by improved control technology, and the limits set by legislators are similarly reduced (Murray 1982), this aspect of measurement of airborne fluoride needs greater attention.

Vegetation is the main immediate sink for deposited fluoride. Measurements show (NAS 1971) that the fluoride content of the differing components of vegetation varies considerably at any one sampling interval and that even in apparently uniform grass swards it varies spatially and temporally (Davison, Blakemore & Craggs 1979). Table 2 shows some of the variation in four plant species sampled from an area of 10 × 10 m downwind of a brickworks in Northumberland. There is quite clear spatial variation between individuals, variation between species and between tissues of a single species. The extent of temporal variation is illustrated in Davison *et al.* (1979) and Davison (1983). Aside from the implications that this variation has for monitoring and legislation, it causes obvious problems in the study of deposition. Deposition must ideally be measured on specific tissues of particular species. Even with the best sampling protocol in relatively uniform vegetation, estimates of the fluoride content of tissue at any one sampling are probably subject to an uncertainty of ± 10% (Davison *et al.* 1979).

In summary, the problems associated with fluoride set limits to the methods that can be used to study deposition and pathways, and they also limit the accuracy of measurements. The same problems also limit the usefulness of much of the published data.

TABLE 1. Comparison of airborne fluoride measured by six air samplers spaced on 2 × 2 m grid, 30 cm above a short grass sward. Total fluoride measured for 29 consecutive 24 h periods varied between 0·76 and 2·72 $\mu$g m$^{-3}$. Data are regression coefficients (standard error) for regression of the sampler at the head of each column on that in each row

| | 1 | 2 | 3 | 4 | 5 |
|---|---|---|---|---|---|
| 2 | 0·83<br>(0·04) | | | | |
| 3 | 0·89<br>(0·05) | 1·03<br>(0·07) | | | |
| 4 | 0·89<br>(0·03) | 1·04<br>(0·05) | 0·94<br>(0·04) | | |
| 5 | 0·76<br>(0·03) | 0·88<br>(0·04) | 0·79<br>(0·04) | 0·82<br>0·43 | |
| 6 | 0·84<br>(0·05) | 1·00<br>(0·05) | 0·91<br>(0·03) | 0·94<br>(0·04) | 1·10<br>(0·05) |

TABLE 2. Mean and variation in fluoride concentration (mg F $kg^{-1}$ dry wt) in four species growing within a 10 × 10 m area near a brickworks. The mean is of three replicate analyses of samples from five randomly collected individuals. Standard deviations represent variation between individuals. Data from Davison, Takmaz-Nisancioglu & Bailey (1985).

| Species | Part of plant analysed | F concentration Mean | Standard deviation |
|---|---|---|---|
| *Chamaenerion angustifolium* | Leaves 80–120 cm above ground | 217 | 56 |
| | Stem 80–120 cm above ground | 9 | 4 |
| | Leaves 20—80 cm above ground | 143 | 68 |
| | Stem 20–80 cm above ground | 5 | 1 |
| *Polygonum cuspidatum* | Leaves 100–150 cm above ground | 32 | 13 |
| | Stem 100–150 cm above ground | 3 | <1 |
| | Leaves 20–80 cm above ground | 57 | 18 |
| | Stem 20–80 cm above ground | 2 | <1 |
| | Flowers | 17 | 2 |
| *Equisetum telmateia* | Stem | 4 | 1 |
| | Branches | 70 | 31 |
| *Plantago lanceolata* | Living leaves | 12 | 6 |
| | Dead leaves | 30 | 14 |

## DEPOSITION, PATHWAYS AND TRANSPORT

### *Rates of deposition*

Airborne fluoride occurs as gas (principally HF) and in a great variety of particles. Particulates vary in chemistry from distinct minerals (e.g. cryolite) to alumina that has HF chemi-sorbed to the surface. The diameters range from $<0 \cdot 1$ $\mu$m to about 10 $\mu$m. Both gaseous and particulate forms are deposited on leaf surfaces while gaseous fluoride is also taken up through stomata. Some fluoride on leaf surfaces may penetrate directly through the cuticle (McCune, Silberman & Weinstein 1977; Ares, Villa & Mondadori 1980).

Pathways of fluoride transport have been reviewed several times (NAS 1971; Weinstein 1977; E.P.A. 1980; Murray 1981) and the literature contains a few reports of airborne fluoride concentrations near sources (Davison, Rand & Betts 1973; McCune, Maclean & Schneider 1976; Davison & Blakemore 1976; EPA 1980; Murray 1981, 1982; O'Conner & Horsman 1982; Davison 1983). There are hundreds of analyses of vegetation (reviewed by EPA 1980), some soil analyses (reviewed by Davison 1983; Murray 1983), and a large number of analyses of animal tissues (reviewed by EPA 1980). However, few of the studies have been directed towards detailed elucidation of pathways or quantitative measurement of transport. A number of publications (McCune *et al.* 1965; Hill 1971; Israel 1974; Less *et al.* 1975; Davison & Blakemore 1976; Schwela 1979; Ares *et al.* 1980; Bonte & Garrec 1980; Garrec & Passera 1980; van der Eerden 1981; Davison 1983) give data on rates and dynamics of fluoride deposition on vegetation but only Murray (1981, 1982) and to a lesser

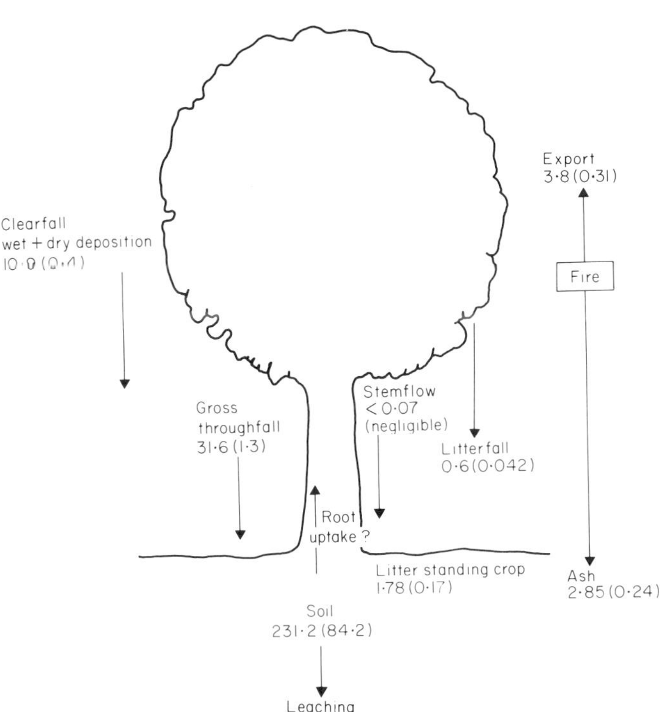


FIG. 1. Fluoride transport in a polluted and unpolluted mixed, open eucalypt forest. Data from Murray (1982). All data are kg F ha$^{-1}$ y$^{-1}$; those in parentheses are for the unpolluted site.

extent Ares (1978) have attempted to quantify cycling from vegetation to soil and animals. Some data of Murray's (1982) for a mixed eucalypt forest are given in Fig. 1.

The deposition velocity $V_g$ of particulate fluoride to grass swards has been measured by a few authors or can be calculated from their data (McCune *et al.* 1965; Less *et al.* 1975; Davison & Blakemore 1976; Garrec & Passera 1980). The available measurements were summarized and compared with data for non-fluoride particulates by Davison (1982). The fluoride data compare well with data from experiments on other pollutants using wind-tunnels under strictly controlled conditions, so it appears that rates of deposition of particulate fluoride to short grass can be calculated provided that the particle size is known. This latter restriction may present a difficulty because there may be a range of sizes emitted from a single source. For example, near a dry-scrubbed aluminium smelter, fugitive fluoridated alumina may range from under 1 to about 5 $\mu$m in diameter (A.W. Davison, unpublished). Over this range $V_g$ changes by a factor of about twelve, so estimates of deposition will depend very much on the accuracy with which particle size can be specified. To give an example of the rate of deposition of particulate fluoride, Davison (1982) calculated that for a grass with a specific leaf weight of 20 g m$^{-2}$, exposed to 1 $\mu$g F m$^{-3}$ at a windspeed 4 m s$^{-1}$ for 24 h, the fluoride deposition due to particles of 1 and 5 $\mu$m diameter respectively would be 0·8 and 17 mg kg$^{-1}$ dry wt of leaf.

Deposition velocities have also been calculated (Davison 1982) for gaseous fluoride using data of Hill (1971), Israel (1974), Davison & Blakemore (1976) and Schwela (1979). The data were obtained in a variety of ways and under different conditions but the values, which ranged from 0·002 to 0·008 m $s^{-1}$, implied that exposure of a short grass sward to 1 $\mu g$ gaseous F $m^{-3}$ leads to the deposition of between about 9 and 35 mg F $kg^{-1}$ dry weight per day. However, the broad range of the estimated deposition rates indicates the major problem in the use of deposition velocity. The concept is not readily amenable to allowance for the variables that affect deposition. Nor does it allow a distinction to be made between deposition on surface and uptake via stomata.

A model of gaseous fluoride deposition based on the well-known (Unsworth, Biscoe & Black 1976) analogy with electrical resistance was examined by Davison (1982, 1983). Using measured or estimated values of the component resistances, the model predicted that over short periods the flux of HF to bean *(Phaseolus vulgaris)* and scarlet oak *(Quercus coccinea)* when exposed to 1 $\mu g$ $m^{-3}$ for 24 h would result in leaf concentrations of 48 and 5 mg F $kg^{-1}$ dry weight respectively. Analysis of the resistances showed how this large difference between the species arose, and the high proportion of fluoride deposited on the outer surface. In bean, surface deposition was predicted to be 60% of the 48 mg F $kg^{-1}$ while in oak it was 40% of the 5 mg F $kg^{-1}$ accumulated. The difference was largely due to the difference in affinity of the surfaces for HF.

At present the main limitation to the use of resistance models is that there are few measurements available of the rate of deposition of HF on surfaces. The data used by Davison (1983) were based on very short-term exposures (Davison 1982) and they leave two important questions unanswered: what happens to the rate of deposition of fluoride over more realistic time periods (days to weeks), as the surface must have a finite capacity for sorption; and what is the basis for differences between surfaces of species such as bean and oak? My final observation, from work currently in progress, is that water droplets tend to spread over the surface of leaves with a relatively high capacity for surface deposition of HF (e.g. bean), whereas they remain spherical and run off rapidly where leaves have a low capacity, i.e. the surface is hydrophobic. Perhaps the affinity of leaf surfaces for water is one of the important determinants of deposition.

### *Dry deposition*

A number of measurements of deposition velocity have been published (Davison 1982, 1983) and they are sufficiently consistent to enable estimates of rates of deposition to grass swards to be made. The data given in Table 3 were calculated for the range of particle sizes commonly encountered near aluminium smelters while the values of deposition velocity of gaseous fluoride cover the range reported by several authors (Davison 1983). In calculating the input, the rate was first calculated on a dry weight of leaf basis then converted to an area basis assuming an above-ground standing crop of 0·3 kg $m^{-2}$. The concentration of fluoride in the air was assumed to be 1 $\mu g$ $m^{-3}$.

Total input (kg $ha^{-1}y^{-1}$) depends upon particle sizes and the proportions of both forms of fluoride. In the vicinity of an aluminium smelter the particulate/gaseous ratio varies with the technology in use, but is in the order of 1:1 (Davison & Blakemore 1980). With a

TABLE 3. Estimates of rate of deposition of 1 $\mu g\ m^{-3}$ gaseous and particulate fluoride to a grass sward using published values of deposition velocity

| | | | |
|---|---|---|---|
| Particulate 1 $\mu g\ m^{-3}$ | | | |
| Particle size ($\mu m$) | 1 | | 5 |
| Deposition velocity ($m\ s^{-1}$) | 0·0002 | | 0·004 |
| Deposition rate* | 0·8 | | 17 |
| Annual deposition to sward of 300 $g\ m^{-2}$ biomass ($kg\ F\ ha^{-1}\ y^{-1}$) | 0·88 | | 18·7 |
| Gaseous 1 $\mu g\ m^{-3}$ | | | |
| Range of values of deposition velocity ($m\ s^{-1}$) | 0·002 | to | 0·008 |
| Deposition rate* | 9 | to | 35 |
| Annual deposition to sward of 300 $g\ m^{-2}$ biomass ($kgF\ ha^{-1}\ y^{-1}$) | 9.9 | to | 38.3 |

* = ($mg\ F\ d^{-1}\ kg^{-1}$ dry weight grass).

total airborne fluoride concentration of 1 $\mu g\ m^{-3}$, particles nearer 1 than 5 $\mu m$ and an average $V_g$ for gaseous fluoride of 0·005 $m\ s^{-1}$ the rate of input to a sward is estimated at around 14 $kg\ ha^{-1}\ y^{-1}$. This is comparable to the data of Murray (1982) for open eucalypt forest.

### *Wet deposition and the influence of rain*

In most studies of pollutant deposition a distinction is made between dry and wet deposition. Pollutants collected by precipitation are deposited on vegetation canopies, but whether rain leads to a net increase in the concentration of fluoride in leaves is open to question. The reason is that rain may have opposing effects, scrubbing and therefore reducing the concentration of fluoride in the air, depositing the pollutant on leaves, leaching it from the same leaves or, by wetting surfaces, increasing the rate of dry deposition.

There are many reports of the concentrations of fluoride found in 'rain' (EPA 1980) but almost without exception they consist of analyses of water collected over long intervals with standard rain-gauges. There is no indication in these reports of whether the collecting bowls were rinsed to remove dry deposition, and the total volume collected is rarely stated, so the data are in fact a measure of an undefinable fraction of the total deposition. What they do show is that rain scrubs the atmosphere. This scrubbing will reduce the concentration of fluoride in the air and therefore reduce deposition and uptake. Whether this effect is sufficient to have a significant influence on deposition will depend on the duration and intensity of the rain event as well as the contribution of emission rate, wind direction and all the other factors that determine the concentration in the air at any one site. Unfortunately, there are no data available that allow a quantitative assessment of rain scrubbing of fluoride but it is known that it is very effective in removing large particles (Gregory 1973).

The fluoride scavenged by rain might be expected to deposit on surfaces and so increase the tissue concentration. Simple calculations can indicate what rate of input might result from wet deposition. Assume 5 mm of rain containing 1 mg F $l^{-1}$ falls on a grass sward with a standing crop of 0·3 kg $m^{-2}$; if all of the fluoride is retained by the grass the increase in concentration in the tissues would be about 12 mg F $kg^{-1}$ dry wt. A concentration of 1 mg F $l^{-1}$ probably occurs only in the most-industrialized areas (EPA 1980) and it is extremely unlikely that vegetation would retain all of the fluoride so 12 mg F $kg^{-1}$ is probably an absolute maximum. It is also unlikely that a sequence of rain events would all have the same fluoride concentration or that deposited fluoride would accumulate with each event. Because of spatial variation in fluoride concentrations (Davison *et al.* 1979), and the consequent sampling problems, it may not be possible to demonstrate whether increases of this order of magnitude occur. The annual input of fluoride to an area with 750 mm of rain containing an average of 1 mg F $l^{-1}$ would be 7·5 kg $ha^{-1}$ $y^{-1}$.

The very high solubility of HF in water suggests that a wet film covering a surface should act as a more effective sink than a dry cuticle so an increased rate of deposition should be expected during and after rain events. In fact, wetting surfaces is known to increase deposition of other pollutants by a factor of about two (Chamberlain & Chadwick 1972; Fowler & Unsworth 1974). However, field observations by a number of authors (NAS 1971; Davison & Blakemore 1976) show that in practice there is usually a small, but significant, negative relationship between the fluoride content of vegetation and rainfall, i.e. the tendency is for low herbage fluoride concentrations to be associated with high rainfall. Whether this is due to scrubbing and reduced deposition or, more likely, leaching cannot be determined. Murray's data (1982) shown in Fig. 1 indicate the effectiveness of throughfall in leaching the fluoride deposited on the canopy of mixed eucalypt forest.

### *Transport of fluoride from vegetation: export from the ecosystem?*

There are several ways in which fluoride is removed or transported from vegetation. Those that lead to direct transport to the soil and to animals are considered later but there is evidence that a considerable proportion of deposited fluoride is exported by other means.

Repeated sampling of vegetation with time shows that fluoride concentrations do not remain constant or rise steadily; they fluctuate up and down (Davison & Blakemore 1976; Davison *et al.* 1979; Davison 1983). Some of this fluctuation is due to sampling errors but much is not, and statistical analysis shows that the variation is largely related to variation in airborne fluoride concentrations (Davison & Blakemore 1976; Davison 1983; Craggs, Blakemore & Davison 1985). An increase in the fluoride content can be readily explained by deposition and uptake but decreases, especially of the magnitude that have been reported by Davison (1983), are more difficult to understand. The fact that whenever simultaneous measurements of fluoride in the air and vegetation have been made there is a significant correlation between the two variables (Davison 1983), is evidence that fluoride must be exported from vegetation at a rate that is comparable to the rate of deposition. A lower rate of export would lead to gradual accumulation and poor or zero correlation. The possible pathways are transport to the roots, litterfall, throughfall leaching, weathering of surface waxes, guttation followed by leaching, and volatilization to the air.

Although some fluoride may be transported to roots (Benedict, Ross & Wade 1964; Garrec & Vavasseur 1978) this would involve entry to the symplast and it is usually considered of minimal importance (Weinstein & Alscher-Herman 1982; Murray 1982). Litterfall could similarly not account for the rapid decreases in concentration that have been observed, even in grasses where leaf turnover is more rapid than in trees. Leaching of fluoride from the canopy is, in view of the work by Murray (1982) an important route of transport to the soil, but rapid decreases in fluoride concentration of leaves have been observed during periods when there was no rain (Davison 1982). Furthermore, statistical analysis of 2 years of weekly monitoring data showed that, over that time period, variation in rainfall accounted for only 9% of the variation in the fluoride concentration in a grass sward (Davison 1982). Weathering of surface waxes is certainly a possibility (Chamberlain 1970) but the rate of such weathering has not been measured. Guttation has recently been investigated by S. Takmaz-Nisancioglu (personal communication) and although fluoride is released from leaves by this pathway, the rate is too low to account for field observations. The remaining route is volatilization to the air. Takmaz-Nisancioglu & Davison (1982) first investigated this route by transplanting contaminated turf from a polluted to a clean ($<0{\cdot}1\ \mu g\ F\ m^{-3}$) site. They showed that, in the absence of rain, measurable growth or leaf turnover, fluoride concentrations fell rapidly. The possibility that this was caused by volatilization was investigated by measuring the loss of fluoride from paper models of leaves that were buffered to different pHs, loaded with fluoride (as NaF) and suspended in the air in a laboratory. Some data are shown in Fig. 2. The papers lost fluoride at a rapid rate that increased at lower pH. The theory was advanced that, although fluoride ions are not significantly volatile, at pH values lower than about 5 a fraction of the fluorine complexes with hydrogen ions to form HF which is sufficiently volatile to be lost. Garrec & Chopin (1983) attempted to confirm this observation by sampling air drawn past a plant of *Abies alba* which had fluoride fed as $NH_4F$ to the roots. They were able to detect fluoride on the filter, which is surprising considering that the source was fed at the roots. They observed that the rate of release was slow but this may have been due to: (i) the high stomatal resistance of conifers; (ii) a high aerodynamic resistance in the unstirred air of the container; (iii) deposition of released fluoride on water surfaces and the walls of the container.

Although it is easy to demonstrate that HF is lost from model leaves, there are still some puzzling features of this pathway. There are, for example, several reports of fluoride being lost from shoots under laboratory conditions (see Davison 1983) but it is not always possible to show rapid loss under those conditions. In one experiment Takmaz-Nisancioglu (personal communication) found rapid loss of fluoride from maize shoots when it was given through the roots yet when *Lolium perenne* was dosed with over 200 mg F $kg^{-1}$ by exposure to HF, no loss could be found over a period of a week in a well ventilated, temperature-controlled greenhouse (A.W. Davison, unpublished). In the field, rapid losses of the same order as rates of deposition are more consistently observed and one reason for this may be the presence of other acid gases. In industrial areas $SO_2$ and $NO_2$ deposited on leaves by dry or wet deposition might decrease the pH of the cuticular surface

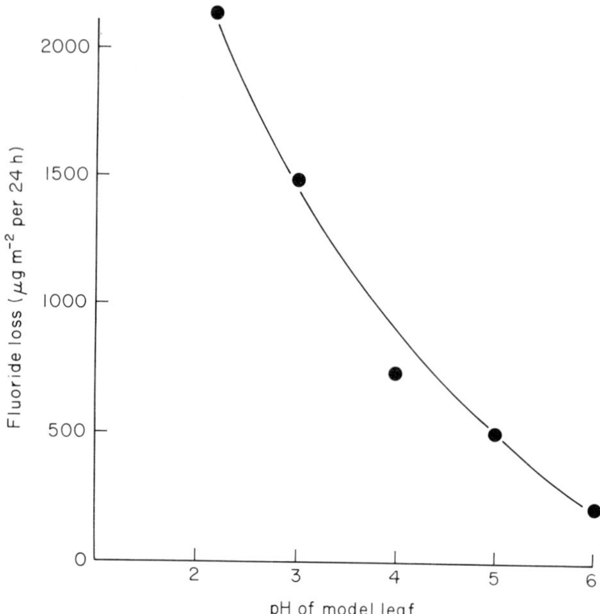


FIG. 2. Rate of loss of fluoride by volatilization from paper models of leaves suspended in air. Models, area 32 $cm^2$, were impregnated with citrate phosphate buffer, dried, and then 0·3 ml of NaF stock solution was applied as evenly as possible to give 60 μg F per 32 $cm^2$. The models were then dried quickly to prevent migration of the fluoride. Data of Takmaz-Nisancioglu & Davision (1982).

and so lead to formation of HF. This is supported by some recent work of T. Davies and J. Cooke (unpublished) at Sunderland who showed that alkali-impregnated papers used as deposit gauges accumulated fluoride at first then, as acid gases deposited, the pH fell and a high proportion of the fluoride was lost. Clearly, the importance of this pathway of export of fluoride needs further investigation.

## *Soil: source and sink*

Soil is the ultimate terrestrial sink for deposited fluoride, receiving it by direct deposition, litterfall and throughfall (Murray 1982). It is also a source of fluoride for vegetation and animals through uptake via the roots and by splash onto shoots.

Direct deposition of airborne fluoride to soil surfaces can be calculated readily using deposition velocity or a resistance measurement, but in most temperate ecosystems the vegetation cover acts as a more important initial sink. The proportion of fluoride transported directly to soil will depend on the surface characteristics of soil and vegetation, the leaf area index and the structure of the standing crop.

Data of Murray (1982) provide estimates of the magnitude and relative contribution of litterfall and throughfall as pathways of transport in eucalypt forest (Fig. 1). The latter route was of much greater importance than the former, accounting for 96% of the fluoride transported from vegetation to soil. The data also provide an interesting estimate of the effect of fire (which is a normal environmental factor in eucalypt forest) in exporting fluoride from the standing crop and litter to the air and soil. After fire the water-soluble fluoride fraction in the soil increased temporarily by a factor of about two.

The chemistry of fluoride in soil was reviewed by Davison (1983) and Murray (1983), and uptake from soil by Davison (1983). In non-saline soils, fluoride is associated with the clay-sized fraction largely because of its occurrence in certain minerals and its reaction with aluminium compounds. The concentration in soils varies from under 20 to several thousand mg $kg^{-1}$ but, in general, it is usually a few hundred mg $kg^{-1}$ (Davison 1983). In undisturbed profiles there is usually a gradient of increasing concentration with depth due to leaching but soils have a great capacity for fixation of fluoride so leaching is usually slow. Fixation also results in deposited fluoride being retained in the upper horizons. Omueti & Jones (1977) showed that much of the fluoride added (in fertilizer) to soil between 1904 and 1924 was still retained in 1955. They calculated that between 1922 and 1944 the average rate of leaching was 2·5 mg F $kg^{-1}$ of soil per year. This is equivalent to about 1·8 kg ha $y^{-1}$. Against this background of export, Davison (1983) calculated that the annual input from typical top dressing of phosphate fertilizer adds 5–28 mg F $kg^{-1}$ to the top 5 cm of soil. Similarly, it was calculated from several field records that the input from an air fluoride concentration of about 1 $\mu g\ m^{-3}$ (i.e. the concentration near an emission source) leads to a comparable annual increase in concentration, i.e. from 4 to 25 mg F $kg^{-1}$ of soil. These rates of input are approximately equivalent to 3–18 kg $ha^{-1}\ y^{-1}$.

The available or labile fluoride in most soils is usually low and this, coupled with the fact that the endodermis acts as a barrier, limits transport from roots to shoot. Fluoride concentrations in leaves that are not exposed to airborne fluoride is usually less than 10 mg $kg^{-1}$ though higher values do occur (Weinstein 1977; E.P.A. 1980; Davison 1983). Because of rapid fixation of deposited fluoride, under most circumstances deposition does not significantly increase transport to the shoot. Certainly all of the considerable body of work by MacIntire and his co-workers (reviewed by Davison 1983) bears this out, so soil not only acts as a sink but as a massive resistance that slows and controls transport to shoots and to aquatic systems. There are, however, a few exceptions where fluoride is or may be rapidly taken up. The report by Murray (1982) of the effects of fire in transporting fluoride to soil via ash is one example where there may be accelerated uptake for the few months that the water-soluble fraction is elevated. A second case is where the soil is saline, because then the high pH and lack of calcium result in high solubility and greater availability of fluoride ions (Davison 1983). Finally, there is evidence that, very close to some aluminium smelters, deposition of particulate fluorides on the surface, and formation of aluminium–fluoride complexes, may lead to increased uptake and transport (Takmaz-Nisancioglu 1983).

Washing field-collected leaves, particularly from pastures, usually reveals that there are mineral soil particles adhering to the surfaces. Healy (1973) commented that vegetation that did not look affected in the field showed contamination after drying out. He also concluded that it is unlikely that well grazed pastures are ever free of soil contamination. The contamination is caused by impaction of wind-blown material, splash by raindrops, physical contact with worm casts and contact brought about by trampling. The importance of animals is shown by the way in which soil contamination increases with stocking rate and by the increase in soil voided in faeces as stocking increases (Healy 1973).

The degree of contamination changes with season. The subject has been studied extensively by a number of authors (Healy 1973) and it appears from their data that soil contributes up to 25–35% of the dry weight of field-collected pasture herbage. An average degree of contamination in a stocked pasture is probably 5–15% on a dry weight basis. As soil contains a high concentration of fluoride, contamination of foilage with soil may provide an important route of transport to large herbivores. Calculation suggests that grass containing 5 mg F $kg^{-1}$ (from uptake) would have the content increased to around 24 mg F $kg^{-1}$ by 10% contamination with a soil containing 200 mg F $kg^{-1}$. When McClenahan & Wiedensaul (1977) incorporated 10% of a soil containing 312 mg F $kg^{-1}$ into dry forage having 7mg F $kg^{-1}$ they found an increase in concentration of only 7 mg F $kg^{-1}$. However, the analytical technique that they used was capable of detecting only about a third of the soil fluoride so the 7 mg F $kg^{-1}$ was roughly what might have been expected. If surface soil is contaminated by top-dressing with phosphate fertilizer or deposited airborne fluoride this route may be of some importance.

### *Transport to animals*

Most of the interest concerning fluorine in animals has revolved around its ingestion by humans, cattle and sheep in relation to fluorosis. Little is known about transport to wild vertebrates and virtually nothing about transport to invertebrates in general or to organisms of detritus chains (Murray 1981).

There are three routes of transport of fluoride to animals: inhalation, deposition on the outer surfaces and ingestion. Inhalation is not usually considered to be an important route (WHO 1984) because the amount taken in and retained during lung ventilation is small in absolute terms and very small relative to other routes. In the case of fattening cattle, an individual weighing 500 kg will inhale about 144 $m^3$ of air a day. Even if all the fluoride were to be retained, the daily input from air containing a background level of 0·05 $\mu$g F $m^{-3}$ would be only 7·2 $\mu$g ($1{\cdot}44 \times 10^{-5}$ mg $kg^{-1}$ body wt). Food intake would be of the order of 10 kg (2% of body weight) and water consumption around 40 l per day. Assuming average fluoride concentrations of 5 mg F $kg^{-1}$ in food and 0·1 mg F $l^{-1}$ in water, the intake from these sources would be 54 mg per day (0·108 mg $kg^{-1}$ body wt) i.e. over 7000 times that inhaled. In a polluted environment the relative contributions would not change. With concentrations of 0·3 $\mu$g F $m^{-3}$ in the air and 40 mg F kg in forage and assuming the water supply is unpolluted, the amounts inhaled and ingested would be 43 $\mu$g and 404 mg respectively ($8{\cdot}6 \times 10^{-5}$ and 0·81 mg $kg^{-1}$ body wt).

The many reports (EPA 1980) of fluoride in animals are of concentrations on a whole-body basis or they are of concentrations in internal tissues (particularly bone). There are no analyses available that enable rates of deposition on wings, skin, shell, etc. to be calculated yet these routes could be important. Surface fluoride may be non-toxic but some particulate fluorides are known to be toxic and they have been used as insecticides (NAS 1971). Deposited fluoride will be consumed by predators. In addition, preening in birds, the collection of pollen from the body by bees and coat-licking by cattle would all transfer deposited fluoride from the surface to within animals. In industrial environments, transfer from the hands and lips to the mouth is an important route of entry of fluoride for man and differences in personal hygiene may be one reason for differences in fluoride loading between individuals. Clearly, more work is needed in this area.

Until more is known about surface deposition it must be assumed that ingestion in food and drink is the main pathway to animals. An example of typical rates of transport has already been given for cattle but there are similar data available for man (EPA 1980; WHO 1984). It is obvious that the daily intake varies greatly in humans with age, the fluoride content of water and the diet, and the amount consumed. In many circumstances, fluorine in water accounts for about half the input to man, which is much more than in cattle. It is impossible to be precise about fluoride intake in humans but an adult typically takes in from about 0·2 to 2·0 mg day$^{-1}$, though it may be higher if the water is fluoridated or if high-fluoride foods (e.g. shellfish, tea) form a significant part of the diet (EPA 1980). In a 65 kg adult this intake is equivalent to 0·003–0·03 mg kg$^{-1}$ body weight. These rates are much lower than the rates previously quoted for cattle. Intake by humans is higher only in poorly-controlled industrial environments and in areas with high fluoride in the water. In the vicinity of industrial emissions, extra intake of fluoride is usually negligible (WHO 1984).

Ingested fluoride may vary in availability for uptake in the digestive system depending on its chemical state, and this affects retention and the route through the animal. Fluoride voided in the urine and faeces may well be of different chemical species (e.g. free ions or complexed with metals) and is transported to different groups of detritus feeders. In cattle which were fed forage supplemented with NaF, about 50% was voided in the faeces but other chemical species containing fluorine have different, usually lower, availability (NAS 1974). Some data of Wright & Thompson (1978) illustrate this point (Table 4). In an

TABLE 4. Effect of source on availability and route of transport of fluoride in rats. Rats of mean weight 140 g were given diets supplemented by fluoride for 1 week. Data from Wright & Thompson (1978)

| | Source of fluoride supplement | | |
|---|---|---|---|
| F(%) | NaF | $Na_3AlF_6$ Cryolite | $AlF_3.H_2O$ |
| Retained | 51 | 55 | 15 |
| Excreted in urine | 48 | 30 | 1 |
| Excreted in faeces | 1 | 15 | 84 |

experiment in which rats were fed diets supplemented with NaF, cryolite ($Na_3AlF_6$, $AlF_3.H_2O$ and two industrial wastes, retention and pathway of excretion varied greatly, especially in relation to the presence of aluminium. With NaF, about 51% was retained and almost all of the rest was excreted in the urine (i.e. > 99% was absorbed through the digestive tract) but in the case of $AlF_3.H_2O$ only 15% was retained and virtually all of the rest was excreted in the faeces. This is of significance in relation to transport of fluorine to animals via soil contamination of herbage because the fluoride in soil is usually strongly bound but there is a high concentration of aluminium present. The strong binding might be expected to reduce availability but no data appear to exist on the effects that the ruminant or any other digestive system has on solubilization of fluoride from recalcitrant materials. With respect to aluminium, Healy (1973) showed that, when soil was added to various digestive liquors from sheep, there was a large increase (3–8 times) in aluminium in solution. If this occurs *in vivo,* the aluminium would complex with a proportion of the fluoride, reduce retention, and change the proportion voided in the faeces. Thus, even if fluoride in soil remains largely non-available to plant roots, aluminium might affect transport of fluoride to the grazing animal.

Within any given area the amount of fluoride encountered by an animal varies greatly because of spatial variation in vegetation (Table 2) and because of differences in feeding strategy, foraging territory and trophic level. This makes it difficult to assess input and to measure transport between trophic levels.

Spatial variation in pasture fluoride, when expressed as the coefficient of variation between samples, is usually of the order of 0·2–0·5 (Davison *et al.* 1979), and over wide limits it is independent of the size of area over which it measured, i.e. an animal foraging over a territory of 1 $m^2$ should encounter a similar degree of variation in fluoride concentration as one foraging over 100 $m^2$.

Variation in the fluoride content of plant species within an area is normal and of a similar degree as that illustrated in Table 2. The variation in leaf fluoride is commonly in the region of 10–20 times but fluoride concentrating in stems tends to be much less variable between species. Within-species variation results in animals with different feeding strategies receiving quite different fluoride loads. Animals feeding on nectar (e.g. adult butterfly), phloem (e.g. aphid) or whole leaf tissues (e.g. caterpillar) of the same plant might be ingesting fluoride at rates differing by an order of magnitude or more. In a pasture where the bulk grass fluoride concentration was in excess of 100 mg $kg^{-1}$, small mammals were feeding on the basal inner sheath tissues of *Dactylis glomerata* that contained less then 5 mg $kg^{-1}$ (Davison, unpublished observations). As the ingested food was so low in fluoride, it is possible that contact with high-fluoride tissues as they stripped off the outer leaf sheaths and preened may have been the major source. The fact that at present it is only possible to speculate on routes of transport to animals is illustrated by some data of Kettle, Port & Davison (unpublished) for whole-body analyses of invertebrates collected near an aluminium smelter and from a control site (Table 5). Sample sizes were small but there were significant differences between froghoppers and aphids, and between those and adult lepidoptera and chewing herbivores. Xylem feeders might be expected to have the lowest intake but the concentration in phloem feeders was surprisingly high in comparison. It is generally accepted that very little fluoride is translocated via the phloem, so aphids might have been predicted to have concentrations similar to or lower than xylem feeders. Is the higher concentration indicative of phloem transport in plants or does much of the 93 mg F $kg^{-1}$ (Table 5) represent surface deposition? If the latter is the case, does moulting reduce whole-body fluoride concentrations? The nymphs of froghoppers are surrounded by a mass of froth which might prevent surface deposition. The fact that the fluoride burden was similar in flower-

TABLE 5. Mean fluoride (mg $kg^{-1}$ air-dry body weight) ± standard error (number of observations in brackets) in animals collected about 500 m from an aluminium smelter and from a control site. Concentration of fluoride in vegetation at the smelter site was 100 to 200 mg $kg^{-1}$ dry wt, and at the control site <5 mg $kg^{-1}$ dry wt. Data from Kettle, Port & Davison (in preparation)

| Organism and feeding type | F concentration Smelter site | Control site |
|---|---|---|
| Froghoppers: xylem fluid | 47 ± 6 (20) | 10 ± 1 (8) |
| Aphids: phloem fluid | 93 ± 3 (5) | 10 ± 1 (10) |
| Lepidoptera larvae: leaf chewing | 186 ± 172 (6) | 29 ± 29 (7) |
| Lepidoptera adults: flower visitors | 145 ± 26 (11) | 14 ± 1 (12) |

visiting adults and leaf-chewing larvae raises further questions, i.e. is it due to surface deposition, and what effect did the difference in size of foraging territory have?

No attempts have been made to answer these questions and clarify routes and rates of transport. Most attention has been paid to vertebrates and as a result there are quite a lot of tissue analyses available (EPA 1980). These are, however, difficult to interpret because of small sample sizes and because of the fact that bone fluoride concentrations are a non-linear function of time. Unless animals can be aged, and the diet assessed over the same time period, bone analyses are of limited use in studies of transport. However, despite the limitations of the data it is usually concluded that fluoride concentration does not increase with trophic level (Murray 1981). Vertebrates have probably been studied most because of ease of identification, emotional appeal and because symptoms of fluorosis are readily recognizable, but in many ways they are less suitable for studies of pathways than are invertebrates. In general, vertebrate foraging areas are larger, longevity varies greatly with trophic level and there is the problem of eliminating the effects of age. A further major drawback to the use of vertebrates is the difficulty of obtaining sufficient animals from the higher trophic levels to make data meaningful.

## CONCLUSIONS

Some aspects of fluoride are now well documented but there remain major gaps. In vegetation, the observation of fluoride export, perhaps by volatilization as HF, at a rate similar to that of deposition needs verification and further measurement. Although the factors controlling deposition of fluorides on vegetation are known, the reasons for the very great differences between species and tissues need to be fully explained and more widely appreciated. Soil is both source and sink. Because of fluoride fixation by soil, soil fluoride has been neglected in recent years but there are circumstances where it is important in transport to plants and animals. Routes of transport to both grazing and detritus-feeding animals need to be clarified, particularly the significance of deposition of fluoride on outer surfaces. The problems raised by different availability of fluorides, differences in feeding strategy, longevity of individuals and foraging area need to be examined. Invertebrates may be more amenable to studies of pathways and rates than vertebrates.

## REFERENCES


**Ares, J.O. (1978).** Fluoride cycling near a coastal emission source. *Journal of the Air Pollution Control Association,* **28**, 344–349.

**Ares, J.O., Villa, A. & Mondadori, G. (1980).** Air pollutant uptake by xerophytic vegetation: fluoride. *Environmental and Experimental Botany,* **20**, 259–269.

**Benedict, H.M., Ross, J.M. & Wade, R.W. (1964).** The disposition of atmospheric fluorides by vegetation. *International Journal of Air and Water Pollution,* **8**, 279–289.

**Bennett, J.H., Hill, A.C. & Gates, D.M. (1973).** A model for gaseous pollutant sorption by leaves. *Journal of the Air Pollution Control Association,* **23**, 957–962.

**Bonte, J. & Garrec, J.-P (1980).** Pollution de l'atmosphere par les composés fluorés et fructification chez *Fragaria* L.: mise en avidence par analyse direct à la microsonde électronique d'une forte accumulation de fluor à la surface des stigmates. *Comptes Rendus Hebdomadaires des Séances de l'Academie des Sciences, Series D: Natural Sciences,* **290**, 815–818.

**Chamberlain, A.C. (1970).** Interception and retention of radioactive aerosols by vegetation. *Atmospheric Environment,* **4**, 57–78.

**Chamberlain, A.C. & Chadwick, R.C.** (1972). Deposition of spores and other particles on vegetation and soil. *Annals of Applied Biology,* **71**, 141–148.

**Craggs, C., Blakemore, J. & Davison, A.W. (1985).** Seasonality in the fluoride concentrations of pasture grass subject to ambient air fluorides. *Environmental Pollution (Series B),* **9**, 163–178.

**Davison, A.W. (1982).** The effects of fluorides on plant growth and forage quality. *Effects of Gaseous Air Pollution in Agriculture and Horticulture* (Ed. by M.H. Unsworth & D.P. Ormrod), pp. 267–292. Proceedings of the 32nd School in Agricultural Science, University of Nottingham School of Agriculture. Butterworths, London.

**Davison, A.W. (1983).** Uptake, translocation and accumulation of soil and airborne fluorides by vegetation. *Fluorides: Effects on Vegetation, Animals and Humans* (Ed. by J.L. Shupe, H.B. Peterson & N.C. Leone), pp. 62–82. Paragon Press Inc., Salt Lake City, Utah.

**Davison, A.W. & Blakemore, J. (1976).** Factors determining fluoride accumulation in forage. *Effects of Air Pollutants on Plants* (Ed. by T.A. Mansfield) pp. 17–30. Society for Experimental Biology Seminar Series, Vol. 1. Cambridge University Press, Cambridge.

**Davison, A.W. & Blakemore, J. (1980).** Rate of deposition and resistance to deposition of fluorides on alkali papers. *Environmental Pollution (Series B),* **1**, 305–320.

**Davison, A.W., Blakemore J. & Craggs, C. (1979).** The fluoride content of forage as an environmental quality standard for the protection of livestock. *Environmental Pollution,* **20**, 279–296.

**Davison, A.W., Rand, A.W. & Betts, W.E. (1973).** Measurement of atmospheric fluoride concentrations in urban areas. *Environmental Pollution,* **5**, 23–33.

**Davison, A.W., Takmaz-Nisancioglu, S. & Bailey, I.F. (1985).** The dynamics of fluoride accumulation by vegetation. *Fluoride Toxity* (Ed. by A.K. Susheela), pp. 30–46. International Society for Fluoride Research, Delhi, India.

**Van der Eerden, L. (1981).** Fluoride-accumulatie in gras. *Voortgangsverslag nr. 1.* Research Institute for Plant Protection, Wageningen.

**EPA (1980).** *Reviews of the Environmental Effects of Pollutants: IX Fluoride.* Report Number EPA-600/1-78-050 prepared for Health Effects Research Laboratory, U.S. Environmental Protection Agency, Cincinatti, Ohio by Oak Ridge National Laboratory, Tennessee.

**Fowler, D. & Unsworth, M.H. (1974).** Dry deposition of sulphur dioxide on wheat. *Nature (Lond.),* **249**, 389–390.

**Garrec, J.-P & Chopin, S. (1983).** Mise en evidence d'un dégagement de fluor gazeux par les végétaux soumis à une pollution fluorée. *Environmental Pollution, (Series A),* **30**, 201–210.

**Garrec, J.-P & Passera, N. (1980).** Uptake of particulate fluorides from an aluminium smelter by plants: Effects of humidity of the air. *Fluoride,* **13**, 105–117.

**Garrec, J.-P & Vavasseur, A. (1978).** Distribution of fluoride in polluted poplars — detection of fluoride accumulation in roots. *European Journal of Forest Pathology,* **8**, 37–42.

**Gregory, P.H. (1973).** *The Microbiology of the Atmosphere.* Leonard Hill, London.

**Healy, W.B. (1973).** Nutritional aspects of soil ingestion by grazing animals. *Chemistry and Biochemistry of Herbage.* Vol. I (Ed. by W.G. Butler & R.W. Bailey), pp. 567–588. Academic Press, London.

**Hill, A.C. (1971).** Vegetation: a sink for atmospheric pollutants. *Journal of the Air Pollution Control Association,* **21**, 341–346.

**Horne, A.W. (1982).** Monitoring and analysis of ambient air hydrogen fluoride. *Fluoride Emissions: Their Monitoring and Effects on Vegetation and Ecosystems* (Ed. by F. Murray), pp. 67–76. Academic Press, Sydney.

**Israel, G.W. (1974).** Deposition velocity of gaseous fluorides on alfalfa. *Atmospheric Environment,* **8**, 1329–1330.

**Jacobson, J.S. & McCune, D.C. (1969).** Interlaboratory studies of analytical techniques for fluorine in vegetation. *Journal of the Association of Official Analytical Chemists,* **52**, 894–899.

**Jacobson, J.S. & McCune, D.C. (1972).** Collaborative study of analytical methods for fluoride in vegetation: Effect of individual techniques on results. *Journal of the Association of Official Analytical Chemists,* **55**, 991–998.

**Less, L.N., McGregor, A., Jones, L.H.P., Cowling, D.W. & Leafe, E.L. (1975).** Fluorine uptake by gas from aluminium smelter fume. *International Journal of Environmental Studies,* **7**, 153–160.

**McClenahan, J.R. & Weidensaul, T.C. (1977).** Geographic distribution of airborne fluorides near a point source in Southeast Ohio. *Research Bulletin 1093.* Ohio Agricultural Research and Development Center, Wooster, Ohio.

**McCune, D.C., Hitchcock, A.E., Jacobson, J.S. & Weinstein, L.H. (1965).** Fluoride accumulation and growth of plants exposed to particulate cryolite in the atmosphere. *Contrib. Boyce Thompson Inst.* **23**, 1–12.

**McCune, D.C., MacLean, D.C. & Schneider, R.E. (1976).** Experimental approaches to the effects of airborne fluoride on plants. *Effects of Air Pollutants on Plants* (Ed. by T.A. Mansfield) pp. 31–46. Society for Experimental Biology Seminar Series Vol. 1. Cambridge University Press, Cambridge.

**McCune, D.C., Silberman, D.H. & Weinstein, L.H. (1977).** Effects of relative humidity and free water on the phytotoxicity of hydrogen fluoride and cryolite. *Proceedings of the International Clean Air Congress,* **4**, 116–119.

**Milne, D.J. (1982).** Evaluation of ambient air fluoride monitoring. *Fluoride Emissions: Their Monitoring and Effects on Vegetation and Ecosystem* (Ed. by F. Murray), pp. 53–66. Academic Press, Sydney.

**Murray, F. (1981).** Fluoride cycles in an estuarine ecosystem. *The Science of the Total Environment,* **17**, 223–241.

**Murray, F. (1982).** Ecosystems as sinks for atmospheric fluorides. *Fluoride Emissions: Their Monitoring and Effects on Vegetation and Ecosystems* (Ed. by F. Murray), pp. 191–205. Academic Press, Sydney.

**Murray, F. (1983).** Fluoride retention by sandy soils. *Water, Air and Soil Pollution,* **20**, 361–367.

**NAS (1971)** *Biologic Effects of Air Pollutants: Fluorides.* US National Academy of Sciences, Washington DC.

**NAS (1974)** *Effects of Fluorides on Animals.* US National Academy of Sciences, Washington, DC.

**O'Conner, J.A. & Horsman, D.C. (1982).** Fluoride levels in vegetation and ambient air in the Portland (Victoria) area. *Fluoride Emissions: Their Monitoring and Effects on Vegetation and Ecosystems* (Ed. by F. Murray), pp. 77–94. Academic Press, Sydney.

**Omueti, J.A.I. & Jones, R.L. (1977).** Fluorine content of soil from Morrow Plots over a period of 67 years. *Soil Science Society of America Proceedings,* **41**, 1023–1024.

**Schwela, D.H. (1979).** An estimate of deposition velocities of several air pollutants on grass. *Ecotoxicology and Environmental Safety,* **3**, 174–189.

**Takmaz-Nisancioglu, S. (1983).** *Dynamics of fluoride uptake, movement and loss in higher plants.* Ph.D. Thesis, University of Newcastle upon Tyne.

**Takmaz-Nisancioglu, S. & Davison, A.W. (1982).** Loss of fluorides from grass swards and other surfaces. *Effects of Gaseous Pollutants in Agriculture and Horticulture* (Ed. by M.H. Unsworth & D.P. Ormrod), p. 465. Butterworth, London.

**Unsworth, M.H., Biscoe, P.V. & Black, V.J. (1976).** Analysis of gas exchange between plants and polluted atmospheres. *Effects of Air Pollutants on Plants* (Ed. by T.A. Mansfield), pp. 5–16. Society for Experimental Botany Seminar Series, Vol. 1. Cambridge University Press, Cambridge.

**Weinstein, L.H. (1977).** Fluoride and plant life. *Journal of Occupational Medicine,* **19**, 49–78.

**Weinstein, L.H. & Alscher-Herman, R. (1982).** Physiological responses of plants to fluoride. *Effects of Gaseous Pollutants in Agriculture and Horticulture* (Ed. by M.H. Unsworth & D.P. Ormrod), pp. 139–167. Proceedings of the 32nd School in Agricultural Science, University of Nottingham School of Agriculture. Butterworths, London.

**WHO (1984).** *Environmental Health Criteria 36 : Fluorine and Fluorides.* World Health Organization, Geneva.

**Wright, D.A. & Thompson, A. (1978).** Retention of fluoride from diets containing materials produced during aluminium smelting. *British Journal of Nutrition,* **40**, 139–147.

# Bone fluoride in four species of predatory bird in the British Isles

D. C. SEEL, A. G. THOMSON AND R. E. BRYANT
*Institute of Terrestrial Ecology, Bangor Research Station,*
*Penrhos Road, Bangor, Gwynedd, LL57 2LQ*

## SUMMARY

**1** Bone fluoride was measured in two species of raptor and two species of owl in the British Isles. Fluoride load was higher in the male compared with female in the raptors, but similar in the sexes in the owls; it increased with age in all four species.

**2** The data for all species were combined for (i) a local comparison with a known fluoride gradient on Anglesey, north Wales, and (ii) an estimate of the regional occurrence of fluoride in predatory birds in the British Isles. Lower than average fluoride loads occurred in most of Scotland, NW England, SW England, and Wales; higher than average loads occurred mostly in NE, central, eastern and southern England, and were attributed to man-made, rather than natural sources.

## INTRODUCTION

Present information on the biological occurrence and effects of fluoride in animals mostly relates to man, to his domestic animals and to standard laboratory experimental animals. Up to *c.* 1960, the information on these species has been summarized by Simons (1965, Vol. 4), who also reviewed fluorine chemistry as a whole (Simons 1954–65). In contrast, the information on wild vertebrates is scarce. Such as there is mostly refers to mammals, in particular primary consumers; experimental work is rare. For fish, see Stewart *et al.* (1974), Wright & Davison (1975), Milhaud, El Bahri & Dridi (1981), Suga, Taki & Wadi (1983); for amphibia, see Macuch *et al.* (1969); for birds, see Macuch *et al.* (1969), Stewart *et al.* (1974), Kay, Tourangeau & Gordon (1975a), Turner *et al.* (1978), van Toledo (1978), Schneppenheim (1980), Bird & Massari (1983), Seel (1983), Seel & Thomson (1984); for mammals, see Karstad (1967), Steward *et al.* (1974), Kay (1974, 1975), Kay *et al.* (1975a, b, 1976), Newman (1976), Shupe & Sharma (1976), Wright, Davison & Johnson (1978), Newman & Murphy (1979), Shupe, Peterson & Olson (1979), Schneppenheim (1980), Andrews, Cooke & Johnson (1982), Walton (1982, 1984). Some of these studies refer specifically to 'uncontaminated' or 'contaminated' environments, but the definition of a 'normal' fluoride level in animal tissues is doubtful (cf. Simons, 1965, p. 517). For the use of animals as biological indicators of air pollution see Newman (1979).

The potential of fluoride for environmental contamination was reviewed by Groth (1975a, b), who stressed that fluoride passes through food chains and that it might accumulate in secondary consumers. Andrews *et al.* (1982) seem to be, as yet, the only authors specifically comparing fluoride loads in prey and predators in potentially the same food chain in a given locality; but no information appears to be available on the transfer of fluoride from prey to predators.

Predatory birds offer an opportunity to investigate the significance of fluoride in predators because they are commonly available from the wild; they can often be kept in captivity; their diets often lend themselves to quantitative study; and, since fluoride is mostly deposited in bone, they are of interest in terms of fluoride acquisition and mobilization (e.g. most, or even all species, reject much of the bone obtained in their diet, while, in the female, body calcium, and hence probably fluoride, is mobilized during egg laying.)

The only studies referring to predatory birds are those of van Toledo (1978) and Seel & Thomson (1984) on the occurrence of fluoride in free-living birds, and Bird & Massari (1983) on the effects of dietary fluoride on a species in captivity. In our previous study (Seel & Thomson 1984), we examined twelve species of predatory bird in the British Isles and found considerable variation in bone fluoride load, and suggested that the immediate principal cause of this variation might lie in the differing diets of these species. The purpose of the present study was to demonstrate further aspects of fluoride contamination in four of those same species, for which more data are now available, and in particular to examine the regional occurrence of fluoride content in the British Isles. The data previously reported for the four species are incorporated in the present report.

The information reported here was obtained as follows. DCS organized the programme and compiled the report; AGT and REB performed laboratory preparation of specimens; the Sub-Division of Chemistry & Instrumentation, Institute of Terrestrial Ecology, Merlewood, made the chemical analyses.

## METHODS

The four species of predatory bird examined were the sparrow-hawk *Accipiter nisus,* kestrel *Falco tinnunculus,* barn own *Tyto alba,* and tawny owl *Strix aluco*. Specimens were obtained via the Ministry of Agriculture, Fisheries & Food, the Monks Wood Experimental Station of the ITE, museums, private taxidermists and the general public.

Both nestlings and fully-grown birds were examined but most of the present paper is based on the latter. Fluoride was measured in the femur. Each femur was dissected out, cleaned in 10% papaine solution for 12 h at 65 °C and then oven-dried for a further 12 h at the same temperature. Fluoride determination was according to the method given by Allen *et al.* (1974) (see also Seel & Thomson 1984).

Actual age could be determined for a few specimens, previously ringed as nestlings and now available by chance for the present purpose. The majority of the specimens were described according to a system of age-classes based on the calendar year (January–December). The depth of this classification in the present context varied between the species owing to differences between them in their external morphology in relation to age. The age-classes were, respectively, 'precise' and 'imprecise'. For *A nisus,* the precise age-classes were 'calendar year 1' (= year of hatching), '2' and '3'; the imprecise classes were simply 'older than' any of the precise classes. For *F. tinnunculus,* the precise classes were '1' and '2'. For *S. aluco* and *T. alba,* there was only one such class,

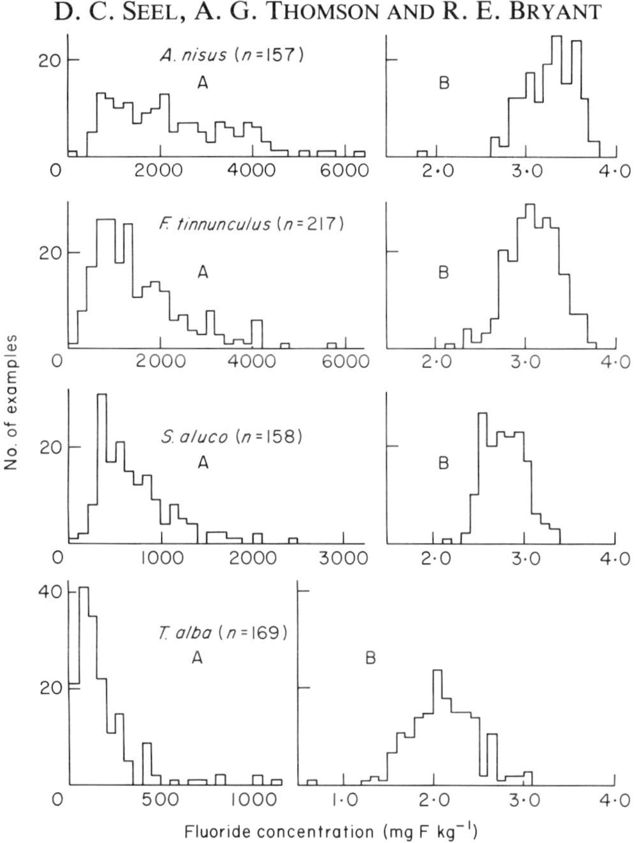


FIG. 1. Statistical distribution of values for fluoride concentrations in four species of predatory bird: sparrowhawk *A. nisus,* kestrel *F. tinnunculus,* tawny owl *S. aluco* and barn owl *T. alba,* in the British Isles: A, original data; B, logarithmically-transformed data.

viz. '1'. In the first year, birds in all species could be separated into 'nestlings' and 'flying-first-year', but sufficient nestlings to justify inclusion in the present report were available for *T. alba* only.

Some previous reports giving fluoride data for wild animals have presented arithmetic means of untransformed data. However, fluoride concentrations in predatory birds in the British Isles, did not follow a normal distribution: extreme, high values, apparently relating to the incidence of pollution, gave rise to assymetrical statistical distribution patterns. This feature was overcome by log transformation of the original data to give normal distribution patterns, from which means and confidence limits were calculated. Back transformation gave geometric means and confidence limits.

## RESULTS

### *Comparison between the species*

#### *Overall patterns of distribution of fluoride*

Fig. 1 shows both untransformed and logarithmically-transformed data for the four species. The untransformed data for *F. tinnunculus, T. alba* and *S. aluco* were mostly

TABLE 1. Summary values for fluoride concentration in femurs of four species of predatory bird* in the British Isles, arranged according to fluoride concentration.

| Species | *n* | Fluoride concentration (mg kg$^{-1}$) Geometric mean | 95% confidence limits of mean | |
|---|---|---|---|---|
| Sparrow-hawk *(A. nisus)* | 157 | 1801·8 | 1621·8 | 2001·7 |
| Kestrel *(F. tinnunculus)* | 217 | 1227·4 | 1124·1 | 1340·3 |
| Tawny owl *(S. aluco)* | 158 | 610·9 | 562·6 | 663·3 |
| Barn owl *(T. alba)* | 169 | 127·4 | 111·9 | 144·9 |

*Fully-grown birds only — no nestlings included. The data in this Table include those previously published in Seel & Thomson (1984).

relatively closely grouped, while those for *A. nisus* were more variable, but in all cases more extreme high values were present, necessitating logarithmic transformation, as indicated above. The geometric means (Table 1) confirm the pattern of relative fluoride load noted previously (Seel & Thomson 1984) in which, from highest to lowest, these four

TABLE 2. Variation in fluoride concentration in femurs of four species of predatory bird* in the British Isles according to sex.

| Species | Male *n* | Male Geometric mean | Male 95% confidence limits of mean | | Female *n* | Female Geometric mean | Female 95% confidence limits of mean | | *P* | Ratio of means |
|---|---|---|---|---|---|---|---|---|---|---|
| Sparrow-hawk *(A. nisus)* | 74 | 2055·4 | 1804·3 | 2341·5 | 80 | 1594·0 | 1355·5 | 1875·0 | 0·016 | 1·3:1 |
| Kestrel *(F. tinnunculus)* | 118 | 1482·2 | 1325·9 | 1656·9 | 95 | 994·3 | 874·0 | 1131·1 | 0·001 | 1·5:1 |
| Tawny owl *(S. aluco)* | 61 | 640·2 | 562·6 | 728·6 | 74 | 606·3 | 539·3 | 681·7 | N.S. | 1·1:1 |
| Barn owl *(T. alba)* | 62 | 138·6 | 114·9 | 167·1 | 74 | 137·0 | 111·2 | 168·6 | N.S. | 1·0:1 |

Fluoride concentration (mg kg$^{-1}$)

*Fully-grown birds only — no nestlings included.

birds were in the order *A. nisus, F. tinnunculus, S. aluco* and *T. alba*. Using the present data, their loads had the ratio 14:10:5:1.

### *Variation in fluoride load in relation to sex*

The two species with the highest load, *A. nisus* and *F. tinnunculus,* showed significant differences between male and female in their fluoride, males containing more than females (Table 2). The other two species showed no significant difference but, none the less, the mean value for male *S. aluco* was also higher than that for the female.

*Variation in fluoride load in relation to age*

Table 3 gives results from all available specimens assigned to the afore-mentioned age-classes (ringed specimens are, of course, included here). In all four species fluoride load increased with age, though in *T. alba* the difference between flying-first-year and post-first-year was not significant. In *A. nisus* and *F. tinnunculus,* where the sexes were

TABLE 3. Variation in fluoride concentration in femurs of four species of predatory bird in the British Isles according to age*.

| Species | Sex | | Fluoride concentration (mg kg$^{-1}$) | | | | | | |
|---|---|---|---|---|---|---|---|---|---|
| | | | Nestling | Flying first year | Post-first year† | Second year | Post-second year† | Third year | Post-third year† |
| Sparrow-hawk | M | $n$ | – | 17 | 49 | 15 | 20 | 11 | 8 |
| *(A. nisus)* | | Geom. mean | – | 1179·2 | 2563·9 | 1625·5 | 3051·4 | 2847·1 | 3288·5 |
| | | 95% conf. | – | 972·1 | 2244·4 | 1290·0 | 2605·0 | 2306·7 | 2373·6 |
| | | limits of $\bar{x}$ | – | 1430·9 | 2928·9 | 248·3 | 3574·4 | 3513·2 | 4555·1 |
| | | $P$‡ | – | | <0·001 | <0·004 | <0·001 | <0·001 | <0·001 |
| | F | $n$ | – | 17 | 50 | 14 | 16 | 6 | 7 |
| | | Geom. mean | – | 848·8 | 1984·7 | 1297·5 | 2822·3 | 2157·2 | 3231·5 |
| | | 95% conf. | – | 680·6 | 1627·0 | 970·5 | 2410·5 | 1719·1 | 2691·5 |
| | | limits of $\bar{x}$ | – | 1058·8 | 2421·0 | 1735·0 | 3304·5 | 2707·7 | 3878·8 |
| | | $P$ | – | | <0·001 | <0·02 | <0·001 | <0·001 | <0·001 |
| Kestrel | M | $n$ | – | 31 | 80 | 16 | 24 | – | – |
| *(F. tinnunculus)* | | Geom. mean | – | 870·6 | 1858·2 | 1132·2 | 2095·6 | – | – |
| | | 95% conf. | – | 749·0 | 1656·5 | 829·1 | 1759·5 | – | – |
| | | limits of $\bar{x}$ | – | 1011·8 | 2085·0 | 1547·7 | 2496·3 | – | – |
| | | $P$ | – | | <0·001 | N.S. | <0·001 | – | – |
| | F | $n$ | – | 23 | 58 | 20 | 9 | – | – |
| | | Geom. mean | – | 744·7 | 1086·9 | 928·3 | 1086·7 | – | – |
| | | 95% conf. | – | 818·3 | 910·8 | 707·9 | 603·0 | – | – |
| | | limits of $\bar{x}$ | – | 897·2 | 1297·2 | 1217·3 | 1958·8 | – | – |
| | | $P$ | – | | <0·04 | N.S. | N.S. | – | – |
| Tawny owl | M + F | $n$ | – | 17 | 97 | – | – | – | – |
| *(S. aluco)* | | Geom. mean | – | 437·8 | 654·8 | – | – | – | – |
| | | 95% conf. | – | 346·6 | 591·3 | – | – | – | – |
| | | limits of $\bar{x}$ | – | 553·2 | 724·9 | – | – | – | – |
| | | $P$ | – | | <0·004 | – | – | – | – |
| Barn owl | M + F | $n$ | 16 | 16 | 106 | – | – | – | – |
| *(T. alba)* | | Geom. mean | 51·1 | 102·8 | 152·8 | – | – | – | – |
| | | 95% conf. | 34·4 | 68·0 | 131·8 | – | – | – | – |
| | | limits of $\bar{x}$ | 76·0 | 155·3 | 177·1 | – | – | – | – |
| | | $P$ | <0·02 | | N.S. | – | – | – | – |

*Age-classes refer to calendar year, viz. January to December.

†This age-class includes both birds whose age was known precisely and birds whose age was known only to the extent of being older than the previous age-class.

‡The probability value is for the comparison between the mean value in the given column and the mean value for 'flying-first-year'.

N.S., not significant.

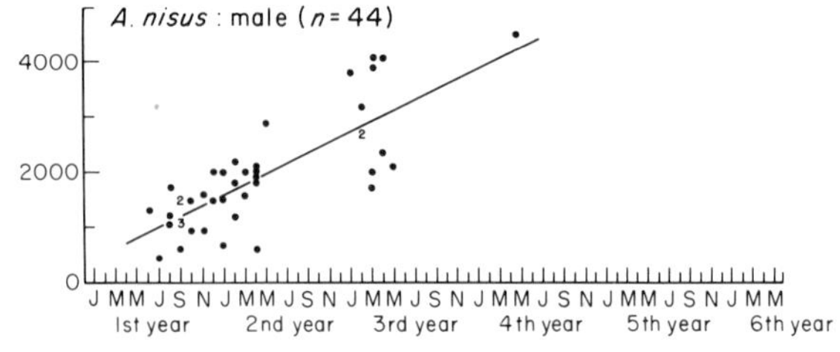


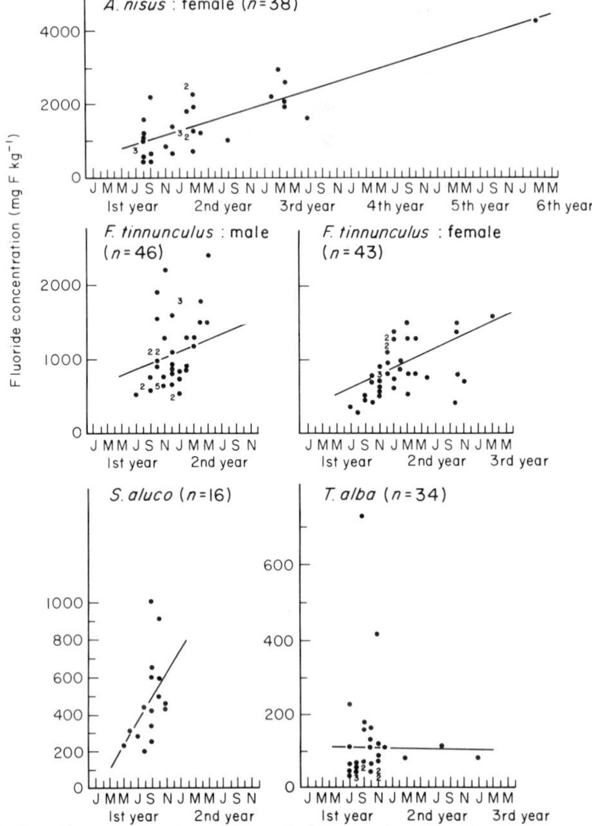


FIG. 2. Variation in fluoride concentration with age in four species of predatory bird: sparrow-hawk *A. nisus,* kestrel *F. tinnunculus,* tawny owl *S. aluco* and barn owl *T. alba,* in the British Isles.

*Footnote to Figure 2*

Regression data for graphs given above.

| Species | Sex | $n$ | Regression constant | Regression coefficient | S.E. | $R^2$, % (adjusted for degrees of freedom, $n$-2) | $P$ |
|---|---|---|---|---|---|---|---|
| Sparrow-hawk *(A. nisus)* | M | 44 | 326·2 | 96·82 | 11·31 | 62·7 | <0·001 |
| | F | 38 | 509·4 | 59·64 | 8·22 | 58·3 | <0·001 |
| Kestrel *(F. tinnunculus)* | M | 46 | 584·0 | 40·95 | 27·35 | 2·7 | N.S. |
| | F | 43 | 319·6 | 45·44 | 14·31 | 17·8 | <0·01 |
| Tawny owl *(S. aluco)* | M + F | 16 | −83·8 | 64·15 | 32·97 | 15·7 | N.S. |
| Barn owl *(T. alba)* | M + F | 34 | 113·0 | −0·434 | 6·508 | −3·1 | N.S. |

N.S., not significant.

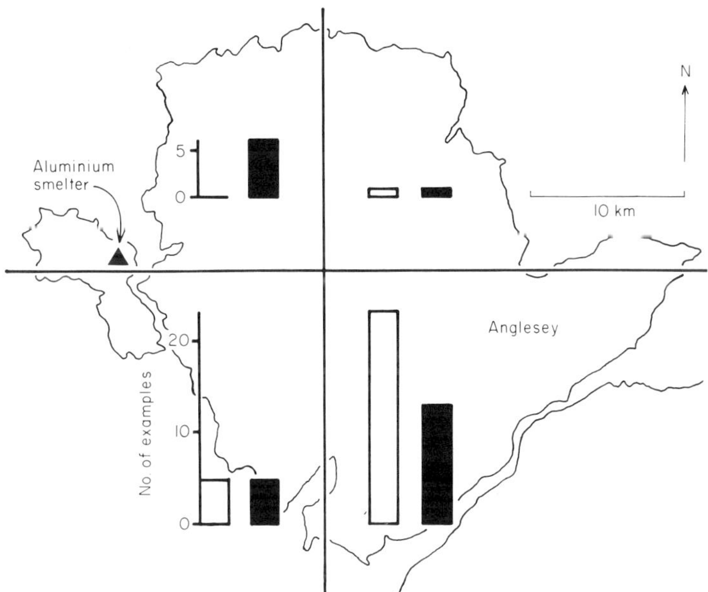


FIG. 3. Local variation in fluoride concentration in four species of predatory bird: sparrow-hawk *A. nisus,* kestrel *F. tinnunculus,* tawny owl *S. aluco* and barn owl *T. alba,* on Anglesey. The data for all four species have been combined (see text): (□) number of examples (in the given sector) with fluoride concentrations less than or equal to the species means; (■) number of examples with fluoride concentrations greater than the species means.

separated for this purpose, the load was higher on average in males than females in each age class. Fig. 2 gives the results for birds of known age, i.e. examples which could be assigned to precise age-classes or which had been ringed (see above). The samples were smaller but show significant increase in fluoride load with age in male and female *A. nisus,* and in female *F. tinnunculus;* non-significant increase in male *F. tinnunculus* and male/female *S. aluco;* but no increase at all in male/female *T. alba.* There is no reason to exclude the one high value obtained for a bird in September of the first year.

## *Geographical occurrence of fluoride*

### *Local occurrence on Anglesey, North Wales*

Several previous studies of fluoride levels in diverse organisms on the island of Anglesey, viz. lichens, magpie *Pica pica,* mammals (Perkins, Millar & Neep 1980a, b; Walton 1982; Seel 1983) have shown that there are local variations in body fluoride which appear to be related principally to emissions from the aluminium smelter in western Anglesey. Broadly speaking, fluoride levels in these organisms were higher in the west of the island than elsewhere, in ways related to the incidence of the prevailing winds.

The number of examples for each of the four predators was both variable and unevenly distributed within Anglesey. It was therefore necessary to construct a single composite map, based on all four taken together, to show the tentative overall distribution of fluoride in predatory birds. Using the original data for each species, two relative levels of fluoride were distinguished: first the concentration less than or equal to the overall geometric mean for the species (as given in Table 1); and secondly the concentration greater than the mean. The combined numbers for all the species in each quadrant of Anglesey, together with the location of the smelter, are shown in Fig. 3.

A $\chi^2$ test on the distribution of birds with higher than average fluoride shows a significant local variation ($P<0\cdot05$): fluoride loads were highest in the north-west, lowest in the south-east, and intermediate elsewhere. This pattern was broadly consistent with those for the other organisms mentioned above. That is, body fluoride in the predatory birds reflected an environmental gradient in fluoride. It was therefore appropriate to examine fluoride load in the four predatory birds in the British Isles as a whole, and to use the relationship in the reverse to estimate the regional occurrence of high environmental fluoride.

### *Regional occurrence in the British Isles*

The number of specimens for each species in the British Isles altogether was fairly large, but, like those for Anglesey, they were unevenly distributed. It was again, therefore, necessary to compile one map based on all four species taken together, as follows. A geometric mean fluoride level was calculated for each 100 km square for each species, and then (where $n \geq 2$ in any given square) the mean was assigned to one or other of the same two categories as before, viz. less than or equal to the overall species mean, and greater than the mean. The results for the four species are presented individually but on the same map in Fig. 4.

Taking the results for all four species collectively, lower than average fluoride levels occurred in most of Scotland, NW England, SW England and in Wales; higher than average fluoride was mostly present in NE, central, eastern and southern England.

## DISCUSSION

It seems unlikely that fluoride load is linked to body size, since the patterns in body size are contrary to the present patterns between the sexes. Thus, males are smaller (i.e. lighter in weight) than females in all species (body weight ratio in *A. nisus* is 1:1·8; in *F. tinnunculus* 1:1·2; in *S. aluco* 1:1·3; and in *T. alba* 1:1·1) (Cramp & Simmons 1980; Hardy, Hirons & Stanley 1981). There is also no consistent pattern between body weight and fluoride load between the species. Thus, any association between fluoride load and sex must be an indirect one. The higher fluoride in males compared with females was the reverse of that found by Bird & Massari (1983) for the American kestrel *F. sparverius* under experimental conditions.

Patterns of fluoride accumulation with age have been observed in mammals, where the identification of age (by anatomical features) is sometimes easier than for birds (e.g. Kay *et al.* 1976; Walton 1984), so in this respect fluoride accumulation in predatory birds is

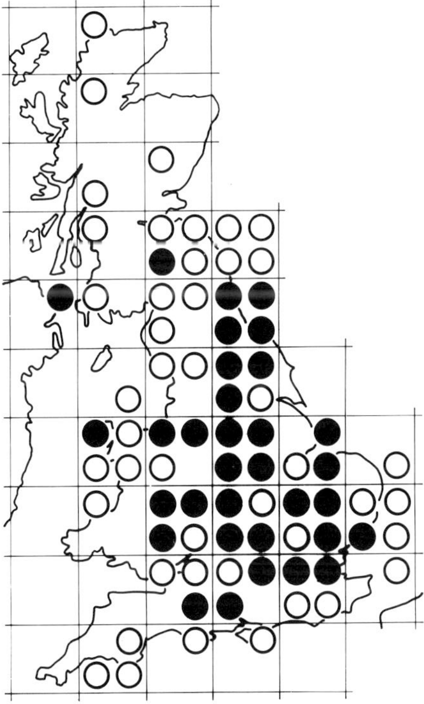

FIG. 4. Regional variation in fluoride concentration in four species of predatory bird: sparrow-hawk *A. nisus*, kestrel *F. tinnunculus*, tawny owl *S. aluco* and barn owl *T. alba* in the British Isles. In each 100 km square, the symbols are arranged as follows: top left, *A.nisus;* top right, *F. tinnunculus;* bottom left, *S. aluco;* bottom right, *T. alba.* (○) Mean fluoride concentration (in the given 100 km square) less than or equal to the species mean; (●) mean fluoride concentration greater than the species mean. These values are shown for each species in any given square, where $n \geq 2$.

similar to that in other organisms. It is not yet evident whether predatory birds eventually achieve a steady state or balance between fluoride intake and excretion (see Simons 1965). What is noteworthy is the different rates of accumulation between the four species, leading to their very different overall fluoride loads. These features may be the consequence of a combination of different diets (which themselves may contain varying amounts of fluoride) and different degrees of retention and rejection of fluoride by the consumer organism.

Studies of man, domestic animals and such studies that have been made on wild animals, suggest that fluoride may enter the body from two main kinds of source: (i) naturally-occurring minerals, probably releasing their fluoride into ground water, which is then taken up in the first instance by vegetation; (ii) from man-made sources, probably mostly industrial, though domestic contributions could be significant, both releasing fluoride into the air. In the British Isles the principal natural source is probably fluospar (fluorite), which is apparently largely restricted to Derbyshire in central England. However, the present data on the regional occurrence of fluoride in the four main species in the British Isles, albeit rather simple in character, suggest that higher fluoride levels occur over a wide area of England. This suggests that fluoride in predatory birds is principally man-made (industrial and domestic), rather than natural, in origin — and so can indeed be described as a 'pollutant'.

The irregularities in the patterns in Fig. 4 could be due in part to small samples for one or other species in given squares; it is evident that the map could be greatly improved by the acquisition of further specimens, which it is hoped this report will encourage.

There is also a need for comparative and linked studies on the movements of fluoride through food chains, viz. on fluoride in taxonomically-related species of predator in all systematic classes, in the prey of those species, and in the sources via the primary producers; while measurements on the same species inhabiting different geographical areas will help to identify environmental variations and their consequences. Such studies need to be backed up by experimental work such as that of Bird & Massari (1983).

## ACKNOWLEDGMENTS

We are pleased to thank the many people who supplied the birds used in this report: many members of the public, museum staff, Dr A. R. Hardy, MAFF, and Dr I. Newton, the Sub-Division of Chemistry & Instrumentation, ITE for performing the chemical analyses; and Dr C. Milner for kindly criticizing the manuscript.

## REFERENCES

**Allen, S.E., Grimshaw, H.M., Parkinson, J.A. & Quarmby, C. (1974).** *Chemical Analysis of Ecological Materials.* Blackwell Scientific Publications, Oxford.

**Andrews, S.M., Cooke, J.A. & Johnson, M.S. (1982).** Fluoride in small mammals and their potential food sources in contaminated grasslands. *Fluoride,* **15**, 56–63.

**Bird, D.M. & Massari, C. (1983).** Effects of dietary sodium fluoride on bone fluoride levels and reproductive performance of captive American kestrels. *Environmental Pollution, (Series A),* **31**, 67–76.

**Cramp, S. & Simmons, K.E.L. (1980).** *Handbook of the Birds of Europe, the Middle East and North Africa.* Oxford University Press, Oxford.

**Groth, E. (1975a).** Along the food chain. *Environment,* **17**, 29–38.

**Groth, E. (1975b).** An evaluation of the potential for ecological damage by chronic low level environmental pollution by fluoride. *Fluoride,* **8**, 224–240.

**Hardy, A.R., Hirons, G.J.M. & Stanley, P.I. (1981).** The relationship of body weight, fat deposit and moult to the reproductive cycles in wild tawny owls and barn owls. *Recent Advances in the Study of Raptor Diseases.* (Ed. by J.E. Cooper & A.G. Greenwood), pp. 159–163. Chirons Publications, Keighley.

**Karstad, L. (1967).** Fluorosis in deer *(Odocoileus virginianus). Bulletin of the Wildlife Disease Association,* **3**, 42–46.

**Kay, E. (1974).** An inquiry into the distribution of fluoride in the environment of Garrison, Montana. *Fluoride,* **7**, 7-31.

**Kay, C.E. (1975).** Flouride distribution in different segments of the femur, metacarpus and mandible of mule deer. *Fluoride,* **8**, 92–97.

**Kay, C.E., Tourangeau, P.C. & Gordon, C.C. (1975a).** Fluoride levels in indigenous animals and plants collected from uncontaminated ecosystems. *Fluoride,* **8**, 125–133.

**Kay, C.E., Tourangeau, P.C. & Gordon, C.C. (1975b).** Industrial fluorosis in wild mule and whitetail deer from western Montana. *Fluoride,* **8**, 182–191.

**Kay, E., Tourangeau, P.C. & Gordon, C.C. (1976).** Populational variation of fluoride parameters in wild ungulates from the western United States. *Fluoride,* **9**, 73–90.

**Macuch, P., Hluchan, E., Mayer, J. & Able. (1969).** Air pollution by fluoride compounds near an aluminium factory. *Fluoride,* **2**, 28–32.

**Milhaud, G., El Bahri, L. &. Dridi, A. (1981).** The effects of fluoride on fish in Gabes Gulf. *Fluoride,* **14**, 161–168.

**Newman, J.R. (1976).** Fluorosis in black-tailed deer. *Journal of Wildlife Diseases,* **12**, 39–41.

**Newman, J.R. (1979).** The effects of air pollution on wildlife and their use as biological indicators. *Animals as Monitors of Environmental Pollutants* (Ed. by S.W. Nielsen, G. Migaki & D.G. Scarpelli), pp. 223–232. National Academy of Science, Washington D.C.

**Newman, J.R. & Murphy, J.R. (1979)** Effects of industrial fluoride on black-tailed deer. *Fluoride,* **12**, 129–135.

**Perkins, D.F., Millar, R.O. & Neep, P.E. (1980a).** Accumulation and effect of airborne fluoride on the saxicolous lichen *Ramalina siliquosa.* Annual Report Institute Terrestrial Ecology 1979, pp. 81–84.

**Perkins, D.F., Millar, R.O. & Neep, P.E. (1980b).** Accumulation of airborne fluoride by lichens in the vicinity of an aluminium reduction plant. *Environmental Pollution (Series A),* **21**, 155–168.

**Schneppenheim, R. (1980).** Concentration of fluoride in Antarctic animals. *Meeresforschung,* **28**, 179–182.

**Seel, D.C. (1983).** Fluoride in the magpie. *Annual Report Institute Terrestrial Ecology, 1982,* pp. 45–48, 73 (plate 8).

**Seel, D.C. & Thomson, A.G. (1984).** Bone fluoride in predatory birds in the British Isles. *Environmental Pollution (Series A),* **36**, 367–374.

**Shupe, J.L., Peterson, H.B. & Olson, A.E. (1979).** Fluoride toxicosis in wild ungulates of the western United States. *Animals as Monitors of Environmental Pollution* (Ed. by S.W. Nielsen, G. Migaki & S.D.B. Scarpelli), pp. 253–266. National Academy of Science, Washington D.C.

**Shupe, J.L. & Sharma, R.P. (1976).** Fluoride distribution in a natural ecosystem and related effects on wild animals. *Trace Substances in Environmental Health.* X. A Symposium. (Ed. by D.D. Hemphill), pp. 137–144. Univ. of Missouri, Columbia.

**Simons, J.H.** (Ed.). **(1950–65).** *Fluorine Chemistry* (5 vols.). Academic Press, New York.

**Stewart, D.J., Manley, T.R., White, D.A., Harrison, D.L. & Stringer, E.A. (1974).** Natural fluorine levels in the Bluff area, New Zealand. 1. Concentrations in wildlife and domestic animals. *New Zealand Journal of Science,* **17**, 105–113.

**Suga, S., Taki, Y. & Wadi, K. (1983).** Fluoride concentration in the teeth of perciform fishes and its phylogenetic significance. *Gyoruigaku Zasshi,* **30**, 81–93.

**Turner, J.C., Solly, S.R.B., Mol-Krijnen, J.C.M. & Shanks, V. (1978).** Organochlorine, fluorine and heavy-metal levels in some birds from New Zealand estuaries. *New Zealand Journal of Science,* **21**, 99–102.

**van Toldeo, B. (1978).** Fluoride content in eggs of wild birds *Parus major* and *Strix aluco* and the common house hen *Gallus domesticus. Fluoride,* **11**, 198–207.

**Walton, K.C. (1982).** Fluoride in small animals. *Annual Report Institute Terrestrial Ecology 1981,* pp. 23–28.

**Walton, K.C. (1984).** Fluoride in fox bone near an aluminium reduction plant in Anglesey, Wales, and elsewhere in the United Kingdom, *Environmental Pollution, (Series B),* **7**, 273–280.

**Wright, D.A. & Davison, A.W. (1975).** The accumulation of fluoride by marine and intertidal animals. *Environmental Pollution,* **8**, 1–13.

**Wright, D.A. & Davison, A.W. & Johnson, M.S. (1978).** Fluoride accumulation by long-tailed field mice *(Apodemus sylvaticus* L.) and field voles *(Microtus agrestis* L.) from polluted environments. *Environmental Pollution,* **17**, 303–310.

# Sources of cadmium discharge to the UK environment

M. HUTTON AND C. SYMON
*Monitoring and Assessment Research Centre (MARC), The Octagon Building, 459A Fulham Road, London SW10 0QX*

## SUMMARY

**1** This paper reviews the sources and discharge rates of cadmium in the UK environment. Estimates are provided for the quantities of cadmium released to the land, atmosphere and water from human activities. Attention is also paid to the sources of cadmium input to sewage, as this acts as a secondary source of the metal.
**2** Published emission data and information provided by the relevant industry or authority were used to compute emission factors. These were applied to recent production or consumption data for the source under consideration.
**3** The major atmospheric sources of cadmium in the UK include refuse incineration, non-ferrous metal production, the iron and steel industries and coal combustion. Waste disposal is responsible for the single largest cadmium input to landfill in the UK and accounts for about one-half of the total input to this compartment. The major sources of cadmium discharge to coastal waters are sewage disposal and the disposal of gypsum from the manufacturing of phosphate fertilizers.
**4** Cadmium inputs to arable land from atmospheric deposition and from the application of phosphate fertilizers and sewage sludge have also been quantified. Phosphate fertilizers represent the largest source of cadmium input with atmospheric deposition next in ranking. Sludge application results in high rates of cadmium input at the local level, but the total input from this source is only about a quarter of the total from phosphate fertilizers.

## INTRODUCTION

Human activities result in the release of large quantities of different contaminants which are dispersed along various pathways through the biosphere. For synthetic chemicals such as the organochlorine pesticides and PCBs, human activities represent the only source of environmental discharge. However, for many other contaminants, there are important natural sources of release and a corresponding 'pre-man' cycle upon which human sources have been superimposed.

In order to determine the significance of human activities on the cycling of such chemicals, it is necessary to identify and quantify the major sources of environmental release. This activity of source quantification has been carried out for several contaminants at the global level. For example, recent estimates of the global fluxes of sulphur to the atmosphere reveal that inputs due to human activities rival or even exceed those from natural sources (Freney, Ivanov & Rodhe 1983). In the case of lead fluxes to the global atmosphere, it has been calculated that inputs from human activities are nearly 20-fold greater than from natural sources (Nriagu 1979). Further evidence for man's perturbation of the global lead cycle is obtained from findings of enhanced lead concentrations in recent ice cores from the Antarctic as well as in surface waters from remote mid-ocean sampling locations (Zoller 1984).

A different picture emerges for cadmium, with an absence of any enrichment in recent ice cores from either the remote Antarctic or the less remote Arctic (Zoller 1984). These findings suggest that human activities have not resulted in a measurable impact on either the global or hemispheric cycling of cadmium. Nevertheless, there are indications from certain parts of the world of a recent increase in the environmental burden of cadmium as reflected by enhanced human exposure. One such region is north-west Europe, an industrialized area which produces and consumes more cadmium than either the USA or the USSR (WBMS 1984). In two European countries it has been reported that there has been a considerable increase in the body burden of cadmium in the general population over the last 100 years, based on the analysis of present-day and preserved kidney specimens (Elinder & Kjellström 1977; Drasch 1983). Furthermore, a study in Belgium is the first outside Japan to link cadmium exposure with signs of renal dysfunction in a group of individuals from the general population (Roels *et al.* 1981).

This paper presents a source inventory of anthropogenic cadmium release for one European country, the UK. The major categories of cadmium discharge are identified and, where possible, estimates of the quantities involved are also provided. These discharges are considered in terms of inputs to the broad environmental compartments: atmosphere, landfill, agricultural land, fresh waters and coastal waters. Particular attention is paid to those sources of cadmium which make a contribution to the soil–plant pathway for human exposure. Attention is also directed to those sources which contribute to the cadmium burden of sewage, as this material acts as a secondary source of cadmium when disposed of.

## METHODS

The first steps in this study were to identify potential sources of environmental cadmium release in the UK and then to assemble the available emission data for each source category. In the absence of published data, requests for information were made to the relevant company or state body. The quantities of cadmium discharged to the major environmental compartments were estimated by two techniques. The first was based on a best judgement emission factor (g cadmium released $t^{-1}$ material consumed or produced) which was applied to the most recent production and consumption data for the source in question. Each emission factor was selected after a critical examination of the available emission studies for the relevant source category. For some sources, the emission factor was based on experimental studies conducted outside the UK. The second technique employed to estimate emission strengths involved the use of specific discharge data which were available for certain categories such as large point source emitters and the marine disposal of certain wastes.

Occasionally, difficulties were experienced in obtaining information on the consumption/production statistics and/or the corresponding cadmium discharge data for a particular source. This was often because the information was considered confidential by the body concerned, or the relevant data were simply not available.

The human sources of environmental cadmium release were divided into three categories. The first included those activities associated with the production and consumption of the metal. For convenience, the mining and production of other non-ferrous metals were also considered under this category. The second group was made up of inadvertent sources, defined as those activities handling materials which naturally contained cadmium as a trace constituent. The third group consisted of those sources associated with the disposal of waste materials which had earlier received cadmium inputs.

## DISCHARGES FROM USE-RELATED SOURCES

### *Mining of non-ferrous metals*

Cadmium is a minor constituent of all the major non-ferrous metal ores, but only the lead–zinc ores contain sufficient cadmium to merit its commercial extraction. Mines which exploit these particular ores can be a significant source of cadmium contamination. Currently, there are no active lead–zinc mines in the UK, but the legacy of contamination from disused sites still remains. Wind-blown dispersal of particulate material from spoil heaps can result in local cadmium contamination. Water-borne dispersal is of greater significance in the UK, with the possibility of agricultural land being contaminated downstream of the mine site. A programme of restoration of mines in Wales has reduced the mobilization of cadmium and other metals from these sites (Johnson & Eaton 1980).

Currently, there is insufficient information available to calculate the quantities of cadmium released by non-ferrous metal mines in the UK. However, the Wolfson Geochemical Atlas of England and Wales gives an insight into the larger areas affected (Applied Geochemistry Research Group 1978).

### *Production of non-ferrous metals*

Trade directories were used to identify the nature, size and location of non-ferrous metal works in the UK. Emission information was requested from about twenty major companies, but only a small number provided useful data. In the absence of relevant information, emission factors were calculated from practical studies of non-ferrous metal works in the USA (EPA 1977, 1979). These reports unfortunately gave no information on effluent discharges to aquatic systems.

#### *The Avonmouth smelter*

The Imperial Smelting furnace at Avonmouth, Bristol, is the only primary zinc smelter in the UK. Lead bullion, copper and cadmium are also produced at this plant. Details of the plant's operational characteristics are given by Coy (1984). It has been estimated from both unpublished and published data that annual stack emissions of cadmium to the atmosphere are currently about 3·5 t while annual effluent discharges to the Bristol

Channel total 2·6 t (DOE 1980; Hutton 1982; Coy 1984). It is understood that about 0·75 t $y^{-1}$ cadmium in furnace slag is retained on site where it is presumably landfilled.

*Other non-ferrous metal works*

Apparently, no other primary smelter in the UK handles ores with a significant cadmium content, and emissions from these plants are considered to be small. In the case of secondary production, the emission factors calculated from American studies were applied to the twelve largest refineries, eight lead plants and four copper works. The resulting totals of cadmium release are, however, considered to be underestimates, with atmospheric emissions of 0·2 t $y^{-1}$ and solid wastes also containing just 0·2 t $y^{-1}$ cadmium.

In the case of solid wastes, another estimate can be obtained from a recent survey which reported that the UK non-ferrous metal industry annually produces about 0·2 × $10^6$ t slag material with a cadmium content of 1–150 mg $kg^{-1}$ (DOE 1984). If an average cadmium concentration of 100 mg $kg^{-1}$ is assumed, then the slag material would contain 20 t $y^{-1}$ cadmium, all of which is probably landfilled.

## *Production and disposal of cadmium-containing products*

Cadmium has four principal uses in the UK: as a protective plating in steel, in nickel-cadmium batteries and in certain pigments and stabilizers used in plastics. Environmental discharges arise both during the manufacture of these articles and as a result of their disposal. The uses of cadmium are dissipative, with only minimal recycling of the metal. This is because most of the cadmium is consumed in the form of compounds which are present in the products at relatively low concentrations. These features pose technical and economic problems for recovery.

Published consumption statistics are not considered to be accurate for some of the uses of cadmium and these have been modified accordingly, as shown in Table 1. The estimated

TABLE 1. Estimated cadmium losses in the UK, arising from the manufacture and disposal of the major cadmium-containing articles (t $y^{-1}$)

| Use | Cadmium consumption (t $y^{-1}$)* | Manufacturing losses | | | Disposal losses | | |
|---|---|---|---|---|---|---|---|
| | | Aqueous wastes to sewers | Solid wastes to landfill | Waste pathway | Scrap steel cycle | Landfill | Recycle |
| Pigments | 630 | 13 | 13 | 606 | – | – | – |
| Electroplating | 323 | 16 | 16 | – | 116 | 175 | – |
| Ni–Cd batteries | 200† | 3 | 11 | 57 | – | – | 10 |
| Stabilizers | 158 | 2 | 2 | 154 | – | – | – |
| Total | 1311 | 34 | 42 | 817 | 116 | 175 | 10 |

*Consumption data for 1983 based on WBMS (1984), but modified according to Cadmium Association (1980) and Tron (1984).

†It is assumed that 144 t $y^{-1}$ cadmium in batteries is exported (Tron 1984). For all other uses, it has been assumed that exports equal imports.

A dash indicates that no cadmium is discharged to the pathway in question.

fate of cadmium associated with the manufacture and disposal of these articles is also depicted in Table 1. The assumptions used in the partitioning of cadmium are based on Yost (1979), Hutton (1982, 1983), DOE (1984) and Tron (1984). For the purposes of this study it has been assumed that equilibrium has been reached between the annual quantity of each product disposed of and the amount consumed.

Inspection of Table 1 reveals an estimated 34 t $y^{-1}$ cadmium discharged to the UK sewage system, principally from electroplating and pigment manufacture. Other discharge pathways of importance include the large cadmium input to the scrap steel cycle and the discharge of over 800 t $y^{-1}$ cadmium to municipal wastes.

## DISCHARGES FROM INADVERTENT SOURCES

### *Iron and steel industries*

The raw materials used in iron and steel production contain some cadmium, but the largest source of input results from the recycling of cadmium-plated steel scrap, and this is mainly used in the electric-arc process. Table 2 summarizes the estimated flow of cadmium in the

TABLE 2. Estimated cadmium discharges to the UK arising from the iron and steel industries in 1982 (t $y^{-1}$)

| Process | Cadmium in products | Atmospheric emissions | Solid wastes | | Aquatic wastes |
|---|---|---|---|---|---|
| | | | Arrested dusts | Slags | |
| Ironmaking | | | | | |
| Cokemaking | – | 0·1 | – | – | 0·2 |
| Sinter production | – | 0·9 | – | – | – |
| Pig iron | – | – | 6·8 | 1·3 | – |
| Steelmaking | | | | | |
| Basic oxygen | 4·5 | 0·1 | 2·0 | 8·1 | 0·8 |
| Electric arc | 2·3 | 1·4 | 19·0 | 3·4 | 0·7 |
| Total | 6·8 | 2·5 | 27·8 | 12·8 | 1·7 |

Consumption data for 1982 based on Iron and Steel Statistics Bureau (1983). Emission estimates based on Hutton (1982), Prater (1982) and DOE (1984).

A dash indicates that insufficient information is available to formulate an estimate, or that there are zero discharges of cadmium.

UK iron and steel industries. Production and consumption data are based on published statistics (Iron and Steel Statistics Bureau 1983), while information on the cadmium concentrations of the various raw materials, products and wastes is based on a variety of sources (Hutton 1982; Prater 1982; DOE 1984). An earlier estimate of cadmium release from the UK iron and steel industries, based on 1979 production data, assigned larger discharges than those shown in Table 2 (Hutton 1982). This difference is due to the recent decline in the output of these industries, together with the phasing out of the outmoded open hearth steelmaking process in the UK after 1979.

Solid wastes from iron and steel manufacture are mainly landfilled in the UK, but some slags are used in the construction industry (DOE 1984). Arrested dusts contain much higher concentrations of cadmium than the slags. This difference is most pronounced for the electric arc process, where the elevated cadmium content of arrested dusts is reflected by the large amount (19 t $y^{-1}$) of cadmium associated with this waste material.

A preliminary study of several UK iron foundries (Simmons 1983) indicates that this previously unquantified sector may be a significant source of cadmium discharge. Once again this relates to the increasing reliance in recent years on scrap steel as a feedstock in UK iron foundries. Further work is needed to provide sufficient data to estimate cadmium release from these facilities at the national level.

It has not been possible to fully quantify discharges to water from iron and steel production, and the values which have been assigned in Table 1 are probably underestimates (Prater 1982). The environmental pathway taken by these wastes has also not been determined, although it is known that some steelworks, such as Teesside, discharge to coastal waters.

## *Fossil fuel combustion*

### *Coal combustion*

The coal consumed in the UK contains low levels of cadmium and a value of 1 mg $kg^{-1}$ is considered to be a representative concentration (Hutton 1982). The majority of coal consumed in the UK is burnt in large power plants which operate at temperatures of up to 1500 °C. Experimental studies of modern power plants conclude that cadmium is vaporized at these temperatures but subsequently condenses onto suspended fly ash particles when the flue gases cool (Davison *et al.* 1974; Gladney *et al.* 1976; Ondov, Ragaini & Bierman 1978; Smith, Campbell & Neilsen 1979). These and other studies also report that volatile metals such as cadmium show an inverse relationship between enrichment and particle size. There is, however, disagreement with regard to the extent of this enrichment on the sub-micron particles (Smith 1980). This issue is of concern because it is the sub-micron particles which are least efficiently retained by the control devices used in modern power plants (Smith 1980). For the purpose of this paper, it has been assumed that there is a collection efficiency of 98% for cadmium in suspended fly ash particles, corresponding to an emission factor of 0·02 g $t^{-1}$.

Little information is available with regard to cadmium emissions from other sectors of coal use. Results from a UK study of one industrial use, coke production, indicates that an emission factor of 0·02 g $t^{-1}$ is also applicable to this process (Prater 1982). In the case of domestic coal-burning appliances, there are no particulate control devices, but combustion occurs at lower temperatures than in power plants and vaporization of cadmium may be less efficient. In the absence of evidence to the contrary, an emission factor of 0·02 g $t^{-1}$ has also been applied to the other sectors burning coal in the UK.

Table 3 presents the estimated partitioning of cadmium between atmospheric emission

TABLE 3. Estimated annual discharges of cadmium arising from fossil fuel combustion in the UK (t $y^{-1}$)

| Sector | Fuel | Consumption ($\times 10^6$ t)* | Atmospheric emissions | Ashes |
|---|---|---|---|---|
| Electricity generation | coal | 82·8 | 1·7 | 81·1† |
| | oil | 3·5 | – | – |
| Domestic and industry | coal | 24·9 | 0·5 | 24·4 |
| | oil | 16·1 | 0·02 | – |
| Total | | | 2·2 | 105·5 |

*Consumption data for electricity generation refer to 1982–83 (CEGB 1983) and for other sectors, 1981 (DOE 1983).
†About one-third of the ashes from this sector is used in the construction industry, the remainder being landfilled.
A dash indicates negligible cadmium burden in this material.

and solid wastes resulting from coal combustion in the UK. The majority of cadmium is retained in ashes and these are generally disposed of to landfill, but about one-third of the boiler ash from power plants is used in the construction industry (DOE 1984).

*Oil combustion*

Little attention has been paid to either the metal content of fuel oils used in the UK or their mobilization resulting from the combustion of these oils. After reviewing the available information on cadmium, Hutton (1982) assigned an average value of 0·05 mg $kg^{-1}$ to European oils and assumed a 98% retention efficiency of cadmium by oil-fired power plants. Application of the corresponding emission factor, 0·001 g $t^{-1}$, to recent consumption data results in atmospheric release of cadmium from all sectors in the UK of under 0·05 t $y^{-1}$ (Table 3). The cadmium contents of ashes from oil combustion are reported to be low and are not considered to be a significant source of discharge (DOE 1984).

### *Cement manufacture*

The raw materials used in cement manufacture — limestone, clays and shales — contain low cadmium concentrations, i.e. $<0·5$ mg $kg^{-1}$ (Brumsack 1977; DOE 1980, 1984). It is considered that the particulate control devices used at cement plants restrict atmospheric cadmium emissions from these processes to less than 1 t $y^{-1}$. In the case of land-based discharges, it is estimated from assumptions given in DOE (1980) that 21 t $y^{-1}$ cadmium are landfilled in cement wastes.

### *Manufacture and use of phosphate fertilizers*

Table 4 presents the 1982 pattern of rock phosphate imports to the UK, together with their estimated cadmium contents. Morocco and Senegal represent the most important sources

TABLE 4. Rock phosphate imports to the UK by source for 1982 together with the associated cadmium burdens

| Source country | Rock phosphate* imports 1982 | | Cadmium in rock phosphate | |
|---|---|---|---|---|
| | ($\times 10^3$ t) | (%) | Content† (mg kg$^{-1}$) | Total quantity (t) |
| USA | 53 | 4 | 6·5 | 0·3 |
| Senegal | 371 | 28 | 71 | 26·3 |
| Morocco | 859 | 65 | 18 | 15·5 |
| Tunisia | 41 | 3 | 18 | 0·7 |
| Totals | 1324 | | | 42·8 |

*Import data from Anon (1983).
†Average values, derived from Wakefield (1980) and Gunnarsson (1983).

of rock phosphates and those from Senegal contain the highest cadmium concentrations. Phosphoric acid, the basic feedstock for fertilizer manufacture, is produced from rock phosphate imports. This process results in a redistribution of cadmium from the rock phosphate to the acid product and the gypsum waste. A study of cadmium in the UK phosphate fertilizer industry (Feenstra 1978) provides sufficient data to estimate both the partitioning of cadmium between the phosphoric acid and gypsum, and to determine the fate of the gypsum waste. Feenstra (1978) reported that the majority of the gypsum was discharged via pipeline to coastal waters while only about 10% was landfilled. Applying the findings of Feenstra (1978) to 1982 data (Table 4) results in an estimated cadmium discharge of 5·7 t y$^{-1}$ to coastal waters and 1·3 t y$^{-1}$ to landfill. It is assumed that the remaining 36·8 t y$^{-1}$ cadmium are retained in the phosphate fertilizers manufactured from the phosphoric acid and that these are applied to UK agricultural land. Further consideration is given later to the relative size and implications of cadmium inputs to agricultural land from phosphate fertilizers and other sources.

## WASTE-RELATED SOURCES

### *Municipal waste disposal*

The cadmium in refuse is mainly derived from disposal of those plastics containing cadmium stabilizers and pigments, and from nickel-cadmium batteries. A survey of the published cadmium values for municipal waste concluded that 15 mg kg$^{-1}$ was a representative concentration for this material (Hutton 1982). In 1982, about 33 $\times$ 10$^6$ t municipal waste were generated in the UK, of which 30 $\times$ 10$^6$ t were landfilled (CIPFA 1983). This practice, therefore, results in an estimated 450 t y$^{-1}$ cadmium entering landfill sites in the UK.

The remaining 3 $\times$ 10$^6$ t refuse are combusted at incinerators which are generally equipped with highly efficient electrostatic precipitators for the control of particulate emissions. Experimental studies of such facilities in both the USA and the UK report a marked enrichment of cadmium on the suspended fly-ash particles and an increasing enrichment with decreasing particle size (Greenberg, Zoller & Gordon 1978; Scott 1979). This association with small particles is of relevance because it is these particles which are retained least efficiently by the emission control devices. Emission factors were calculated from these experimental studies by Hutton (1982), and an average value of 1·5 g $t^{-1}$ was assigned to this source. Application of this emission factor to the 1982 incineration data results in an atmospheric emission of 4·5 t cadmium.

The remaining 40·5 t $y^{-1}$ cadmium are retained in the fly ash and furnace clinker and both these wastes are landfilled. The cadmium in these wastes is considered to be of greater potential significance than the 450 t $y^{-1}$ cadmium discharged by direct landfill of refuse. This is because the cadmium is not only present at much higher concentrations in the ashes but is also likely to be more soluble, and therefore show a greater potential for biological uptake.

## *Sewage*

Since the sewage system receives considerable cadmium inputs, the subsequent disposal of sludge and sewage effluents is responsible for 'secondary' discharges of environmental cadmium. The cadmium in the sewage system originates from both domestic and industrial inputs as well as from road surface run-off. It is difficult to quantify the relative importance of these sources and there are currently no published estimates for the UK. However, initial estimates made by the Monitoring and Assessment Research Centre (MARC, unpublished) assign a value of 9 t $y^{-1}$ cadmium from domestic sources and 1·5 t $y^{-1}$ from road run-off. The remaining burden was presumed to originate from industrial discharges.

### *Sewage sludge disposal*

Sewage sludge production in the UK currently totals about $1 \cdot 5 \times 10^6$ t dry solids (DOE/NWC 1983). Table 5 presents the estimated disposal pattern of sludge, together with the corresponding discharges of cadmium. It is apparent that the majority of sludge is disposed of to land, and a considerable proportion of the total cadmium is applied to arable land. The relative importance of this input is considered later. The term 'non-agricultural land' refers mainly to landfill sites for sewage sludge. Sludge dumping at sea takes place at a small number of licensed sites around the UK coast. In 1982, the two major sites were the outer Thames estuary, which received 3·9 t $y^{-1}$ cadmium, and Liverpool Bay, which received 1·5 t $y^{-1}$ cadmium (Oslo Commission 1984). Table 5 shows that pipeline disposal of sludge to coastal waters is a minor route of discharge in the UK.

Incineration is a means of sludge treatment which reduces the bulky nature of this material to a low volume ash. This process redistributes the cadmium between ashes and

TABLE 5. The current pattern of sewage sludge disposal in the UK and corresponding cadmium inputs*

| Disposal route | Quantity of sludge (× $10^3$ t dry solids) | Cadmium input (t) |
|---|---|---|
| Agricultural land | | |
| Arable and horticultural land | 473·1 | 4·7 |
| Pasture | | 2·8 |
| Non-agricultural land | 318·7 | 8·5 |
| Dumped at sea | 321 | 7·3 |
| Pipeline | 19 | 0·3 |
| Incineration | 44·6 | 2·3 |
| Totals | 1176·4 | 25·9 |

*Source DOE/NWC (1983); Oslo Commission (1984).

atmospheric emissions. In the absence of any published emission data from sludge incineration, reliance is placed on the corresponding data for refuse incineration. Applying the appropriate partition characteristics of cadmium for the two waste streams to sludge incineration data for 1980 results in an estimated atmospheric emission of 0·2 t $y^{-1}$ and an input to landfill of 2·1 t $y^{-1}$.

### *Sewage and sewage-effluent disposal*

Certain sewage works in the UK, particularly those in coastal areas, discharge sewage-effluent and untreated sewage directly to adjacent aquatic systems. A recent survey (Mance & O'Donnell 1984) revealed that this activity results in major inputs of metals to coastal waters and that an estimated 24 t cadmium entered these waters in 1980 as a result of this activity.

Sewage works also release metals to fresh waters from the discharge of effluent produced by sewage treatment. This disposal pathway is estimated (MARC, unpublished) to result in an annual cadmium input of about 3 t.

## DISCUSSION

### *Inventory of cadmium release in the UK*

Estimated quantities of cadmium released to the major environmental compartments by human activities in the UK are summarized in Table 6. This inventory reveals that, in contrast to the situation for lead (Nriagu 1979), there is no single predominant atmospheric source of the metal. Thus, although the largest emissions arise from refuse

TABLE 6. Inventory of current cadmium release in the UK arising from human activities (t $y^{-1}$)

| Source | Atmosphere | Agricultural land | Landfill | Fresh waters | Estuarine and coastal waters |
|---|---|---|---|---|---|
| Non-ferrous metal production | 3·7* | – | 20·8* | – | 2·6* |
| Production and disposal of cadmium-containing materials | – | – | 217 | – | – |
| Iron and steel production | 2·5 | – | 40·6 | – | 1·7⁺ |
| Fossil fuel combustion | 2·2 | – | 78·5 | – | – |
| Cement manufacture | 1 | – | 21 | – | – |
| Manufacture and application of phosphate fertilizers | – | 35·8 | 1·3 | – | 5·7 |
| Municipal waste disposal | 4·5 | – | 490·5 | – | – |
| Sewage disposal | 0·2 | 7·5 | 10·6 | 3 | 31·6 |
| Totals | 14·1 | 43·3 | 880·3 | 3 | 41·6 |

*These values are considered to be underestimates.
⁺A portion of this input probably enters fresh waters.

incineration, these are closely followed by those from non-ferrous metal production, with ferrous metal production and coal combustion next in ranking.

It should be stressed that the broad approach adopted in this study conceals large differences in the relative environmental impact of such emissions. For example, the largest atmospheric source of cadmium, refuse incineration, takes place at over forty plants in the UK (DOE 1984). In contrast, the zinc smelter at Avonmouth is responsible for the majority of cadmium emissions from non-ferrous metal production. Increased atmospheric inputs of cadmium and associated environmental impacts in the vicinity of this smelter are well-documented (Little 1974; Coughtrey *et al.* 1979), while there is no published evidence of any increase in cadmium deposition in the vicinity of UK incinerator plants. Despite these shortcomings, the values of atmospheric cadmium release can be used to evaluate the relative contribution of different sources to the deposition of cadmium in areas removed from local sources of contamination. This approach has been used to estimate the likely changes in atmospheric cadmium deposition in background agricultural areas of Europe resulting from hypothetical controls on the major sources of atmospheric cadmium in the region (Hutton 1982).

The generalized nature of the inventory in Table 6 exaggerates the significance of landfilled cadmium in those high-volume wastes with relatively low cadmium contents, such as the ashes from coal combustion. Another example, noted earlier, is the large tonnage of cadmium (450 t $y^{-1}$) associated with landfilled refuse; it is considered that the cadmium in this material is of less environmental significance than the input of 40·5 t $y^{-1}$ to landfill from the disposal of incinerator ashes.

Several discrepancies are apparent in the inventory presented, the most notable being the important cadmium inputs associated with the disposal of cadmium-containing materials. For example, a total of 817 t $y^{-1}$ cadmium were estimated to enter the municipal waste pathway resulting from the disposal of plastics and batteries (Table 1). However, the total burden of cadmium in UK refuse itself was estimated to be only 495 t $y^{-1}$. This discrepancy may reflect an underestimate of the cadmium content of UK refuse or, more likely, that production and disposal of the cadmium-containing materials is not in equilibrium, as was assumed in this study. Another example, the disposal of cadmium-plated items, was responsible for an estimated cadmium input of 116 t $y^{-1}$ to the scrap steel cycle, yet the total burden of cadmium in the UK steel industry was given as only 42·3 t $y^{-1}$ (Table 2). One explanation for this disparity, in addition to those given above for the refuse cycle, is that iron foundries also utilize large quantities of scrap steel. Unfortunately, cadmium discharges from this sector could not be quantified.

## *Cadmium inputs to specific compartments*

### *Inputs to the UK sewage system*

There is continuing concern over environmental inputs of cadmium resulting from the disposal and utilization of sewage (Davis & Coker 1980; Oslo Commission 1984). Nevertheless, surprisingly little is known about the relative contribution of different sources of cadmium input to the sewage system. Such information would facilitate the implementation of effective control strategies, if a reduction in the cadmium content of sewage was considered necessary. This study has provided initial UK estimates for two of the three source categories, domestic inputs (9 t $y^{-1}$) and road run-off (1·5 t $y^{-1}$). Certain industrial discharges were also quantified and these totalled 34 t $y^{-1}$. The total industrial input can therefore be calculated after estimating the total cadmium burden in the UK sewage system. This is considered to be currently about 53 t $y^{-1}$, made up of 25·9 t $y^{-1}$ in sewage sludge (Table 5), 24 t $y^{-1}$ in direct coastal discharges of sewage and 3 t $y^{-1}$ in sewage works effluents. Subtraction of domestic inputs, road run-off and those industrial inputs already quantified, leaves an additional 8 t $y^{-1}$ unaccounted for. Presumably, this value represents other industrial inputs which still require identification.

### *Cadmium inputs to UK coastal waters*

Mance & O'Donnell (1984) estimated that land-based industrial discharges of cadmium to UK coastal waters totalled 22 t $y^{-1}$, based on unpublished monitoring data supplied by the regulatory authorities. An input to coastal waters from specific industrial sources of only 10 t $y^{-1}$ is accounted for in the present study, suggesting that over half of the total input remains to be identified. Unfortunately, further comparison is not possible because Mance & O'Donnell (1984) did not disaggregate the estimated total discharge into specific industrial activities.

Another finding from the Mance & O'Donnell study used in this paper was the large, and hitherto unreported, cadmium input of 24 t $y^{-1}$ arising from direct discharges of sewage to coastal waters. Previous assessments of cadmium release in the UK (DOE 1980; Norton 1982) have focused exclusively on the much smaller cadmium inputs arising from the dumping of sewage sludge in UK coastal waters.

*Cadmium inputs to UK agricultural land*

The quantities of cadmium entering UK agricultural land from phosphate fertilizer and sewage sludge applications were estimated to be 35·8 and 7·5 t $y^{-1}$. Of greater relevance for human exposure are the quantities and rates of cadmium entering arable land from these sources. This is because the soil–crop plant pathway of dietary exposure in humans is more susceptible to increases in soil cadmium than the soil–livestock pathway (Ryan, Pahren & Lucas 1982). To estimate the relevant values for phosphate fertilizers, it is first necessary to calculate an average cadmium content for this material, expressed in terms of phosphorus pentoxide ($P_2O_5$). This is estimated to be 78 g Cd $t^{-1}$, based on the total agricultural use of phosphate fertilizers in 1981–82 of $4{\cdot}57 \times 10^5$ t $P_2O_5$ (FMA 1982) and a total cadmium burden of 35·8 t. Multiplication of this value by the average application rate of phosphate fertilizers to UK arable land, 54·5 kg $P_2O_5$ $ha^{-1}$ (Church & Leech 1983), results in an annual cadmium input of 4·3 g $ha^{-1}$. This may be compared with the estimated cadmium input of 3 g $ha^{-1}$ from atmospheric deposition, a value considered to be representative of rural areas in Europe (Hutton 1982).

The total cadmium input to UK arable land from phosphate fertilizers is estimated to be 22 t $y^{-1}$, based on the UK arable land area of $5{\cdot}12 \times 10^6$ ha (FMA 1982) and the unit area cadmium input of 4·3 g $ha^{-1}$. The corresponding total cadmium input to arable land from atmospheric deposition is 15·4 t $y^{-1}$. In comparison, the total cadmium discharge to arable land from sludge application is 4·7 t $y^{-1}$ (Table 5). In contrast with the situation for phosphate fertilizers, only about 1·2% of UK arable land receives sewage sludge (Davis 1984). Sewage sludge application to this small land area at the typical rate of 5 t $ha^{-1}$ (Davis 1984) results in an annual cadmium input of about 80 g $ha^{-1}$.

In conclusion, phosphate fertilizers represent the largest source of cadmium discharge to UK arable land. Atmospheric inputs are next in ranking, but this source category receives contributions from a variety of industrial activities (Table 6). Sludge application practice in the UK results in high input rates of cadmium, but these currently affect only about 1% of the arable land area.

## REFERENCES

**Anon. (1983).** World Trends: Phosphate Rock. *Phosphorus and Potassium,* **123**, 5–7.

**Applied Geochemistry Research Group (1978).** *The Wolfson Geochemical Atlas of England and Wales.* Clarendon Press, Oxford.

**Brumsack, H.J. (1977).** Potential metal pollution in grass and soil samples around brickworks. *Environmental Geology,* **2**, 33–41.

**Cadmium Association (1980).** *Cadmium Plating in the United Kingdom.* Cadmium Association, London.

**CEGB (1983).** *CEGB Statistical Yearbook 1982–83.* Central Electricity Generating Board, London.

**Church, B.M. & Leech, P.K. (1983).** *Survey of Fertilizer Practice - Fertilizer Use on Farm Crops in England and Wales 1982.* (SS/CH/11). ADAS, Rothamsted Experimental Station, The Fertilizer Manufacturers' Association.

**CIPFA (1983).** *Waste Disposal Statistics 1981–82 Actuals.* Statistical Information Service ref. 63–83. The Chartered Institute of Public Finance and Accountancy, London.

**Coughtrey, P.J., Jones, C.H., Martin, M.H. & Shales, S.W. (1979).** Litter accumulation in woodlands contaminated by Pb, Zn, Cd and Cu. *Oecologia (Berlin),* **39**, 51–60.

**Coy, C.M. (1984).** Control of dust and fume at a primary zinc and lead smelter. *Chemistry in Britain,* **20**, 418–20.

**Davis, R.D. (1984).** Cadmium in sludge used as a fertilizer. *Experientia,* **40**, 117–126.

**Davis, R.D. & Coker, E.G. (1980).** *Cadmium in Agriculture with Special Reference to the Utilization of Sewage Sludge on Land.* Water Research Centre, Report TR 139 Stevenage, Herts.

**Davison, R.L., Natusch, D.F.S., Wallace, J.R. & Evans, C.A., Jr. (1974).** Trace elements in fly ash: dependence of concentration on particle size. *Environmental Science Technology,* **8**, 1107–1113.

**DOE (1980).** *Cadmium in the Environment and its Significance to Man.* Department of the Environment, Pollution Paper No. 17 HMSO, London.

**DOE (1983)** *Digest of Environmental Pollution and Water Statistics No. 5, 1982.* Department of the Environment HMSO, London.

**DOE (1984).** *Waste Management Paper No. 24 Cadmium-bearing Wastes.* Department of the Environment HMSO, London.

**DOE/NWC (1983).** *Sewage Sludge Survey 1980 Data.* Standing Committee on the Disposal of Sewage Sludge, Department of the Environment/National Water Council.

**Drasch, G.A. (1983).** An increase of cadmium body burden for this century — an investigation on human tissues. *Science Total Environment,* **26**, 111–119.

**Elinder, C-G. & Kjellström, T. (1977).** Cadmium concentration in samples of human kidney cortex from the 19th century. *Ambio,* **6**,270–272.

**EPA (1977).** *Assessment of Industrial Hazardous Waste Practices in the Metal Smelting and Refining Industry.*EPA/530/SW-145C, US Environmental Protection Agency, Office of Solid Waste, Washington D.C.

**EPA (1979).** *Sources of Atmospheric Cadmium.* EPA-450/5-79-006, US Environmental Protection Agency, Office of Air Quality Planning and Standards, North Carolina.

**Feenstra, J.F. (1978).** *Control of Water Pollution due to Cadmium Discharged in the Process of Producing and Using Phosphate Fertilizers.* ENV/223/74-E REV 2, Commission of the European Communities, Brussels.

**FMA (1982).** *Fertilizer Statistics 1982.* The Fertilizer Manufacturers' Association Ltd., London.

**Freney, J.R., Ivanov, M.V. & Rodhe, H. (1983).** The sulphur cycle. *The Major Biogeochemical Cycles and their Interactions* (Ed. by B. Bolin & R.B. Cook), pp 56–65. John Wiley and Sons, Chichester.

**Gladney, E.S., Small, J.A., Gordon, G.E. & Zoller, W.H. (1976).** Composition and size distribution of in-stack particulate material at a coal-fired power plant. *Atmospheric Environment,* **10**, 1071–1077.

**Greenberg, R.R., Zoller, W.H. & Gordon, G.E. (1978).** Composition and size distributions of particles released in refuse incineration. *Environmental Science Technology,* **12**, 566–573.

**Gunnarsson, O. (1983).** Heavy metals in fertilizers. Do they cause environmental and health problems?. *Fertilizers and Agriculture,* No. 85, 27–42.

**Hutton, M. (1982).** *Cadmium in the European Community. MARC Report No. 26.* Monitoring and Assessment Research Centre, Chelsea College, University of London.

**Hutton, M. (1983).** A prospective atmospheric emission inventory for cadmium — the European Community as a study area. *Science Total Environment,* **29**, 29–47.

**Iron and Steel Statistics Bureau (1983).** *Iron and Steel Industry: Annual Statistics for the United Kingdom 1982.* Iron and Steel Statistics Bureau, London.

**Johnson, M.S. & Eaton, J.W. (1980).** Environmental contamination through residual trace metal dispersal from a derelict lead-zinc mine. *Journal Environmental Quality,* **9**, 175–179.

**Little, P. (1974).** *Airborne zinc, lead and cadmium pollution and its effects on soils and vegetation.* PhD Thesis, University of Bristol.

**Mance, G. & O'Donnell, A.R. (1984).** The magnitude of land-based discharges of heavy metals entering coastal waters of the United Kingdom and the North Sea. *Environmental Contamination, International Conference 1984,* pp. 542–548. CEP Consultants Ltd., Edinburgh.

**Norton, R.L. (1982).** *Assessment of Pollution Loads to the North Sea.* Water Research Centre, Report TR 182, Medmenham, Bucks.

**Nriagu, J.O. (1979).** Global inventory of natural and anthropogenic emissions of trace metals to the atmosphere. *Nature (London),* **279**, 409–411.

**Ondov, J.M., Ragaini, R.C. & Bierman, A.H. (1978).**Elemental particle size emissions from coal-fired power plants: Use of inertial cascade impactor. *Atmospheric Environment* **12**, 1175–1185.

**Oslo Commission (1984).** *Eighth Annual Report.* Oslo Commission, London.

**Prater, B.E. (1982).** The environmental significance of cadmium in UK steelmaking operations. *Environmental Science Technology Letters,* **3**, 179–182.

**Roels, H.A., Lauwerys, R.R., Buchet, J-P. & Bernard, A. (1981).** Environmental exposure to cadmium and renal function of aged women in three areas of Belgium. *Environmental Research,* **24**, 117–130.

**Ryan, J.A., Pahren, H.R. & Lucas, J.B. (1982).** Controlling cadmium in the human food chain: a review and rationale based on health effects. *Environmental Research,* **28**, 251–302.

**Scott, D.W. (1979).** *The Fate of Cadmium in Municipal Refuse Incinerators.* Warren Spring Laboratory Report, 1710 (AP). Warren Spring, Stevenage, Herts.

**Simmons, S.A. (1983).** Estimating heavy metal emissions from iron foundry cupola furnaces. *Clean Air,* **12**,127–137.

**Smith, R.D. (1980).** The trace element chemistry of coal during combustion and the emissions from coal-fired plants. *Progressive Energy and Combustion Science,* **6**,53–119.

**Smith, R.D., Campbell, J.A. & Nielsen, K.K. (1979).** Concentration dependence upon particle size of volatilized elements in fly ash. *Environmental Science Technology,* **13**, 553–558.

**Tron, A.R. (1984).** *Battery-related Wastes in the U.K. and their Disposal.* Warren Spring Laboratory Report, LR 430 (MR). Warren Spring, Stevenage, Herts.

**Wakefield, Z.T. (1980).** *Distribution of Cadmium and Selected Heavy Metals in Phosphate Fertilizer Processing.* Tennessee Valley Authority, Tennessee.

**WBMS (1984).** *World Metal Statistics,* 37, No. 12, World Bureau of Metal Statistics, London.

**Yost, K.J. (1979).** Some aspects of cadmium flow in the US. *Environmental Health Perspectives,* **28**, 5–16.

**Zoller, W.H. (1984).** Anthropogenic perturbation of metal fluxes into the atmosphere. *Changing Metal Cycles and Human Health* (Ed by J.O. Nriagu), pp. 27–41. Springer Verlag, Berlin.

# Physico-chemical speciation and chemical transformations of toxic metals in the environment

R. M. HARRISON
*Department of Chemistry, University of Essex, Colchester CO4 3SQ*

## SUMMARY

**1** The commoner experimental techniques for speciation of trace metals such as Pb, Cd, Cu and Zn in environmental media are described. The shortcomings inherent in the various methodologies are discussed and selected results of speciation investigations are reviewed.

**2** Where precise chemical speciation information is obtainable, it is possible to infer chemical transformation reactions occurring within the environment, and some such reactions are proposed.

## INTRODUCTION

In studies of the behaviour of trace metals in natural systems, there are many instances in which the ecologist will wish to know the precise physico-chemical form (known as speciation) of the metal. This is because speciation profoundly influences the environmental mobility and toxicity of the metal. For example, if it is intended to dose a soil with lead to simulate a soil polluted by industrial, or automotive lead, the chemical form of the added lead may influence appreciably its availability for plant uptake. It is thus desirable to have knowledge of the physico-chemical forms in which pollutant lead enters soils, and also of how rapidly these chemical forms alter within the soil, and to what other forms they change.

This problem extends far beyond soils. In another example, suppose an animal is exposed to lead salts by ingestion. It is well established that the gastro-intestinal absorption of lead is influenced by the chemical form of the ingested lead (Chamberlain *et al.* 1978). For inhaled lead, absorption is far less dependent upon the chemistry of the lead (Chamberlain *et al.* 1978), but the proportion of inhaled lead subject to tracheo-bronchial as opposed to alveolar deposition is a function of aerodynamic particle size (another aspect of physico-chemical speciation). Lead deposited in the tracheo-bronchial region is elevated by the muco-ciliary escalator mechanism, swallowed and may then be available for gastro-intestinal absorption (O'Neill, Harrison & Williams 1982).

Similar considerations apply in the aquatic environment. Metal associated with rather large-grained suspended sediment will rapidly deposit and then be relatively immobile providing a source of exposure for organisms which live within the bottom sediment. Metal in smaller size associations will remain in the water column for longer periods, but the toxicity is crucially dependent upon chemical form (Laxen 1983); generally, complexed forms of metals are relatively unavailable and present a far lesser toxic hazard than free metal ions.

TABLE 1. Compounds of lead identified in the atmosphere by X-ray powder diffraction

| Source | Compounds |
|---|---|
| Automotive lead | $PbSO_4.(NH_4)_2SO_4$ |
| | $PbSO_4$ |
| | PbBrCl |
| | $PbBrCl.(NH_4)_2BrCl$ |
| | $\alpha$-$2PbBrCl.NH_4Cl$ |
| | $PbBrCl.2NH_4Cl$ |
| Industrial (smelter) lead | PbS |
| | $PbSO_4$ |
| | $PbO.PbSO_4$ |

## METAL AEROSOLS IN THE ATMOSPHERE

A variety of experimental techniques exist which have the potential for use in physico-chemical speciation of atmospheric aerosols (de Mora & Harrison 1984). Two methods have been used most frequently: electron microscopy with X-ray analysis and X-ray powder diffraction. The two techniques are essentially complementary in that the former examines individual particles, indicating, but not fully quantifying, chemical composition. The latter method is applied to the bulk sample and allows unequivocal identification of major crystalline components from their diffraction patterns; amorphous and poorly crystalline substances are not identifiable.

Our work has centred mainly upon lead because of the interest which it has stimulated as a pollutant, and because of its relative abundance in polluted air (concentrations of the order of 1 $\mu g\ m^{-3}$ are very commonly encountered). Our greatest success in speciation of lead in air has been with X-ray powder diffraction, and the compounds identified in our work and the related studies of other workers are shown in Table 1. By far the most frequently encountered compound is the mixed salt $PbSO_4.(NH_4)_2SO_4$. This had not been reported as present in the environment when first detected by our group (Biggins & Harrison 1979), but has since been found in the USA (O'Connor & Jaklevic 1981; Tani *et al.* 1983). Since it is well accepted that the major vehicle-emitted lead compound is lead bromochloride, PbBrCl, the question arose of how the ammonium lead sulphate is formed. Laboratory studies indicated that the likely mechanism is via coagulation of sub-micrometre PbBrCl with ambient ammonium sulphate aerosol, with reaction as below in aqueous solution (the $(NH_4)_2SO_4$ is hygroscopic):

$$2PbBrCl + 2(NH_4)_2SO_4 \rightarrow PbSO_4.(NH_4)_2SO_4 + PbBrCl.(NH_4)_2BrCl$$

The second reaction product has also been identified in some air samples by X-ray powder diffraction (Table 1).

Most other toxic metals occur most commonly in air at lower concentrations than lead and are not readily identified by any presently available speciation method. Zinc has been reported in several studies of ambient air to be present in the form of zinc ammonium sulphate $ZnSO_4.(NH_4)_2SO_4.6H_2O$, and metal ammonium sulphate mixed salts appear to be

TABLE 2. Measurements of specific tetraalkyllead compounds in the atmosphere (from Hewitt & Harrison 1985)

| Time | Compound* (ng $m^{-3}$) | | | | |
|---|---|---|---|---|---|
| | TML | TMEL | DMDEL | MTEL | TEL |
| Town centre | | | | | |
| 02.00–05.00 | 3·2 | 0·4 | 0·2 | 0·4 | 1·5 |
| 08.00–11.00 | 50·7 | 3·4 | 5·4 | 3·8 | 78·2 |
| 17.00–20.00 | 22·6 | 1·4 | 1·6 | 1·2 | 30·1 |
| Semi rural | | | | | |
| 23.00–02.00 | 4·6 | — | — | — | 0·6 |
| 08.00–11.00 | 1·6 | — | — | — | 1·1 |
| 17.00–20.00 | 5·0 | — | — | — | 1·1 |

– Represents below detection limit (approx. 0·2 ng $m^{-3}$).
* TML, tetramethyllead; TMEL, trimethylethyllead; DMDEL, dimethyl - diethyllead; MTEL, methyltriethyllead; TEL, tetraethyllead.

of significance for a number of metals in the air (de Mora & Harrison 1984). In a study of a major zinc–lead smelter, from which cadmium is produced as a by-product, α-ZnS*, ZnO and CdO were identified in stack emissions, whilst β–ZnS was found in the surrounding atmosphere. The large particle sizes of the β-ZnS suggested that blowage of dusts from ore stockpiles was the major source at this sampling site. Cadmium is commonly present in air at concentrations of only a few ng $m^{-3}$ which is insufficient for speciation work with current techniques (Williams & Harrison 1984).

## METAL VAPOURS AND ALKYLS IN THE ATMOSPHERE

Volatile forms of metals and their compounds may exist in air in the vapour phase, when their behaviour will be rather different from that of particulate metals and the problems of speciation are entirely different.

Mercury is an element with a number of volatile forms including elemental vapour, mono- and dialkyl species and inorganic salts of high vapour pressure. At present there is no entirely satisfactory means of identification of specific chemical forms at atmospheric concentrations, although some useful fractionation schemes do exist (de Mora & Harrison 1984).

One focus of our own work has been the development of a sensitive gas chromatography–atomic absorption spectrophotometry system which allows separation of individual tetraalkyllead compounds and their analysis in picogram quantities (de Mora, Hewitt & Harrison 1984). This has been applied with considerable success to the analysis of tetraalkyllead in both urban and rural atmospheres (Hewitt & Harrison 1985) and has the potential for application to volatile forms of other metals and metalloids. Some results from tetraalkyllead analysis by this technique are presented in Table 2. The method is applicable also to tetraalkyllead compounds extracted from other environmental media, and to ionic alkyllead after butylation with a Grignard reagent (Harrison & Radojević 1985).

* α-ZnS and β-ZnS are different crystal forms of zinc sulphide.

## METALS IN SEDIMENTS AND SOILS

Generally, toxic trace metals exist even in polluted sediments and soils at concentrations rarely exceeding 1000 mg $kg^{-1}$ (0·1%). Their existence as discrete mineral phases is unlikely, other than when they arise by dispersion from an area of mineralization and, even if present in this form, their abundance is too low to enable identification by presently available techniques. The best approaches to speciation currently available depend upon sequentially dissolving different major fractions of a sediment or soil, and measuring the trace metal released during each sequential stage.

Salomons & Förstner (1980) elaborated five mechanisms for metal accumulation on sedimentary particles:

(a) adsorptive bonding on fine-grained substances;
(b) precipitation of discrete metal substances;
(c) co-precipitation of metals with hydrous iron and manganese oxides and carbonates;
(d) association with organic components;
(e) incorporation in crystalline material (i.e. mineral lattices).

There are a number of sequential extraction schemes designed to separate the species indicated above, although the availability of selective reagents is very limited and only a partial separation can be achieved. The classification of metal forms, and types of extracting reagent utilised are listed in Table 3. In the most commonly utilized scheme, due to Tessier, Campbell & Bisson (1979), the 'soluble' and 'exchangeable' fractions are not distinguished.

There is much evidence to suggest that sequential extractions do not cleanly remove 100% of metal in a given association and, in any case, there will be many minor chemical forms not described within Table 3. The results obtained should therefore be regarded as essentially 'operationally-defined', i.e. fractions are determined by the extraction reagent used, as much as by the speciation of the metal in the sample. However, the results obtained with such schemes make a great deal of qualitative sense and give confidence in

TABLE 3. Classification of metal associations in sediments, soils, and deposited dusts (after Harrison *et al.* 1981)

| Classification | Form of association | Extraction technique |
|---|---|---|
| Soluble | Metal ppt, pore water | Release to pore water or river water |
| Exchangeable | Specifically adsorbed, exchangeable | Exchange with excess cations |
| Carbonate phase | Ppt or co-ppt | Release by mild acid |
| Fe–Mn oxide phase | Specifically adsorbed co-ppt | Reduction |
| Organic phase | Complex, adsorbed | Oxidation |
| Residual | In mineral lattices | Digestion with strong acids |

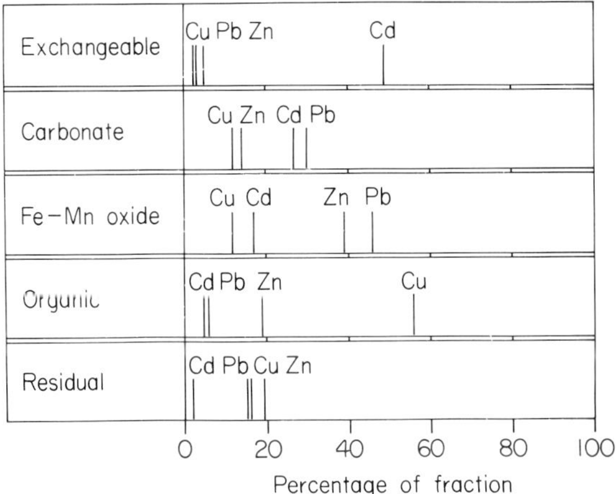


FIG. 1. Distribution of trace metals between different chemical fractions of suspended sediment from road drainage waters (mean of two samples) as a percentage of total metal.

the methodology. Fig. 1 shows mean results obtained from the analysis of suspended sediment collected from two samples of road drainage water. Certain features, such as the high level of 'exchangeable' cadmium, and the abundance of 'organic' copper show up frequently and are consistent with our knowledge of the chemical behaviour of these metals in sediments and soils. Cadmium shows an elevation of the 'exchangeable' fraction relative to the other metals in soils also, and there is a clear indication from the analysis of soil cores of the relative mobility of this element.

Rather than evaluating the chemical association of metals with the solid matter in soils, other workers have studied the potentially mobile material contained within the pore waters. Examples of this work are the studies of Tills & Alloway (1983) and Haswell, O'Neill & Bancroft (1983). The former workers used ion exchange techniques to separate four fractions: neutral, cationic, anionic and low polarity organic complexed. Whilst clearly contributing to our understanding of metal associations in soil solutions, this tells us rather little about specific chemical forms and indeed the authors did no more than make suggestions based upon inappropriate inorganic chemical modelling studies. Haswell, O'Neill & Bancroft (1983) used reverse phase high performance liquid chromatography to separate associations of lead with polar dissolved organic compounds in soil pore waters. Both u.v. absorbance and metal concentration were measured in the HPLC effluent and it was concluded that the majority of the soluble lead was associated with citric, malic and formic acids. However, the use of orthophosphoric acid at pH 2·6 as eluent may well have profoundly altered the speciation of metals in the original soil solution. It is clear that methods of this type will need considerable further development before confidence may be placed in the results obtained.

## METALS IN ROADSIDE DUSTS

There has been a great interest in street dusts (i.e. sedimented dusts collected from pavements and gutters) arising from their possible significance as a route of exposure of

TABLE 4. Lead phases identified in street dusts by X-ray diffraction (from Biggins & Harrison 1980)

| |
|---|
| $PbSO_4$ |
| $Pb^{\circ}$ |
| $PbSO_4.(NH_4)_2SO_4$ |
| $Pb_3O_4$ |
| $PbO.PbSO_4$ |
| $2PbCO_3.Pb(OH)_2$ |

children to lead. Initial approaches to speciation involved metal enrichment followed by X-ray powder diffraction. Metal phases identified in our work (Biggins & Harrison 1980) are shown in Table 4 and include compounds originating both from motor vehicles and leaded paint. Other studies (Olson & Skogerboe 1975) have reported $PbSO_4$ as the most abundant phase of lead in street dusts, and this compound appeared commonly in our work. However, crude quantification of the abundance of $PbSO_4$ indicated that it could account for only, at most, a few per cent of the lead present in the dust (Biggins & Harrison 1980) and we therefore assumed that non-crystalline forms of the metal were dominant. The sequential extraction scheme of Tessier *et al.* (1979) was therefore applied and showed a distribution of Pb, Cd, Cu and Zn very similar to that in local roadside soils (Harrison, Laxen & Wilson 1981) and a behaviour for lead quite different to that given by finely ground $PbSO_4$ added to the soil.

Since the predominant form of automotive lead in the atmosphere is believed to be $PbSO_4.(NH_4)_2SO_4$, the presence of $PbSO_4$ and non-crystalline forms of lead required some explanation. A laboratory study showed that water-leaching of ammonium lead sulphate gave high yields of $PbSO_4$. The latter is significantly water soluble and presumably weathers to give the broad distribution of chemical associations of lead shown by the sequential extraction scheme.

$$PbSO_4(NH_4)_2SO_4 \xleftarrow{\text{water leaching}} PbSO_4 \xrightarrow{\text{weathering}} \text{non-crystalline forms of lead}$$

## METALS IN NATURAL WATERS

Natural waters provide a very severe challenge in speciation studies because of the highly variable chemical matrix which they provide, the variety of metal forms which may exist in a natural water, and the generally low overall concentrations of metal present.

There have been three major approaches to elucidation of the speciation of metals in natural waters:

(a) modelling studies based upon knowledge of stability constants of ion pairs and metal complexes;
(b) laboratory studies on simplified systems;
(c) analytical studies of real water samples.

Although the first two approaches have been widely used, they cannot account adequately for the complexity of natural waters (e.g. the presence of humic substances of poorly defined composition, or of colloidal metal-oxide species) and hence generate results of true value only in rather simple systems. The third approach, i.e. the actual analytical determination of species in a real water sample, is fraught with problems, but some attempts have been made to follow this route. Since no technique is available for unequivocal characterization of any specific metal form at the concentrations normally encountered in natural waters, it is necessary to fall back upon the concept of operational definition of species outlined earlier in the context of sediments.

Table 5 indicates some of the range of metal species which may be expected to occur in a natural water. A number of speciation schemes have been proposed (de Mora & Harrison 1984); one devised in our own laboratories (Laxen & Harrison 1981a) is fairly typical and is used here to exemplify the methodology adopted. The scheme includes four major components, which are as follows.

(a) Parallel filtration of the sample through five Nuclepore filters of different pore diameter. This utilizes the fact that different physico-chemical species are associated with different physical sizes (Table 5) and hence useful information may be gained from such a fractionation.

(b) Analysis of selected size fractions by differential pulse anodic stripping voltammetry (DPASV) at the natural pH of the sample. This approach is based upon the fact that the DPASV measurement is sensitive only to certain chemical forms of the metal — termed 'ASV-labile' — generally assumed to be free metal ions and simple ion pairs. Recent work in our laboratory has indicated that the ASV response is a complex function of the pH and ionic strength of the solution (Davison *et al.* 1985) and that quantitative interpretation of DPASV measurements in terms of metal speciation is fraught with problems.

(c) Irradiation with ultra-violet light, followed by DPASV measurement. The

TABLE 5. Metal species occurring in natural waters in relation to size association

| Typical size range | Metal species | Example | Phase state |
|---|---|---|---|
| < 1 nm | Free metal ions | $Pb^{2+}$, $Cu^{2+}$ | Dissolved |
| 1–10 nm | Inorganic ion pairs, inorganic complexes, low MW organic complexes | $CdCl_4^{2-}$ Pb–fulvates | Dissolved |
| 10–100 nm | High MW organic complexes | Cu-humates | Colloidal |
| 100–1000 nm | Adsorbed onto inorganic colloids (or complexation by surface-adsorbed humics); associated with detritus | Co-$MnO_2$ Pb-FeOOH | Colloidal |
| < 1000 nm | Adsorbed onto living cells; associated with mineral solids and precipitates | Cd-clay $2PbCO_3.Pb(OH)_2$ | Particulate |

irradiation decomposes organic complexing agents, freeing complexed metal into an ASV-labile form. This technique may work with saline waters, but in freshwaters the UV irradiation destabilizes inorganic colloidal species causing them to precipitate, scavenging a considerable proportion of the trace metal of the sample in the process. Valuable qualitative information may be gained from use of the method (Harrison & Wilson 1982).

(d) Equilibration with Chelex exchange resin. The Chelex removes metal from thermodynamically weakly complexed forms, allowing estimation of this fraction by difference.

Other techniques, such as ultrafiltration and gel chromatography can complement these schemes, but none is without major inherent problems.

In our experience, the most useful component of the speciation scheme is the determination of size-association. For example, the final effluent from a sewage treatment works contained cadmium in an organically-complexed form predominantly within the dissolved and colloidal size ranges (Laxen & Harrison 1981b). The effluent was discharged to a river where it elevated Cd concentrations from $< 1\ \mu g\ l^{-1}$ to $9\ \mu g\ l^{-1}$, but showed no appreciable change in speciation. Its size-association suggests appreciable mobility within the river, but low aquatic toxicity is indicated due to the strong complexation.

In another study of a lead-acid battery works effluent which contained $4\ mg\ l^{-1}$ Pb (Laxen & Harrison 1983), the lead was present in two major fractions, firstly free metal ions and ion pairs, and secondly particulate $PbSO_4$ in the range 1–10 $\mu m$ diameter. On entering a turbulent river, the effluent lead became associated rapidly with large particles of $> 12\ \mu m$ and would be expected to be of low bioavailability and amenable to sedimentation in a slow-running section of the river.

## CONCLUSIONS

Progress is being made in the development of experimental techniques of metal speciation. Much remains to be done, however, before fully satisfactory methodology is available for all environmental media. In particular, amorphous material in atmospheric aerosols, and particulate and colloidal metal associations in natural waters, are areas requiring a great deal of research effort. Full understanding of chemical transitions will come about only through improved knowledge of metal speciation.

## REFERENCES

**Biggins, P.D.E. & Harrison, R.M. (1979).** Atmospheric chemistry of automotive lead. *Environmental Science & Technology,* **13**, 558–565.

**Biggins, P.D.E. & Harrison, R.M. (1980).** Chemical speciation of lead compounds in street dusts. *Environmental Science & Technology,* **14**, 336–339.

**Chamberlain, A.C., Heard, M.J., Little, P., Newton, D., Wells, A.C. & Wiffen, R.D. (1978).** Investigations into lead from motor vehicles. *AERE Report R9198,* HMSO, London.

**Davison, W., Harrison, R.M., de Mora, S.J. & Wilson, S. (1985).** pH and ionic strength dependence of the ASV response of cadmium, lead and zinc in solutions which simulate natural waters. *Science of the Total Environment* (in press).

**De Mora S.J. & Harrison, R.M. (1984).** Physico-chemical speciation of inorganic compounds in environmental media. *Hazard Assessment of Chemicals: Current Developments, Vol. III* (Ed. by J.I. Saxena), pp. 1–61. Academic Press, Orlando, Florida.

**De Mora, S.J., Hewitt, C.N. & Harrison, R.M. (1984).** Environmental applications of gas chromatography-atomic absorption spectroscopy. *Analytical Proceedings,* **21**, 415–418.

**Harrison R.M., Laxen D.P.H. & Wilson, S.J. (1981).** Chemical associations of lead, cadmium, copper and zinc in street dusts and roadside soils. *Environmental Science & Technology,* **15**, 1378–1383.

**Harrison, R.M. & Wilson, S.J. (1982).** Physico-chemical speciation of trace metals in environmental samples *Analytical Techniques in Environmental Chemistry 2* (Ed. by J. Albaiges), pp. 301–314. Pergamon Press, Oxford.

**Harrison, R.M. & Radojević, M. (1985).** Determination of tetraalkyl and ionic alkyllead compounds in environmental samples by butylation and gas chromatography-atomic absorption. *Environmental Technology Letters,* **6**, 129–136.

**Haswell, S.J., O'Neill, P. & Bancroft, K.C.C. (1983).** Association of lead with polar dissolved organic compounds in soil pore water. *Proc. International Conference: 'Heavy Metals in the Environment', Heidelberg,* CEP Consultants, Edinburgh.

**Hewitt, C.N. & Harrison, R.M. (1985).** A sensitive and specific method for the determination of tetraalkyllead in air by gas chromatography-atomic absorption spectroscopy. *Analytica Chimica Acta,* **167**, 277–287.

**Laxen, D.P.H. (1983).** The chemistry of metal pollutants in water. *Pollution: Causes, Effects and Control* (Ed. by R.M. Harrison), pp. 104–123, Royal Society of Chemistry, London.

**Laxen, D.P.H. & Harrison, R.M. (1981a).** A scheme for the physico-chemical speciation of trace metals in freshwater samples. *Science of the Total Environment,* **19**,59–82.

**Laxen, D.P.H. & Harrison, R.M. (1981b).** The physico-chemical speciation of Cd, Pb, Cu, Fe and Mn in the final effluent of a sewage treatment works and its impact on speciation in the receiving river. *Water Research,* **15**, 1053–1065.

**Laxen, D.P.H. & Harrison, R.M. (1983).** The physico-chemical speciation of selected metals in the treated effluent of a lead-acid battery manufacturer and its effect upon metal speciation in the receiving water. *Water Research,* **17**,71–80.

**O'Connor, B.H. & Jaklevic, J.M. (1981).** Characterization of ambient aerosol particle samples from the St Louis area by X-ray powder diffractometry. *Atmospheric Environment,* **15**,1681–1690.

**Olson, K.W. & Skogerboe, R.K. (1975).** Identification of soil lead compounds from automotive sources. *Environmental Science & Technology,* **9**, 227–230.

**O'Neill, I.K., Harrison, R.M. & Williams, C.R. (1982).** Characterisation of airborne heavy metal particulate in a zinc-lead smelter. Potential importance of gastrointestinal absorption in occupational exposure. *Transactions of the Institute of Mining and Metallurgy,* **91** C84–90.

**Salomons, W. & Förstner, U. (1980).** Trace metal analysis on polluted sediments, Part II. Evaluation of environmental impact. *Environmental & Technology Letters,* **1**, 506–517.

**Tessier, A., Campbell, P.C.G. & Bisson, M. (1979).** Sequential extraction procedure for the speciation of particulate trace metals. *Analytical Chemistry,* **51**, 844–851.

**Tani B., Siegel A., Johnson S.A. & Kumar, R. (1983).** X-ray diffraction identification of atmospheric aerosols in the 0·3-1·0 μm aerodynamic size range. *Atmospheric Environment,* **17**, 2277–2283.

**Tills, A.R. & Alloway, B.J. (1983).** The speciation of cadmium and lead in soil solutions from polluted soils, *Proc. International Conference: 'Heavy Metals in the Environment', Heidelberg.* CEP Consultants, Edinburgh.

**Williams, C.R. & Harrison, R.M. (1984).** Cadmium in the atmosphere. *Experientia,* **40**, 29–36.

# Trace elements in saxicolous lichens

D. A. JENKINS

*Department of Biochemistry and Soil Science, University College of North Wales, Bangor, Gwynedd, LL57 2UW*

## SUMMARY

1 Analyses of saxicolous lichens from Snowdonia indicate the relative importance of atmospheric deposition in their trace element content as compared to contributions from the underlying rock substrate. This relationship is more clearly expressed with foliaceous species on rock types which are slow weathering and with inherently low trace element contents, being most obvious for vein quartz: it is explicable in terms of the peculiar physiology of lichens and their slow growth rates.

2 It is suggested that the composition of carefully selected samples of saxicolous lichens can be used as a monitor of atmospheric deposition. On this basis, analyses of *Parmelia* species (particularly *P. omphalodes* (L.) Ach.) from a number of sites in Britain and other parts of the globe reveal interesting trends. It would appear that most parts of Britain's land surface, even where nominally 'unpolluted', receive significant atmospheric inputs of such trace elements as Pb, Sn, Cu, Ag, Zn, Mo, Be, etc., the development of this trend over the first half of this century being evident in analyses of samples from herbarium collections.

3 Only in remoter parts of the Southern hemisphere is the effect not detectable, and even there localized trends of increasing trace element deposition are now to be found around urban centres. This geochemical pattern reflects man's accumulative extraction of trace elements from localized concentrations in ore deposits, their use, and their redispersion back to the environment, in part via the atmosphere.

## INTRODUCTION

Modern industrial society is heavily dependent on the use of metals. This dependence can be traced back over 10 000 years to the copper artefacts found in the eastern Mediterranean, marking the beginning of the Chalcolithic period. Since then man's use of, and dependence on, trace elements has expanded to include a significant proportion of the periodic table. This has been possible because metals and other trace elements, whilst only present within the earth's crust at average concentrations measured in mg $kg^{-1}$ (or ppm), have been geochemically concentrated in localized ore deposits to the extent where their extraction has proved possible. However, an important corollary to this useage has been the redispersion of these same elements back to the earth's surface, often via the atmosphere and thence to the biosphere. The scale of the imbalance which this has caused in the natural cycling of trace elements has recently been quantified by Nriagu (1979): from a detailed consideration of available data he has calculated that anthropogenic contributions to the atmosphere now exceed those from natural sources by a factor of around twofold in the case of Ni and Cu, to tenfold for Cd and Zn, and up to twentyfold in the case of Pb. Such major changes have imposed on the biosphere significant burdens whose potentially harmful effects are yet to be evaluated fully.

The biosphere derives trace elements from various sources. For most terricolous plants the dominant source is the soil, this being the basis for geobotanical and biogeochemical prospecting (Brooks 1972). In certain circumstances, however, atmospheric deposition can contribute a significant component of the inorganic composition of the plant. This is especially true in the case of lichens for a number of reasons. First, lichens are a symbiotic association of fungus and alga which lack the rooting and vascular system of higher plants whereby nutrients can be readily extracted from the substrate and transmitted through the organism. Instead, being 'ectohydrates' they absorb nutrients over the whole of their surfaces (Hale 1983) by mechanisms, including ion exchange, which have been investigated and reviewed by Brown (1976) and Nieboer, Richardson & Tomassini (1978). Second, they lack the protective excluding cuticle of higher plants, having a porous surface which solutions can readily penetrate and within which fine particulate material can be entrapped (Fig. 1a). Third, they are relatively slow-growing plants, rates for saxicolous species varying from 10 mm per year (foliaceous species) down to 0·1 mm per year (crustaceous species) such that some individual plants may achieve ages of the order of hundreds to thousands of years (Hale 1983): atmospheric material can thus be accumulated over long periods. Finally, they would appear to be tolerant of the high levels of otherwise toxic elements which can build up within their tissues. Although most species are susceptible to damage by the higher levels of $SO_2$ in industrial and urban areas (Ferry, Badeley & Hawksworth 1973), and although instances of heavy metal toxicity have been recorded (Nieboer *et al.* 1978), tolerance is evident in the colonization by some species of the actual sulphide ore minerals of Pb, Cu, Zn and similar elements. It is possible that this tolerance is achieved by the immobilization of potentially toxic elements as extracellular precipitates. For example, recent studies have demonstrated the presence in certain lichen species of copper oxalate (Purvis 1984) and manganese oxalate crystals (Wilson & Jones 1984) as well as Pb compounds including hydrocerussite (Jones, Wilson & Launden 1982).

For these various reasons the inorganic composition of lichens can be particularly responsive to atmospheric material. The possibility of its use to monitor trace elements in atmospheric deposition has long been recognized (Jenkins & Davies 1966; Nieboer *et al.* 1972; Seward 1973) as for similar reasons has that of bryophytes (Goodman & Roberts 1971) and — in the marine environment — of algae (Phillips 1977). This realization achieved a more immediate significance in the context of radioactive fallout where the accumulation of radionuclides by arctic lichens browsed by reindeer pointed to the dangers of enhanced levels in those people subsequently eating the reindeer (Tuominen & Jaakkola 1974). The general interest in lichens and pollution is now expressed in a specific bibliography on this topic which is published regularly in *The Lichenologist* and in the reviews by Ferry *et al.* (1973) and Hawksworth & Rose (1976). In this paper an account is first presented of a study of trace elements in saxicolous lichens in relation to their substrates in Snowdonia which raised the possibility of using analyses of carefully selected lichens to monitor atmospheric deposition. Some analyses are then presented to illustrate variations in relatively unpolluted areas both within the UK and globally, and also with

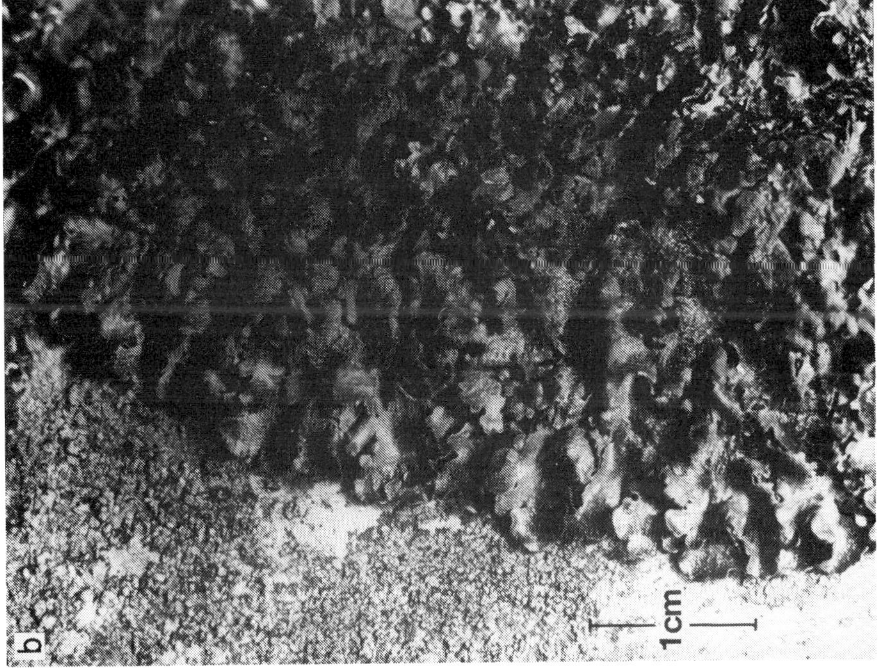


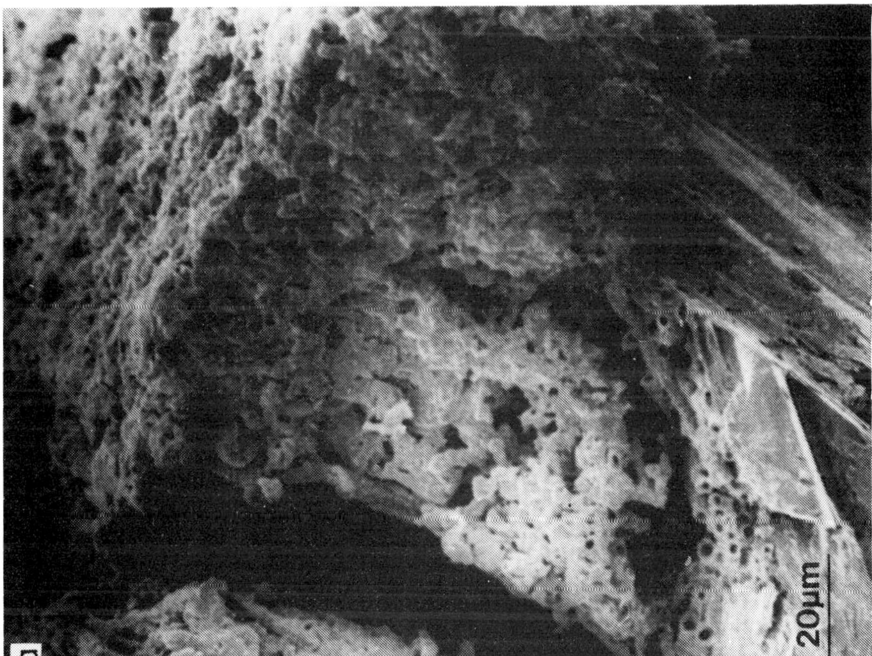


FIG. 1. (a) *Pertusaria* species colonizing amphibolite. (b) *Parmelia omphalodes* on rhyolitic ash.

time. Finally, the significance of the perceived trends is assessed in terms of the input of trace elements to the biosphere from atmospheric sources.

## SAMPLING AND ANALYTICAL PROCEDURES

Lichen plants were sampled from sites on the crest of a rock outcrop or boulder of identifiable petrology, such that they received no run-off from either adjacent surfaces or overhanging vegetation. Obvious bird perches were also avoided. Whole plants amounting, where possible, to several grams dry weight were detached, stored in paper envelopes, and then dried at 105 °C prior to subsampling for analysis. Samples were not washed or pretreated in any way as it was the total inorganic composition that was of interest. For the initial studies on rock weathering, samples of the rock were detached with the lichen plant intact, and these were subsampled in the laboratory for the unaltered fresh rock material, and for the weathered zone in contact with the lichen, by careful crushing and hand picking. Selected lichen/rock samples were also resin-impregnated and standard petrological thin-sections prepared of the total system for microscopic examination.

A semiquantitative arc-spectrographic method was employed as a convenient means of investigating trends in the trace element contents. This has the advantage of providing simultaneous results for a dozen or so elements, diminishing the inevitable preselection that occurs when a limited number of elements (e.g. Cu, Pb, Zn) are estimated by, for example, atomic-absorption spectroscopy. Some 1–2 g of the oven-dried lichen was ashed overnight at 470 °C in silica crucibles. To minimize matrix effects, 11 mg of the ash was diluted with 9 mg of standard base (i.e. 2 units on the scale described below) and then diluted 1:1 with carbon, prior to arcing using a modification of the method described by Mitchell (1964). First and second minutes were recorded separately over the spectral range 245–360 nm on Kodak SA1 plates. Visual comparison was then made in a projection comparator against standards over the range 1–10 000 mg kg$^{-1}$ on the series $10^{n/2}$ (i.e. . . . 1·0,3·2,10,32,100 . . .) By interpolation, three intermediate values can be recognized so that the results presented are limited to the logarithmic series of $10^{n/8}$ (i.e. . . . 10,13,18,24,32,42,56,75,100,130,180 . . .). This logarithmic scale has the advantage of results approximating to a normal distribution and of producing more homogeneous variance leading to easier statistical treatment (cf. Erdman & Gough 1977): 'averages' therefore take the form of geometric means.

All results are presented in terms of ashed rather than air-dried material, as this removes the effects of dilution by organic components and is a more useful basis for geochemical comparisons with rock and soil material. Results can be readily converted to the air-dried basis by subtracting the relevant number of units on the $10^{n/8}$ scale above as determined by the % ash values (e.g. 20·6–14·4% 6 units; 15·4–11·6% 7; 11·6–8·4% 8; 8·4–6·3% 9; etc.). Samples checked by wet ashing and atomic absorption spectroscopy for Cu, Pb and Zn agreed within 2 units or better on the $10^{n/8}$ scale and reproducibility between replicates was of the same order (i.e. × 1·8 to × 0·56 or better). The spectral lines used and their respective lower sensitivity limits are given in Table 1, together with the range of values encountered in lichen ashes and their geometric means.

TABLE 1. Summary of analytical conditions and results. Values in mg kg$^{-1}$ (ppm) in the lichen ash

| | λnm (1st/2nd min) | Detection limit (*S*) | % >*S* | Range | Geometric mean |
|---|---|---|---|---|---|
| Ag | 313·1 (1) | 1·8 | 44 | < - 13 | 1·8 |
| Be | 324·2 (1) | 10 | 8 | < - 18 | < |
| Co | 345·4 (2) | 7·5 | 80 | < - 75 | 13 |
| Cr | 267·7 (2) | 5·6 | 98 | < - 1000 | 75 |
| Cu | 327·4 (1) | 18 | 82 | < - 1000 | 100 |
| Ga | 294·4 (1) | 1·8 | 100 | 1·8 - 100 | 13 |
| Ge | 265·1 (1) | 18 | 6 | < - 18 | < |
| Mn | 280·1 (2) | 5·6 | 100 | < - >3·2% | 1300 |
| Mo | 317·0 (2) | 10 | 16 | < - 24 | < |
| Ni | 341·5 (2) | 1·8 | 100 | 1·8 - 320 | 42 |
| Pb | 283·3 (1) | 18 | 82 | < - 1·0% | 320 |
| Sn | 284·0 (1) | 18 | 54 | < - 24 | 32 |
| Ti | 324·2 (2) | 5·6 | 100 | 1300 - >3·2% | 5600 |
| V | 318·3 (2) | 32 | 92 | < - 750 | 180 |
| Y | 332·8 (2) | 5·6 | 90 | < - 100 | 18 |
| Yb | 328·9 (2) | 1·8 | 82 | < - 7·5 | 3·2 |
| Zn | 334·5 (1) | 420 | 76 | < - 3200 | 750 |
| Zr | 343·8 (2) | 5·6 | 100 | 32 - 2400 | 240 |

< Signifies not detectable (i.e. <*S*).

## TRACE ELEMENTS IN SAXICOLOUS LICHENS AND THEIR ROCK SUBSTRATES

As part of a more general study of rock weathering designed to elucidate the interactions between saxicolous lichens and their substrates, trace element analyses were made of selected plant/rock systems from Snowdonia. The lichen species investigated were mainly *Parmelia omphalodes* (L) Arch. and *Lecanora gangaleoides* Nyl. but also included *P. saxatilis* (L) Arch., *P. conspersa* (Ehrh. ex. Ach) Ach., *Haematomma ventosum* (L) Massal., and *Rhizocarpon geographicum* (L) DC., and the rock types involved were shale, dolerite, basic tuff, rhyolite and vein quartz. Where possible, the lichen, weathered rock surface and fresh unaltered rock were sampled and analysed separately. Clearly, there is the possibility of incorporation of detached weathered rock material into the lichen sample, especially for crustaceous species; this will be reflected in increased ash content and would accordingly 'dilute' the lichen composition. To allow for such contamination to some extent in the graphical presentation of the results, in which fresh-rock/weathered-rock/lichen-ash values are linked by tie-lines, the lichen results are dispersed horizontally on a logarithmic scale of ash %.

The results obtained showed interesting and consistent trends for the two lichen species investigated, as illustrated in Fig. 2. For certain elements (i.e. Mn, Y, Ba, Rb) there is a depletion in the weathered zone of rocks of low trace element content, but enhancement in those of high content, resulting in an apparent convergence toward some 'preferred' value within the lichen ash irrespective of rock type. For other elements (i.e. Li, Co, Ga, Sr) the lichen ash value is higher than that of many of the rocks, resulting in tie-lines concave

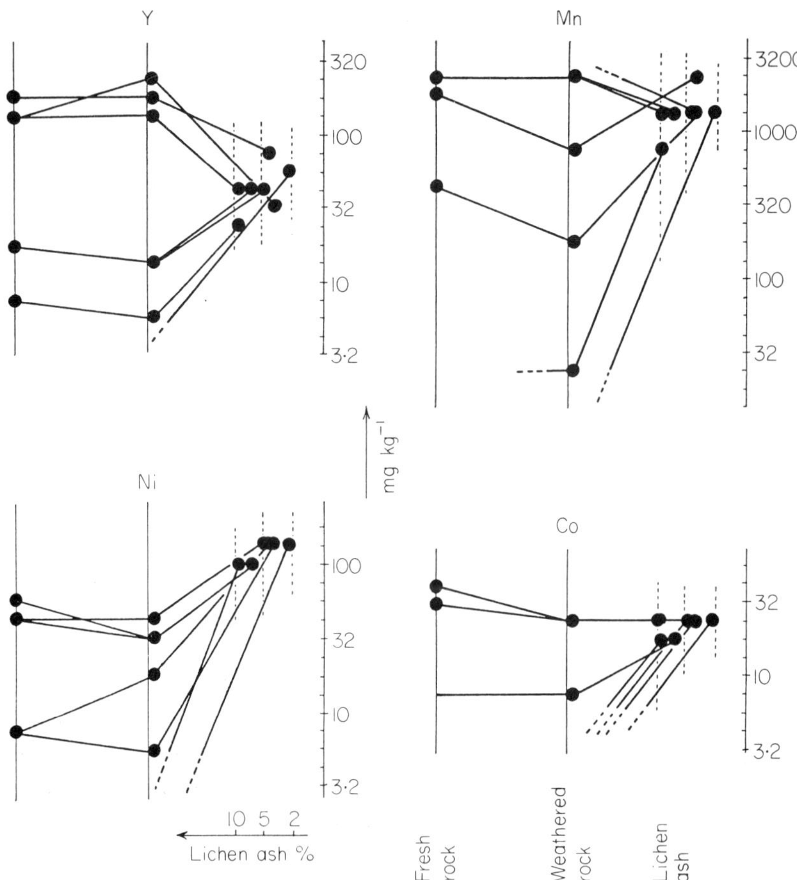


FIG. 2. Trace element trends in *Parmelia omphalodes* colonizing different rock types in Snowdonia.

upwards, whilst for others the lichen ash values are higher than in any rock substrate (i.e. Cr, Ni, Cu, Sn, Zn, Pb). It thus became clear that some control mechanism was operating to determine the trace element levels, either internally (metabolic) or imposed by an external factor.

The impracticality of an internal mechanism is apparent in the pronounced accumulation of specific elements in lichens relative to their crustal abundance. This is most marked for rock substrates with inherently low levels of these trace elements, such as silicic igneous rock (i.e. rhyolitic) and, more dramatically, vein quartz which (other than Li) contains no trace elements detectable by the analytical methods employed. The likelihood of an external factor was thus indicated by the poor correlaton of trace element composition between lichen ash and rock substrate considered either individually (Table 2) or generally in terms of crustal abundances (Krauskopf 1979; Fig. 3). It is also interesting to note (Table 2) that the correlation with substrate is marginally better for the crustaceous *(L.gangaleoides)* as compared to the foliaceous *(P.omphalodes)* species, presumably due

TABLE 2. Correlation coefficients between metal composition of ashed lichen samples and various potential sources of metal. Metals included in the analysis were Cr, Mn, Co, Ni, Cu, Zn, Ga, Y, Zr, Mo, Sn and Pb

| | | Snowdonian samples | |
|---|---|---|---|
| | | *Parmelia omphalodes* (7)† | *Lecanora gangaleoides* (4) |
| *Parmelia omphalodes* | (Finland; Lounamaa 1956) | 0·93** | - |
| *Lecanora gangaleoides* | (Snowdonia) | 0·96** | - |
| Atmospheric deposits | (Warren Spring Laboratory) | 0·95** | 0·89* |
| Average of rock substrates | | 0·41 | 0·70 |

*$P<1\cdot0\%$
**$P<0\cdot1\%$
†Values in brackets denote the number of samples.

in part to greater 'contamination' of the former as reflected in their higher average ash contents (13·7 cf. 5·4%). This poor correlation stems from the consistent enrichment of specific elements relative to their crustal abundance and is most marked for Pb (500×), Sn and Ag (50×), Zn and Cu (10×): it is also evident for elements such as Mo, Ge, and Be which are at levels near to their respective sensitivity limits by the arc-spectrographic method used.

By contrast, correlation with data then available for the composition of atmospheric deposits (unpublished from Warren Springs Laboratory; Stocks, Commings & Aubrey 1961) proved to be highly significant (Table 2 and Fig. 3) being marginally more so for the foliaceous as compared to the crustaceous species. Further support for an atmospheric source for trace elements comes from the fact that there is a marginally better correlation between different lichen species in the same area than for the same species in different areas (Snowdonia & Finland: Lounamaa 1956), implying a regional component to the external factor.

The balance between atmospheric and substrate as a source of trace elements in lichen plants will vary with the petrology of the substrate and the nature of the lichen plant. There is an observable association of many lichen species with specific rock types, e.g. silicic or mafic igneous, serpentines or – in particular – limestones (Brodo 1974). There is also an observable acceleration of the weathering of rock surfaces, both physical and chemical, when colonized by lichens (e.g. Syers & Iskander 1974; Jones, Wilson & McHardy 1981) and the components liberated from the rock, including trace elements, will become available to the lichen plant and may thus be reflected in its composition. However, it is of interest that, whilst Yarilova (1950) and Lounamaa (1956) emphasized rock substrates as a

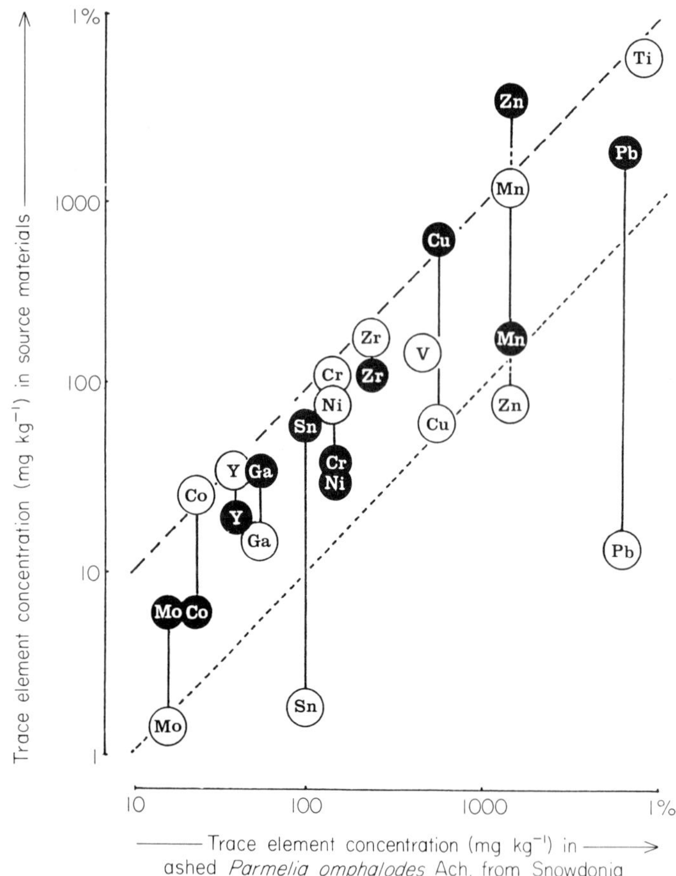


FIG. 3. Correlations between the trace element contents of lichen ash with atmospheric deposits (●) or with crustal abundances (©).

source of the inorganic constituents of lichens, Stahlecker (1906) had earlier stressed the importance of atmospheric sources. The influence of substrate will clearly decrease for those rock types which are slow weathering and of inherently low trace element content, such that the effect of atmospheric deposition is enhanced: the extreme situation will arise with vein quartz which is effectively pure $SiO_2$ and very stable. With regard to lichen species, the actual content of trace elements derived from the atmosphere will be determined by growth rate, areal exposure and surface morphology, whilst adhesion will affect ease of separation and thus degree of contamination by detached rock particles. Crustaceous species are the slowest growing, but foliaceous species present a large absorbing surface and are more readily detached.

Taking these factors into account, a strategy was devised for sampling those saxicolous lichens which would reflect most closely the trace element composition of atmospheric material. Foliaceous *Parmelia* species were selected, preferably the distinctive dark brown

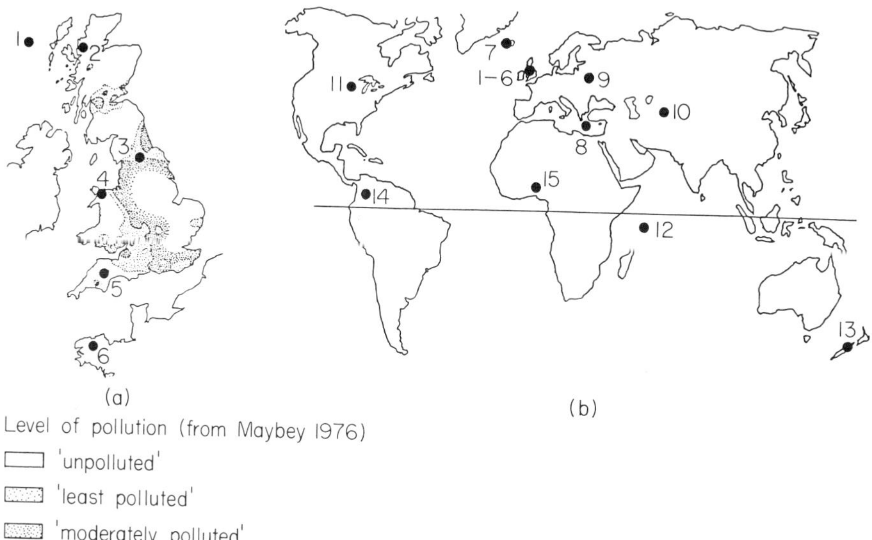


FIG. 4. Sampling sites for saxicolous lichens from (a) the UK area and (b) other parts of the world.

*P. omphalodes* (L) Ach. (Fig. 1b): where this was unavailable, species of comparable morphology were selected (e.g. *P. conspersa* (Ehrh. ex Ach.) and *P. saxatilis* (L) Ach.) to minimize any effects of different patterns of entrapment. Ideally, the substrate chosen was vein quartz from the crest of outcrops or boulders and free from canopy drip, in order to minimize substrate contributions, but where unavailable a quartzitic or silicic rock was chosen.

## TRENDS IN SAMPLES FROM THE UK AND ELSEWHERE

Much interest has been shown in Britain in recent years over the effects of atmospheric pollution on lichens. This interest was stimulated by Gilbert (1965) who mapped 'lichen deserts' around industrial areas and ascribed them to the effects of enhanced $SO_2$ levels, and it led to the national pollution maps produced by Maybey (1974) and Hawksworth & Rose (1976). In this study analyses are presented from six sites in or adjacent to the UK, all but one (No. 3) being from mountainous areas within the zone deemed 'unpolluted' (Fig. 4a). The plants sampled were all *P. omphalodes* on quartz (4,6) quartzitic (2,3) or granitic (1,2,5) rocks, although mafic igneous rocks were also involved in one instance (No. 1). The results are presented in Table 3 and tend to show the same pattern of enhanced Pb, Sn, Zn, Cu (and possibly Ag, Mo, Cr, Ni) levels as discerned in Snowdonia. This enhancement is in fact surpassed by the sample from the 'relatively pure/polluted' site in Yorkshire where Be, Ge and Mo are also detectable, and it decreases to the North and South though still manifested in the enhanced levels of Pb and Zn in Brittany (No. 6) and even in the remote St Kildas (No. 1). Superimposed upon these trends are local influences of rock substrate such as the somewhat higher levels of Cr, Cu and Ni due to a mafic igneous substrate in St

TABLE 3. Trace element content of *Parmelia omphalodes* samples from the UK area

| | | St Kildas (2)† (80 km W of 2) | Torridon (3) NW Scotland | Whernside N. Yorks. | Snowdonia (7) N. Wales | Kestor Devon | Trevezel (3) Brittany |
|---|---|---|---|---|---|---|---|
| | | 1 | 2 | 3 | 4 | 5 | 6 |
| | %Ash | 8·6 | 7·9 | 4·6 | 6·3 | 3·6 | 14·8 |
| | *S* (mg kg$^{-1}$) | (mg kg$^{-1}$ of sample ashed at 470 °C) | | | | | |
| Ag | 1·8 | < | < | 5·6 | 3·2 | 2·4 | < |
| Be | 10 | < | < | 10 | < | 18 | < |
| Co | 7·5 | 32 | 10 | 24 | 24 | 7·5 | < |
| Cr | 5·6 | 320 | 42 | 320 | 130 | 75 | 42 |
| Cu | 18 | 100 | 130 | 1000 | 560 | 130 | < |
| Ge | 18 | < | < | 18 | < | < | < |
| Ga | 1·8 | 24 | 18 | 24 | 56 | 10 | 4·2 |
| Mn | 5·6 | 1000 | 240 | 1800 6 | 1300 | 1000 | 320 |
| Mo | 10 | < | < | 24 | 18 | < | < |
| Ni | 1·8 | 180 | 42 | 130 | 130 | 56 | 24 |
| Pb | 18 | 180 | 1000 | 7500 | 5600 | 1300 | 100 |
| Sn | 18 | < | 24 | 180 | 100 | 56 | < |
| Ti | 5·6 | 18000 | 3200 | 4200 | 7500 | 3200 | 4200 |
| V | 18 | 420 | 180 | 420 | 420 | 320 | 56 |
| Y | 5·6 | 5·6 | 18 | 24 | 42 | 42 | < |
| Yb | 1·8 | 1·8 | 1·8 | 4·2 | 3·2 | 7·5 | 2·4 |
| Zn | 420 | 420 | 560 | 2400 | 1300 | 1300 | 420 |
| Zr | 10 | 100 | 240 | 420 | 240 | 750 | 240 |

†Values in brackets denote the number of samples.
< Signifies not detectable (i.e. < *S*: lower sensitivity limit).

Kilda, and of Be and Zr in the Devon sample (No. 5) which was colonizing granite.

Analyses of lichens from a limited number of sites from other parts of the globe (Fig. 4b) are presented in Table 4. Although obviously too sparse to allow discussion of global trends, they do highlight certain features. For example in the samples from the more populous Crete, (No. 8), Western USSR (No. 9) and Midwest USA (No. 11), moderately enhanced levels of Pb and (in Nos 8, 11) of Cu are evident, although neither Sn or Zn are detectable. In the Midwest USA sample, the values of Co, Mn, Ni and Pb are higher than those reported for *Parmelia chlorochroa* Tuck. from the less populous Montana/ Wyoming area by Erdman & Gough (1977). In contrast to the composition of these samples, no enhancement of Pb is evident in the Icelandic samples (No. 7), although the Cu (and possibly Co) levels might reflect the specific influence of particulate volcanic emissions as discussed by Weiss *et al.* (1975). Neither Pb nor Cu is detectable in the lichens from the more remote part of the USSR (The Parmirs: No. 10) nor in those from the two sites in the southern hemisphere, the Seychelles (No. 12) and New Zealand (No. 13). The former has an extreme composition which could be accounted for by that of the rock substrate (granite) alone, while the latter has lower values for Cu and Pb than those reported by Ward, Brooks & Roberts (1977) in bryophytes in ‘unpolluted background’ sites from a mining area in New Zealand.

Analyses from two further series of sites from equatorial regions are presented in Table

TABLE 4. Trace element content of *Parmelia* (10–13) and other (7–9) species from various parts of the world

| | | Iceland (3)† | Crete (2) | Zhigouli USSR | Varzob, USSR (2) | Wisconsin USA | Seychelles Indian Oc. | S. Island NZ (3) |
|---|---|---|---|---|---|---|---|---|
| | | 7 | 8 | 9 | 10 | 11 | 12 | 13 |
| | %Ash | 10·0 | 27·6 | 14·4 | 13·3 | 14·0 | 26·7 | 19·8 |
| | *S* (mg kg$^{-1}$) | (mg kg$^{-1}$ of sample ashed at 470 °C) | | | | | | |
| Co | 7·5 | 18 | < | 7·5 | 7·5 | 13 | < | 13 |
| Cr | 5·6 | 75 | 24 | 100 | 75 | 42 | < | 32 |
| Cu | 18 | 75 | < | 24 | < | 42 | < | < |
| Ga | 1·8 | 7·5 | 1·8 | 3·2 | 4·2 | 5·6 | 18 | 7·5 |
| Mn | 5·6 | 1800 | 240 | 560 | 420 | 750 | 130 | 1800 |
| Ni | 1·8 | 32 | 10 | 42 | 24 | 32 | 2·4 | 18 |
| Pb | 18 | < | 130 | 56 | < | 420 | < | < |
| Ti | 5·6 | 18000 | 2400 | 5600 | 4200 | 1300 | 5600 | 13 |
| V | 18 | 180 | 130 | 180 | 130 | 100 | < | 180 |
| Y | 5·6 | 10 | 32 | 24 | 18 | 13 | < | 180 |
| Yb | 1·8 | 3·2 | 4·2 | 3·2 | 2·4 | < | < | < |
| Zr | 10 | 180 | 56 | 75 | 75 | 180 | 130 | 320 |

Not detected in any sample (i.e. $<S$ mg kg$^{-1}$): Ag (<1·8); Be (<10); Ge (<18); Mo (<10); Sn (<18); Zn (<420).

†Values in brackets denote the number of samples.

<Signifies not detectable (i.e. $<S$: lower sensitivity limit).

5. These series were sampled to test for localized effects around densely populated areas in developing countries. Those from Nigeria (No. 15) do indeed show predictable trends of increasing levels of Pb, to a lesser extent of Cu, and also of Cr, Ni and Ga in progressing from the remote mountain village of Idanre (Pb not detectable) to the densely populated city of Ibadan. These trends parallel those recorded by Onianwa & Eguyomi (1983) in bryophytes in this same area and are detectable despite the high ash levels (18–30%) which presumably result from the influence of the dust-laden 'Harmattan' which blows down from the Sahara. The series from Colombia (No. 14) shows a similar trend of increase with proximity to the city of Bogota for Pb, and possibly for Zn, but the results for other elements such as Ni, Cr and Cu are confused by the relatively high levels of Co, Cr, Cu, Ni and V in the remote (least polluted?) sample which, although on vein quartz itself, was from a site with serpentine outcrops nearby.

A final set of analyses is presented in Table 6 to illustrate temporal variations in the trace element content of saxicolous lichens from Snowdonia, North Wales. The lichen material analysed in any one sample will have probably ranged in age from 0 to 10 years with a weighted average of around 3 years. This calculation is based on growth rates of the order of 5 mm per year for *Parmelia* species and a life span of 25–35 years (Hale 1983), with the plants sampled in this study being of the order of 10 cm in diameter. It was possible to supplement analyses of *P. omphalodes* collected in 1961 and 1984 with samples from the Herbarium in the School of Plant Biology, UCNW, collected in 1929. A further sample of '*Parmelia fuliginosa*' collected in 1897 was also analysed to provide four sample points

TABLE 5. Trace element trends in *Parmelia* species from Nigeria and Colombia

| | | 14. Colombia, SA | | | 15. Nigeria | | |
|---|---|---|---|---|---|---|---|
| | | Bogota | 30 km NE, El Dorado | 400 km SW, Rosas | Ibadan | 80 km E, Ile Ife | 120 km E, Idanre |
| | %Ash | 27·8 | 22·5 | 18·3 | 29·6 | 22·3 | 17·7 |
| | *S* (mg kg$^{-1}$) | (mg kg$^{-1}$ of sample ashed at 470 °C) | | | | | |
| Co | 7·5 | < | < | 75 | 18 | < | < |
| Cr | 5·6 | 130 | 100 | 1000 | 130 | 24 | 10 |
| Cu | 18 | 24 | 18 | 75 | 18 | < | < |
| Ga | 1·8 | 7·5 | 10 | 5·6 | 7·5 | 5·6 | 1·8 |
| Mn | 5·6 | 56 | 180 | 2400 | 1800 | 750 | 1000 |
| Ni | 1·8 | 32 | 5·6 | 560 | 32 | 18 | 5·6 |
| Pb | 18 | 130 | 56 | < | 130 | 100 | < |
| Ti | 5·6 | 3200 | 2400 | 3200 | 5600 | 2400 | 3200 |
| V | 18 | < | < | 100 | 75 | 100 | 100 |
| Y | 5·6 | 24 | 18 | 18 | < | 75 | 10 |
| Yb | 1·8 | 3·2 | < | 1·8 | 3·2 | 5·6 | 2·4 |
| Zn | 420 | 750 | < | < | < | < | < |
| Zr | 10 | 2400 | 320 | 420 | 560 | 2400 | 320 |

Not detected in any sample (i.e. $<S$ mg kg$^{-1}$): Ag (< 1·8); Be (< 10); Ge (< 18); Mo (< 10); Sn (< 18); T1 (< 18).
< Signifies not detectable (i.e. $<S$: lower sensitivity limit).

23–32 years apart, although this last sample is of unspecified substrate, different morphology (more compact), higher ash content, and unknown contamination since collection. Nevertheless, from the results it is apparent that up to 1961 there have been pronounced increases in the levels of Pb (50 ×), Cu, Ni, Sn, V, Zn (10 ×), Ga, Y (6 ×) and Cr (3 ×): in contrast the levels of Co and Yb have remained relatively constant whilst those of Mn, Ti and Zr have fluctuated irregularly. These trends have continued through to 1984 for Mn, Ti and Zr (erratic), Co and Yb (constant), but of those with rising levels up to 1961 only Cr and Zn continue this trend to still higher levels whilst the values for Cu, Ga, Ni, Sn, Y an Pb show a distinct fall (× 0·4–0·2) over the last 23 years.

## DISCUSSION

In this paper spectrographic analyses for eighteen trace elements in over fifty saxicolous lichen samples are presented. These results have been assessed initially in terms of the relative inputs from substrate and atmospheric sources, and of devising a sampling strategy to isolate as far as possible the latter source. Subsequently, the results of applying this strategy to a limited number of samples from the UK and elsewhere have been presented and the features and trends which they reveal noted. These trends in both space and time are considered below in more general terms and their implications discussed in the context of the atmospheric content of trace elements.

TABLE 6. Trace element contents of *Parmelia* species from Snowdonia

| | Ash% | *P. fuliginosa* | *P. omphalodes* | *P. omphalodes* | *P. omphalodes* |
|---|---|---|---|---|---|
| | | 1897 | 1929 | 1961 | 1984 |
| | | (1)[†] | (3) | (3) | (4) |
| | | 28·5 | 8·4 | 5·1 | 5·5 |
| | (*S*) | mg kg$^{-1}$ (ppm) in sample ashed at 470 °C | | | |
| Ag | 1·8 | < | < | 2·4 | 2·4 |
| Co | 7·5 | 10 | 10 | 24 | 13 |
| Cr | 5·6 | 42 | 36 | 130 | 240 |
| Cu | 18 | 56 | 75 | 560 | 180 |
| Ga | 1·8 | 5·6 | 13 | 32 | 13 |
| Mn | 5·6 | 420 | 2400 | 1300 | 1000 |
| Mo | 10 | < | < | 24 | < |
| Ni | 1·8 | 13 | 32 | 130 | 56 |
| Pb | 18 | 130 | 420 | 56000 | 1300 |
| Sn | 18 | < | 56 | 100 | 42 |
| Ti | 5·6 | 4200 | 3200 | 7500 | 2400 |
| V | 18 | 42 | 42 | 420 | 240 |
| Y | 5·6 | 7·5 | 24 | 56 | 24 |
| Yb | 1·8 | 3·2 | 4·2 | 3·2 | 4·2 |
| Zn | 420 | < | 1000 | 1300 | 1800 |
| Zr | 10 | 420 | 56 | 180 | 240 |

Not detected in any sample (i.e. <*S*, mg kg$^{-1}$): Be (<10); Ge (<18); Tl (<18); Cd (<320).

<Signifies not detectable (i.e. <*S*: lower sensitivity limit).

[†]Values in brackets denote the number of samples.

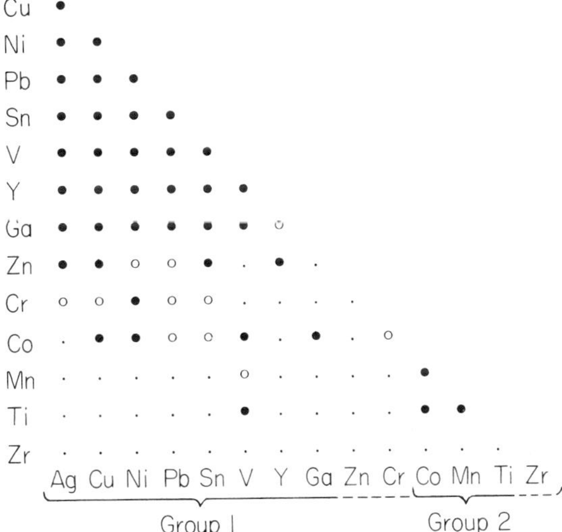


FIG. 5. Correlation matrix for trace elements in saxicolous lichens. ● denotes $P < 0{\cdot}1\%$, ○ denotes $P < 1{\cdot}0\%$, $n \pm 50$.

Considering first the overall results in isolation, the various trace elements show well defined inter-relationships as evident from their correlation matrix (Fig. 5). One distinct group comprises Ag, Cu, Ni, Pb, Sn, V and Y which all show mutual positive correlation coefficients significant at the 0·1% level: Ga, Zn, Co and Cr are also associated with this group though less consistently and, by inspection of their occasional detectable values, so too are Be, Ge and Mo. By contrast, Co, Mn and Ti form an independent second group with mutual positive correlation coefficients again significant at the 0·1% level, while Zr stands alone and is not significantly correlated with any of the other elements. In seeking the provenance of these elements it has already been shown (Fig. 3) that the concentrations of the first group correlate with atmospheric sources whilst those of the second group (and Zr) reflect their average crustal abundances. This relationship is also borne out by negative correlations with lichen ash % values (i.e. 'substrate contamination') which are significant at the 0·1% level for Ag, Cu, Ni, Pb, Sn and Zn, and at the 1% level for Y. It can therefore be concluded that, for the plant/substrate systems sampled, Co, Mn, Ti and Zr can generally be accounted for by normal terrigenous (i.e. rock/soil) sources, directly or via the atmosphere, whilst non-terrigenous atmospheric sources dominate for Ag, Cu, Ni, Pb, Sn, V, Y and, to a lesser extent, Ga, Zn and Cr. This pattern is generally similar to that shown by Co, Mn, Cr, V, Zn, Cu and Pb in terms of their 'enrichment factors' (atmosphere/crust) in atmospheric samples from over the Atlantic, reported by Duce, Hoffman & Coller (1975), and by Pierson, Cawse & Cambray (1974) from UK sites.

In focusing on the atmospheric input to the lichen plant, it would be convenient to be able to allow for terrigenous material by 'normalizing' trace element values according to a reference value for a typical terrigenous element(s) by the calculation of 'enrichment factors'. For example, Nieboer *et al.* (1978) suggested the Ti/Fe ratio as a reference value, but Ti — and Zr in particular — occur in rocks and soils as discrete oxide/silicate minerals and it is therefore unlikely that their presence would be other than sporadic and variable when sampled from different geological terrains; this is borne out by their lack of mutual correlation ($T = -0 \cdot 12$, Fig. 5). Unfortunately, the major element Al used by Duce *et al.* (1975) was not estimated in this study and Sc, which is suitably dispersed in ferromagnesian minerals and has been used as a potential reference element by Peirson *et al.* (1974), lies just at, or below, the detection limit (18 mg $kg^{-1}$) of the spectrographic method used. In the correlation analysis none of the trace elements shows a significant positive correlation with lichen ash % such as would suggest its possible use as a correcting factor, Ti being the nearest but with an insignificant coefficient of $+0 \cdot 13$.

Trace element relationships within source materials will be subject to modification once within the lichen plant. This will vary according to the mode of input (particulate/solution), manner or tenacity of binding, and consequent mobility of the element. Compared to Pb, for example, Zn is potentially more mobile (Brown 1976) which could account for its poorer correlations with the other elements of its group: Zn is also seen (Fig. 3) to be the only trace element apparently depleted in lichen ash as compared to atmospheric deposits. Such differential depletion could be through leaching from the plant or possibly by microscopic particulate losses back to the atmosphere, the loss of Zn (from plants) having been shown to be two orders of magnitude greater than that of Pb (Beauford, Barber & Barringer 1977). Information for many other elements in lichens is at present lacking, and this raises many questions of detail. For example, the specific association of Co with Mn in the sedimentary environment is well known, as is the susceptibility of Mn to redistribution and concentration by fluctuations in redox potential. It would be interesting to know whether the strong correlation of these two elements in the lichens ($P < 0 \cdot 1\%$) is inherited via particulate manganese oxides or is due to subsequent adsorption of aqueous Co onto precipitated solid Mn phases.

The trace element content of the atmosphere has been a subject of much interest in recent years with regard to contributing sources, methods of monitoring, and spatial and temporal variations. A valuable collation of available quantitative data for Cd, Cu, Ni, Pb and Zn has been presented by Nriagu (1979) who emphasized the massive burden imposed by man's industrial activities upon the cycling of these elements through the biosphere. As noted in the introduction, the anthropogenic contributions to atmospheric material are calculated to exceed those from natural sources by factors ranging from twofold (Ni) to twentyfold (Pb). Natural sources are apparently dominated by eroded soil materials of 'normal' composition (60–80%) except in the case of Cd where sporadic volcanic sources dominate (60%). Vegetative 'exudates' are considered important for Zn (20%), Cd and Cu (10–15%) whilst smaller, but important, localized additions will arise from forest fires and from the concentration of trace elements in sea spray.

Contributions to the atmosphere by man are dominated by the mining/processing industry for Cd, Cu and Zn (40–65%) but by petrol and diesel additives respectively for Pb (60%) and Ni (55%), with smaller contributions from the combustion of wood, coal and refuse, and from other industrial processes. Clearly the importance of these various sources will vary with proximity to industry, the sea, extensive forests, etc: industrial sources are more important in the Northern hemisphere which is estimated to receive 90% of the world's particulate pollutants. Although these now appear to be dispersed so as to produce a general elevation of background levels in addition to localized effects, atmospheric residence times are apparently too short to permit transfer to the Southern hemisphere (Duce *et al.* 1975). This is reflected in the trends of increasing trace element content in ice cores from the arctic, enrichments having increased two to threefold for Pb and Zn over the last century (Herron *et al.* 1977), as compared to the antarctic where no marked increases have been found for Ag, Cu, Mn, Pb or Zn over the last 60 years (Boutron & Lorius 1979).

Viewed in this context, the lichen analyses reported in this paper conform generally to the various trends noted for similar plants and atmospheric materials. Thus, the pattern of elemental enrichment is similar with Pb pre-eminent, due to its extensive use as an anti-knock additive, followed by Sn, Ag, Zn and Cu and probably also by Mo, Be and Ge. There are also suggestions in some of the spatial and temporal trends that Cr, Ga, Ni, V and Y may be implicated. In recent years much information has been acquired about the environmental distribution and cycling of elements such as Pb, Zn, Cu and Cd, but conspicuously little is available for elements of less popular interest such as Sn, Be and Ge. Unfortunately, this concentration of attention by analytical preselection has an inherent tendency to both self-perpetuation, in terms of follow-on studies, and self-fulfilment, in terms of interpretation of causal relationships. Hopefully the advent of improved analytical techniques with lower sensitivity limits will encourage more catholic trace element studies to be undertaken in the future.

The spatial trends reported here confirm a generally enhanced background level of atmospheric contamination in the Northern Hemisphere. They also indicate that localized effects are already building up to detectable levels in the otherwise cleaner Southern hemisphere. The temporal trends in Snowdonia are interesting in that they reveal increasing levels over the period 1897–1961. This parallels the twentyfold increase of Pb levels in foliaceous lichens in north-eastern USA over the period 1907–78, found by Lawrey & Hale (1981), and the rising levels of Cu, Zn, Ni, Cr and Pb in bryophytes from Sweden over the period 1870–1969 recorded by Ruhling & Tyler (1968, 1969). However, the values for Cu, Ga, Mo, Ni, Pb, Sn and Y in lichens sampled from Snowdonia appear to have declined between 1961 and 1984, perhaps reflecting the implementation of the Clean Air Act of 1956: a similar trend in air and deposit concentrations is recorded by Cawse (p.89). It will be interesting to see whether further studies substantiate this trend and, if so, whether it will be maintained.

Although most obvious in lichens, mosses and similar plants, it should be noted that atmospheric deposition of trace elements applies to all exposed systems. For example, it can also be detected in contemporary as compared to buried organic top-soils (Jenkins & Bower 1974). It is difficult to predict the long-term effects of this additional input of trace elements into the biosphere. Some of the elements concerned are essential micronutrients (Cu, Zn, Mo) yet are toxic at elevated levels; in some instances the atmospheric input could alleviate subclinical deficiences, whilst in others it could tip the balance towards toxicity. For elements such as Pb, Sn, Ni and Be, for which no essentiality has yet been established, there only remains the potential for toxicity. Despite their apparent tolerance, this might even arise for lichens themselves with their slow growth rates and consequent accumulation, although such effects are likely to be pre-empted by the more immediate problem of rising $SO_2$ levels.

It is concluded from this study that, when judiciously chosen for location, species, and substrate, analysis of saxicolous lichens can provide a convenient monitor for atmospheric deposition of trace elements. This can of course only be an approximation insofar as factors of differential interception, retention and dilution will be involved; nor can such analyses take the place of carefully planned and executed programmes of direct sampling and analysis of atmospheric material (e.g. Beavington & Cawse 1979). Nevertheless, lichen analyses do provide the possibility of quick, inexpensive monitoring for reconnaissance surveys or for areas where inaccessibility and cost are limiting factors.

## ACKNOWLEDGMENTS

I have pleasure in recording my thanks to a number of colleagues who kindly collected samples from various localities, namely: Dr M. Hornung (ITE, Bangor) from St Kildas (No. 1); Mr C. Orgil (UCNW Bangor) from Hofn, Iceland (No. 7); Dr C.P. Burnham (Wye College, Kent) from the Zhigouli Reserve, USSR (No. 9); Mr F. Collier (Hunting Technical Services) from the Seychelle Islands (No. 12); Dr P. MacKenzie (DSIR, Soil Bureau, NZ) from Lake Wakatipu, South Island, NZ, (No. 13); and finally Dr A. Smith (UCNW, Bangor) for permission to use samples from the Herbarium, School of Plant Biology, UCNW (Table 6; 1897 and 1929). Further details as to lichen species, petrology of substrate, and location, are available from the author. I am also grateful to Dr A. Kershaw and Dr A. Pentecost for assistance in identification of the lichen species.

## REFERENCES

**Beauford, M., Barber, J. & Barringer, A.R. (1977).** Release of particles containing metals from vegetation into the atmosphere. *Science,* **195**, 571–573.

**Beavington, F. & Cawse, P.A. (1979).** The deposition of trace elements and major nutrients in dust and rainwater in Northern Nigeria. *The Science of the Total Environment,* **13**, 263–274.

**Boultron, C. & Lorius, C. (1979).** Trace metals in antarctic snows since 1914. *Nature (London),* **277**, 551–554.

**Brodo, I.M. (1974).** Substrate ecology. *The Lichens* (Ed. by V. Ahmadjian, M.E. Hale), pp. 401–441. Academic Press, New York.

**Brooks, R.R. (1972).** *Geobotany and Biogeochemistry in Mineral Exploration.* Harper & Row, New York.

**Brown, D.H. (1976).** Mineral uptake by lichens. *Lichenology: Progress and Problems* (Ed. by D.H. Brown, D.L. Hawksworth & R.H. Bailey), pp. 419–439, Academic Press, London.

**Duce, R.A., Hoffman, G.L. & Zoller, W.B. (1975).** Atmospheric trace metals at remote northern and southern hemisphere sites: pollution or natural? *Science,* **187**, 59–61.

**Erdman, J.A. & Gough, L.P. (1977).** Variation in element content of *Parmelia chlorochoa* from the Powder River basin of Wyoming and Montana. *The Bryologist,* **80**, 292–303.

**Ferry, B.W., Baddeley, M.S. & Hawksworth, D.L.** (Ed.) **(1973).** *Air Pollution and Lichens.* The Athlone Press, London.

**Gilbert, O.L. (1965).** Lichens as indicators of air pollution in the Tyne Valley. *Ecology and the Industrial Society* (Ed. by G.T. Goodman, R.W. Edwards & J.M. Lambert), pp. 35–47. Blackwell, Oxford.

**Goodman, G.T. & Roberts, T.M. (1971).** Plants and soils as indicators of metals in the air. *Nature (London),* **231**, 287–292.

**Hale, M.E. (1983).** *The Biology of Lichens,* 3rd edn. Arnold, London.

**Hawksworth, D.L. & Rose, F. (1976).** *Lichens as Pollution Monitors.* Arnold, London.

**Herron, M.M., Langway, C.C., Weiss, H.V. & Cragin, J.H. (1977).** Atmospheric trace metals and sulphate in the Greenland ice sheet. *Geochimica & Cosmochimica Acta,* **41**, 915–920.

**Jenkins, D.A. & Davies, R.I. (1966).** Trace element content of organic accumulations. *Nature (London),* **210**, 1296–1297.

**Jenkins, D.A. & Bower, R.P. (1974).** The significance of the atmospheric contribution to the trace element content of soils. *Transactions of the 10th International Congress of Soil Science, Moscow,* **6**(11) 446–472.

**Jones, D., Wilson, M.J. & McHardy, W.J. (1981).** Lichen weathering of rock forming minerals: application of scanning electron microscopy and microprobe analysis. *Journal of Microscopy,* **124**, 95–104.

**Jones, D., Wilson, M.J. & Laundon, J.R. (1982).** Observations on the location and form of lead in *Stereocaulon vesuvianum. Lichenologist,* **14**, 281–286.

**Krauskopf, K.R. (1979).** *Introduction of Geochemistry,* 2nd Edn. McGraw Hill, New York.

**Lawrey, J.D. & Hale, M.G. (1981).** Retrospective study of lichen lead accumulation in the north eastern United States. *The Bryologist,* **84**, 449–456.

**Lounamaa, J. (1956).** Trace elements in plants growing wild on different rocks in Finland. A semiquantitative spectrographic survey. *Annales bot.Soc.Zool-bot.Fenn. Vanamo,* **29**, 1–196.

**Maybey, R. (1974).** *The Pollution Handbook.* Penguin, London.

**Mitchell, R.L. (1964).** The spectrochemical analyses of soils, plants and related materials. *Commonwealth Bureau Soil Science Technical Communication,* No. 44a.

**Nieboer, E., Ahmed, H.M., Puckett, K.J. & Richardson, D.H.S. (1972).** Heavy metal content of lichens in relation to distance from a nickel smelter in Sudbury, Ontario. *Lichenologist,* **52**, 292–304.

**Nieboer, E., Richardson, D.H.S. & Tomassini, F.D. (1978).**Mineral uptake and release by lichens: an overview. *The Bryologist,* **81**, 226–246.

**Nriagu, J.O. (1979).** Global inventory of natural and anthropogenic emissions of trace metals to the atmosphere. *Nature (London),* **279**, 409–411.

**Onianwa, P.C. & Egunyomi, A (1983).** Trace metal levels in some Nigerian mosses used as indicators of atmospheric pollution. *Environmental Pollution (Series B),* **5**, 71–81.

**Phillips, D.J.H. (1977).** The use of biological indicator organisms to monitor the metal pollution in marine and estuarine environments — a review. *Environmental Pollution,* **13**, 281–317.

**Pierson, D.H., Cawse, P.A. & Cambray, R.S. (1974).** Chemical uniformity of airborne particulate material, and a maritime effect. *Nature (London),* **251**, 675–679.

**Purvis, D.W. (1984).** The occurrence of Cu-oxalate in lichens growing on copper sulphide bearing rocks in Scandinavia. *Lichenologist,* **16**, 197–207.

**Rühling, A. & Tyler, G. (1968).** An ecological approach to the lead problem. *Botaniska Notiser,* **121**, 34–42.

**Rühling, A. & Tyler, G. (1969).** Ecology of heavy metals: a regional and historical study. *Botaniska Notiser,* **122**, 248–259.

**Seaward, M.R.D. (1973).** Lichen ecology of the Scunthorpe heathlands. 1: Mineral accumulation. *Lichenologist,* **5**, 423–433.

**Stahlecker, E. (1906).** Untersuchungen über Thallusbildung und Thallusbau in ihren Bezeihungen zum Substrat bei siliciseden Krustenflechten. *Fünstucks Beitrage fur Wissenshaft Botanik,* Bd. **5**, 405–451.

**Stocks, P., Commins, B.T. & Aubrey, K.V. (1961).** A study of polycyclic hydrocarbons and trace elements in smoke in Merseyside and other northern localities. *International Journal of Air and Water Pollution,* **4**, 141–153.

**Syers, J.K. & Iskander, I.K. (1974)** Pedogenetic significance of lichens. *The Lichens* (Ed. by V. Ahmadjian & M.E. Hale), pp. 225–248. Academic Press, New York.

**Tuominen, Y. & Jaakkola, T. (1974).** Absorption and accumulation of mineral elements and radioactive nuclides. *The Lichens* (Ed. by V. Ahmadjian & M.E. Hale), pp. 185–233. Academic Press, New York.

**Ward, N.I., Brooks, R.R. & Roberts, R. (1977).** Heavy metals in some New Zealand bryophytes. *The Bryologist,* **80**, 304–312.

**Weiss, H., Bertine, K., Koide, M. & Goldberg, E.D. (1975).** The chemical composition of a Greenland glacier. *Geochemica & Cosmochimica Acta,* **39**, 1–10.

**Wilson, M.J., Jones, D. & McHardy, W.J. (1981).** The weathering of serpentinite by *Lecanora atra. Lichenologist,* **13**, 167–176.

**Wilson, M.J. & Jones, D. (1984).** The occurence and significance of manganese oxalate in *Pertusaria corallina* (Lichenes). *Pedobiologia,* **26**, 373–379.

**Yarilova, E. (1950).** *Trudy Pochvovedenie Institute Dokuchaeva,* **34**, 110–142.

# Plant availability of heavy metals in soils

B. E. DAVIES*, J.M. LEAR AND N.J. LEWIS
*Geography Department, University College of Wales,*
*Penglais, Aberystwyth, Dyfed SY23 3DB*

## SUMMARY

**1** This paper draws on data from a field and a greenhouse experiment to illustrate the relationships between leaf and soil concentrations of Cd, Cu, Pb and Zn. Radish *(Raphanus sativus)* was grown in forty-six garden soils, characterized by varying degrees of metal contamination, and also in pots of contaminated soil where liming was used to modify soil pH. *Cyperus esculentus* was also grown in pots of contaminated soil and redox potential was lowered by keeping one set of pots flooded with tap water.

**2** The Zn contents of radish leaf were best predicted by extracting soil with 1M $NH_4NO_3$ and liming greatly reduced leaf Zn concentrations. Evidence is presented which suggests that the pH effect is exercised through a modification of the release of Zn from other soil compartments to the exchange sites.

**3** Liming was less effective in reducing leaf Cd levels but lowering redox potential greatly reduced them, possibly by the formation of the nearly insoluble CdS. Both EDTA and $CH_3COOH$ are satisfactory soil extractants for Cd.

**4** Neither changes in pH nor redox potential were effective in modifying tissue Pb concentrations and several soil extractants or even total Pb could be used to predict plant availability. None of the extractants used successfully predicted Cu uptake.

## INTRODUCTION

Metals whose densities exceed 6 g $cm^{-3}$ are conveniently referred to as heavy metals. Some are essential for life processes (Co, Cr, Cu, Fe, Mn, Mo, Sn, V, W, Zn) whereas others (e.g. Cd, Pb, Hg) have no known beneficial biological role; all of them are injurious to plants or animals at higher concentrations. All or most of these metals are released to the environment when their compounds are mined, smelted or later used industrially. When the heavy metal compounds reach the soil they tend to accumulate and become available for absorption by crop and food plants. The Royal Commission on Environmental Pollution in their Seventh Report (RCEP 1979) reviewed agriculture and pollution and commented that Cd was then the element of greatest concern due both to its toxicity to man and to the way it was readily taken up by plants. In the Ninth Report (RCEP 1983) dealing with Pb, it was noted that food, though not a primary source, is a most important pathway for the transport of Pb from a multiplicity of sources, including soil. Davies & White (1981) reported that vegetables grown in soils contaminated by Pb and other metals as a consequence of ore-processing contain unusually elevated concentrations of these metals. Subsequently, an investigation was made of blood Pb concentrations of women resident in

*Present address: School of Environmental Science, University of Bradford, Bradford, West Yorkshire, BD7 1DP.

one metal-contaminated Pb mining area. Women who were heavily dependent on home-grown vegetables had blood Pb concentrations that were 28% higher than those of women who consumed no locally-grown vegetables. Although the significance of these results is not clear it would seem prudent to seek to understand, and hence minimize, the uptake of metals by plants. This paper presents an account of the relationships between plant metal concentrations and soil composition and conditions.

## EXPERIMENTAL

The data used in this paper were taken from two experiments. The first one involved field experimentation, using radish (*Raphanus sativus* cv: French Breakfast) as the experimental plant, in order to investigate the control of soil composition on heavy metal uptake. Radish was sown in shallow drills in forty-six private gardens in England and Wales. The experimental sites were chosen to cover a wide range of heavy metal concentrations. The selected locations were: the Winster area of Derbyshire, eleven plots; the Halkyn Mountain area of Clywd, nineteen plots; the environs of Aberystwyth in Dyfed, sixteen plots; and Chepstow in Gwent, one plot. None of the plots was fertilized or limed but the plots were maintained free of weeds, commercial slug pellets were deployed to protect the plants and the owner of each garden carried out any necessary watering using domestic tap water.

The second experiment took place in a greenhouse and was designed to examine the effects of liming and redox potential on the uptake of metals from contaminated soils under controlled conditions. Five test soils were used; the control was an uncontaminated soil from the University College of Wales Frongoch Farm and the others derived from the old metal mining areas of Halkyn Mountain in Clywd; Parys Mountain in Anglesey; the Ystwyth Valley in Ceredigion; and Shipham in Somerset. All treatments were triplicated and all the soils received a basal fertilizer treatment. Except for the Somerset soil where pH was already near neutral, one-half of the soils were limed so as to raise their pH values to 6·5–7. The effect of redox potential was investigated by growing yellow nutsedge *(Cyperus esculentus)* on soils which were either maintained flooded, i.e. under 2 cm water, or were maintained aerobic and watered whenever soil moisture tension registered 50 bar on a tensiometer. Redox potential (Eh) was monitored on selected pots. The effects of liming were investigated on several agricultural or horticultural plants and the data for radish (*Raphanus sativus* var. Webb's French Breakfast) are considered in this paper.

Harvested plants were washed carefully, dried at 80–100 °C and milled. Subsamples were dry ashed at 450 °C and the ash dissolved in 0·1M $HNO_3$. Soils were analysed by conventional methods. Metal analysis was by flame atomic absorption spectrometry. Full details of this work and the experimental methods used have been published elsewhere (Davies, Ginnever & Lear 1981; Davies & Lear 1984; Davies & Houghton 1984).

## DISCUSSION

### *Soil metal content and plant uptake*

There is a broad, direct relationship between the concentration of a given metal in plants and in the soil. However, only a small fraction of the total soil metal content is available for plant uptake. Furthermore, availability to the root is influenced by soil factors among which pH and redox potential are widely regarded as important. The assessment of this plant-available fraction is usually done by equilibrating the soil with a variety of mild extractants. Many of the extractants used are those which have been found reliable in soil-testing for agricultural advisory work. Nevertheless, they were developed for an entirely different purpose, i.e. 'Soil analysis is used principally to assess the nutrient status of soils in relation to crop responses to applied nutrients and as a diagnostic aid in plant nutritional problems' (Needham 1983). There is no basis for supposing therefore that such extractants will predict metal uptake by plants nor that one extractant will serve for all metals or all plants.

The choice of suitable extractants was investigated by extracting the soils from the radish field experiment with several reagents. Total metals were determined by boiling the soil with refluxing concentrated nitric acid. Ethylene diamine tetraacetic acid (EDTA) (diammonium salt) was used to extract the soil solution + exchangeable + humus-bound forms. Ammonium nitrate (1M solution) extracted the exchangeable + soil solution forms. Acetic acid (5%) was used to extract the soil solution + exchangeable + specifically sorbed forms.

For this discussion the composition of the radish leaf was chosen for comparison and correlation with the soil values and the analytical data from the radish experiment are summarized in Tables 1 and 2. It was noted that all the Pb, Zn, Cu and Cd arithmetic

TABLE 1. Summary analytical data for soil and radish from forty-six plots. EDTA = 0·05M diammonium EDTA extractable; Acetic = 5% acetic acid extractable; Ammon = 1M ammonium nitrate extractable, < implies below detection limit.

| | Dry material (mg kg$^{-1}$) | | | |
|---|---|---|---|---|
| Soil data | Pb | Zn | Cu | Cd |
| Total: Mean | 1126 | 899 | 53 | 6·3 |
| Geometric mean | 577 | 535 | 41 | 1·6 |
| Maximum | 3780 | 5929 | 158 | 62 |
| Minimum | 25 | 79 | 10 | 0·05 |
| EDTA: Mean | 566 | 274 | 22 | 3·6 |
| Geometric mean | 282 | 127 | 17 | 1·5 |
| Maximum | 2100 | 1275 | 75 | 24 |
| Minimum | 5·5 | 1·5 | 2·0 | 0·1 |
| Acetic: Mean | 72 | 3·8 | 3·2 | 2·2 |
| Geometric mean | 22 | 133 | 2·8 | 0·8 |
| Maximum | 500 | 2000 | 16 | 23 |
| Minimum | 2·0 | 2·0 | 2·0 | 0·1 |
| Ammon: Mean | 1·3 | 9·5 | < | 0·2 |
| Geometric mean | 1·2 | 5·0 | < | 0·1 |
| Maximum | 3·5 | 70 | < | 1·3 |
| Minimum | 1·0 | 1·0 | < | 0·1 |
| Leaf data | | | | |
| Mean | 50 | 299 | 6·7 | 2·8 |
| Geometric mean | 24 | 201 | 5·8 | 1·1 |
| Maximum | 606 | 1184 | 21 | 25 |
| Minimum | 2·3 | 17 | 2·6 | 0·1 |

TABLE 2. Summary analytical data for soils from forty-six experimental plots. CEC = cation exchange capacity. K and Mg were extracted with 1M $NH_4NO_3$

| | pH | Ignition loss (%) | CEC (meq 100 $g^{-1}$) | K | Mg |
|---|---|---|---|---|---|
| | | | | (mg $kg^{-1}$) | |
| Mean | 6·0 | 10·3 | 28·4 | 314 | 131 |
| Maximum | 7·1 | 17·4 | 54·4 | 1125 | 365 |
| Minimum | 4·6 | 4·9 | 11·8 | 75 | 34 |

means were appreciably greater than their geometric means which was taken as evidence that the metal concentration data were all positively skewed. The metal concentrations were
therefore log-transformed to normalize the data before statistical analyses were carried out.

The data were first examined by simple correlation analysis, the results of which are presented in Table 3. The values for the correlation coefficient *(r)* in the cases of Cd and Pb

TABLE 3. Pearson correlation coefficients for soil and radish leaf metal concentrations. All metal data were log-transformed prior to analysis. EDTA = diammonium EDTA extractable (0·05M), Acetic = acetic acid extractable (5%), Ammon. = ammonium nitrate extractable (1M). Data from forty-six plots for which critical values of *r* are: 5% probability, *r* = 0·29; 1% probability, *r* = 0·37; 0·1% probability, *r* = 0·46

| | Radish leaf | | | |
|---|---|---|---|---|
| Soil | Pb | Zn | Cu | Cd |
| Total | 0·88 | 0·59 | 0·39 | 0·84 |
| EDTA | 0·89 | 0·57 | 0·38 | 0·86 |
| Acetic | 0·83 | 0·59 | 0·32 | 0·86 |
| Ammon. | 0·15 | 0·74 | - | 0·37 |

suggest that any fairly strong extractant satisfactorily predicts the radish leaf concentration of these metals. In contrast, ammonium nitrate provides a poor prediction. None of the extractants predicted Cu uptake satisfactorily, including EDTA, and this is perhaps surprising since EDTA has proved widely useful in identifying copper responsive soils. In the case of Zn, ammonium nitrate was the best extractant.

Besides their use in predicting metal uptake by plants these extractants have also been used to fractionate soils chemically. By comparing the various geometric means given in Table 1, it can be seen that EDTA extracts a large part of the total Pb, Cu and Cd (49%, 41% and 94%, respectively) whereas it extracts only 24% of the total Zn. EDTA is assumed to extract the humus-bound, readily exchangeable and specifically sorbed fractions. In contrast, ammonium nitrate mostly extracts the readily exchangeable forms and the proportions of the total metal extracted were Pb = 0·2%, Zn = 0·9% and Cd = 7·5%. Acetic acid extracts the readily and specifically sorbed fractions and the proportions extracted were, Pb = 4%, Zn = 25%, Cu = 7% and Cd = 47%. Thus, the metals differ in their soil behaviour. Lead and Cu appear to be bound mostly in the organic fraction whereas the specifically sorbed forms are more important for Cd and Zn. The correlation coefficients (Table 3) imply that the readily exchangeable Zn fraction is the plant-available form whereas humus-bound forms are more important for Pb and Cd.

### *Effects of lime and pH on metal uptake*

Evidence for a pH control on metal uptake by plants can be derived from the greenhouse study. The heavy metal contents of the experimental soils are presented in Table 4. None of

TABLE 4. Total metal contents (mg $kg^{-1}$ dry soil) of experimental soils used in greenhouse study

| Soil | Pb | Zn | Cu | Cd |
|---|---|---|---|---|
| Ystwyth | 488 | 256 | 12 | 0·42 |
| Parys Mountain | 597 | 114 | 488 | 0·30 |
| Halkyn Mountain | 419 | 782 | 151 | 0·40 |
| UCW Frongoch | 23 | 46 | 6·6 | 0·21 |
| Somerset | 445 | 3406 | 8·7 | 109 |

the soils was unusually high in Pb although all, except the control, were contaminated (Davies 1983). The Cu content of the Parys Mountain soil is consistent with its derivation from near an old copper mine and the Cd and Zn contents of the Somerset soil are very high since the soil was collected from an old zinc mining area.

Pb, Zn and Cd contents of the radish leaves and soil pH values in the limed and unlimed treatments are summarized in Table 5. Liming raised the pH of all soils; in the case of the

TABLE 5. Mean values (mg $kg^{-1}$ dry matter) for radish (leaf) grown on four experimental soils: UL = unlimed, L = limed

| | Soil pH | | Pb | | Zn | | Cd | |
|---|---|---|---|---|---|---|---|---|
| Soil | UL | L | UL | L | UL | L | UL | L |
| Y | 5·7 | 7·1 | 8·4 | 6·8 | 117 | 41 | 1·2 | 0·74 |
| PM | 5·9 | 6·4 | 9·3 | 5·7 | 138 | 45 | 1·5 | 0·34 |
| HM | 5·7 | 7·4 | 2·7 | 3·2 | 129 | 128 | 0·22 | 0·36 |
| F | 6·6 | 7·2 | 3·0 | 2·9 | 37 | 30 | 0·46 | 0·35 |

Parys Mountain material the calculated amount of lime was not sufficient to raise the pH to 7 whereas in the case of Halkyn Mountain soil the pH exceeded 7. The most marked liming effect in reducing leaf metal concentrations was seen for Zn except on the overlimed

Halkyn Mountain soil where no significant differences were observed between the treatments. Liming caused a smaller reduction in tissue Cd concentrations, and raised the leaf Cd level on the Halkyn soil. The evidence for a control of leaf tissue Pb content by liming is equivocal.

The leaf metal concentration changes recorded in Table 4 could be caused either by the increase in soil pH which resulted from liming or simply due to immobilization of the metals in the soil because of the presence of calcium carbonate. The data from the radish field experiment were examined to determine whether pH was a significant factor in controlling metal uptake. This was done by multiple regression analysis.

The Pearson correlation coefficients in Table 3 indicate that leaf zinc concentrations were best predicted by ammonium nitrate extractable, i.e. exchangeable, Zn. The relationship was further investigated by entering pH as a second dependent variable. This resulted in a small increase in the value of the correlation coefficient, from 0·74 to 0·86 and the multiple regression was computed as:

$$\log \text{leaf-Zn} = 0{\cdot}86 \log \text{soil-Zn} + 0{\cdot}3\ \text{pH} - 0{\cdot}106$$

This is a slightly curious result since it suggested that increasing pH would increase uptake of Zn. The regression analysis was repeated using soil total Zn. The correlation coefficient increased from 0·59 to 0·79 and the regression was:

$$\log \text{leaf-Zn} = 0{\cdot}93 \log \text{soil total-Zn} - 0{\cdot}42\ \text{pH} + 2{\cdot}28$$

In this case, increasing pH leads to a decrease in leaf Zn concentration. For example, taking the geometric mean soil Zn of 535 mg $kg^{-1}$ (Table 1), at pH 5 the leaf Zn is 522 mg $kg^{-1}$ and at pH 7 it reduces to 75 mg $kg^{-1}$. The strength of the association between leaf and soil total zinc is comparable with that for exchangeable Zn. The possible explanation is that ammonium nitrate extracts the plant available Zn, i.e. readily exchangeable Zn, and it is the transfer of Zn from other soil compartments to the exchange complex which is controlled by pH. This latter hypothesis was tested by investigating the effects of pH on the correlation between exchangeable Zn and total and EDTA-extractable Zn. In the case of soil total and exchangeable Zn, including pH as an independent variable increased the value of $r$ from 0·19 to 0·86 and the resulting regression was:

$$\log \text{exchangeable-Zn} = 0{\cdot}99 \log \text{total-Zn} - 0{\cdot}79\ \text{pH} + 2{\cdot}78.$$

For EDTA-extractable Zn similar results were found: including pH increased the value of $r$ from 0·49 to 0·83 and yielded the regression:

$$\log \text{exchangeable-Zn} = 0{\cdot}61 \log \text{EDTA-Zn} - 0{\cdot}71\ \text{pH} + 3{\cdot}7$$

These results are consistent with a hypothesis that the effect of pH on tissue Zn concentrations is exerted through a control of the supply of Zn to the cation exchange complex rather than on uptake by the root from the soil solution. The major pool of plant available Zn appears to be readily exchangeable Zn.

It was observed above that although increasing pH lowered radish leaf Cd concentrations the effect was not as marked as for Zn. The data in Table 3 suggest that either EDTA or acetic acid extractable Cd are both good predictors of leaf Cd concentrations. Allowing for the effect of pH made no significant difference to the correlation for EDTA and it increased the correlation with exchangeable Cd from 0·37 to 0·58. The correlation for exchangeable Cd vs. total Cd was 0·29 and allowing for pH yielded a multiple correlation of 0·50. This suggests that pH can also control the transfer of Cd to the cation exchange sites but that the plant derives its Cd from other soil compartments where pH control is less important.

Multiple regression analysis for Cu and Pb revealed no role for pH, or organic content and cation exchange capacity.

### *Control by redox potential*

In normal agriculture large alternations in oxidizing and reducing conditions, i.e. markedly changing redox (Eh) values, in the soil are of no great importance since most agricultural soils are free draining and wet soils are artificially drained. Only in the production of padi rice does redox become important. Redox is important for contaminated dredged sediments which may be disposed under water, in the tidal zone to create artificial marshes or on land, for reclamation where the sediment drains and oxidizes. It is also important for natural wetlands which may have been contaminated by heavy metals. The metals under consideration here, namely Pb, Zn, Cu and Cd are generally present in natural systems as their divalent cations, although cuprous $Cu^{+}$ is a possibility in very acid and reducing environments. Redox potential therefore cannot control the availability of these metals through changes in their oxidation states but rather through sulphur species derived from organic matter. When redox falls sulphate reduces to sulphide and the metals discussed here are all chalcophilic, i.e. they have a strong affinity for sulphur. Solubility products for heavy metal sulphides are listed in Table 6 and it is

TABLE 6. Reducing conditions and metal solubility (Data from Lindsay 1979)

| | $pK_s$* | pe + pH† |
|---|---|---|
| Pb S | 27·5 | 4·66 |
| Zn S | 24·7 | 4·28 |
| Cu S | 36·1 | 4·73 |
| Cd S | 27·1 | 4·73 |

*Solubility product.
†Critical value for precipitation when pH = 7·0, $SO_4^{2-}$ = $10^{-3}$ M, $CO_2$ = 0·0003 atm.

clear that they are all nearly insoluble, so that reducing conditions may markedly affect their solubilities and consequent plant availabilities.

Redox potential is traditionally represented as Eh (in volts) but there are advantages in converting it to pe (i.e. − log electron activity in moles $l^{-1}$ and Eh = 59·2 pe) which makes it comparable and combinable with pH. In Table 6 are listed the critical values of (pe + pH) at which metal sulphides precipitate when pH = 7 (Lindsay 1979).

Cd and Pb concentrations of *C. esculentus* leaves when grown on oxidized or reduced soils together with (pe + pH) values are summarized in Table 7. Comparison with Table 6

TABLE 7. Metal concentrations of leaves of *Cyperus esculentus* grown on contaminated soils under oxidizing (O) and reducing (R) conditions

| | | pe + pH | | Pb | | Cd | |
|---|---|---|---|---|---|---|---|
| | | | | mg kg$^{-1}$ dry matter | | | |
| Soil | pH | O | R | O | R | O | R |
| Y | 7·1 | 18·6 | 2·0 | 1·01 | 1·62 | 0·87 | 0·02 |
| PM | 6·4 | 17·4 | 4·7 | 5·88 | 2·34 | 0·36 | 0·15 |
| HM | 7·4 | 17·5 | 3·3 | 2·39 | 1·32 | 0·08 | 0·08 |
| F | 7·2 | 19·7 | 5·0 | 0·97 | 1·89 | 0·04 | 0·01 |
| S | 7·1 | 19·8 | 2·4 | 0·94 | 1·23 | 5·04 | 0·54 |

demonstrates that, when the experimental soils were maintained flooded, reducing conditions were induced which were severe enough to reduce sulphate to sulphide and therefore to precipitate heavy metals. A marked contrast between Cd and Pb is observed so far as uptake by *C. esculentus* is concerned. In only two soils (Parys Mountain and Halkyn Mountain) did reducing conditions result in a reduction in tissue Pb, by 60% and 45% respectively. For Cd, with the exception of the Halkyn Mountain soil, leaf concentrations were reduced by 58–98%, mean 80%.

The applicability of simple chemical criteria to marsh plants is uncertain. Radial oxygen loss from the root is likely to modify redox conditions in the rhizosphere and availability for uptake may depend on the balance between increased solubility due to raised redox potential and decreased availability because of adsorption on or co-precipitation in the ferric oxide which precipitates near the root.

## CONCLUSIONS

Increasing soil concentrations of the metals discussed here are accompanied by increasing plant contents of the same metals. Available Zn is best assessed by extracting the exchangeable fraction, e.g. with $NH_4NO_3$. Soil total Zn values can also be used to predict plant uptake providing an allowance is made for soil pH since pH appears to modify the release of Zn from other soil compartments to the cation exchange sites. Liming a soil is an effective way of reducing Zn availability and overcoming phytotoxicity.

The consequences of liming on Cd uptake are less certain. Humus-bound and specifically sorbed Cd appear to contribute to the plant-available pool and both EDTA and $CH_3COOH$ can provide a satisfactory measure of the availability of Cd. The Cd content of plants is also markedly affected by changes in redox potential. Therefore, draining Cd contaminated marsh soils or the disposal of contaminated dredged sediment in conditions where it can drain and aerate, should be avoided.

Plant uptake of Pb does not appear to be greatly controlled by changes in redox potential nor would raising soil pH by liming be certain to reduce its availability. Any reasonably vigorous soil extractant can be used to predict uptake of Pb and total soil Pb content is also a satisfactory guide.

Copper remains an enigma. None of the extractants reviewed here is satisfactory in predicting the uptake of the metal.

## ACKNOWLEDGMENTS

JML thanks the Welsh Office for funds to support the field experiment and the greenhouse experiment (NJL) was performed under contract DAJA-37-82C-0195 from the US Army Corps of Engineers.

## REFERENCES

**Davies, B.E. (1983).** A graphical estimation of the normal lead content of some British soils. *Geoderma,* **29**, 67–75.

**Davies, B.E., Ginnever, R.C. & Lear, J.M. (1981).** Cadmium and lead contaminated soils in some British metal mining areas. *Trace Substances in Environmental Health. XV* (Ed. by D.D. Hemphill), pp. 323–332. University of Missouri.

**Davies, B.E. & Houghton, N.J. (1984).** The use of radish as a monitor crop in heavy metal polluted soils. *Environmental Contamination.* pp. 327–332. Proceedings International Conference, London 1984. CEP, Edinburgh.

**Davies, B.E. & Lear, J.M. (1984).** Heavy metal uptake by radish in relation to soil fertility and chemical extractability of metals. *Being Alive on Land.* (Ed. by N.S. Margaris, M.A. Farragitaki & W. Oechel), pp. 307–312. W. Junk, Holland.

**Davies, B.E. & White, H.M. (1981).** Trace elements in vegetables grown on soils contaminated by base metal mining. *Journal of Plant Nutrition,* **3**, 387–488.

**Lindsay, W.L. (1979).** *Chemical Equilibria in Soils.* John Wiley & Sons, New York.

**Needham, P. (1983).** The occurrence and treatment of mineral disorders in the field. *Diagnoses of Mineral Disorders in Plants, Chapter 4.* HMSO, London.

**RCEP (1979).** Agriculture and Pollution; 7th Report, Royal Commission on Environmental Pollution. HMSO, London.

**RCEP (1983).** *Lead in the Environment; 9th Report, Royal Commission on Environmental Pollution.* HMSO, London.

# Experimental investigations on the relative contribution of atmosphere and soils to the lead contents of crops

ROY M. HARRISON* AND W.R. JOHNSTON†

*Department of Chemistry, University of Essex, Wivenhoe Park, Colchester CO4 3SQ, United Kingdom*

## SUMMARY

**1** Three independent experimental approaches have been used to estimate the separate air and soil-derived contribution of lead to a range of crop plants.
**2** Values of the soil lead concentration factor (lead in plant (mg $kg^{-1}$ dry wt) ÷ total lead in soil (mg $kg^{-1}$ dry wt) in the absence of an air lead contribution) mostly fall within the range $1–30 \times 10^{-3}$.
**3** In most instances, both air and soil contribute a significant proportion of the lead content of the plant tissues.
**4** Some species, but not others, give an indication of translocation of air-derived lead within the plant to tissues not directly exposed to the atmosphere.

## INTRODUCTION

For most individuals, the diet comprises the major route of lead intake (Harrison & Laxen 1981). Plant material makes up an appreciable proportion of the diet, whether in fresh, cooked or processed form. In his excellent review of fallout of lead and uptake by crops, Chamberlain (1983) identified two key questions:

(a) What proportion of dietary lead comes from basic food, as distinct from that in handling, manufacture or cooking?
(b) What proportion of lead in consumer crops and other foods is derived from current fallout of lead and not taken up from the soil?

The latter point is particularly salient in view of current governmental policies to limit or reduce population exposure to lead by reduction of the lead level in gasoline. Such a reduction is likely to be effective in reducing the atmospheric fallout (or deposition) contribution of lead to crops, but not in influencing the soil-derived component.

Lead is present naturally in soils at concentrations typically within the range 10–200 mg $kg^{-1}$ in the absence of local mineralization. Chamberlain (1983) estimated that atmospheric fallout due to leaded gasoline combustion has raised UK soil lead levels by approximately 3 mg $kg^{-1}$ since its introduction in 1946. This may be viewed in the light of an estimated median total lead content of 42 mg $kg^{-1}$ in air-dried agricultural soils sampled in England and Wales (Archer 1980). Even allowing for a possibly greater availability of

*To whom correspondence should be addressed.
†Present address: Shirley Institute, Didsbury, Manchester.

fallout-derived soil lead for plant uptake, the incremental gasoline-derived contribution to crop lead arising from the soil is rather small.

It is far less clear how great a contribution direct deposition of atmospheric lead to plant surfaces can make to the ultimate lead content of the crop. Such deposition may take the form of dry particle deposition, or of rainwater. It is presently unclear whether the net effect of rainwater is as a net source or cleanser of lead from plant surfaces (or perhaps either, dependent upon conditions).

The literature reviewed by Chamberlain (1983) shows a great disparity of estimates of both soil and atmospheric contributions to lead in crops. In this paper, we describe measurements made by three experimental approaches to the elucidation of these contributions. The approaches were as follows.

(a) Labelling of soil with $^{210}$Pb tracer and growth of plants in the field.
(b) Growth of plants in a two compartment chamber in which one side is ventilated with ambient air and the other with filtered air which is essentially lead-free.
(c) Growth of plants in a common soil at field sites with differing air lead levels. Atmospheric deposition was measured by means of moss bag collectors, and extrapolation of plant lead concentrations to zero deposition allowed estimation of soil lead in the plant.

## EXPERIMENTAL

### *$^{210}$Pb tracer experiments*

Two soils were amended with $^{210}$Pb tracer, initially at a level of 3 $\mu$Ci kg$^{-1}$ and subsequently also to 21 $\mu$Ci kg$^{-1}$. The $^{210}$Pb was derived from a high specific activity of $^{210}Pb(NO_3)_2$ in dilute nitric acid (Amersham International). This was diluted to *c.* 1 $\mu$Ci ml$^{-1}$ with deionized water. Oven-dried soil (200 g) was dried and ground to a fine powder using a pestle and mortar, and was added to a beaker containing 1 $\mu$Ci ml$^{-1}$ $^{210}$Pb solution (50 ml) and a further 50 ml of deionized water was added. The resultant slurry was mixed

TABLE 1. Characteristics of the soils used

| Name | pH* | Total lead content (mg kg$^{-1}$ dry wt) | Available stable lead (%) | Available $^{210}$Pb(%) |
|---|---|---|---|---|
| Hazlerigg | 7·1 | 39·1 | 0·30 | 0·90 |
| Nateby | 6·5 | 49·7 | 0·85 | 0·54 |
| John Innes No. 1 | 6·5 | 15·5 | ND | ND |
| Thornton House Farm (THF) | 6·9 | 9·0 | ND | ND |
| M6 | 7·5 | 150 | ND | ND |
| Sellerley (Sel) | 7·5 | 41·5 | ND | ND |

*pH values determined from a mixture of air-dried soil (3 g) and deionized water (15 ml).
ND = not determined.

TABLE 2. Results of $^{210}Pb$ field plot experiments

| Plant material | Soil | Plant Pb (mg $kg^{-1}$ dry wt) | CF* | Soil contribution to plant lead (%) |
|---|---|---|---|---|
| Lettuce | | | | |
| Outer leaves | Nateby | 34·2 | 0·188 | 27 |
| | | 3·81 | 0·016 | 21 |
| | | 5·08 | 0·008 | 8 |
| Heart | Nateby | 1·32 | 0·014 | 52 |
| Outermost leaves | Hazlerigg | 24·9 | 0·465 | 56 |
| Outer leaves | | 11·5 | 0·206 | 54 |
| Intermediate leaves | | 6·5 | 0·062 | 30 |
| Inner leaves | | 1·8 | 0·013 | 22 |
| Heart | | 0·5 | 0·005 | 31 |
| Root | | 11·0 | 0·274 | 75 |
| Turnip | | | | |
| Dead leaves | Hazlerigg | 5·45 | 0·020 | 14 |
| | | 3·60 | 0·011 | 12 |
| | | 14·0 | 0·006 | 2 |
| Leaves | Hazlerigg | 3·50 | 0·010 | 11 |
| | | 2·00 | 0·003 | 6 |
| | | 4·26 | 0·005 | 5 |
| Dividing stalk | Hazlerigg | 5·04 | 0·028 | 22 |
| | | 2·19 | 0·033 | 59 |
| | | 5·70 | 0·038 | 26 |
| | | 15·0 | 0·010 | 3 |
| Inner root | Hazlerigg | 0·72 | 0·015 | 81 |
| | | 0·61 | 0·003 | 19 |
| | | 1·36 | 0·015 | 42 |
| | | 0·24 | 0·005 | 76 |
| | | 0·60 | 0·005 | 33 |
| Outer peelings | Hazlerigg | 0·73 | 0·020 | 107 |
| | | 2·24 | 0·021 | 37 |
| | | 0·44 | 0·011 | 94 |
| | | 2·20 | 0·020 | 35 |
| | | 0·35 | 0·009 | 123 |
| Pea | | | | |
| Dead leaves | Hazlerigg | 13·5 | 0·077 | 22 |
| Leaves | | 13·3 | 0·013 | 4 |
| Stalk | | 2·19 | 0·019 | 34 |
| Pod | | 3·93 | 0·014 | 14 |
| Peas | | 0·25 | 0·002 | 36 |

*CF values refer only to $^{210}Pb$ measurements.

thoroughly, dried at 80 °C for 24 h, ground finely and mixed into air-dried soil (2·4 or 16·7 kg) in a sealed rotating box. The labelled soil was placed in free draining glass-reinforced plastic tanks (each about 0·4 × 0·4 × 0·4 m) which were sunk into the ground so as to be almost flush with the surrounding soil surface. The tanks contained *c.* 50 kg of labelled soil derived from Hazlerigg and Nateby (Table 1) and plants were grown during 1981 and 1982 within the tanks and in the surrounding soil so as to avoid excessive isolation of individual plants after adding a top-dressing of lime and N-P-K fertilizer and ageing prior to use. The surface soil was covered with fine gravel to limit splashing of plant leaves by soil during rain storms. The field plot was located at Hazlerigg, near Lancaster. The site is described by Harrison & Williams (1982) who measured a mean air lead level of 132 ± 51 ng $m^{-3}$ (7-day samples) during 1978–79.

Plants listed in Table 2 were grown from commercial seed (Suttons). Since the plots were exposed to the atmosphere, no additional watering was carried out and no further fertilizer was added. Plants were harvested at the end of the growth season, sealed in plastic bags and frozen prior to analysis. Plant leaves were analysed without washing, having been dissected from the plant with an acid-washed stainless steel scalpel. The turnip root was washed with deionized distilled water prior to dissection into outer and inner portions, the latter being entirely internal material without contact with soil.

Soil and plant material was digested twice in nitric acid prior to taking up the residue in 5% nitric acid, filtering and determination of total lead by atomic absorption spectroscopy. Standard addition procedures were used routinely and matrix modification procedures were used when necessary in graphite furnace analysis of plant material.

The $^{210}$Pb content of plant and soil material was evaluated by drying an aliquot of the above 5% nitric acid solution onto a filter paper and counting the 47 keV emission on a $\gamma$-spectrometer in comparison with standards prepared in the same manner. This procedure gave good precision (*c.* 5% relative standard deviation) and a detection limit of *c.* 10 pCi $^{210}$Pb. 'Plant available' lead was evaluated by extraction of soil with 1M ammonium acetate at pH 7.

### *Filtered air cabinet*

The design of the cabinet has been described elsewhere (Johnston & Harrison 1983). Although the cabinet design permits growth of plants in soil, or nutrient solution culture, only the former technique was consistently successful. A number of soils were used. Results in this paper refer to THF and John Innes soils (see Table 1).

When mature, plants were harvested and dissected with an acid-washed stainless steel scalpel. They were dried and analysed without prior freezing. Twofold nitric acid digestion and graphite furnace atomic absorption analysis were used as described above.

Air lead levels in the two sides of the chamber were determined by continuous sampling through Millipore filters type HAWP 04700, followed by nitric acid extraction and atomic absorption analysis.

### *Field site experiments*

During 1983 plants were grown in pots in close array at each of four field sites, chosen for their differing air lead levels. Moss bags comprising *c.* 1 g of washed air-dried moss (*Sphagnum squarrosum*) contained in a nylon hairnet were suspended above the plants for the entire growth periods. In 1984, plants were grown in plastic tubs using a different set of sites. In all, three soils were used, in 1983 an agricultural soil and a soil taken from the M6 verges, and in 1984 a different agricultural soil. The soil surface was covered with a thin layer of gravel to avoid soil splashage, and plants were watered with deionized water at all sites during extended periods of dry weather.

Analytical procedures for plants, moss bags and soils were as described above. Results for plant tubers refer to well washed inner and outer material; onions were separated from their roots and analysed whole, whilst leaves, stems, pods and peas were analysed without washing. Both turnips and onions showed very slow growth and were harvested at sizes well below normal.

Results were plotted in the form of plant lead concentration vs. moss bag lead concentration (corrected for background) as well as plant total lead vs. moss bag total lead.

## RESULTS AND DISCUSSION

Details of the various soils used are summarized in Table 1. The results of the three experiment types are discussed separately.

### *$^{210}$Pb tracer experiments*

The soils were well aged (by 6 months or 18 months) when plants were grown in them to enable the $^{210}$Pb tracer to equilibrate with the soil. Measurements of available lead ratios to total lead for both stable and tracer lead (Table 1) are indicative of a reasonably similar availability for the label as for stable lead. The results for 'available' lead showed poor consistency between replicates, due presumably to soil inhomogeneity. Since the plant roots extend through an appreciable amount of soil, these should act as integrators and average out much of the variability. In interpretation of the results, no correction for differential tracer availability has been applied.

Representative results are presented in Table 2. In these,the concentration factor (CF) defined by Chamberlain (1983) has been calculated in respect of $^{210}$Pb.

$$\text{CF} = \frac{\text{Lead in plant (ppm dry wt)}}{\text{Total lead in soil (ppm dry wt)}}$$

Additionally, the percentage soil contribution to plant lead has been estimated from:

$$\frac{\text{mg kg}^{-1}\text{ Pb in soil (dry wt)}}{\text{mg kg}^{-1}\text{ Pb in plant (dry wt)}} \times \frac{^{210}\text{Pb g}^{-1}\text{ in plant (dry wt)}}{^{210}\text{Pb g}^{-1}\text{ in soil (dry wt)}} \times 100$$

There are few literature studies which allow an estimate of CF since most have not controlled for air-derived metal. Chamberlain (1983) used data from Davies (1978) to estimate a CF value of about 0·07 for radish grown on a soil of 100 mg (Pb) kg$^{-1}$, and using data from Warren (1973) estimates highly variable CF values from 0·02 to 5·3, with a median of 0·5, for potatoes grown on a variety of soils.

CF values are also estimable from other studies with $^{210}$Pb tracer. Karamamos, Bettany & Stewart (1976) gave CF values for alfalfa and bromegrass of 0·19 for 20 mg (Pb) kg$^{-1}$

and 0·09 for 100 mg (Pb) $kg^{-1}$ lead in soil. Their study differed from that described here in that sufficient stable lead was added to the soil to appreciably elevate the total lead content from its original 6–9 mg (Pb) $kg^{-1}$. In our work, tracer addition did not significantly influence the total lead content of the soil. With a soil lead content of 14 mg (Pb) $kg^{-1}$, Tjell, Hovmand & Mossbaek found a CF value of 0·0027 for ryegrass, highly consistent with earlier work by Wilson & Cline (1966) who reported CF values of 0·002–0·005 for young barley shoots grown on three different soils. Unpublished data from Heard, reported by Chamberlain (1983) indicate similarly low CF values of 0·001–0·002 for cabbages grown on soil amended with $^{210}Pb$.

The CF values reported here (Table 2) vary considerably from 0·002 to 0·465. We cannot entirely rule out soil splashage of outer leaves, nor soil contamination of outer root portions. Taking values for inner portions of plants, such as the lettuce heart, turnip and pea, gives CF values in the range 0·002–0·015 with a median value of 0·005. This behaviour is consistent with all but one of the $^{210}Pb$ tracer studies reported above.

### *Filtered air cabinet experiments*

Data from the filtered air cabinet are summarized in Table 3. Interestingly, the CF values are in remarkably good agreement with those from the $^{210}Pb$ tracer experiments (Table 2). Values range from $< 0·001$ to 0·036. Spinach appears to be a relatively efficient absorber of soil lead; peas and beans absorb rather inefficiently into the edible portion of the crop. The absence of values of CF greater than 0·036 in these experiments is presumably a reflection of complete absence of soil splashage (pots were watered from the bottom, and soil was covered with vermiculite), as well as a possibly lower availability of lead in the THF and John Innes soils relative to the Nateby and Hazlerigg.

### *Field site experiments*

The type of data generated by the field site experiments is exemplified by Figs 1 and 2. Due to the limited number of data points, mathematical curve-fitting procedures did not appear to be warranted, and best-fit lines were fitted by eye, as exemplified in the figures. Some data were a reasonable fit to a straight line, whilst others were best fitted by a curve, and others were apparently rather random in nature. The intercept on the plant lead axis should indicate the plant lead content for zero atmospheric exposure (i.e. the soil contribution) and where possible the intercept has been estimated. The data are summarized in Table 4, which includes an estimate of the soil lead contribution as a percentage of total lead for the rural Hazlerigg site, which allows comparison with the $^{210}Pb$ experimental data. The graphs have been plotted in two ways: (i) from lead concentrations in plants and moss bags, (ii) from total lead contents of plants and moss bags, which takes some account of differing plant yield. Results of both approaches are included in Table 4. The CF values for peas are low and consistent with the other sources of data. Many of the other CF values are relatively high. This, we believe is a result of rather slow growing conditions in which absorbed lead is diluted only slowly by the increase in biomass. In general, the lead content of the plants grown in the field site experiment was

TABLE 3. Results from plants in filtered air cabinet experiment

| Plant material | Soil | Unfiltered air Air Pb (ng m$^{-3}$) | Unfiltered air Plant Pb (mg kg$^{-1}$ dry wt) | Filtered air Air Pb (ng m$^{-3}$) | Filtered air Plant Pb (mg kg$^{-1}$ dry wt) | CF | % soil | S (m$^3$ g$^{-1}$) |
|---|---|---|---|---|---|---|---|---|
| Lettuce ($n=4$) | John Innes | 86 | | 2 | | | | |
| Outer leaves | No. 1 | | 1·04 ± 0·16 | | 0·08 ± 0·04 | 0·005 | 8 | 11 |
| Inner leaves | | | 0·75 ± 0·34 | | 0·04 ± 0·02 | 0·003 | 5 | 8 |
| Heart | | | 0·19 ± 0·12 | | 0·02 ± 0·01 | 0·001 | 11 | 2 |
| Radish ($n=2$) | John Innes | 83 | | <1 | | | | |
| Dead leaves | No. 1 | | 0·40 ± 0·15 | | <0·01 | <0·001 | <3 | 5 |
| Live leaves | | | 0·25 ± 0·10 | | 0·12 ± 0·00 | 0·008 | 48 | 2 |
| Tuber | | | 0·85 ± 0·56 | | 0·02 ± 0·10 | 0·001 | 2 | 10 |
| Spinach ($n=3$) | THF | 58 | | <1 | | | | |
| Leaves | | | 0·60 ± 0·13 | | 0·21 ± 0·06 | 0·023 | 20 | 7 |
| Peas ($n=3$) | THF | 198 | | 11 | | | | |
| Leaves & stalk | | | 2·6 ± 0·70 | | 0·29 ± 0·08 | 0·032 | 11 | 12 |
| Pods | | | 0·26 ± 0·11 | | 0·07 ± 0·03 | 0·008 | 27 | 1 |
| Peas | | | 0·01 ± 0·00 | | 0·01 ± 0·00 | 0·001 | 100 | 0 |
| Lettuce ($n=3$) | THF | 190 | | 10 | | | | |
| Outer leaves | | | 1·6 ± 0·41 | | 0·27 ± 0·08 | 0·030 | 17 | 7 |
| Inner leaves | | | 0·67 ± 0·08 | | 0·20 ± 0·06 | 0·022 | 30 | 3 |
| Heart | | | 0·25 ± 0·09 | | 0·14 ± 0·08 | 0·016 | 56 | 0·6 |
| Spinach ($n=3$) | THF | 165 | | 2 | | | | |
| Leaves | | | 1·8 ± 0·70 | | 0·28 ± 0·06 | 0·031 | 16 | 9 |
| Stalk | | | 0·52 ± 0·23 | | 0·32 ± 0·15 | 0·036 | 62 | 1 |
| Dwarf beans ($n=3$) | THF | 105 | | 7 | | | | |
| Leaves & stalk | | | 0·33 ± 0·12 | | 0·01 ± 0·00 | 0·001 | 3 | 3 |
| Pods | | | 0·67 ± 0·29 | | 0·08 ± 0·00 | 0·009 | 12 | 6 |
| Beans | | | 0·01 ± 0·00 | | 0·01 ± 0·00 | 0·001 | 100 | 0 |

higher than for corresponding plants grown in the filtered air cabinet, whilst growth rates were far slower.

### *Foliar uptake and translocation*

Having established that air lead does contribute appreciably to lead in plants, it is useful to examine the efficiency of lead incorporation. Chamberlain (1983) defined an accumulation ratio, $S$, where

$$S = \frac{\text{Lead in foliage (mg kg}^{-1}\text{ dry wt)}}{\text{Lead in air } (\mu\text{g m}^{-3})}$$

The only experiments in which we have measured directly lead levels in air are the filtered

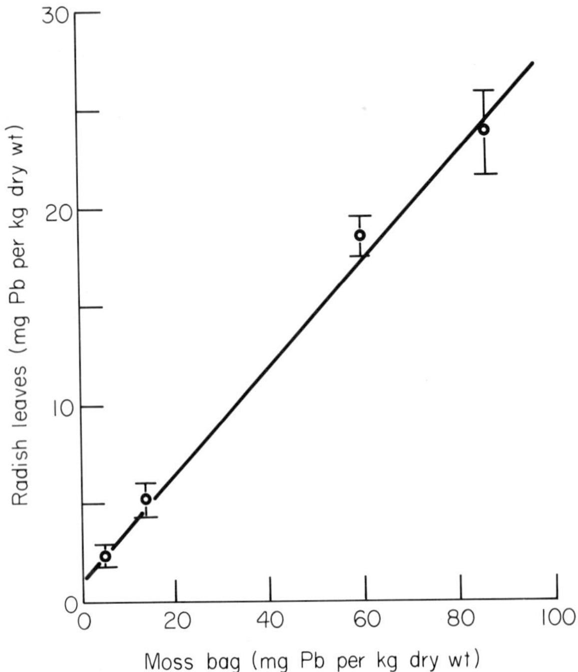


FIG. 1. Lead content of radish leaf in relation to that of moss bags for THF soil in 1983.

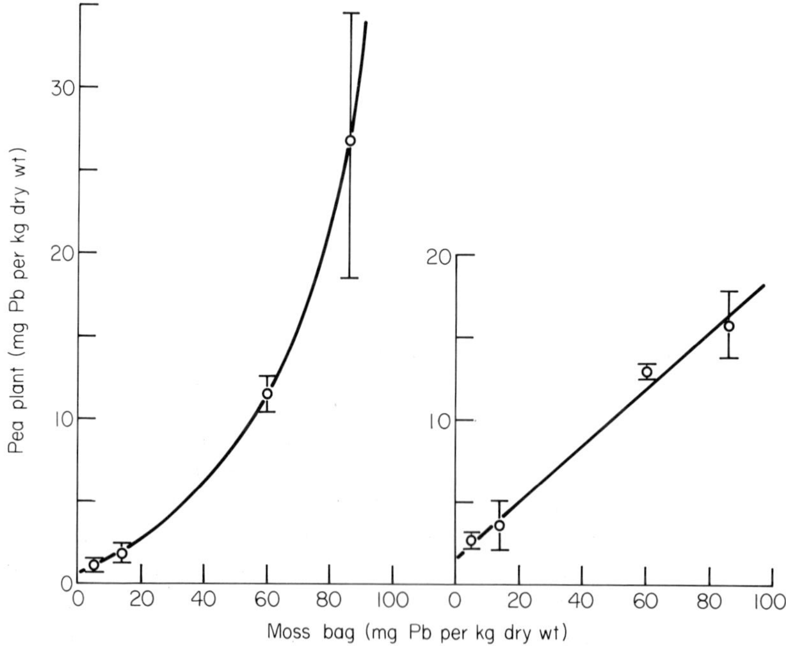


FIG. 2. Lead content of pea foliage in relation to that of moss bags for THF and M6 soils in 1983.

TABLE 4. Results of field plot experiment: graphs based on (a) lead concentrations and (b) total lead

| Plant material | Soil | (a) Lead concentrations: Form of curve* | Estimated intercept† ($mg\ kg^{-1}$ dry wt) | Approx % soil lead for Hazlerigg | CF | (b) Total lead: Form of curve* | Estimated intercept† ($mg\ kg^{-1}$) | Approx % soil cont for Hazlerigg | CF |
|---|---|---|---|---|---|---|---|---|---|
| *1983 Experiment* | | | | | | | | | |
| Radish: leaves | THF | L | 1·0 | 19 | 0·111 | C | 2 | 33 | 0·080 |
| | M6 | C | 2·0 | 55 | 0·013 | C | 0·71 | 30 | 0·005 |
| Radish: tubers | THF | C | 0·35 | 78 | 0·039 | C | 0·45 | 71 | 0·050 |
| | M6 | C | 0·80 | 60 | 0·005 | C | 1·20 | 100 | 0·008 |
| Peas: pods | THF | C | 0·20 | 21 | 0·022 | S | NC | – | – |
| | M6 | C | 0·30 | 26 | 0·002 | C | 0·6 | 36 | 0·004 |
| Pea: pea | THF | S | NC | 100 | 0·004‡ | S | NC | – | – |
| | M6 | S | NC | 100 | 0·001‡ | S | NC | – | – |
| Pea: leaves & stalk | THF | C | 0·6 | 33 | 0·067 | C | 1·11 | 18 | 0·120 |
| | M6 | L | 1·4 | 39 | 0·009 | C | 1·42 | 35 | 0·009 |
| Lettuce: leaves | THF | C | 0·6 | 34 | 0·067 | C | 0·24 | 16 | 0·03 |
| | M6 | I | NC | – | – | I | NC | – | – |
| Turnip: leaves | THF | C | 2·0 | 65 | 0·222 | C | 2·5 | 60 | 0·28 |
| | M6 | I | NC | – | – | I | NC | – | – |
| Turnip: roots | THF | C | 1·4 | 100 | 0·156 | C | 1 | 100 | 0·111 |
| | M6 | I | NC | – | – | I | NC | – | – |
| Onion | THF | C | 0·6 | 60 | 0·067 | C | 0·7 | 67 | 0·08 |
| | M6 | I | NC | – | – | I | NC | – | – |
| *1984 experiment* | | | | | | | | | |
| Radish: leaves | Sel | L | 1·2 | 69 | 0·029 | L | 0·60 | 38 | 0·014 |
| Radish: tubers | Sel | S | NC | – | – | S | NC | – | – |
| Lettuce: inner leaves | Sel | S | NC | – | – | S | NC | – | – |
| Lettuce: outer leaves | Sel | S | NC | – | – | S | NC | – | – |
| Onion: leaves | Sel | S | NC | – | – | S | NC | – | – |
| Onion: bulb | Sel | L | 0·1 | 91 | 0·002 | S | NC | – | – |
| Pea: pea | Sel | L | 0·05 | 100 | 0·001 | S | NC | – | – |
| Pea: pods | Sel | S | NC | – | – | C | 0·03 | 26 | 0·001 |
| Pea: stalks and leaves | Sel | S | NC | – | – | C | 0·16 | 36 | 0·004 |

*L = linear; C = curvilinear; S = scattered; I = insufficient data.
†NC = non-calculable.
‡CF values calculated from mean lead concentration of all pea samples.

air cabinet experiments. These are arguably inappropriate to use for calculation of $S$ values relevant to field conditions due to the absence of rainwater influences. Nonetheless, the total lead values for the plant tissues are similar to those for plants grown in the field, and indeed lettuce removed from the chamber when small, and grown subsequently in a suburban garden, showed lead levels within the range of those grown in the non-filtered side of the cabinet. $S$ values, based upon the air-derived lead in the plants only, are included in Table 3. Chamberlain (1983) estimated an $S$ value of 20 $m^3\ g^{-1}$ for grass and corn leaves and 2·6–4 $m^3\ g^{-1}$ for lettuce based upon data sets available in the literature.

Our own values of $S$ (Table 3) are generally very similar for the exposed parts of plants. Rough and exposed leaves show the highest values (e.g. outer lettuce leaves, spinach leaves and pea leaves and stalks).

The data from the filtered-air cabinet and field site experiments indicate that lead in the fruit of peas and dwarf beans does not derive from the atmosphere. The one measurement for peas in the $^{210}Pb$ experiment does indicate an atmospheric contribution, but this result was obtained only by bulking a number of samples and is considered less reliable than the other measurements. In contrast, lead in the heart of the lettuce and the inner tuber of the radish and turnip does appear to have a significant contribution from atmosphere, despite having no direct exposure route. This is contrary to the findings of Schuck & Locke (1970) for cabbage, but nonetheless appears quite clearly in our data, especially in regard to lettuce hearts (see Tables 2 and 3). The lettuce were large and well hearted, especially those grown in the filtered air cabinet, and this observation can be explained in terms of foliar absorption of lead, either via the stomata, or more probably through breaks in the cuticle, with subsequent translocation within the plant. No explanation is offered of the mechanism of translocation, nor of the reason why it appears to happen in some species and not in others.

## CONCLUSIONS

The data obtained, in terms of both plant lead levels and CF values generally, fall within the range reported in the literature. CF values appear to vary according to the particular plant and soil and, ignoring values which may have been influenced by soil contamination, mostly fall within the range $1–30 \times 10^{-3}$. This range is too great for convenient prediction of soil-derived plant lead, but is not surprising in view of the great variability which can occur within both plants and soils.

Whilst the parts of certain plants which are not exposed to the atmosphere do not appear to accumulate atmospheric lead (most notably peas and beans), in some other plants (i.e. lettuce and radish) it appears that lead may be translocated to unexposed tissues.

Rural air lead levels in the UK are generally of the order of 100 ng $m^{-3}$, and thus the levels at our Hazlerigg site and in the filtered air cabinet are reasonably reflective of typical agricultural sites. The estimated percentage soil contributions to plant lead indicate that for most species examined, both air and soil contribute significantly to the ultimate lead content of the plant grown at such a location.

## ACKNOWLEDGMENTS

We are grateful to Mrs J.M. Walsh and Mrs I. Dootson who carried out the majority of the plant growth and analytical work.

## REFERENCES

**Archer, F.C. (1980).** Trace elements in soils in England and Wales. *Inorganic Pollution and Agriculture*, MAFF Reference Book 326, pp. 184–190. HMSO, London.

**Chamberlain, A.C. (1983).** Fallout of lead and uptake by crops. *Atmospheric Environment,* **17**, 693–706.

**Davies, B.E. (1978).** Plant-available lead and other metals in British garden soils. *Science of the Total Environment,* **9**, 243–262.

**Harrison, R.M. & Laxen, D.P.H. (1981).** *Lead Pollution: Causes and Control.* Chapman & Hall, London.

**Harrison, R.M. & Williams, C.R. (1982).** Airborne cadmium, lead and zinc at urban and rural sites in north-west England. *Atmospheric Environment,* **16**, 2669–81.

**Johnston, W.R. & Harrison, R.M. (1983).** A design for a filtered air cabinet used in the study of foliar uptake of metals by crop plants. *Environmental Technology Letters,* **4**, 291–296.

**Karamanos, R.E., Bettany, J.R. & Stewart, J.W.B. (1976).** The uptake of native and applied lead by alfalfa and bromegrass from soil. *Canadian Journal of Soil Science,* **56**, 485–494.

**Schuck, E.A. & Locke, J.K. (1970).** Relationship of automotive lead particulates to certain consumer crops. *Environmental Science & Technology,* **4**, 324–330.

**Tjell, J.C., Hovmand, M.F. & Mossbaek, H. (1979).** Atmospheric lead pollution of grass grown in a background area of Denmark. *Nature (London),* **280**, 425–426.

**Warren, H.V. (1973).** Some trace element concentrations in various environments. *Environmental Medicine* (Ed. by G.H. Howe & J.A. Lorraine), pp. 2–24. Heinemann, London.

**Wilson, D.O. & Cline, J.F. (1966).** Removal of plutonium-239, tungsten-185 and lead-210 from soils. *Nature (London),* **209**, 941–942.

# Partitioning and transport of copper in various components of Kenyan *Coffea arabica* stands

N. W. LEPP AND N. M. DICKINSON
*Department of Biology, Liverpool Polytechnic,*
*Byrom Street, Liverpool L3 3AF*

## SUMMARY

**1** Recent studies on Kenyan coffee *(Coffea arabica)* stands with differing histories of copper fungicide treatment are reviewed. Excess copper accumulates in various components of coffee stands as a result of routine regular application of copper-based inorganic fungicides. Selective accumulation in certain plant tissues, soil and litter fractions is found together with some transient accumulation in certain seasonal tissues.
**2** Current husbandry of coffee in Kenya includes two practices which affect the accumulation and retention of copper within a stand. Multiple stem pruning and mulching are discussed in this context and the relative contribution of each to the overall pattern of copper circulation is assessed. In addition, the traditional wet-processing method of separating the coffee beans from the fruits is related to the overall circulation of copper within the plantation as a whole.
**3** Copper budgets have been constructed for different-aged coffee stands, and their significance in the present and future cultivation of coffee is evaluated from both biological and economic viewpoints.

## INTRODUCTION

Coffee is the most important commodity crop in world trade (Cannell 1983). Several species of *Coffea* are cultivated in the humid tropics of Africa, Asia, the Americas and the South Pacific. Of these, the most important commercial species is *Coffea arabica* L. In many of the producing nations, coffee exports account for a highly significant proportion of foreign exchange (36–40% of all exports in Kenya). This makes the health and productivity of the plantations a matter of prime national importance.

In common with other monocultures, *Coffea* is subject to attack by a wide variety of insect pests and plant diseases. Historically, the most important plant pathogen is coffee rust *(Hemilea vastatrix)* which, if left unchecked, can cause dramatic crop losses. This disease was responsible for the complete collapse of coffee cultivation in Sri Lanka between 1860 and 1870. Today, coffee rust is widespread in Africa and Asia, but only of local occurrence in the Neotropical regions. In East Africa, two other diseases are of equal, if not greater, importance: coffee berry disease (CBD *Colletotrichum coffeanum)* and elgon die-back *(Pseudomonas syringae)* (Purseglove 1974; Cannell 1983). CBD is a pathogenic strain of a common fungus, known to occur as a saprophyte on *Coffea* in other regions of production; the emergence of this pathogenic race in East Africa has been attributed to frequent protective and 'tonic' spraying with copper in the past (Acland

1975).

In Kenya, the above diseases, particularly leaf rust, are routinely controlled with copper fungicides, normally copper oxychloride applied as one of a variety of proprietary formulations. Input usage levels of copper in the smallholder and estate coffee sectors in 1982–83 were 7·8 and 25·4 kg $ha^{-1}$ respectively (Whittaker 1984): 60% of coffee is produced by smallholders and the remainder is grown on large-scale estates, which maintain approximately a 70% higher yield per hectare (Anon 1984).

Coffee stands are mulched annually; on the estate under study this consists of mixtures of grasses from adjacent areas of poorer soils and the current-season's prunings and thinnings from the bushes. In addition to conserving moisture, this treatment also increases the organic matter content of the upper soil horizons.

Established bushes are subjected to a system of multiple stem pruning which involves the removal of stems; new shoots are established from suckers on the main trunk every 4–7 years (Cannell 1983).

The finest quality coffee is produced by the wet processing method of separating the seeds (beans) from the remainder of the components of the ripe fruit (Acland 1975). Ripe berries are pulped (epicarp removed) and the beans are fermented in water overnight to remove their mucilage. The beans are sun-dried, and the testa and endocarp are removed by milling.

The synthesis presented here was derived from a series of sampling excercises designed to evaluate the consequences of routine heavy applications of inorganic copper fungicides on stands of coffee bushes, and to relate the flow of this copper through not only the coffee bushes and their associated soils and litter, but also within the coffee estate as a whole, as related to current practices in coffee husbandry and processing. The conclusions drawn from the present investigations are then used to produce recommendations for future developments in fungicide applications in the cultivation of coffee.

## STUDY AREA, SITES AND MANAGEMENT

The study area was located on a large (1140 ha) coffee estate in the Kiambu district, 15 km north of Nairobi. The climate of this area is classified as semi-humid, with a mean annual temperature of 18–20 °C. Precipitation is largely confined to two discrete rainy seasons. The soils are classified as eutric nitisols overlying Tertiary basic igneous rocks on volcanic footridges (Sombroek, Braun & Van der Pouw 1982). In the study area, soils are shallower than usual in this coffee-growing belt, having a mean depth of 1·2 m. Coffee has been cultivated in this region since 1915; since then, numerous new stands have been established, usually on virgin soil.

The main portion of the present study was conducted on coffee stands planted on virgin soil in 1959, 1969 and 1979; 4, 14 and 24-years old respectively. In the 4 and 14-year old stands plant spacing was 1 × 1·5 m (6767 trees $ha^{-1}$) but trees in the 24-year old stand were spaced more traditionally at 3·2 × 2·2 m (1350 trees $ha^{-1}$). The closer spacing of coffee trees is a relatively recent innovation, having been shown to increase fruit yields (Clowes & Allison 1982). As the trees mature, alternate rows are removed; in the 14-year old stand this was carried out, rather later than usual, after the present study was complete.

The following pruning technique has been adopted on the study stands: trees are initially capped at a height of 30 cm, then 1–3 stems maintained in a 4–7 year rotation, after which time they are replaced by selected suckers. The crop is borne on a framework of lateral branches. At the time of the present study, trees in the 4-year old stand had not been capped; trees in the 14-year old stand had been capped and a single stem established; trees in the 24-year old stand contained two 7-year old stems. These latter were pruned and replaced by selected suckers at the completion of the present study, for the third time since the trees were established. In addition, each stand received one round of pruning and two rounds of handling (shoot thinning) per year (Cannell 1983; Clowes & Allison 1982).

Coffee beans are extracted from the fruit by the wet-processing method in the estate factory. Waste pulp is removed and some is re-spread between the bushes as part of the mulch. Typically, 3 kg of fruit will yield 1–1·5 kg coffee beans. The river and borehole water from the wet-processing tanks is led away from the factory by a series of drainage channels and settlement ponds to a river.

Mulching is used as a means of water conservation and it also provides a handy repository for the various waste products from the processing of the fruit, which are produced in considerable quantities. In this estate, areas of poorer soils are set aside for the permanent cultivation of grasses for eventual harvest and use as mulch.

## METHODS

### *Plant and soil samples*

Samples of stem and foliar material were collected as required, using stainless-steel secateurs. Samples of root tissue were collected following the uprooting of selected bushes. Samples of the by-products of fruit-processing were obtained from the plantation factory, and could not be assigned to any particular stand. Samples of mulch grasses were collected on site and aged mulch and litter samples were hand-sampled from between the rows of bushes. Soil samples were taken using a trowel for surface and litter material and a stainless steel auger for deeper samples.

### *Sample preparation and analysis*

Plant samples were washed thoroughly in deionized water prior to oven drying to constant weight (90 °C). Soil samples were oven dried (90 °C) and finely ground. Dried plant samples were ground, and weighed subsamples were digested under pressure at 70 °C with 71% $HNO_3$ (Williams 1978). Digested samples were analysed for copper content by flame atomic absorption spectroscopy (AAS). Dried soil and sediment samples were ground, then wet-digested with $HNO_3$ prior to analysis by flame AAS.

## COPPER ACCUMULATION AND CIRCULATION

Initial investigations on copper accumulation in coffee bushes and their associated soils and litter were made on samples from bushes originally planted in 1915. A stand dating from this period was uprooted in 1982, and samples of above and below-ground tissues obtained. Details of plant and soil analyses are given in Table 1. From these data, it was

TABLE 1. Copper content of tissues of 68-year old coffee bushes, their associated soils and litter. Values (mg $kg^{-1}$ dry wt) represent means of three replicates ± SD

| | |
|---|---|
| (a) Plant material | |
| Axial roots | 5·1 ± 0·2 |
| Medium roots | 21·6 ± 4·1 |
| Fine roots | 154·0 ± 4·3 |
| Trunk wood | 6·7 ± 2·3 |
| Trunk bark | 1122·5 ± 128·5 |
| Whole stems | 9·4 ± 1·1 |
| Young bark | 74·5 ± 4·5 |
| Foliage (unwashed) | 409·4 ± 51·2 |
| (b) Soils and litter | |
| Depth (cm) | |
| 0–10 | 236·1 ± 98·5 |
| 10–20 | 68·5 ± 12·6 |
| 20–30 | 64·8 ± 19·1 |
| 30–40 | 31·6 ± 6·4 |
| Litter | 2035·0 ± 152·3 |

Data from Dickinson, Lepp & Ormand (1984).

evident that certain of the perennial tissues of the plant showed accumulation of copper, as did the litter and upper soil layers. Internal tissues, such as stem and trunk wood showed little evidence of copper contamination, a feature shared by the main and axial roots. Only the fine roots showed significant copper accumulation, not unexpected when the distribution of copper in the soil profile is considered.

Investigations were also made into the copper content of annual tissues, and those which had a finite lifespan on the bush, such as the lateral branches (routinely removed in the pruning programme). Table 2 illustrates the copper content of some of the annual

TABLE 2. Copper content (mg $kg^{-1}$ dry wt) of some annual components of the coffee bush. Values represent means of three replicates ± SD

| Tissue | Copper content |
|---|---|
| Mature foliage | |
| washed | 461·6 ± 48·8 |
| unwashed | 129·1 ± 3·3 |
| Flowers | 172·5 ± 5·3 |
| Fruit | |
| Epicarp (pulp) | 74·5 ± 3·9 |
| Hull | 48·6 ± 0·1 |
| Silverskin | 68·7 ± 3·7 |
| Beans (firsts) | 17·6 ± 0·5 |

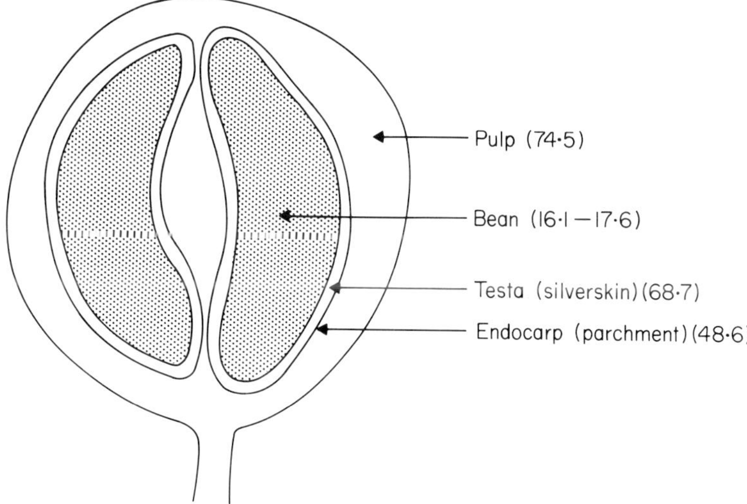


FIG. 1. TS coffee berry to show the various components of the fruit, together with their copper content (mg $kg^{-1}$ dry wt).

components of the coffee bush, including the various constituents of the fruit (Fig. 1). Elevated levels are found in foliage, flowers and those tissues of the fruit in direct contact with the fungicide sprays. Copper levels in the beans are unchanged from reports in the past (Taylor 1949). Changes in copper content in the various tissues of the lateral shoots are illustrated in Table 3. As these shoots grow and age, their copper content changes in a

TABLE 3. Copper content (mg $kg^{-1}$ dry wt) of the internodes of developing coffee shoots. Values represent means of three replicates ± SD

| Tissue | Copper content |
|---|---|
| Whole shoots | |
| Young (internodes 1,2) | 90·9 ± 2·7 |
| 'Green' (internodes 3–6) | 65·0 ± 6·5 |
| 'Brown' (internodes 6+) | 303·4 ± 16·5 |
| Cuticle/bark | |
| Young | 142·3 ± 3·5 |
| 'Green' | 156·5 ± 7·0 |
| 'Brown' (bark + cuticle) | 1156·3 ± 40·8 |
| 'Brown' (bark only) | 4186·1 ± 185·3 |
| Inner Wood | |
| Young | 1001·4 ± 7·5 |
| 'Green' | 89·0 ± 2·7 |
| 'Brown' | 67·5 ± 2·9 |

characteristic manner. Increases in cuticle and bark copper content are attributable primarily to the development of bark tissue, with its high cation sorption capacity. Conversely, the copper content of the woody tissue declines with age and increasing biomass. This, taken with the previous data for copper content of the internal tissues of the coffee bush, indicates that the majority of the copper in the shoot system is derived from spraying. The lack of a steady increase in the copper content of the shoot wood with age indicates an absence of the regular contribution via the transpiration stream, either from the root system or via the phloem from the sprayed foliage.

Data on the copper content of the fruit and its component tissues show a significant accumulation in the pulp (Table 2). Approximately 50% of the harvest is copper-contaminated waste. Not surprisingly, the sediments of the drainage channels and lagoons which conduct the water from the wet-processing of the fruit contain elevated levels of copper (Table 4). These watercourses are dredged frequently; some of the sediments are

TABLE 4. Copper content of sediments from watercourses recieving waste water from 'wet' coffee processing. Values refer to $HNO_3$-extractable levels (mg $kg^{-1}$ dry wt) and are means of three replicates ± SD

| Location | Copper content |
|---|---|
| Drainage channel (adjacent to factory) | 605·2 ± 2·5 |
| Settlement lagoon | 435·5 ± 12·2 |

spread on the areas of grassland used for mulching purposes. This practice does not appear to influence unduly the copper content of fresh mulch (Table 5); this only increases

TABLE 5. Copper content of mulch, as related to age of sample. Values (mg $kg^{-1}$ dry wt) represent means of three replicates ± SD

| Mulch age (time after application) | Copper content |
|---|---|
| Fresh grass | 13·6 ± 0·5 |
| 8·5 weeks | 421·0 ± 25·7 |
| 15 months | 943·9 ± 18·1 |

following exposure to fungicide sprays after it has been spread between the bushes. The apparent concentration of copper in the ageing mulch may also be the consequence of the close association between this element and the more recalcitrant organic fraction of the litter and mulch. The flow of the sprayed copper around the plantation is summarized in Fig. 2.

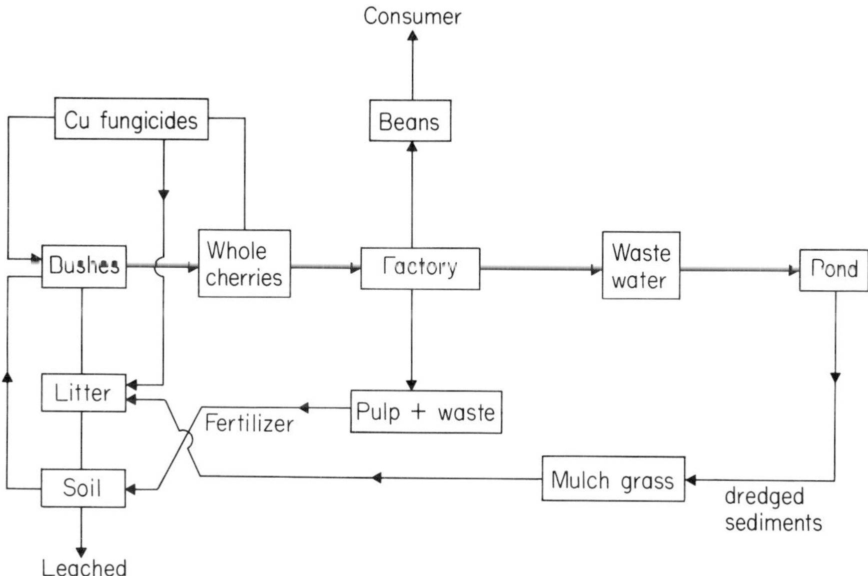


FIG. 2. Flow of copper through the various components of the coffee plantation. Arrows indicate the direction of flow between compartments.

## COPPER BUDGETS

Three different-aged stands of coffee bushes were investigated in order to produce a copper budget for each stand and to indicate the effects of increased fungicide usage and altered cultural practices on the pattern of copper loss and retention. Data for the copper content of selected soils, litter and tissues from each stand are given in Table 6. Age-dependent accumulation is clearly evident in all these components.

A model for the distribution, retention and loss of copper from the various components of each coffee stand was constructed in the following manner. The dimensions of the shoot systems of five replicate trees were measured in each stand. Using a series of volume: dry weight regressions obtained from subsamples, dry weight values were derived for trunks, stems, laterals, sublaterals, leaves and bark. In addition, to provide data for a growth model of the stems of mature trees, similar measurements were made of 3, 5, and 7-year old stems in other stands. This made possible an estimate of dry matter accumulation in branches previously pruned and removed from the 1959 stand. Cannell (1971) reported detailed dry weight measurements on the root systems of 3·5–5 year old coffee bushes in this area, and these measurements were used for the 1979 stand. Representative specimens from the two older stands were uprooted, and dry weight data were derived as for the shoot system. A correction factor for roots left in the soil was obtained by predicting losses from the size of the broken ends of the larger roots. Even using this, it is probable that the root biomass values for the two older stands are underestimates.

Dry weight measurements of litter and soil were made, and yields of coffee beans were

TABLE 6. The distribution of copper in soil, litter and plant samples from different-aged stands of coffee. Values represent the mean of three replicates ± SD, and are expressed as mg Cu $kg^{-1}$ dry wt sample. Soil values are derived from $HNO_3$ extractions (data from Lepp, Dickinson & Ormand (1984))

| | Stand age (years) | | | |
|---|---|---|---|---|
| (a) Soils | | | | |
| Soil depth (cm) | 24 | 14 | 4 | Control |
| 0–20 | 136·1 ± 4·73 | 48·0 ± 4·81 | 24·5 ± 0·67 | 11·1 ± 0·53 |
| 20–40 | 56·6 ± 1·57 | 25·2 ± 1·34 | 15·9 ± 0·31 | 10·1 ± 0·95 |
| 40–60 | 22·8 ± 1·15 | 25·8 ± 0·57 | 19·4 ± 0·34 | 9·8 ± 0·46 |
| 60–80 | 30·2 ± 0·33 | 27·2 ± 0·89 | 13·5 ± 0·39 | 9·7 ± 0·08 |
| (b) Litter | | | | |
| Litter fraction | 24 | 14 | 4 | |
| Leaf | 424·2 ± 38·3 | 211·9 ± 2·5 | 170·3 ± 5·6 | |
| Twig | 220·4 ± 14·5 | 100·8 ± 5·33 | 158·4 ± 14·5 | |
| Total* | 883·7 ± 397·5 | 320·4 ± 150·1 | ** | |
| (c) Plant material | | | | |
| Plant tissue | 24 | 14 | 4 | Control |
| Fine root | 332·8 ± 15·1 | 118·7 ± 5·51 | 46·4 ± 1·08 | 15·8 ± 0·52 |
| Young leaves | 36·2 ± 0·34 | 38·5 ± 8·7 | 54·8 ± 1·75 | 13·9 ± 1·3 |
| Mature leaves | 103·9 ± 3·2 | 121·5 ± 2·6 | 129·1 ± 3·3 | 14·7 ± 0·93 |
| Lateral branches | 300·8 ± 15·9 | 151·6 ± 13·5 | 33·1 ± 1·14 | 14·7 ± 0·79 |
| Bark† | 415·4 ± 51·8 | 283·8 ± 15·0 | 97·0 ± 3·5 | 17·3 ± 0·56 |
| Wood† | 10·6 ± 1·4 | 17·3 ± 2·0 | 8·3 ± 0·6 | 4·6 ± 0·4 |

* Total litter including amorphous fraction.
**Insufficient litter had accumulated to obtain samples for analysis.
†From main trunk.

provided by the estate managers. The yields of different components of the fruit were then estimated using the relative proportions reported by Bressani (1979).

Dry weight data were converted to estimates of the standing crop of copper using the copper values reported in Table 6. Prior to this, values representing the copper content of control trees with no history of spraying were subtracted from the copper values from the three stands. Accurate records of the total quantity of fungicide application per annum were made available by the estate managers.

The total amount of copper applied as a fungicide has increased substantially in recent years (Fig. 3); current rates are four times greater than in the mid 1960s. The accumulation of copper in the above-ground components of the stand increases with age (Table 7); even allowing for under-estimation of biomass, root levels do not show a similar trend.

Overall copper budgets for each stand are given in Fig. 4. In the 1979 stand, less than 50% of the copper input could be accounted for; it seems reasonable to assume that this has been lost from the system, as a consequence of leaching. In the 1969 stand, 60% of the applied copper could be identified, and in common with the younger stand the majority of the residue had accumulated in the soil. Relatively small amounts had been removed with

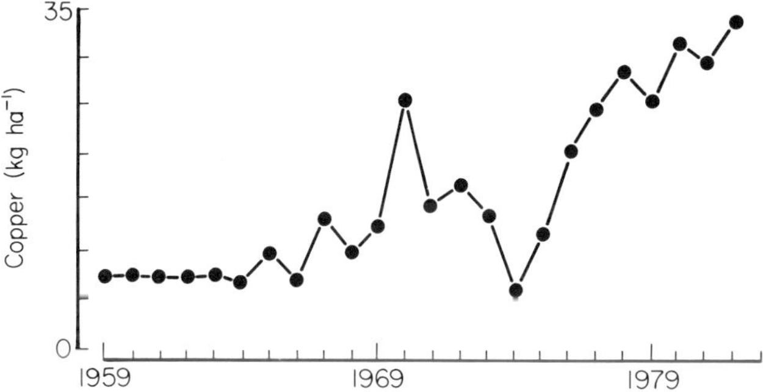


FIG. 3. Annual rates of copper fungicide application (kg $ha^{-1}$) 1959–83. Data from Dickinson & Lepp (1984).

TABLE 7. Total amount of copper in tissues of bushes from different-aged stands. values are expressed as g Cu $tree^{-1}$ ± SD. Figures in brackets represent % distribution with respect to tissue (data from Dickinson & Lepp (1984))

| Tissue | Stand age (years) 4 | 14 | 24 |
|---|---|---|---|
| Leaves | 0·19 ± 0·05 (81·9) | 0·21 ± 0·1 (45·6) | 0·60 ± 0·21 (33·0) |
| Trunk | 0·003 ± 0·001 ( 1·3) | 0·03 ± 0·007 ( 6·5) | 0·07 ± 0·02 ( 3·9) |
| Stems | 0·012 ± 0·003 ( 5·2) | 0·12 ± 0·014 (26·1) | 0·09 ± 0·017 ( 5·0) |
| Branches* | 0·006 ± 0·003 ( 2·6) | 0·06 ± 0·008 (13·0) | 1·02 ± 0·32 (56·0) |
| Large roots† | 0·003 ± 0·0003 ( 1·3) | 0·02 ± 0·002 ( 4·4) | 0·002 ± 0·004 ( 1·1) |
| Fine roots | 0·018 ± 0·0004 ( 7·8) | 0·02 ± 0·0004 ( 4·4) | 0·02 ± 00·001 ( 1·1) |
| Bark‡ | 0·006 ± 0·001 ( 2·6) | 0·05 ± 0·013 (10·9) | 0·39 ± 0·062 (21·4) |

*All laterals and sublaterals.
†Stump and roots > 3mm diam.
‡From trunk, branches and stem.

the fruit and in pruning, and the total copper content of the litter was 50% of that found in the plants. In the 1959 stand, there was a completely different picture; all of the sprayed copper could be accounted for, and in excess of 98% of this had accumulated in the litter and soil.

Copper is an element which shows a strong affinity for organic matter, which is known to dominate its interactions with soils, litter and plant materials (Lepp 1981; Thornton 1979). When the 1959 stand was planted, application rates of copper fungicides were low (Fig. 3); it is probable that the rate of input of organic matter to the stand kept pace with copper inputs from routine spraying. In contrast, the two younger stands received much higher copper inputs from an early age, prior to an accumulation of organic material sufficient to restrict the leaching of excess copper down the soil profile.

These data may be used to predict the possible consequences of the continued use of copper-based fungicides in coffee cultivation. A substantial proportion of the cost of coffee production is accounted for in the use of fungicides (Foxall 1983) and the rates of application are unlikely to increase at the rate of the last few years. Assuming that application rates are maintained at the levels of the past 5 years, a further 297 kg $ha^{-1}$ of

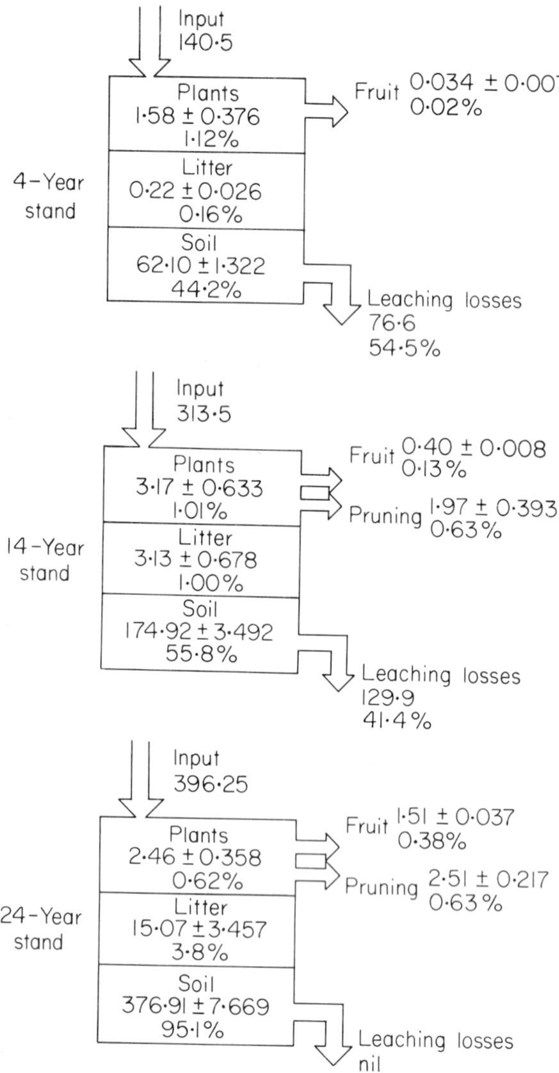


FIG. 4. Overall copper budget for the three stands of coffee. Values are expressed in kg Cu ha$^{-1}$ ± SE. Leaching losses obtained by subtracting all other values from input. Data from Dickinson & Lepp (1984).

copper will have been applied to each stand in the next 10 years. A regression line based on the data in Fig. 3 suggests that annual rates of application will have increased by 38·6% over this time, producing a corrected value of 341·2 kg ha$^{-1}$ in the 10-year period. There is the potential for an 86% increase in the copper content of each of the stands during this period; total copper in fractions such as trunk bark, lateral branches and soil show the potential for an exponential increase, with a doubling time of 3–13 years.

At present, the coffee crops show no sign of deterioration, in either quality or yield, but the potential trends highlighted to date strongly suggest that efforts should be made within the next decade to switch from copper-based fungicides for disease control in coffee.

## ACKNOWLEDGMENTS

The authors would like to thank the management of SOCFINAF, especially Mr R Hemmings and Mr J. Enoch, for their assistance at all stages of this project.

## REFERENCES

**Acland, J.C. (1975).** *East African Crops.* Longmans, London.

**Anon. (1984).** *Annual report, balance sheets and accounts for the year ended 30 September 1983.* Coffee Board of Kenya.

**Bressani, R. (1979).** The by-products of Coffee berries. *Coffee Pulp. Composition, Technology and Utilisation.* (Ed. by J.E. Braham & R. Bressani), pp. 5–10, IDRC, Ottowa, Canada.

**Cannell, M.G.R. (1971).** Production and distribution of dry matter in trees of *Coffea arabica* L. in Kenya as affected by seasonal climatic differences and the presence of fruits. *Annals of Applied Biology,* **67**, 99–120.

**Cannell, M.G.R. (1983).** Exploited crop — Coffee. *Biologist,* **30**, 257–263.

**Clowes, M.St J. & Allison, J.C.S. (1982).** A review of the coffee plant (*Coffea arabica.* L.), its environment and management in relation to coffee growing in Zimbabwe. *Zimbabwe Journal of Agricultural Research,* **20**, 1–19.

**Dickinson, N.M. & Lepp, N.W. (1984).** Pollution of tropical plantation crops by copper fungicides: a copper budget for a Kenyan coffee plantation. *Proceedings of International Conference on Environmental Contamination,* pp. 341–346. CEP Consultants, London.

**Dickinson, N.M., Lepp, N.W. & Ormand, K.L. (1984).** Copper contamination of a 68-year old coffee (*Coffea arabica* L.) plantation. *Environmental Pollution (Series B),* **7**, 223–231.

**Foxall, C. (1983).** Pesticides in Kenya. *Swara,* **6**, 25–27.

**Lepp, N.W. (1981).** Copper. *Effects on Heavy Metal Pollution on Plants. 1. Effects of Trace Metals on Plant Function.* (Ed. by N.W. Lepp), pp. 111–143. Applied Science Publishers, London.

**Lepp, N.W., Dickinson, N.M. & Omand, K.L. (1984).** Distribution of fungicide-derived copper in soils, litter and vegetation of different-aged stands of Coffee (*Coffea arabica.* L.) in Kenya. *Plant and Soil,* **77**, 263–270.

**Purseglove, J.W. (1974).** *Tropical Crops: Dicotyledons.* Longmans, London.

**Sombroek, W.G., Braun, H.M.K. & Van der Pouw, D.J.A. (1982).** Exploratory soil maps and agro-climatic zone map of Kenya 1980; scale 1:1,000,000. *Exploratory Soil Survey report No. E.1.* Kenya soil survey, Nairobi.

**Taylor, A. (1949).** The copper content of Coffee and Coffee products. *Chemical Industry,* **43**, 737–738.

**Thornton, I. (1979).** Copper in soils and sediments. *Copper in the Environment. 1. Ecological Cycling* (Ed. by J.O. Nriagu), pp. 171–216. J. Wiley, New York.

**Whittaker, M.J. (1984).** Study of inputs to smallholder coffee farmers through the co-operative distribution system in Kenya from 1981 to 1983. *Kenya Coffee,* **49**, 227–238.

**Williams, E.V. (1978).** New techniques for the digestion of biological materials: applications to the determination of tin, iron and lead in canned foods. *Journal of Food Technology,* **13**, 367–384.

# Long-term studies of metal transfers following application of sewage sludge

S. P. McGRATH
*Soils and Plant Nutrition Department, Rothamsted Experimental Station, Harpenden, Herts AL5 2JQ*

## SUMMARY

**1** There is a lack of information on the persistence and long-term effects of metal accumulation in the soil resulting from applications of sewage sludge on which guidelines for sludge disposal on land can be based. Most field investigations of the problem have lasted for less than a decade. Results are reported for an experiment which received metal-contaminated sludge treatments from 1942 to 1961 and inorganic fertilizers thereafter.
**2** Additions of sewage sludge and a sludge-straw compost increased considerably the concentrations of Zn, Cu, Ni, Cd, Cr and Pb in the plough-layer soil, and these have persisted for approximately 40 years.
**3** Both 'total' and 0·05 M EDTA-extractable metal concentrations in the soil have fallen with time, but at the same rate since sludge applications ceased in 1961.
**4** Metal analyses of soil profile samples taken during the experiment showed no evidence of significant movement below the depth to which the soil was cultivated. Concentrations measured in soil solutions were not great enough to suggest much leaching of metals in drainage water.
**5** The removal of metals in crops harvested over 20 years was small, ranging from 0·57 to 0· 03% for zinc and chromium respectively.
**6** Using the known concentrations of metals in sludges, the loadings applied and crop metal-offtakes, metal balances for the experiment were calculated. For the soils sampled from individual plots only 75 and 38% of the added metals could be accounted for in the topsoil in 1960 and 1980, respectively. Analyses of soils from a transect across sludged and non-sludged plots showed that movement of metals had occurred and it is concluded that this was due to soil movement during cultivations. This was similar for all six metals tested. If soil movement could be allowed for then a much larger proportion of the metals would be accounted for in the topsoil. Such soil movement may explain low apparent recoveries of metals reported in other field experiments.

## INTRODUCTION

Potentially toxic metals accumulate in ecosystems both as a result of natural processes and as a result of man's activities. There has been an upsurge in the latter during the last 200 years, both in the development of industry and increasing intensity of agriculture which has included increased use of metal-containing pesticides and fertilizers. Metalliferous mining, smelting and refining release metals to the atmosphere, water and sediments, as do electricity generating plants, incinerators and residues from leaded petrol. Although the amounts of metals entering ecosystems can decline rapidly if emissions are stopped or their sources are removed, there is increasing evidence that metals accumulate in the biologically-active layers of soils and sediments and remain there for very long periods. The responses of plants and animals to increased metal concentrations in the environment have been studied and the mechanisms by which some organisms detoxify metals have been identified. However, the long-term implications of metal pollution on ecosystems, particularly in relation to possible medical and veterinary effects are by no means certain at present and await more complete epidemiological studies.

Sewage sludge is a highly organic waste material which is produced during the cleaning up of wastewater from industrial and domestic sources, and includes direct runoff from roads and other surfaces. During water treatment metals are strongly bound to the solid, largely organic phase, and the resulting metal-loaded sludge is separated from the clean water. Currently, approximately 60% of the sewage sludge produced in the UK is deposited on land, amounting to three-quarters of a million tonnes of dry solids annually (about eighteen million tonnes wet weight). In the future, European Economic Community regulations may limit dumping at sea and increased spreading on land or alternative methods for disposal may have to be found. High transport costs for the bulky sludge and mounting costs for landfill sites increase the need for the sludges to be spread on land close to sewage treatment works.

The risks of metal toxicity from highly-polluted sludges have been studied for a number of years and guidelines (Department of the Environment 1981) have been developed for sludge application to land which aim to prevent metal toxicity to plants, animals and humans during a notional 30-year period. However, there is a lack of information on the long-term effects of heavy metals deposited on soil on which to base such guidelines. The two most important questions which need to be answered are (i) whether metals are mobile in soils, and to what extent, and (ii) what changes in bioavailability occur over long periods of time? To answer these, experiments must include plots with no metals added to serve as background during the monitoring of metal mobility. Whilst changes in metal uptake by plants, total soil contents and chemical extractability of elements can be determined from regular applications of metal-contaminated sludges, it is difficult to determine from such experiments what might happen in the long-term if no more sludge was applied. One hypothesis is that, when the added organic matter breaks down, the metals held by it will be released and become more mobile and more phytotoxic (e.g. Berrow & Burridge 1980). Another hypothesis, called 'reversion', is that, after the metals are released from the sludge, they subsequently interact with the mineral soil and are occluded in oxides or precipitated as hydroxides, leading to a reduction in metal activity and therefore toxicity (e.g. Lewin & Beckett 1980). To test these hypotheses, fresh inputs must cease and be followed by chemical and biological investigations into the residual effects of the applications over a substantial period of time.

Field investigations of metal transfers following sludge addition have usually lasted for less than a decade in total, including studies of residual effects of less than 4 years in most cases (Table 1). In many of these reports, there was maintained plant uptake and/or extractability of metals from soil and in only two were decreases recorded (in one of these

TABLE 1. Published evidence on metal mobility, 'availability' and recovery from field experiments of long duration on arable soils treated with sewage sludge

| | Duration* (years) | Metals studied | Metal movement | Changes in extractibility (*e*) from soil and plant uptake (*u*) | Recovery (%) |
|---|---|---|---|---|---|
| Baker, Amacher & Leach (1979) | 6 | Cd | - | 4 year residual: *u* & *e* constant | - |
| Berrow & Burridge (1980) | 9 | Zn, Cu, Ni, Cr | - | 8 year residual: *u* & *e* constant | - |
| Chaney & Hornick (1978) | 6(?) | Cd | - | 5 year residual: *u* & *e* constant | - |
| Chang, Page & Bingham (1982) | 8 | Zn, Cu, Ni, Cd, Pb | None below 15 cm (dc)† | 4 year residual: *u* constant | 67 |
| Chang *et al.* (1984) | 6 | Zn, Cu, Ni, Cd, Pb, Cr | Little below 15 cm (dc) | 3 year residual: not reported | >90 |
| Dowdy *et al.* (1978) | 4 | Zn, Cu | - | 3 year residual: *u* constant | - |
| Giordano, Mortvedt & Mays (1979) | 9 | Zn, Cd | - | 4 year residual: *u* constant | - |
| Hinsley *et al.* (1979) | 7 | Zn, Cd | None below 15 cm (dc) | 4 year residual: *u* decreased | 65 |
| Kelling *et al.* (1977) | 3 | Zn, Cu, Ni, Cd, Cr | None below 15 cm | 2 year residual *u* & *e*‡ decreased | - |
| Larsen (1984) | 7 | Cd | None below 25 cm | 6 year residual: *u* constant | - |
| Lönsjö (1984) | 9 | Cd | - | 8 year residual: *u* constant | - |
| Schaaf & Boguslawski (1982) | 13 | Zn, Cu, Ni, Cd, Pb, Cr, Hg | None below 20 cm | No residual study | - |
| Sidle *et al.* (1977) | 9 | Zn, Cu, Ni, Cd, Pb | None below 30 cm | No residual study | - |
| Webber & Beauchamp (1979) | 6 | Cd | None below 20 cm (dc) | 3 year residual: *u* constant | 88 |
| Williams *et al.* (1984) | 6 | Zn, Cu, Cd, Pb | None below 30 cm 20 cm (dc)§ | No residual study | - |

*Includes period of residual study; the number of sludge applications varied: see references for details.
†dc = depth of incorporation.
‡Large pH changes occurred.
§Sludge addition acidified the soil.

the soil pH changed greatly and the results are therefore not conclusive). Movement of metals down the soil profile could also affect bioavailability and some of the experiments in Table 1 give information on this without, often, giving the depth to which sludges were incorporated. Despite this, it seems clear that in the short to medium-term, hardly any leaching of metals below the plough layer occurred over a large range of soils and sludge combinations. This confirms findings from laboratory tests in which soil columns treated with sludge or inorganic metal salts or derived from sites with metals deposited from the air have been leached intensively with amounts of water equivalent to up to 10 years rainfall. In these tests, no movement of Zn, Cu, Ni, Cd, Pb and Cr was detected (Giordano & Mortvedt 1976; Emmerich *et al.* 1982; Miller, McFee & Kelly 1983).

The percentage recovery of metals added to a site is an important quantity to establish, since it represents what remains in the soil pool and shows how much has been lost over a period of time. Based on only a few field experiments, recoveries for various metals have ranged from 65 to 90% (Table 1), whilst one laboratory leaching experiment gave a recovery of 100% of the added metals (Emerich *et al.* 1982). Explanations for apparent low recoveries in the short-term of metals added to field experiments has been a cause of some controversy. Low recoveries are most likely to result from errors in sampling, chemical analysis and conversion of concentrations of metals in soil to total loadings (Chang *et al.* 1984; McGrath 1984). Results of recent analyses of samples from a long-term experiment including applications of metal-contaminated sewage sludge are given in this paper. The metal inputs considered below are from sludges and manures, inorganic fertilizers and lime. The losses investigated were: solute leaching, removal of metals with harvested portion of crops and mass transport of soil due to cultivation. The aims were to determine the metal balances and mobilities of Zn, Cu, Ni, Cd, Pb and Cr remaining in the soil.

## EXPERIMENTAL

The Market Garden Experiment at Woburn, started in 1942, included various bulky organic manure treatments which were compared with additions of inorganic fertilizers to assess the manurial value of the organic manures. Included was an anaerobically-digested, lagoon-dried sludge from West Middlesex Drainage works at Isleworth, London which was later shown to contain large amounts of heavy metals. The treatments were: (i) sewage sludge, single (S1) and double (S2) dressing, (ii), sludge-straw compost, single (SC1) and double (SC2) dressing, (iii) farmyard manure, single (FYM1) and double (FYM2) dressing, (iv) inorganic NPK fertilizers only (Fert). Each organic treatment was replicated on four plots and fertilizer on eight; the whole experiment was replicated in two blocks. Results from one block and for four control plots only are given here. An additional treatment of vegetable compost at two rates gave similar effects to FYM and is also not reported here. Sludge and sludge-composts were applied from 1942 to 1961 then discontinued; FYM treatments continued until 1967 and all plots received inorganic fertilizers after organic treatments ended. The manures were applied on the basis of fresh weights, 37·5 te $ha^{-1}$ and 75 te $ha^{-1}$ $year^{-1}$, but, because of differences in dry matter contents of the manures, the total amounts of dry matter added varied amongst treatments (Table 2). After each application the manures were incorporated to 23 cm depth by cultivation, though the soil is now completely mixed to 27cm due to occasional variation in ploughing depth. The plots were not cultivated between 1974 and 1982 when they were down to grass.

The soil was a sandy loam with 9% clay (Cottenham Series) which has been maintained at a pH (in 1:2·5 w/v soil:$H_2O$) of 6·4–7·0 since 1944 by small maintenance dressings of lime. Soil samples were taken throughout the experiment with a 5 cm diameter stainless

TABLE 2. Total amounts of organic manures added to the Market Garden experiment, Woburn, 1942–67

| Manure | Rate | Amount added (te $ha^{-1}$) Fresh weight | Dry weight |
|---|---|---|---|
| Sewage sludge (to 1961) | 1 | 713 | 383* |
| | 2 | 1426 | 766 |
| Sludge-straw compost (to 1961) | 1 | 713 | 273* |
| | 2 | 1426 | 546 |
| Farmyard manure | 1 | 888 | 226† |
| | 2 | 1777 | 452 |

*In twenty-five applications.
†In thirty-two applications.

steel semicylinder auger, air-dried, ground and 'total' concentrations of Zn, Cu, Ni, Cd, Pb and Cr determined after extraction of soils and manures in 4:1 (v/v) $HCl:HNO_3$ (McGrath & Cunliffe 1985) and extractable Zn, Cu, Ni, Pb and Cd in soils were determined after shaking with 0·05 M EDTA (Ministry of Agriculture, Fisheries and Food 1981). Plant samples were washed briefly in tap water then digested in $HNO_3$ and $HClO_4$ as in Thompson & Walsh (1983). The concentrations of metals in the above solutions were determined using an ARL 34000 inductively-coupled plasma emission spectrometer, with appropriate corrections for spectral interferences or on a Perkin-Elmer atomic adsorption spectrometer fitted with a heated graphite analyser. All samples have been analysed since 1982 using the same methods.

## RESULTS AND DISCUSSION

### *Metals applied in the organic manures*

Complete sets of samples of each sewage sludge and farmyard manure applied were available for chemical analysis. Fig. 1, taken from McGrath (1984), shows how the concentrations of six metals in sewage sludge changed during the period 1942–61. The clearest trends were increases in nickel and cadmium concentrations which reflect greater use of these metals by the industries served by the sewage works. The mean concentrations of metals in these sludges (Table 3) were typical of sludges from urban industrial areas, with the exception of cadmium, which was present at relatively large concentrations (McGrath 1984).

These concentration data were used to calculate the total loadings of each metal applied during 1942–61. The loadings for Zn, Cu, Ni and particularly Cd in sewage-sludge and sludge-composts exceeded current DOE guidelines for maximum permissible additions of these metals (Table 4). Metal loadings from additions of sludge-straw composts were calculated by McGrath (1984). The amounts of metals added in FYM treatments were small in comparison with sludge and sludge-compost treatments (Table 4). The superphosphate fertilizer applied to all plots contained an average of 15 mg Cd $kg^{-1}$ which would have increased the total soil Cd by not more than 0·06 mg $kg^{-1}$.

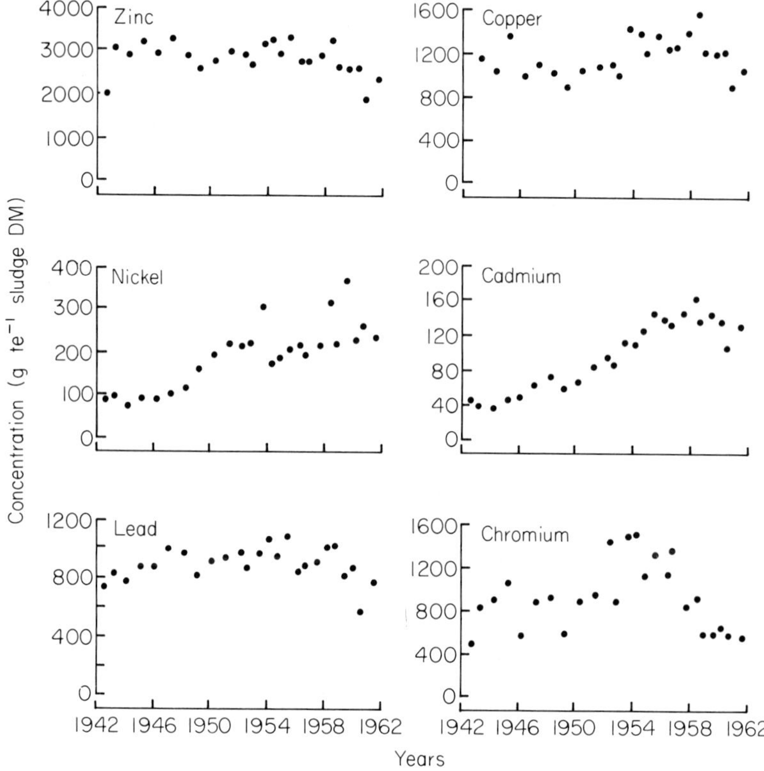


FIG. 1. Concentrations of metals in sewage sludges applied to the Market Garden Experiment, Woburn, 1942–61.

TABLE 3. Mean concentrations of metals in sewage sludges applied to the Market Garden Experiment, Woburn, 1942–61

| Concentration (g metal te$^{-1}$ dry matter) | | | |
|---|---|---|---|
| Zinc | 2780 | Cadmium | 99 |
| Copper | 1138 | Lead | 900 |
| Nickel | 188 | Chromium | 919 |

Cadmium concentrations in the soil from plots receiving only inorganic fertilizers remained at background concentrations. The other inorganic fertilizers and lime used probably contained only very small concentrations of metals (Stojkovaska & Cooke 1958; Chater & Williams 1974) in comparison with sludge and no attempt has been made to include their contribution.

TABLE 4. Total amounts of metals in farmyard manure and sewage sludges applied to the Market Garden Experiment, Woburn 1942–61

| | Amounts of metals added (kg ha$^{-1}$) | | | | | | DOE guidelines on maximum permissible additions of elements (kg ha$^{-1}$) |
|---|---|---|---|---|---|---|---|
| | FYM1 | FYM2 | SC1 | S1 | SC2 | S2 | |
| Zn | 68·0 | 136·0 | 626 | 1079 | 1252 | 2158 | 560* |
| Cu | 10·8 | 21·6 | 251 | 432 | 502 | 864 | 280 |
| Ni | 3·0 | 6·0 | 39 | 67·5 | 78 | 135 | 70 |
| Cd | 0·3 | 0·6 | 20·4 | 35·1 | 40·8 | 70·2 | 5 |
| Pb | 16·7 | 33·4 | 201 | 347 | 402 | 694 | 1000 |
| Cr | 6·8 | 13·6 | 204 | 352 | 208 | 704 | 1000 |
| Mo | 0·5 | 1·0 | 1·7 | 2·9 | 3·4 | 5·8 | 4 |

*For arable land with non-calcareous soil of pH >6·5; overall addition of 'zinc equivalent' ([Zn] + [Cu × 2] + [Ni × 8]) should not exceed 560 kg ha$^{-1}$ (DOE 1981).

### *Metal concentrations in the soil*

The additions of sludge and sludge-compost considerably increased the concentrations of Zn, Cu, Ni, Cd, Cr and Pb in the soils, as illustrated by data for the single and double rates of sludge (Table 5). The balance between metal additions and amounts of metal in the soil

TABLE 5. Concentrations of metals in 0–23 cm soils of the Market Garden Experiment, Woburn, in 1960

| | Total metal concentration* (mg kg$^{-1}$) | | |
|---|---|---|---|
| | Fertilizer | S1 | S2 |
| Zn | 74·5 | 340·0 | 635·4 |
| Cu | 22·8 | 125·4 | 239·4 |
| Ni | 14·3 | 29·6 | 42·3 |
| Cd | 0·2 | 8·8 | 19·4 |
| Pb | 30·3 | 130·6 | 209·1 |
| Cr | 40·8 | 154·3 | 258·4 |

*Mean of four replicate plots.

was estimated from the relationships presented in Fig. 2. For each metal there was a straight line relationship with all treatments lying close to the line. The slope of each regression line is a measure of the recovery of the added metal in 1960 and 1980 (Table 6). Recoveries ranged from 55 to 85% in 1960 and from 32 to 42% in 1980. Nickel was the least 'recoverable' metal in 1960. The average recovery of the other five metals was 75% in 1960 and this fell by 50% in the 20 years following 1960 after sludge dressings ceased in 1961.

The relationships in Fig. 2 for 1960 differ from those in McGrath (1984) because differences in soil bulk density due to treatment have been taken into account. The carbon content of the soils treated with bulky organic manures changed with time, increasing until

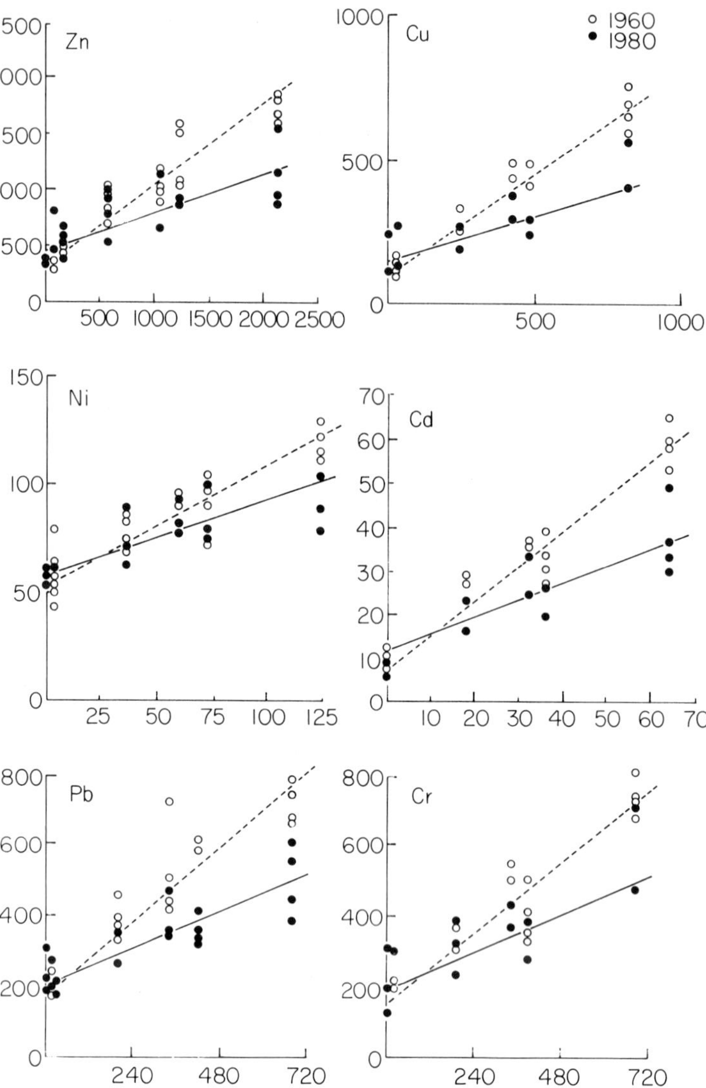


FIG. 2. Amounts of metals measured in the topsoil in 1960 and 1980 compared with the amounts of metals added in the organic manures. Lines fitted by linear regression.

1961 and then declining by a factor of two (Johnston 1975) so that bulk density also varied significantly. Bulk density for each treatment in 1960 and 1980 was calculated from known %C using the equation:

$$BD = 1{\cdot}46 + (-0{\cdot}126 \times \%\ C) \quad r = -0{\cdot}87,\ P < 0{\cdot}001$$

Where BD = bulk density (g ml$^{-1}$) and %C is the percentage carbon in soil measured by the Walkley-Black method (Walkley 1935). This regression was derived from data from this experiment given by Williams (1971). The estimated recoveries of the metals in 1960

TABLE 6. Slopes of linear regression of amounts of metals added in organic manures against observed total metal content of the soils of the Market Garden Experiment, Woburn, in 1960 and 1980

| | 1960 | 1980 |
|---|---|---|
| Zn | 0·74 | 0·35 |
| Cu | 0·71 | 0·32 |
| Ni | 0·55 | 0·40 |
| Cd | 0·81 | 0·40 |
| Pb | 0·85 | 0·42 |
| Cr | 0·82 | 0·42 |

from the slopes of the lines in Fig. 2 are less than those given by McGrath (1984) by an average of 10%, after allowing for changes in bulk density of the soil.

Total metal concentrations in the soils changed with time, the changes in Zn, Cu, Ni, Cd, Cr and Pb for both the double rate of sludge and inorganic fertilizers are shown Fig. 3.

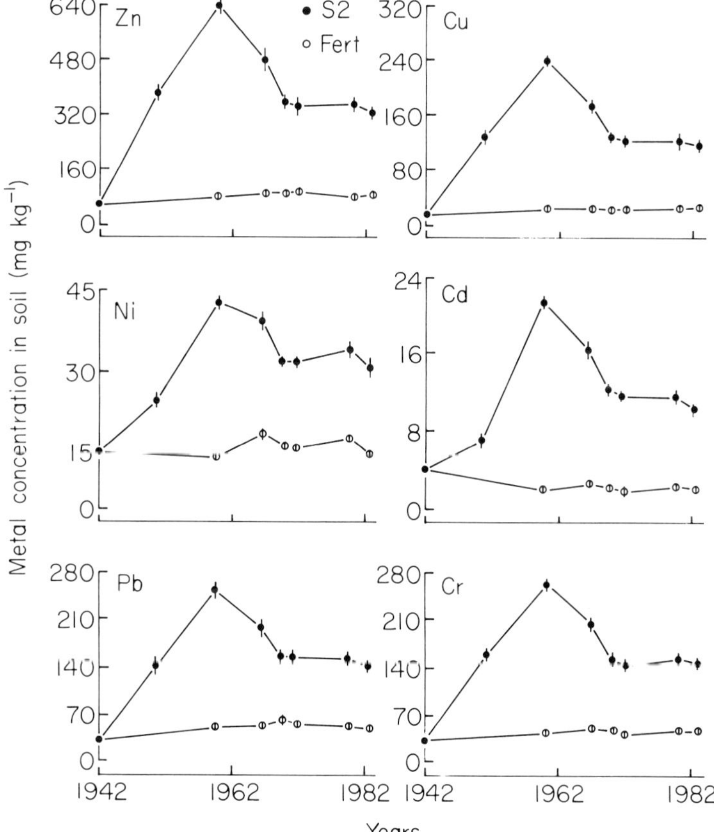


FIG. 3. Mean concentrations of metals in the topsoils from the double rate of sewage sludge and fertilizer only treatments of the Market Garden Experiment, Woburn, 1942–83. Bars represent standard errors.

McGrath (1983) showed that the concentrations of EDTA-extractable Zn, Cu and Ni also decreased in these soils after 1960. It is possible to interpret these results as evidence of large losses of metals from the topsoils, especially in the 10–12 years after sludge applications ceased. Before discussing further the current investigations into these losses the relationships between extractable and total metal concentrations in the soils needs to be examined.

### *Extractability of metals from the soils*

In the UK, 0·05 M EDTA is used to monitor potential bioavailability of metals in sludge-treated soils (MAFF 1981; DOE 1981); this extractant was used on soils from the experiment and the results expressed as percentages of total metal concentrations (by aqua regia digests). The EDTA:total-metal percentages changed little between 1960 and 1983 (Table 7), suggesting that the EDTA-extractable and total concentrations in sludged soils

TABLE 7. Extractability of metals in 0·05 M EDTA as a percentage of total metal concentrations in soils of three treatments of the Market Garden Experiment, Woburn

| | Year | | | | | |
|---|---|---|---|---|---|---|
| | 1960 | 1967 | 1970 | 1972 | 1980 | 1983 |
| Zinc | | | | | | |
| Sludge | 48 | 48 | 52 | 50 | 52 | 50 |
| FYM | 38 | 46 | 47 | 47 | 41 | 46 |
| Fertilizer | 34 | 37 | 37 | 37 | 38 | 38 |
| Copper | | | | | | |
| Sludge | 58 | 61 | 64 | 65 | 66 | 62 |
| FYM | 55 | 54 | 59 | 63 | 61 | 56 |
| Fertilizer | 56 | 52 | 59 | 57 | 58 | 55 |
| Nickel | | | | | | |
| Sludge | 30 | 28 | 29 | 26 | 26 | 24 |
| FYM | 13 | 16 | 17 | 18 | 18 | 16 |
| Fertilizer | 10 | 11 | 13 | 13 | 12 | 13 |
| Cadmium | | | | | | |
| Sludge | 54 | 55 | 58 | 56 | 58 | 56 |
| FYM | 39 | 46 | 51 | 50 | 51 | 47 |
| Fertilizer | 32 | 38 | 46 | 42 | 41 | 37 |
| Lead | | | | | | |
| Sludge | 43 | 43 | 47 | 43 | 49 | 46 |
| FYM | 38 | 34 | 39 | 42 | 43 | 40 |
| Fertilizer | 30 | 35 | 33 | 33 | 38 | 40 |

were falling at approximately the same rate. Also, these percentages differ for each of the manurial treatments, with inorganic fertilizer giving the lowest extractability. This is to be expected as, in this treatment, more of the metals would be present in the mineral matrix and therefore unavailable to the extractant (Beckett, Warr & Brindley 1983). Another important feature of the data in Table 7 is that, with the exception of nickel, the

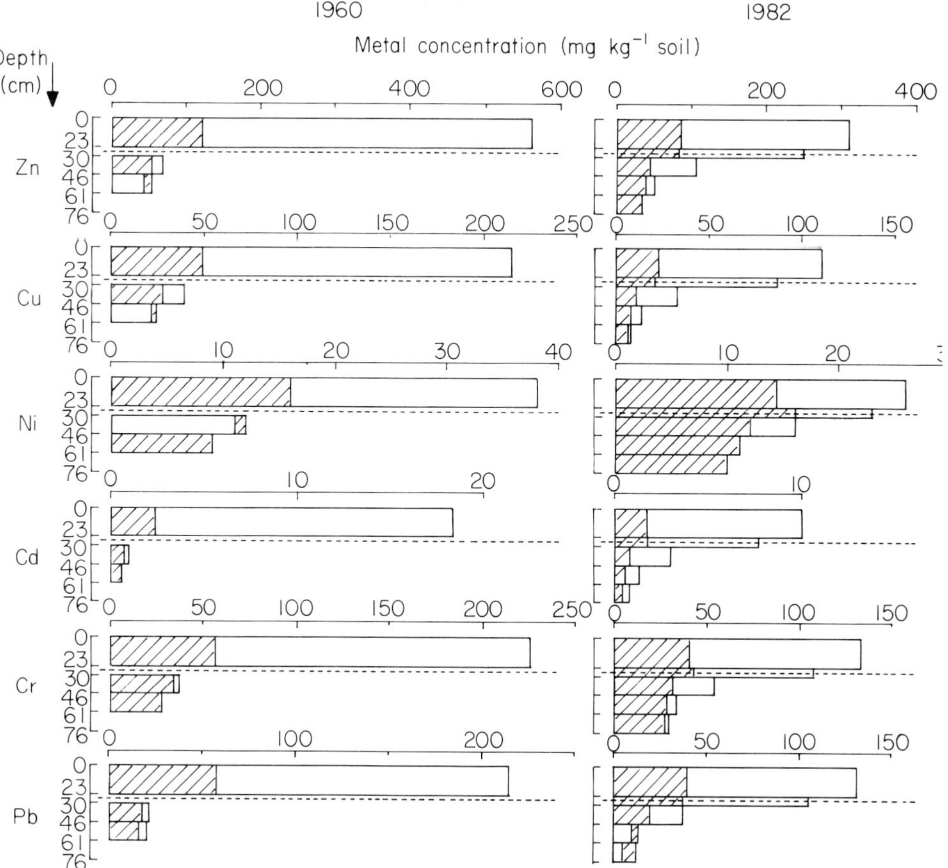


FIG. 4. Concentrations of metals in soil profile samples taken from the double rate of sewage sludge (□) and fertilizer only treatments (▨) in 1960 and 1982; pecked line represents the depth of cultivation. No samples were taken at the 23–30cm and 61–76 cm depths in 1960.

proportions of metal extracted by EDTA from sludged soils are in the range 43 to 66%. The assumption on which the current guidelines for extractable metal concentrations in sludge-treated soils are based is that 100% of the added metals remain EDTA-extractable (DOE 1981). If EDTA extracts provide an estimate of the maximum long-term potential bioavailability of metals, this means that the guidelines are conservative, at least for the soil studied (a sandy loam with a pH above 6·5).

## *Metal losses*

### *Leaching*

Soil cores to 61 cm were taken in 1960 with a semicylinder auger and divided into various depths then stored air-dry; this procedure was repeated in 1982 to 76 cm. The metal contents of the soils treated with the double rate of sludge or inorganic fertilizer are given in Fig. 4. On both occasions much of the added metal was found in the 0–27 cm plough layer into which the sludge had been incorporated. Additionally, the metal concentrations in the topsoil were less in 1982 than in 1960 with little evidence of increased metal concentrations in layers below 30 cm. It is of course possible that metals had been leached completely out of the soil profile, however, this is unlikely. The amount of water which percolates through this soil (i.e. precipitation minus evapotranspiration) averages 30 cm year$^{-1}$. If the amount of each metal apparently lost from an S2 plot between 1960 and 1980 had dissolved in the amount of water that passed through the soil, assuming that constant amounts of metals have leached with time, the concentrations of metals in drainage water would be those given in Table 8. In order to determine whether such concentrations could be attained in soil water, soils from the experiment sampled in 1961 and 1984 were incubated in moist condition for 10 days and the concentrations of Zn, Cu, Ni, Cd and Pb were measured in the soil water extracted by centrifugation. All the metal concentrations except nickel proved to be lower than the predicted concentrations (Table 8). This makes it

TABLE 8. Apparent amount of metal loss between 1960 and 1980, theoretical concentrations in drainage water and measured concentrations in soil water

| | Metal loss (kg ha$^{-1}$) | Theoretical metal concentrations in drainage water (mg l$^{-1}$) | Metal concentrations in soil water (mg l$^{-1}$) | |
|---|---|---|---|---|
| | | | 1961 sample | 1984 sample |
| Zn | 846 | 14·1 | 0·58 | 0·27 |
| Cu | 536 | 8·9 | 1·19 | 0·36 |
| Ni | 23·8 | 0·4 | 0·38 | 0·10 |
| Cd | 30·1 | 0·5 | 0·015 | 0·004 |
| Pb | 301 | 5·0 | 0·0015 | 0·0014 |

unlikely that leaching has been a major route of metal loss from the cultivated topsoil of this experiment, especially as there is little evidence of metal enrichment in the subsoil. Subsoils, unless very acid or devoid of finer particles, have a potentially large capacity to re-sorb any metals which percolate from the topsoil (Miller, McFee & Kelly 1983; Brümmer *et al.* 1983).

### *Removal in crops*

Some of the crops which were grown between 1960 and 1980 were stored in air-dry conditions at Rothamsted and the metal concentrations in them were determined recently. In addition, some crops not now available were analysed by Le Riche (1968). From these data and recorded yields, estimated metal offtakes have been calculated for the 20-year period (Table 9). There may be some errors in these data but even if the errors were as large as 100% the data summarized in Table 10 show that only a small fraction of added metal has been removed in the crops; the range varied from 0·57% (for Zn) to 0·03% (for Cr). This finding is supported by those of Kelling *et al.* (1977) and Chang, Page & Bingham

TABLE 9. Crops, mean yields, and metal offtakes from plots treated with double rate of sewage sludge 1942–61

| | Crop | Part | Yield ($t\ ha^{-1}$) | Zn | Cu | Ni | Cd | Cr | Pb |
|---|---|---|---|---|---|---|---|---|---|
| | | | | | | ($mg\ kg^{-1}$) | | | |
| 1960* | Red beet | Roots | 2·35 | 588 | 42 | 31 | – | 1·9 | 3·8 |
| | | Tops | 2·29 | 1145 | 23 | 34 | – | 2·3 | 6·9 |
| 1961* | Potatoes | Tubers | 4·03 | 113 | 38 | 2·4 | – | 0·1 | 0·8 |
| 1961 (1960*)[†] | Leeks | Tops | 1·70 | 230 | 27 | 29 | – | 0·9 | 2·7 |
| 1962 (1963) | Red beet | Tops (only) | 1·19 | 364 | 20 | 10 | 15 | 16 | 17 |
| 1963 | Carrots | Roots | 1·60 | 66 | 11 | 7·2 | 2·2 | 0·8 | 3·4 |
| | | Tops | 2·06 | 122 | 13 | 8·2 | 4·5 | 1·6 | 5·8 |
| 1963 (1960*) | Leeks | Tops | 1·03 | 139 | 17 | 18 | – | 0·6 | 1·6 |
| 1964 (1963) | Red beet | Roots | 3·04 | 398 | 40 | 16 | 7·6 | 14 | 8·8 |
| | | Tops | 2·19 | 670 | 37 | 19 | 27 | 29 | 31 |
| 1965 (1963) | Carrots | Roots | 2·79 | 114 | 19 | 13 | 3·9 | 1·4 | 5·9 |
| | | Tops | 3·95 | 233 | 24 | 16 | 8·7 | 3·2 | 11 |
| 1966 (1963) | Red beet | Roots | 2·92 | 383 | 38 | 15 | 7·3 | 14 | 8·5 |
| | | Tops | 2·58 | 790 | 44 | 22 | 32 | 34 | 36 |
| 1967* | Carrots | Roots | 2·41 | 101 | 11 | 4·8 | – | 0·2 | 0·2 |
| | | Tops | 4·59 | 454 | 46 | 14 | – | 4·0 | 7·8 |
| 1968[‡] | Beans | Seed | 3·81 | 95 | 21 | 24 | 0·6 | – | 8·4 |
| 1969[†] | Beans | Seed | 2·37 | 59 | 13 | 15 | 0·4 | – | 5·2 |
| 1970 | Sugar beet | Roots | 10·45 | 784 | 303 | 12 | 11 | 20 | 22 |
| | | Tops | 6·91 | 1506 | 90 | 42 | 35 | 46 | 66 |
| 1971 | Barley | Grain | 5·58 | 329 | 78 | 3·3 | 1·1 | 3·3 | 12 |
| | | Straw | 3·90 | 179 | 12 | 2·3 | 1·6 | 1·2 | 8·2 |
| 1972 | Potatoes | Tubers | 14·26 | 242 | 106 | 8·6 | 4·3 | 5·7 | 30 |
| 1973 (1971) | Barley | Straw | 3·90[§] | 179 | 12 | 2·3 | 1·6 | 1·2 | 8·2 |
| 1974–80 (1982) | Ryegrass | Cut twice per year = | 45·48[¶] | 2456 | 246 | 77 | 18 | 14 | 96 |

* Values from Le Riche (1968) are for two replicate plots, otherwise four.

[†] Date in brackets indicates that analyses from a different year were used.

[‡] Metal concentrations estimated from results of Davis & Carlton-Smith (1980).

[§] Yields not taken, grain eaten by birds; assumed same straw yield as 1971.

[¶] Yields not taken 1974–80; assumed same yield as 1982.

(1982) for shorter cropping sequences and demonstrates that metal removal by crops in an unimportant pathway for metal loss from sludged soils. Indeed, even for Zn which had the largest percentage removed, it would take approximately 3700 years to remove all of the metal added in the sludge through offtake in crops. Lead and chromium, which are known to be less mobile from the soil through the plants (Koeppe 1981; McGrath 1982) would require even longer periods for removal by crops (Table 10). Although the rankings of crop metal-offtakes which can be seen in Table 10 are in broad agreement with rough estimates of crop metal removals from uncontaminated soils (Allaway 1968; Bowen 1979), the time periods calculated for the sludged soils of this experiment are longer because of the large amounts of metals added.

TABLE 10. Amounts of metals removed in harvested crops, related to amounts of metals added in sewage sludge and their residence times

| | Metal offtake 1960–80 (kg ha$^{-1}$) | Offtake as % of total amount of metals added in sludges | Residence time* (years) |
|---|---|---|---|
| Zinc | 11·74 | 0·57 | 3700 |
| Copper | 1·33 | 0·16 | 13100 |
| Nickel | 0·45 | 0·37 | 5700 |
| Cadmium | 0·18 | 0·28 | 7500 |
| Chromium | 0·22 | 0·03 | 70000 |
| Lead | 0·41 | 0·06 | 35000 |

* Number of years of same cropping regime required to remove 100% of the added metals, given to the nearest 100 years.

*Soil movement*

As shown above, only relatively small proportions of the metals lost between 1960 and 1980 were accounted for by estimates of leaching or crop offtakes. Nevertheless, the difference in the amount of each metal between sludged and non-sludged soil was much less than the amount added. This loss may have occurred through soil movement.

In the large and long-lasting experiment described the plots are, of necessity, cultivated by standard farm machinery. The plots were cultivated in directions which alternated between each ploughing in order to minimize bias due to constantly ploughing in a single direction. Soil and plant samples have always been taken from an area of the plot 1 m inside the boundaries of the paths surrounding each plot to minimize 'edge effects'.

To check for possible soil movement and to estimate the magnitude of any movement, an existing set of plough-layer soil samples taken from a transect across some of the plots in 1960 and another set from a slightly longer transect across the same plots in 1984 were analysed. The results of the 1984 transect show that, within the plough layer, there was a smooth gradation of metal concentrations with distance, approximating a Gaussian distribution (Fig. 5). This indicates a reduction of metal concentrations in the middle of a sludged plot and a concomittant increase in the surrounding soil. This increase outside the sludged plot declined gradually towards background approximately 5 m from the edge of the plot. This movement had occurred over 42 years, but for the last 7 years there was no ploughing or cultivation as the plots were down to grass. During this period, no decrease in the concentrations of any metal was observed (Fig. 3). This strongly suggests that in field experiments where adjacent plots have very contrasted treatments, leading to accumulation of elements in soil, there can be dispersion of metals due to the physical movement of soil during cultivations over many years. Of the studies listed in Table 1 only four reported metal balances and none gave details of how the plots were cultivated. It seems likely that the low recoveries reported could be due to soil movement if methods to prevent this (e.g. physical barriers or hand-digging) were not used.

If the movement of soil across plot boundaries can be modelled then the amount of metal both outside and inside the treated plot boundaries could be calculated. Recoveries would then be much larger and it would become clearer whether significant losses were occurring from the crop–soil system and under what conditions these losses occur.

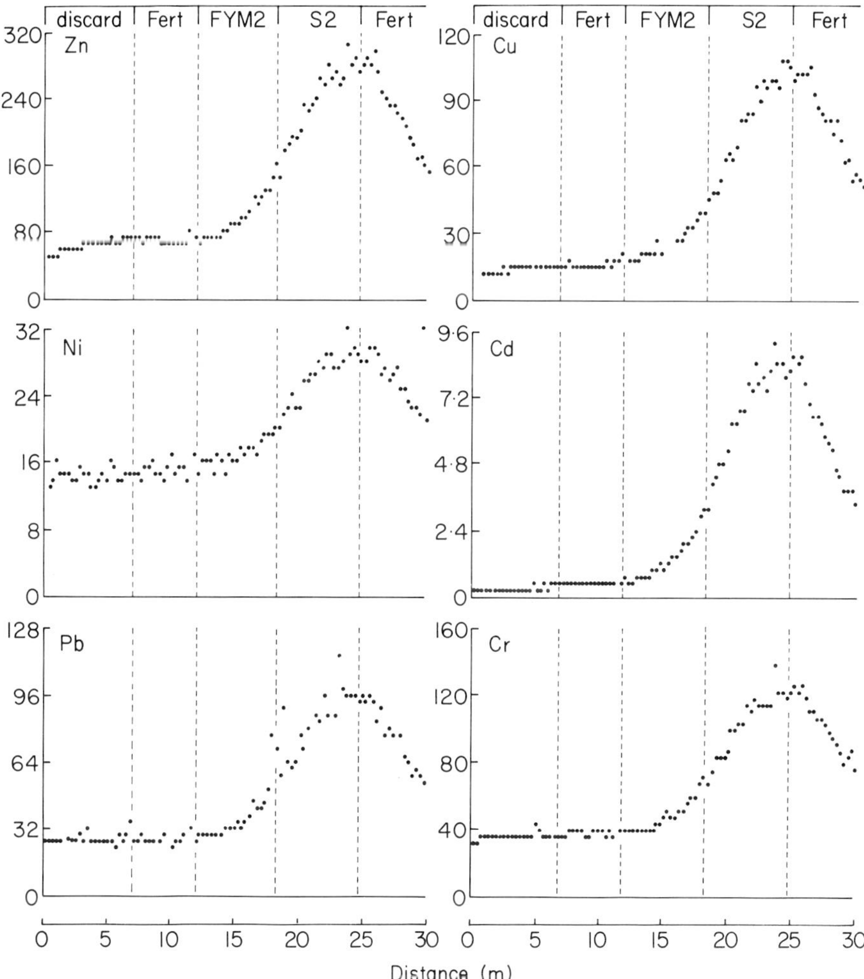


FIG. 5. Concentrations of metals in soil samples taken in a transect across plots in 1984; for details of treatments see Experimental section. 'Discard' is area outside the experiment without treatment or cropping.

## ACKNOWLEDGMENTS

I thank P.R. Hellon, N.J. Freeman and R. Brunsdon for their assistance, and A.E. Johnston for useful discussion.

## REFERENCES

**Allaway, W.H. (1968).** Agronomic controls over environmental cycling of trace elements. *Advances in Agronomy,* **20**, 235–274.

**Baker, D.E., Amacher, M.C. & Leach, R.M. (1979).** Sewage sludge as a source of cadmium in soil-plant-animal systems. *Environmental Health Perspectives,* **28**, 45–49.

**Beckett, P.H.T., Warr, E. & Brindley, P. (1983).** Changes in extractabilities of the heavy metals in water-logged soils. *Water Pollution Control,* **82**, 107–113.

**Berrow, M.L. & Burridge, J.C. (1980).** Trace element levels in soils: effects of sewage sludge. *Inorganic Pollution and Agriculture,* pp. 159–183. Ministry of Agriculture, Fisheries and Food Reference Book 326, HMSO, London.

**Bowen, H.J.M. (1979).** *Environmental Chemistry of the Elements.* Academic Press, London.

**Brümmer, G., Tiller, K.G., Herms, U. & Clayton, P.M. (1983).** Adsorption-desorption and/or precipitation-dissolution processes of zinc in soils. *Geoderma,* **31**, 337–354.

**Chaney, R.L. & Hornick, S.B. (1978).** Accumulation and effects of cadmium on crops. *Proceedings of the 1st Cadmium Conference,* San Francisco, pp. 125–140. Metal Bulletin Ltd, London.

**Chang, A.C., Page, A.L. & Bingham, F.T. (1982).** Heavy metal absorption by winter wheat following termination of cropland sludge applications. *Journal of Environmental Quality,* **11**, 705–708.

**Chang, A.C., Warneke, J.E., Page, A.L. & Lund, L.J. (1984).** Accumulation of heavy metals in sewage sludge-treated soils. *Journal of Environmental Quality,* **13**, 87–91.

**Chater, M. & Williams, R.J.B. (1974).** The chemical composition of British agricultural liming materials. *Journal of Agricultural Science,* **82**, 193–205.

**Davis, R.D. & Carlton-Smith, C. (1980).** *Crops as indicators of the significance of contamination of soil by heavy metals: Technical Report TR140.* Water Research Centre, Stevenage, U.K.

**Department of the Environment (1981).** Report of the sub-committee on the disposal of sewage sludge to land. *Standing Technical Committee Reports No. 20.* National Water Council, London.

**Dowdy, R.H., Larson, W.E., Titrud, J.M. & Latterell, J.J. (1978).** Growth and metal uptake of snap beans grown on sewage sludge-amended soils: a four-year field study. *Journal of Environmental Quality,* **7**, 252–257.

**Emmerich, W.E., Lund, L.J., Page, A.L. & Chang, A.C. (1982).** Movement of heavy metals in sewage sludge-treated soils. *Journal of Environmental Quality,* **11**, 174–178.

**Giordano, P.M. & Mortvedt, J.J. (1976).** Nitrogen effects of mobility and plant uptake of heavy metals in sewage sludge applied to soil columns. *Journal of Environmental Quality,* **5**, 165–168.

**Giordano, P.M., Mortvedt, J.J. & Mays, D.A. (1979).** Residual effects of municipal wastes on yield and heavy metal content of sweet corn and bush beans. *Agronomy Abstracts,* p.29.

**Hinsley, T.D. Ziegler, E.L. & Barrett, G.L. (1979).** Residual effects of irrigating corn with digested sewage sludge. *Journal of Environmental Quality,* **8**, 35–38.

**Johnston, A.E. (1975).** The Woburn Market Garden Experiment, 1942–69. II. The effects of the treatments on soil pH, soil carbon, nitrogen, phosphorus and potassium. *Rothamsted Report for 1974,* Part 2, pp. 102–132.

**Kelling, K.A., Keeney, D.R., Walsh, L.M. & Ryan, J.A. (1977).** A field study of the agricultural use of sewage sludge: III. Effect on uptake and extractability of sludge-borne metals. *Journal of Environmental Quality,* **6**, 352–358.

**Koeppe, D.E. (1981).** Lead: understanding the minimal toxicity of lead to plants. *Effects of Heavy Metal Pollution on Plants.* Vol. 1 (Ed. by N.W. Lepp), pp. 55–76. Applied Science Publishers, London.

**Larsen, K.E. (1984).** Cadmium content in soil and crops after use of sewage sludge. *Utilisation of Sewage Sludge on Land* (Ed. by S. Berglund, R.D. Davis & P. L'Hermite), pp. 157–165. D. Reidel Publishing Company, Dordrecht.

**Le Riche, H.H. (1968).** Metal contamination of soil in the Woburn Market Garden Experiment resulting from the application of sewage sludge. *Journal of Agricultural Science,* **71**, 205–208.

**Lewin, V.H. and Beckett, P.H.T. (1980).** Monitoring heavy metal accumulation in agricultural soils treated with sewage sludge. *Effluent and Water Treatment Journal,* **20**, 217–221.

**Lönsjö, H. (1984).** Isotope studies on crop uptake of cadmium under Swedish field conditions. *Utilisation of Sewage Sludge on Land* (Ed. by S. Berglund, R.D. Davis & P. L'Hermite), pp. 135–145. D. Reidel Publishing Company, Dordrecht.

**McGrath, S.P. (1982).** The uptake and translocation of tri- and hexavalent chromium and effects on the growth of oat in flowing nutrient solution and in soil. *New Phytologist,* **92**, 381–390.

**McGrath, S.P. (1983).** The metal concentrations in the soil of a long-term sewage sludge experiment and the decline in extractability of the metals after sludge applications ceased. *Heavy Metals in the Environment,* Heidelberg, 1983, pp. 397–400. CEP, Edinburgh.

**McGrath, S.P. (1984).** Metal concentrations in sludges and soil from a long-term field trial. *Journal of Agricultural Science,* **103**, 25–35.

**McGrath, S.P. & Cunliffe, C.H. (1985).** A simplified method for the extraction of the metals Fe, Zn, Cu, Ni, Cd, Pb, Cr, Co and Mn from soils and sewage sludges. *Journal of the Science of Food and Agriculture,* **36**, 794–798.

**Miller, W.P., McFee, W.W. & Kelly, J.M. (1983).** Mobility and retention of heavy metals in sandy soils. *Journal of Environmental Quality,* **12**, 579–584.

**Ministry of Agriculture, Fisheries and Food. (1981).** *Tha Analysis of Agricultural Materials.* Reference Book 427. HMSO, London.

**Schaaf, H. & Boguslawski, E. von (1982).** Schwermetallanreicherung in Boden und Pflanze bei langjährige Anwendung von Klärschlamm. *Landwirtschaftliche Forschung, Sonderheft,* **39**, 224–237.

**Sidle, R.C., Hook, J.E. & Kardos, L.T. (1977).** Accumulation of heavy metals in soils from extended wastewater irrigation. *Journal of the Water Pollution Control Federation,* **49**, 311–318.

**Stojkovaska, A. & Cooke, G.W. (1958).** Micronutrients in fertilizers. *Chemistry and Industry, p. 1368.*

**Thompson, M. & Walsh, J.N. (1983).** *A Handbook of Inductively Coupled Plasma Spectrometry.* Blackie, Glasgow.

**Walkley, A. (1935).** An examination of methods for determining organic carbon and nitrogen in soils. *Journal of Agricultural Science,* **25**, 598–609.

**Webber, L.R. & Beauchamp, E.G. (1979).** Cadmium concentration and distribution in corn (*Zea mays* L.) grown on a calcareous soil for three years after annual sludge applications. *Journal of Environmental Science and Health,* **B14**, 459–474.

**Williams, D.E., Vlamis, J., Pukite, A.H. & Corey, J.E. (1984).** Metal movement in sludge-treated soils after six years of sludge addition: I. Cadmium, copper, lead and zinc. *Soil Science,* **137**,351–359.

**Williams, R.J.B. (1971).** Relationships between the composition of soils and physical measurements made on them. *Rothamsted Report for 1970,* Part 2, pp. 5–35.

# Cycling and fate of heavy metals in a contaminated woodland ecosystem

M. H. MARTIN
*Department of Botany, The University, Bristol BS8 1UG*

P. J. COUGHTREY
*Associated Nuclear Services, Eastleigh House, East Street, Epsom, Surrey*

## SUMMARY

**1** The metal concentrations in litter and soil profiles of aerially-contaminated woodland have been determined over a period of 10 years. Data for lead, zinc and cadmium at various depths in the soil profile have been used to determine total contents per unit area and to compare distributions within the profile with time.
**2** The results obtained have been used for the development of a simple mathematical model describing observed distributions in the profile with time at the woodland. Direct observations and model predictions demonstrate that the mobility of lead, zinc and cadmium is considerably greater than that which would be inferred from existing literature.
**3** Model predictions suggest that the total loss of metals from the 0–20 cm layer of the profile could amount to 2–3 g $m^{-2}$ for cadmium, 150–200 g $m^{-2}$ for zinc and 10–50 g $m^{-2}$ for lead over the next 10–100 years. In the case of cadmium, partial equilibrium in that layer would be reached within a period of about 10 years and this would be associated with the re-establishment of a 'classic' concentration–depth profile generally interpreted as representing immobility in soil at aerially-contaminated sites.
**4** An important factor in current and potential future movement of metals in soil is the rate of loss from organic layers. Possible causes of the apparent recent increase in mobility are discussed. It is concluded that a major factor is the considerable reduction in pH of litter and soil over the past 5–10 years and it is suggested that this reflects acid deposition in the area.

## INTRODUCTION

There are relatively few detailed, published examples of the cycling of heavy metals through forest ecosystems. Studies of heavy metal cycling in relatively unpolluted forest ecosystems include Van Hook, Harris & Henderson (1977), Denaeyer-de Smet (1974), Mayer (1981) and Heinrichs & Mayer (1977); studies from more heavily polluted sites include Martin & Coughtrey (1981), Martin, Duncan & Coughtrey (1982), Tyler (1970, 1972), Parker, McFee & Kelly (1978) and Denaeyer-de Smet & Duvigneaud (1974). Hughes (1981) and Hughes, Lepp & Phipps (1980) have reviewed the general field of cycling of heavy metals in woodland ecosystems.

There is general agreement that the soil and organic litter layer components of woodland ecosystems are the ultimate sinks for heavy metals in aerially-contaminated areas. Davies (1983) for example, in reviewing the literature on soils as heavy metal sinks, stated that once soils have become contaminated they are unlikely to be depleted of lead and other metals by leaching. Davies (1980) drew attention to the dearth of information concerning the distribution of heavy metals in contaminated soil profiles. Numerous authors have described work relevant to this topic and the evidence for relative immobility and long-term retention of heavy metals in most soils comes mainly from a variety of observations and considerations.

(a) The major part of the total ecosystem burden of heavy metals is found in the soil and organic litter components (see Table 1 for examples).
(b) The distribution of heavy metal concentrations in soil profiles generally shows high levels of contamination in the surface horizons only. For example, Friedland *et al.* (1984), Harmsen (1977), Mayer (1981), Chang *et al.* (1984), Rutherford & Bray (1979), Davies (1980, 1983), Williams, Vlamis & Pukite (1980), Fleming & Ryan (1964), Kreutzer *et al.* (1983), Sharma & Shupe (1977), John, Chuah & Vanlaerhoven (1972), Scokart, Meeus-Verdinne & Deborger (1983), Greszta *et al.* (1979).
(c) Experimental evidence (including differential removal by chemical extractants) of the solubility and mobility of heavy metals in soil. For example, Korte *et al.* (1975, 1976), Khan, Nandau & Khan (1982), Harmsen (1977), Adams & Sanders (1984), Brummer & Herms (1983), Anderson & Christensen (1983), Herms & Brummer (1980), Marinsky, Wolf & Bunzl (1980), Zunino *et al.*(1979). Levi-Minzi, Soldatini & Riffaldi (1976), Harter (1983).
(d) Calculation of residence times and depletion times of heavy metals in soil. For example, Bowen (1975, 1977), Frissel & Pennders (1983), Van Enk (1983), Alloway (1968), Tyler (1978).
(e) Artificial or laboratory leaching experiments using spiked soils or field soils. For example, Tyler (1978), Miller, McFee & Kelly (1983), Jurinak & Santillon-Medrano (1974), Frissel, Poelstra & Van der Klugt (1974), Chang & Broadbent (1980), Alesii, Fuller & Boyle (1980).
(f) Repeated measurements of metal concentrations in soil profiles over a period of years. For example, Sidle & Kardos (1977), Sidle, Kardos & Genutchen (1977), Sidle, Hook & Kardos (1977).
(g) Analysis of ground-water leaching from soils and lysimeter studies. For example, Sopper & Kerr (1980), Sidle & Kardos (1977), Sidle, Hook & Kardos (1977), Harmsen (1977), Mayer (1981, 1983), Heinrichs & Mayer (1977), Smith & Siccama (1981).

It is clear from the types of investigations listed above that mobility differs between metals and according to the type of soil and type and degree of contamination being considered. There is general agreement that cadmium and zinc are relatively more mobile in soils than elements such as lead and copper, and that soil characteristics which influence mobility include pH, cation exchange capacity, organic matter content, free-lime content, clay content, manganese, aluminium and iron oxide content, soil texture, redox potential and leaching rate.

TABLE 1. Percentage distribution of certain heavy metals in biomass, organic litter and mineral soil of forest ecosystems. (a) and (b) are for woodland fairly heavily contaminated by aerial deposition of metals; (c), (d), (e) and (h) for relatively unpolluted forests; (f) is for a roadside site; (g) is for unpolluted forests

| | | Biomass | Litter | Soil |
|---|---|---|---|---|
| (a) Hallen Wood | Pb | 3·92 | 37·25 | 58·83 |
| Oak/hazel | Zn | 1·08 | 9·50 | 89·41 |
| Martin & Coughtrey (1981) | Cd | 1·31 | 10·75 | 87·94 |
| | Cu | 1·93 | 14·37 | 83·71 |
| (b) Haw wood | Pb | 23.86 | 37·50 | 38·64 |
| Oak/hazel | Zn | 4·24 | 18·08 | 77·67 |
| Martin *et al.* (1982) | Cd | 5·05 | 30·26 | 64·69 |
| | Cu | 4·53 | 5·96 | 89·51 |
| (c) Mixed deciduous | Pb | 0·30 | 0·75 | 98·95 |
| Van Hook *et al.* (1977) | Zn | 0·32 | 0·57 | 99·11 |
| | Cd | 3·32 | 1·72 | 94·96 |
| (d) Solling B | Pb | 1·18 | 10·06 | 88·76 |
| Beech | Zn | 1·36 | 2·60 | 95·74 |
| Mayer (1981)* | Cd | 9·86 | 10·90 | 79·24 |
| | Cu | 8·41 | 2·00 | 89·59 |
| | Ni | 5·03 | 1·15 | 93·82 |
| | Cr | 0·96 | 0·51 | 98·54 |
| (e) Solling F | Pb | 1·60 | 10·08 | 88·32 |
| Spruce | Zn | 4·40 | 3·45 | 92·15 |
| Mayer (1981)* | Cd | 26·67 | 8·33 | 65·00 |
| | Cu | 7·36 | 2·49 | 90·15 |
| (f) Spruce | Pb | 5·00 | 50·00 | 45·00 |
| Tyler (1970) | | | | |
| (g) Spruce† | Pb | 1·52 | 4·51 | 93·96 |
| Beech | Pb | 1·71 | 6·11 | 92·18 |
| Denaeyer-de Smet (1974) | | | | |
| (h) Open savannah | Cd | 0·74 | 1·03 | 98·23 |
| Black oak | Zn | 0·75 | 0·92 | 98·34 |
| Parker *et al.* (1978) | Pb | 1·04 | 2·32 | 96·63 |
| | Cu | 1·24 | 1·73 | 97·04 |

*Also published in Heinrichs & Mayer (1980).
†Data calculated from tabulation by Hughes (1981).

There are few, if any, field studies of soil profiles over extended periods of time which examine the extent to which various heavy metals move in soils that have been aerially-contaminated.

In this paper we present field data collected from a woodland ecosystem, Hallen Wood, close to Avonmouth, near Bristol, which is subjected to relatively heavy aerial deposition of lead, zinc, cadmium and, to a lesser degree, copper. Estimates of deposition to the woodland in the late 1970s were 9·2 mg $m^{-2} y^{-1}$ cadmium, 285 mg $m^{-2} y^{-1}$ lead, 600 mg $m^{-2} y^{-1}$ zinc and 26·3 mg $m^{-2} y^{-1}$ copper (Martin & Coughtrey 1981). The woodland is 3 km downwind of an industrial complex which includes a major primary lead, zinc, cadmium smelter (see Coy 1984 for details of emissions), and a domestic incinerator; the area is also traversed by the M5 motorway. The woodland is an old oak–hazel *(Quercus robur, Corylus avellana)* woodland with a field layer which includes *Endymion non-scriptus, Milium effusus* and *Holcus lanatus.* The soil is a heavy clay, developed from Rhaetic Clays. Previous studies have described the concentrations of metals in woodland plants and soils (Martin *et al.* 1982; Martin & Coughtrey 1981), in invertebrates (Martin & Coughtrey 1976; Coughtrey 1978; Hopkin & Martin 1982, 1983). Coughtrey *et al.* (1979) showed that substantial accumulation of organic litter occurred and that this appeared to be correlated with the degree of zinc and cadmium contamination. Hutton (1984) has also discussed the data for this woodland.

## METHODOLOGY

### *Field studies*

During our investigations on the effects of heavy-metal pollution in the Avonmouth area we have measured the concentrations and total amounts of the heavy metals lead, zinc and cadmium in soil profiles of Hallen Wood at regular intervals. Initial, but limited data were obtained by Denny in 1974, and subsequently more detailed profiles were analysed in 1975, 1979 and 1983. Profiles were sampled from the side of a soil pit, having first removed the organic litter layer. Sampling depths for the 1974 profile were at 6 cm depth intervals to 18 cm and, for later profiles, at 0–1 cm and then at 2·5 cm intervals to 21 cm, with a 10 cm deep interval to 31 cm. Samples of soil from the upper and lower parts of the profile were collected for bulk density measurements; soil pH at each depth was measured with a glass electrode in a 1:2·5 soil:distilled water suspension; loss-on-ignition was measured by weight, loss-after-ignition at 375 °C. Metal concentrations were measured by atomic absorption spectrophotometry after wet digestion in boiling, concentrated, Analar nitric acid. All concentration data are expressed on an oven-dry weight basis.

### *Mathematical modelling studies*

Having obtained data for soil profiles in 1975, 1979 and 1983, it has been possible to determine rate coefficients for transfer of metals between successive layers of soil and to use these coefficients to predict future trends in the soil profile with time. The conceptual model used is

$$I \rightarrow \boxed{q} \xrightarrow{\lambda}$$

which is represented mathematically by the differential equation

$$\frac{dq}{dt} = I - \lambda q \tag{1}$$

where $q$ = compartment content, $I$ = the rate of input to the compartment, and $\lambda$ = the rate constant for loss from the compartment (with units of $t^{-1}$).

The mathematical solution of eqn (1) is

$$q(t) = \frac{I}{\lambda}(1 - e^{-\lambda t}) + q(o)e^{-\lambda t} \tag{2}$$

where $q(t)$ = the compartment content at time $t$, and $q(0)$ = the original compartment content.

In the absence of any continuing input to the system ($I = 0$) eqn (2) reduces to

$$q\ (t) = q(o)e^{-\lambda t}$$

In these circumstances, the half-life (the time taken for the contents of the compartment to be reduced by a factor of two, $t\frac{1}{2}$) is defined as ln 0·5/$\lambda$. The residence time, mean lifetime, or turnover time for material in the compartment (often denoted as $\Upsilon$) assuming a single pathway for loss is defined as $1/\lambda$. Values of $\lambda$, $t\frac{1}{2}$ and $\Upsilon$ can be derived from systems relatively easily when $I$ is small compared to $\lambda$ or $q$ or is zero. Given $\lambda$ and $I$, the equilibrium content of the compartment is derived from eqn (2) which reduces to $I/\lambda$ as $t \rightarrow \infty$.

For the present study, the soil profile can be regarded as a series of compartments with transfers represented as net one-way flows, i.e.

$$\boxed{q_1} \xrightarrow{\lambda_1} \boxed{q_2} \xrightarrow{\lambda_2} \cdots \xrightarrow{\lambda_{n-1}} \boxed{q_n}$$

In reality, there are refluxes between each pair of compartments (e.g. as a result of upwards percolation caused by evaporation at the surface, mass movement by burrowing animals, etc.). The advantage of representing the transfers as net one-way flows is that the system is amenable to an explicit solution, frequently termed the Bateman equations. A derivation of these equations is given by Whicker & Schultz (1982); for a detailed discussion of techniques in compartmental analysis, Jacquez (1972) should be consulted. In order to derive values of $\lambda$ to $\lambda_n$, an iterative fitting technique was used in conjunction with the metal contents of various layers in the soil profile at different years. This technique provides a set of rate coefficients (Table 2) which, when inserted into the Bateman equations, provides results for changes in compartmental contents with time that are consistent with observation. The technique was applied to changes in normalized metal contents of each compartment over the periods 1975–79, 1979–83 or 1975–83 and results

TABLE 2. Rate coefficients ($year^{-1}$) for transfer of metals between layers of the Hallen Wood soil profile based on 1975 to 1983 measurements. All coefficients shown are negative.

| Soil depth | Cadmium ($\times 10^{-1}$) | Zinc ($\times 10^{-1}$) | Lead ($\times 10^{-1}$) |
|---|---|---|---|
| L - 1 | 2·1814 | 1·3394 | * |
| 1 - 2 | 15·4770 | 5·6019 | 1·5162 |
| 2 - 3 | 6·9643 | 3·9435 | 2·5171 |
| 3 - 4 | 7·5715 | 5·4234 | 6·2425 |
| 4 - 5 | 10·1130 | 6·9568 | 3·9041 |
| 5 - 6 | 9·8280 | 5·1710 | 3·0147 |
| 6 - 7 | 11·0220 | 5·1710 | 3·3375 |
| 7 - 8 | 12·4390 | 5·1710 | 1·8323 |
| 8 - 9 | 15·3870 | 5·1710 | 1·8323 |
| 9 - 10 | 19·1690 | 5·1710 | 1·8323 |

*Litter component showed accumulation of lead over the period of measurement, therefore transfer coefficients based on losses not calculated.

Soil depths: L = litter layer; 1 = 0–1 cm; 2 = 1-3·5 cm; 3 = 3·5-6 cm ; 4 = 6-8·5 cm; 5 = 8·5-11 cm; 6 = 11-13·5 cm; 7 = 13·5-16 cm; 8 = 16-18·5 cm; 9 = 18·5-21 cm; 10 = 21-31 cm.

are illustrated in Table 3. The results obtained on the basis of data for 1975–1983 were used to predict compartmental contents in 1979 and compared with observed contents to provide a test of the method and data. For lead it was not possible to include the litter layer in calculations because the total and percentage contents of this compartment increased with time. Calculations were therefore restricted to soil layers alone (i.e. to contents in layers other than litter, normalized to the total content in soil alone excluding litter).

Other predictions described in this paper were obtained by using a technique developed and implemented at Associated Nuclear Services for the purpose of solving a set of first order differential equations with constant rate coefficients. Further details are given by Jackson & Smith (p.385). The technique uses a fixed-step, second-order predictor-corrector method in conjunction with a fourth-order Runge-Kutta method to increase or decrease the step length as convergence is being achieved or not being achieved respectively. Results obtained using this technique in respect of a simple series of compartments were confirmed, where appropriate, against similar results obtained with the direct solution using the Bateman equations.

Rate coefficients were obtained with the assumption that input rates to the upper layer could be neglected, whereas predictions based on the derived model were obtained both for systems in which no inputs were assumed and for those in which inputs were assumed. Appropriate input rates ( 10 mg Cd $m^{-2} y^{-1}$, 600 mg Zn $m^{-2} y^{-1}$ and 300 mg Pb $m^{-2} y^{-1}$) for these investigations were derived from various published and unpublished investigations at

TABLE 3. Observed and predicted percentage distributions of cadmium, zinc and lead in Hallen Wood soil profiles using 1975 – 83 transfer rate coefficients applied to 1975 date to predict 1979 data. Obs. = data measured in 1979; Pred. = data predicted

| | Cadmium | | Zinc | | Lead | |
|---|---|---|---|---|---|---|
| Soil depth (cm) | Obs. | Pred. | Obs. | Pred. | Obs. | Pred. |
| L | 10·9 | 10·9 | 9·59 | 11·1 | – | – |
| 1 | 5·96 | 1·84 | 4·54 | 4·74 | 37·7 | 22·0 |
| 2 | 10·6 | 8·38 | 7·19 | 12·2 | 24·4 | 16·9 |
| 3 | 12·3 | 12·1 | 7·74 | 10·6 | 5·79 | 7·01 |
| 4 | 11·7 | 10·8 | 9·50 | 8·63 | 4·35 | 9·79 |
| 5 | 11·1 | 11·4 | 9·11 | 10·8 | 3·73 | 9·22 |
| 6 | 10·1 | 9·3 | 8·38 | 9·15 | 3·97 | 6·27 |
| 7 | 6·05 | 7·09 | 6·56 | 7·31 | 3·33 | 6·86 |
| 8 | 5·31 | 4·83 | 6·94 | 5·72 | 3·25 | 4·71 |
| 9 | 4·14 | 3·30 | 6·62 | 4·50 | 2·62 | 3·67 |
| 10 | 11·80 | (19.9) | 23·8 | (15·3) | 11·10 | (13·62) |

N.B. Soil depths as in Table 2. Parentheses emphasize that the model does not allow for movement out of layer 10 whereas actual measurements will reflect loss from this layer.

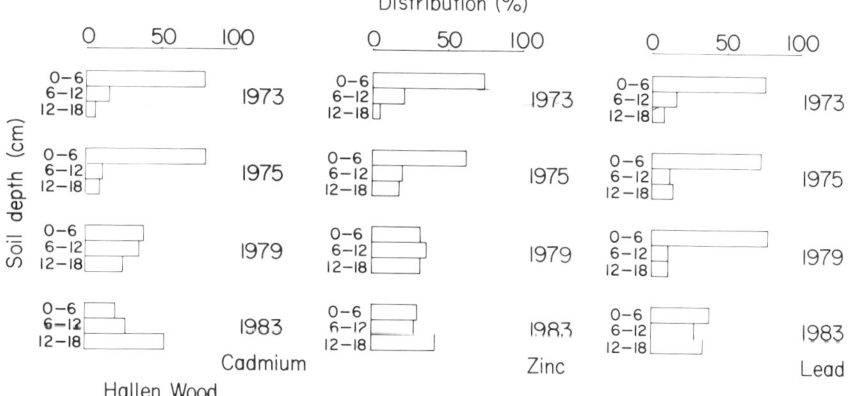


FIG. 1. Percentage distribution of cadmium, zinc and lead in Hallen Wood soil at three mineral depths over the 10 year period 1973–83. Note that the litter layer was not included in the calculations.

Hallen Wood (e.g. Denny 1974; Manning 1978; Coughtrey 1978; Martin & Coughtrey 1981).

## RESULTS

Data describing the change in metal distribution with soil depth are presented here in three ways: concentrations of metals with soil depth, percentage total metal content with soil depth, and percentage total content normalized to unit depth against soil depth.

Percentage distributions of cadmium, zinc and lead in 3 mineral layers (i.e. excluding litter) of the soil profile at Hallen Wood over a 10-year period are shown in Fig. 1. At the

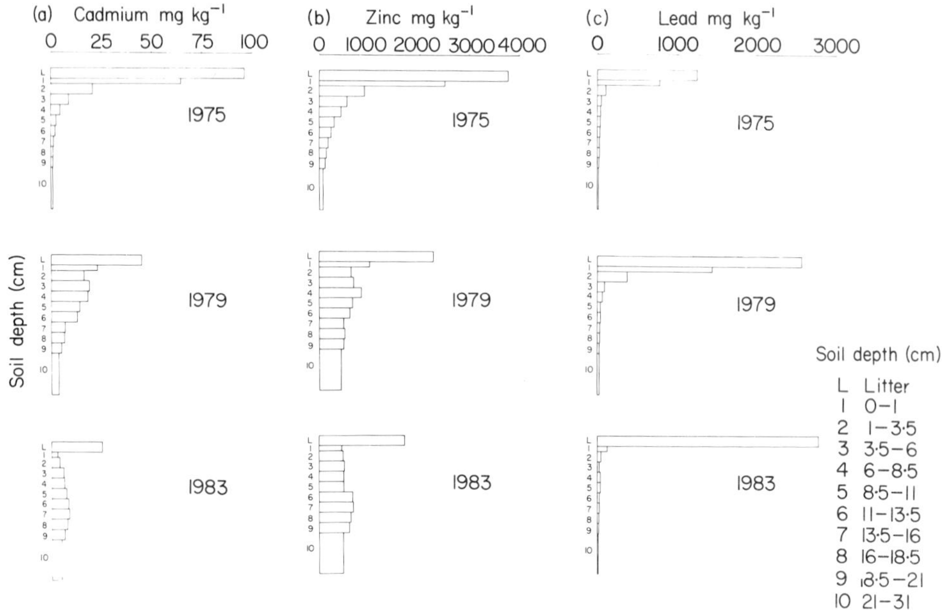


FIG. 2. Depth profiles of metal concentrations (mg kg$^{-1}$ dry weight) in Hallen Wood soils at three sampling times; (a) cadmium, (b) zinc and (c) lead.

1973 and 1975 sampling dates the profiles of all three metals were essentially similar, with markedly higher proportions of the total metal content in the top 6 cm of soil. However, progressive redistribution of zinc and then cadmium and finally (in 1983) some lead, appears to have occurred (the changes in lead concentration in the mineral soil (below) are rather small). Thus, by the 1983 sampling date, the 12–18 cm depth layer contains a greater proportion of zinc and cadmium than the 0–6 or 6–12 cm layer.

More detailed soil profiles for metal concentration data in the years 1975, 1979 and 1983, but including the more heavily contaminated organic litter layer, are shown in Fig. 2 a, b, c. These data show that the concentrations of lead with soil depth have remained substantially unchanged during the period 1975–83 although, apparently, concentrations in the litter have increased. The changes in the 0–1 cm depth layer, i.e. the higher concentrations in 1979 compared with 1983, may be related to sampling errors and to the effectiveness with which all organic matter was scraped from the mineral soil surface in 1979 (and would also contribute substantially to the apparent change in percentage distribution of lead shown in Fig. 1 for 1983; these do, however, represent rather small changes in concentration). These data also show that, even in 1975, the concentration gradient with soil depth was in the order of Pb>Cd>Zn, suggesting that zinc and cadmium were more mobile than lead. This is confirmed by the data for 1979 and 1983 which show evidence, particularly in the case of zinc, of a 'wave' of concentration moving down the profile. It is emphasized that the data presented (Fig. 2) reflect single profiles at each sampling date. Nevertheless, the general pattern shown by these profiles is supported both by replicate sampling at Hallen Wood (e.g. 1975) and by data obtained at nearby

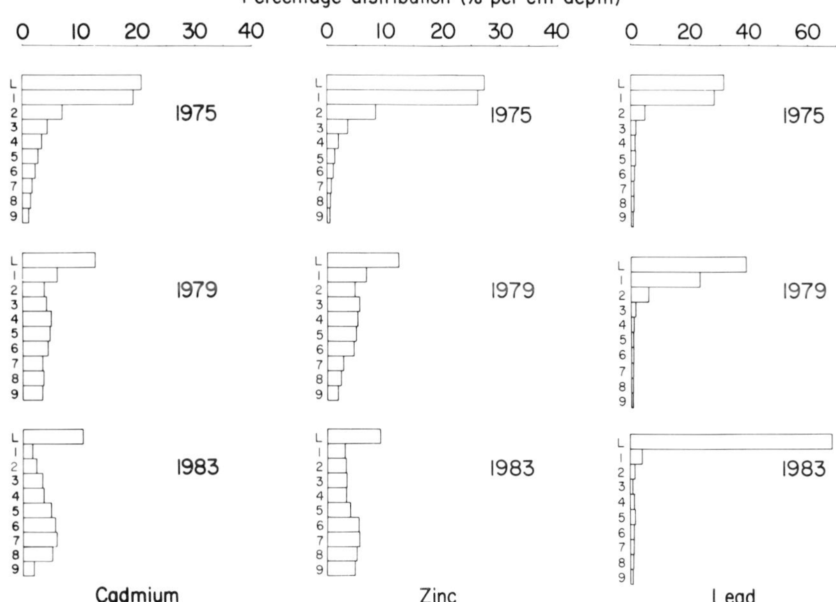


FIG. 3. Percentage distribution, expressed as percentage per centimetre depth, of cadmium, zinc and lead in Hallen Wood soils at three sampling times. Soil depths (L to 9) as in Fig. 2.

woodlands in 1981 and 1983. In Fig. 3 the data have been transformed to percentage of total metal (including 0–31 cm soil and litter layers) normalized for depth of each layer

TABLE 4. Soil pH and loss-on-ignition (LOI) data for Hallen Wood soils

| Soil depth | pH | | | | LOI |
|---|---|---|---|---|---|
| | 1973 | 1975 | 1979 | 1984 | 1979 |
| L | | 4·2 | 3·8 | – | 64·2 |
| 1 | | 4·65 | 4·0 | 4·05 | 12·6 |
| 2 | 5·01 | 4·95 | 4·1 | 4·00 | 5·63 |
| 3 | | 4·72 | 4·15 | 4·15 | 3·87 |
| 4 | | 4·55 | 4·5 | 4·00 | 2·70 |
| 5 | 4·96 | 4·27 | 4·7 | 3·95 | 2·01 |
| 6 | | 4·25 | 4·85 | 4·10 | 2·09 |
| 7 | 5·29 | 4·45 | 4·70 | 4·15 | 2·12 |
| 8 | | 4·40 | 4·65 | 4·15 | 2·04 |
| 9 | | 4·58 | 4·50 | 4·15 | 1·51 |
| 10 | | 4·93 | 4·75 | 4·25 | 1·69 |
| 31 cm + | | – | 5·15 | 4·70 | 1·30 |
| 46 cm | | – | – | 6·30 | |

Soil depths L – 10 as in Table 2.

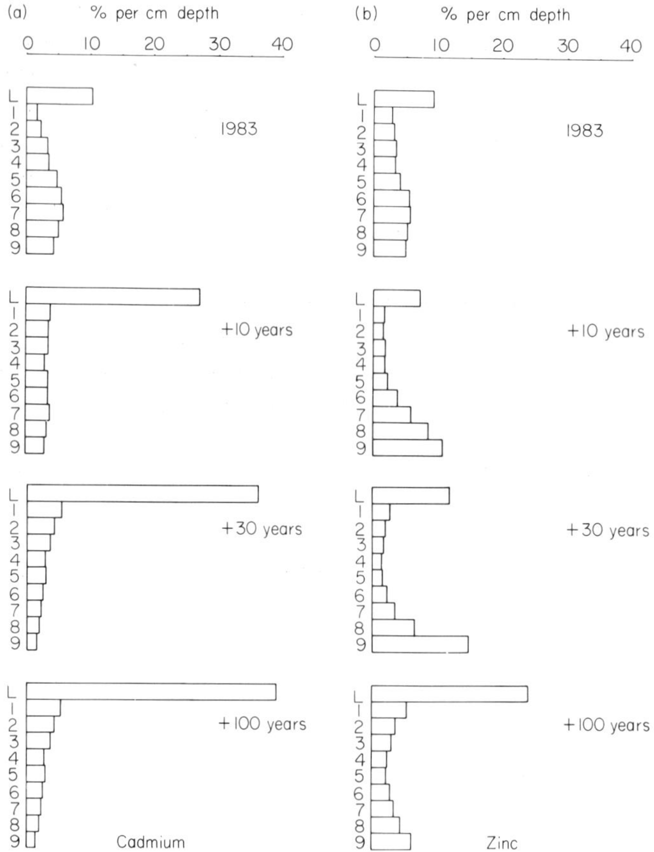


FIG. 4. Percentage of total metal content per unit depth in each soil layer in 1983 and model predictions for 10, 30 and 100 years after 1983 for cadmium and zinc. Soil depths (L to 9) as in Fig. 2. The model predictions assume an input, subsequent to 1983, of 10 mg Cd $m^{-2} y^{-1}$, 600 mg Zn $m^{-2} y^{-1}$ and 300 mg Pb $m^{-2} y^{-1}$.

sampled. Such data clarify the situation for cadmium and zinc and suggest that these two metals were moving relatively quickly through the soil profile, indeed to such an extent that sampling to a depth of 31 cm can now be regarded as inadequate. The data for lead further emphasize the relative immobility of this metal and the importance of the organic layers in its retention. The progressive movement of cadmium and zinc with time is also shown in Figs 5 and 6 (see next Section). Loss-on-ignition and pH data for these profiles are given in Table 4. These data show a general trend of increasing acidification both in the surface layers and at greater soil depths with time.

## *Mathematical modelling*

Data for the percentage distribution of cadmium and zinc with soil depth in 1983 and the predicted soil profile data for 10, 30 and 100 years after 1983 are shown in Fig. 4. They show a 'wave' of cadmium moving down the soil profile which, as it penetrates to depths deeper than those shown, changes the distribution profile to a shape similar to that of the earlier 1975 profiles. The re-establishment of the surface-horizon dominated profiles occurs within 10–30 years in the case of cadmium and rather more than 100 years in case of zinc. Figs 5 and 6 depict the movement of cadmium and zinc respectively in particular soil layers over the period 1975–83 and 100 years from 1983 and confirm the 'wave' of metal moving down the profile as observed in the original concentration data (Fig. 2).

Fig. 7 gives the predicted quantities of metal in g $m^{-2}$ which would enter the basal layer of the profile (as sampled, i.e. the 21–31 cm horizon) from 1983 onwards. This layer is not depicted in Fig. 4 because it represents the compartment of accumulation in the model, whilst in the field it would also have an output to the next layer down the profile and in reality is not subject to accumulation. Fig. 7 indicates that the relative rates of movement are Cd > Zn > Pb with lead, unlike cadmium and zinc, not reaching an equilibrium over the 100 years.

Table 5 shows the computed residence times of the three metals based on the transfer rate coefficients derived for the 1975–83 sampling period. These data also confirm that the mobility of the metals, as indicated by residence times is in the order Cd > Zn > > Pb. For all these metals the rate-limiting transfer in the system is the loss from litter to surface soil.

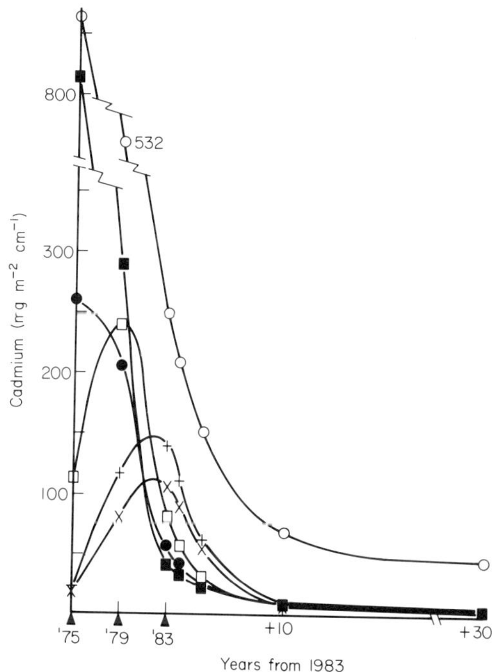


FIG. 5. Cadmium content per unit soil depth (g $m^{-2}$ $cm^{-1}$) of Hallen Wood soils observed in different soil layers in 1975, 1979 and 1983 and model predictions for up to 30 years from 1983. Soil depths shown are ○ litter, ■ 0–1 cm, ● 1–3·5 cm, □ 3·5–6 cm, + 13·5–16 cm, × 18·5–21 cm.

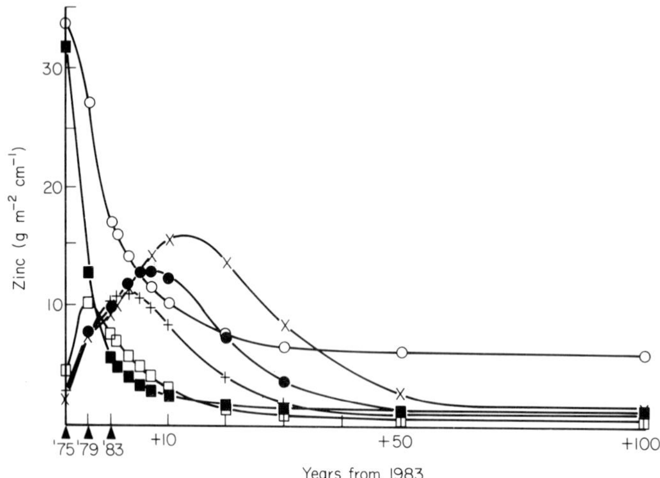


FIG. 6. Zinc content per unit soil depth (g $m^{-2}$ $cm^{-1}$) of Hallen Wood soils observed in different soil layers in 1975, 1979 and 1983 and model predictions for up to 100 years from 1983. Soil depths shown are ○ litter, ■ 0–1 cm, □ 8·5–11 cm, + 13·5–16 cm, • 16–18·5 cm, × 18·5–21 cm.

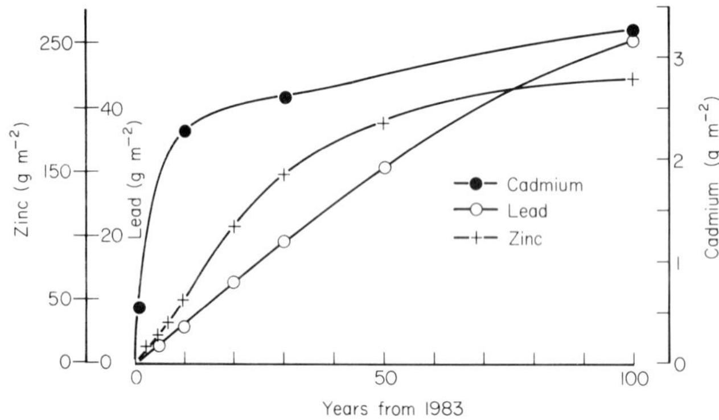


FIG. 7. Cumulative amounts (g $m^{-2}$) of cadmium, zinc and lead predicted to move from the 18·5–21 cm layer to the 21–31 cm layer with time from 1983.

## DISCUSSION

In situations of no input of heavy metals to soils, the depletion times for metals, by uptake and removal in crops, is extremely long. Alloway (1968) for example estimated times of $10^2$–$10^4$ y for cadmium and lead respectively. Bowen (1975, 1977) studied the residence times of heavy metals in Thames Valley soils and estimated 280 y for cadmium, 860 y for copper, 2100 y for zinc and 3000 y for lead. Bowen pointed out that there were very few studies of residence times of metals in specific soils; it is also clear that residence times may

TABLE 5. Residence times (years) for cadmium, zinc and lead in Hallen Wood soil profiles based on transfer rate coefficients. See text for details of calculation of residence times.

| Soil depth | Cadmium | Zinc | Lead |
|---|---|---|---|
| L – 1 | 4·6 | 7·5 | – |
| 1 – 2 | 0·65 | 1·8 | 6·6 |
| 2 – 3 | 1·44 | 2·5 | 4·0 |
| 3 – 4 | 1·32 | 1·8 | 1·6 |
| 4 – 5 | 0·99 | 1·4 | 2·6 |
| 5 – 6 | 1·02 | 1·9 | 3·3 |
| 6 – 7 | 0·91 | 1·9 | 3·0 |
| 7 – 8 | 0·8 | 1·9 | 5·5 |
| 8 – 9 | 0·65 | 1·9 | 5·5 |
| 9 – 10 | 0·52 | 1·9 | 5·5 |

Soil depths as in Table 2.

well change with the degree of contamination of the soils. Bowen's data seem to be for normal, unpolluted, soils. Enk (1983) however, in a specific study of cadmium, based predominantly on a mathematical-modelling approach suggested that the average concentration of cadmium in the top 20 cm of European soils increased over the period 1900–70 and that the average residence time for cadmium has reduced from 210 y to 53 y over the same period. In the case of agricultural soils, residence times were estimated to have decreased from 877 y in 1750 to 390 y in 1980. Enk further suggested that, in more heavily-polluted soils, the time for 50% depletion of cadmium is between 2 and 11 y.

The data presented in Table 4 are clearly of great contrast to Bowen's data and seem to confirm Enk's suggestion of very short residence times in heavily-contaminated soils. Tyler (1978) leached a purely organic mor layer from a spruce forest close to a brass mill in south-eastern Sweden with artificial rainwater of different acidities. He calculated the number of years necessary to cause 10% depletion of the original metal concentrations in the sample. The results were closely related to the acidity of the leaching water; at pH 4·2 the times were 9, 20 and $>200$ y for Zn, Cd and Pb, respectively; at pH 3·2 they were 3, 4–5 and $>200$ y, respectively; and at pH 2·8 they were 1·2, 1·7 and 17 y, respectively. Comparison with the date for Hallen Wood shown in Table 4 for the litter layer, shows that residence times for zinc are longer than those of cadmium, whilst Tyler's data show cadmium to be more mobile than zinc. Lead is relatively immobile in both situations.

The field data for Hallen Wood collected over the periods 1973, 1975, 1979 and 1983 provide clear evidence for movement of zinc and cadmium through the soil profile. Many authors have investigated metal concentration profiles in soils (see Introduction for examples) at single points in time and have concluded a lack of mobility if the concentrations of metals in the profile show high concentrations at the surface and low concentrations at depth. It must be emphasized, however, that, per se, such concentration data do not necessarily imply a lack of mobility of metals. Thus, if the transfer rate coefficients change with soil depth and if those rate coefficients were higher in the deeper parts of the soil profile or increased with soil depth, it would be possible to maintain a steep concentration profile provided input to the surface was sufficient to maintain and drive the system. In Hallen Wood it seems that the transfer rate coefficients do not increase with depth for lead but the position for cadmium and zinc is less clear (Table 2). Thus, to ascertain the degree of mobility, it becomes necessary either to measure loss from the soil profile in leachates, or to conduct a total inventory of metals per unit area of surface at intervals of time. The data presented in Figs 1 and 3 show changes in the percentage of the total metal load in the soil profile (to 31 cm depth) in the different soil layers, and confirm that zinc and cadmium are being redistributed in the soil profile.

Very few authors have described heavy metal concentrations in soil profiles in the same site over a period of time, but Sidle, Hook & Kardos (1977) and Sidle & Kardos (1977) reported metal concentrations in a variety of soils measured in successive years during the application of sewage sludge or wastewater to a hardwood forest ecosystem. In soil treated with sewage sludge, small amounts of copper, zinc and cadmium (in the series Cd > Zn > Cu) moved out of the 120 cm deep layer of the soil profile, but chromium, lead, cobalt and nickel showed little penetration into the profile. In the absence of published work describing metal concentrations in soil profiles over a period of years in aerially-polluted sites, data for soils showing different degrees of aerial-contamination become important. Buchauer (1973) showed that the distribution of zinc concentrations with soil depth changed with distance from a zinc smelter. She found that, close to the smelter, substantial contamination occurred even in the deeper parts of the soil profile, whilst further from the smelter the contamination became increasingly restricted to the surface layers. Scokart *et al.* (1983) investigated highly-polluted soils near zinc smelters in Belgium and described the mobility of lead, zinc, cadmium and copper in two contrasting soil types, i.e. an acidic (pH 4) sandy podsolic soil, and a neutral loamy soil. Based on concentration data with soil depth these authors showed that, in the neutral loamy soil, there was little movement of metals, although there were indications that cadmium became mobile at pH 6 and below. Zinc became mobile below pH 5 whilst lead and copper remained in the upper layers and pH had no effect. In contrast, in the acid soil considerable movement occurred in association with organic matter with a series in mobility of Cd > Zn > Cu > Pb. This is in agreement with the observations for Hallen Wood (Figs 1, 2 and 3) and with experimental work reported by other authors, e.g. Adams & Sanders (1984), Brummer & Herms (1983), Herms & Brummer (1980). The sites investigated by Scokart *et al.* had a wide range of contamination which encompasses that at Hallen Wood. There are several points of similarity between Scokart *et al.*'s sites and Hallen Wood, e.g. type of contamination and soil pH. However, whilst Hallen Wood soils have pH values comparable to the mid-upper range of the sandy soils, and substantially below the pH of the loamy soils of Scokart *et al.*, the Hallen Wood soils are, in contrast, heavy clays.

A feature which affects the degree to which clear-cut concentration profiles develop is the degree of soil mixing by earthworms (Van Hook *et al.* 1976; Bouma *et al.* 1981). However, in the case of the Hallen Wood soils this is negligible because earthworms of the burrowing type are virtually absent (Hopkin *et al.* 1985) and incorporation and redistribution of organic matter is extremely restricted.

The computer modelling studies (Figs 4, 5, 6, 7) show how the percentage distributions of cadmium and zinc can be expected to change as the 'wave' of metal, observed to be moving down the profile in the 1979 and 1983 data, moves deeper into the soil. This 'wave'-like movement of metals through the profile has a number of possible causes. In the mid 1970s a tip of smelter waste which had been burning internally for some years was moved and this coincided with an input to our experimental site at Hallen Wood of about 90 mg $m^{-2}$ $y^{-1}$ for cadmium and 5·7 g $m^{-2}$ $y^{-1}$ for zinc (Manning 1978). These values are about ten times higher than input to the same site for 1977–78 measured by Coughtrey (1978). However, an input of about 4 g $m^{-2}$ cadmium (forty times that observed) would have been necessary to cause a wave of the magnitude observed. A 'wave' could also be generated by an increased rate of release of metals from litter by, for example, enhanced decomposition. However, lead is observed to be accumulating in the litter layer whilst zinc and cadmium are being lost, and there is no indication that litter decompositon has been enhanced. Also, the rate coefficients for cadmium and zinc loss from litter are relatively unchanged over the period 1975–83, whilst the rate coefficients for the mineral horizons change substantially and support enhanced movement.

We consider that contamination of Hallen Wood soils with heavy metals would have commenced at or soon after the commissioning of the first zinc smelter at Avonmouth in 1928 (Coy 1984). Evidence from nearby woodlands on similar soil parent material but less heavily-contaminated suggests that the soil pH would have been near neutral in the surface mineral layers. Hallen Wood is downwind of a major sulphuric acid plant and close to major nitric acid plants. We believe that acid deposition has been the major cause of acidification of the soil profile (see Table 5) and that progressive acidification to pH values below 6 to substantial soil depths has resulted in the observed movement of cadmium and zinc. Present-day pollution levels in the area are certainly lower than in the early to mid-1970s, both with respect to heavy metal deposition and airborne sulphur dioxide and nitrogen oxides. Coy (1984) reports current $SO_3$ emissions from a 91 m stack of 400 kg $h^{-1}$ from the smelting complex, and this is but one of several sources in the area.

The extent to which the transfer rate coefficients measured over the period 1975–83 will continue to apply cannot be predicted precisely, and the extent of both acidification and metal movement through the deeper (greater than 31 cm) layers of these profiles is uncertain. However, the potential fates of metals in this woodland ecosystem are: (i) to remain associated with surface organic material in the case of lead; (ii) to be precipitated or adsorbed in deeper soil layers of higher pH in the cases of cadmium and zinc until these layers too become acidified; and/or (iii) continued movement through the soil with ultimate contamination of ground-water and entry to other ecosystems or local water sources. The time-course of these movements, and the extent to which they have occurred and may occur in nearby woodland ecosystems, is worthy of further study, because those forms of metals which are most mobile in soil are also likely to be the most bioavailable. The rates at which these forms of metals are removed from the rooting zones of the species of the woodland biomass, and the rate of entry to water courses and hence availability to other systems, are important considerations especially in the case of cadmium.

ACKNOWLEDGMENTS

The authors wish to thank Dr M. C. Thorne for assistance with the mathematical modelling, Mr D. Jackson for commenting on the manuscript, and the Natural Environment Research Council for financial support during part of this investigation.

## REFERENCES

**Adams, T.McM. & Sanders, J.R. (1984).** The effect of pH on the release to solution of zinc, copper and nickel from metal-loaded sewage sludge. *Environmental Pollution Series B,* **8**, 85–99.

**Alesii, B.A. Fuller, W.H. & Boyle, M.V. (1980).** Effect of leachate flow rate on metal migration through soil. *Journal of the Environmental Quality,* **9**, 119–126.

**Alloway, W.H. (1968).** Agronomic controls over the environmental cycling of trace elements. *Advances in Agronomy,* **20**, 235–274.

**Anderson, P.R. & Christensen, T.H. (1983).** Paremeters controlling the distribution of Cd, Co, Ni and Zn in soils. *International Conference on Heavy Metals in the Environment.* Heidelberg, Sept. 1983. Vol. 2, pp. 1187–1190. CEP Consultants, Edinburgh.

**Bouma, J., Belman, C.F.M. & Dekker, L.W. (1982).** Water infiltration and redistribution in a silt loam subsoil with vertical worm channels. *Soil Science Society of America Journal,* **46**, 917–921.

**Bowen, H.J.M. (1975).** Residence times of heavy metals in the environment. *International Conference on Heavy Metals in the Environment.* Toronto 1975, Vol. 1 pp. 1–19. Toronto, Canada.

**Bowen, H.J.M. (1977).** Natural cycles of the elements and their perturbation by man. *Environment and Man,* Vol. 6. *The Chemical Environment* (Ed. by J. Lenihan & W.W. Fletcher), pp. 1–37. Blackie, Glasgow.

**Brummer, G. & Herms, U. (1983).** Influence of soil reaction and organic matter on the solubility of heavy metals in soils. *Effects of Air Pollutants in Forest Ecosystems* (Ed. by B. Ulrich & J. Pankrath), pp. 233–243. Reidel Publishing Co, Dordrecht.

**Buchauer, M.J. (1973).** Contamination of soil and vegetation near a zinc smelter by zinc, cadmium, copper and lead. *Environmental Science & Technology,* **7**, 131–135.

**Chang, F.H. & Broadbent, F.E. (1980).** Effect of nitrification on movement of trace metals in soil columns. *Journal of Environmental Quality,* **9**, 587–592.

**Chang, A.C., Warneke, J.E., Page, A.L. & Lund, L.J. (1984).** Accumulation of heavy metals in sewage sludge-treated soils. *Journal of Environmental Quality,* **13**, 87–91.

**Coughtrey, P.J. (1978).** *Cadmium in terrestrial ecosystems: a case study at Avonmouth, Bristol, England.* Ph.D. thesis, University of Bristol.

**Coughtrey, P.J., Jones, C.H., Martin, M.H. & Shales, S.W. (1979).** Litter accumulation in woodlands contaminated by Pb, Zn, Cd and Cu. *Oecologia (Berlin),* **39**, 51–60.

**Coy, C.M. (1984).** Control of dust and fume at a primary zinc and lead smelter. *Chemistry in Britain,* May 1984, pp. 418–420.

**Davies, B.E. (1980).** Trace element pollution. *Applied Soil Trace Elements, Chapter 9.* (Ed. by B.E. Davies), pp. 287-351. Wiley, New York.

**Davies, B.E. (1983).** Heavy metal contamination from base metal mining and smelting: implications for man and his environment. *Applied Environmental Geochemistry.* Chapter 14. (Ed. by I. Thornton), pp. 425–462. Academic Press, London.

**Denaeyer-de Smet, S. (1974).** Cycle biologique annuel et distribution du plomb dans une pessiere (Piceetum) et une hetraie (Fagetum) etablies sur meme roche-mere. *Bulletin, Sociétér. de botanique de Belgique,* **107**, 115–125.

**Denaeyer-de Smet, S. & Duvigneaud, P. (1974).** Accumulation de metaux lourds toxique dans divers ecosyustemes terrestres pollue par des retombees d'origine industrielle. *Bulletin, Sociétér. de botanique de Belgique,* **107**, 147–156.

**Denny, J.A. (1974).** *A study of some aspects of heavy metal contamination, with special reference to the Avonmouth industrial area of Bristol.* Unpublished B.Sc. thesis, Depts of Geography and Botany, University of Bristol.

**Enk, R.H. van, (1983).** Trends in environmental cadmium concentrations in the European Community. *European Applied Research Report,* **2**, 129–220.

**Fleming, G.A. & Ryan, P. (1964).** Trace element distribution in whole soils and their mineral fractions. *8th International Congress Soil Science, Bucharest, Rumania,* **4**, 297–308.

**Friedland, A.J., Johnson, A.H., Siccama, T.G. & Mader, D.L. (1984).** Trace metal profiles in the forest floor of New England. *Soil Science Society America Journal,* **48**, 422–425.

**Frissel, M.J. & Pennders, R. (1983).** Models for the accumulation and migration of $^{90}$Sr, $^{137}$Cs, $^{239}$, $^{240}$Pu and $^{241}$Am in the upper layer of soil. *Ecological Aspects of Radionuclide Release (*Ed. by P.J. Coughtrey *et al.*) pp. 63–72. Blackwell Scientific Publications, Oxford.

**Frissel, M.J., Poelstra, P. & Van der Klugt, N. (1974).** The contamination of Dutch soils with Hg and a few other heavy metals. *Geologie en Mijnbouw,* **53**, 163–170.

**Greszta, J., Braniewski, S., Marczynska-Galkowska, K. & Nosek, A. (1979).** The effect of dusts emitted by non-ferrous metal smelters on the soil, soil microflora and selected tree species. *Ecologia Polska,* **27**, 397–426.

**Harmsen, K. (1977).** Behaviour of heavy metals in soils. *Agricultural research report (Versl. Landbouwk. Onderz) 866.* Centre for agricultural publishing and documentation, Wageningen.

**Harter, R.D. (1983).** Effect of soil pH on adsorption of lead, copper, zinc and nickel. *Soil Science Society America Journal,* **47**, 47–51.

**Heinrichs, H. & Mayer, R. (1977).** Distribution and cycling of major and trace elements in two central European forest ecosystems. *Journal Environmental Quality,* **6**, 402–407.

**Herms, U. & Brummer, G. (1980).** Einfluss de bodenreaktion auf loslichkeit und tolerierbare gesamtgehalte an nickel, kupfer, zink, cadmium und blei in boden und kompostierten siedlungsabfallen. *Landwirtschaftliche Forschung,* **33**, 408–423.

**Hopkin, S.P. & Martin, M.H. (1982).** The distribution of zinc, cadmium, lead and copper within the woodlouse *Oniscus asellus* (Crustacea, Isopoda). *Oecologia (Berlin),* **54**, 227–232.

**Hopkin, S.P. & Martin, M.H. (1983).** Heavy metals in the centipede *Lithobius variegatus* (Chilopoda). *Environmental Pollution (Series B),* **6**, 309–318.

**Hopkin, S.P., Watson, K., Martin, M.H. & Mould, M.L. (1985).** The assimilation of heavy metals by *Lithobius variegatus* and *Glomeris marginata* (Chilopoda: Diplopoda). *Bijdragen tot der Dierkunde,* **55**, 88–94.

**Hughes, M.K. (1981).** Cycling of trace metals in ecosystems. *Effect of Heavy Metal Pollution on Plants,* Vol. 2. *Metals in the Environment,* Chapter 3 (Ed. by N.W. Lepp), pp. 95–118. Applied Science Publishers, London.

**Hughes, M.K., Lepp, N.W. & Phipps, D.A. (1980).** Aerial heavy metal pollution and terrestrial ecosystems. *Advances in Ecological Research,* **11**, 218–327.

**Hutton, M. (1984).** Impact of airborne metal contamination on a deciduous woodland ecosystem. *SCOPE 22. Effects of Pollutants at the Ecosystem Level,* (Ed. by P.J. Sheehan, D.R. Miller, G.C. Butler & Ph. Bordeau), pp. 365–375. Wiley, New York.

**Jacquez, J.A. (1972).** *Compartmental Analysis in Biology and Medicine.* Elsevier, Amsterdam.

**John, M.K., Chuah, H.H. & Vanlaerhoven, C.J. (1972).** Cadmium contamination of soil and its uptake by oats. *Environmental Science & Technology,* **6**, 555–557.

**Jurinak, J.J. & Santillon-Medrano, J. (1974).** *The chemistry and transport of lead and cadmium in soils.* Research Report 18, Utah Agricultural Experimental Station, Logan Utah.

**Khan, S., Nandau, D. & Khan, N.N. (1982).** The mobility of some heavy metals through Indian red soil. *Environmental Pollution (Series B)* , **119–125.**

**Korte, N.E. Skopp, J. Niebla, E.E. & Fulle, W.H. (1975).** A baseline study on trace metal elution from diverse soil types. *Water, Air and Soil Pollution,* **5**, 149–156.

**Korte, N.E., Skopp, J., Fuller, W.H., Niebla, E.E. & Alesii, B.A. (1976).** Trace element movement in soil : influence of soil physical and chemical properties. *Soil Science,* **122**, 350–359.

**Kreutzer, K., Knorr, A., Brosinger, F. & Kretzschmar, P. (1983).** Scots Pine — dying within the neighbourhood of an industrial area. *Effects of Accumulation of Air Pollutants in Forest Ecosystems.* (Ed. by B. Ulrich & J. Pankrath), pp. 343–357. Reidel Publishing Co., Dordrecht, Holland.

**Levi-Minzi, R., Soldatini, G.F. & Riffaldi, R. (1976).** Cadmium adsorption by soils. *Journal of Soil Science,* **27**, 10–15.

**Manning, J.C.A. (1978).** *Effects of cadmium on plant growth, its relationship with zinc and their cycling in a contaminated woodland.* Ph.D. thesis. University of Bristol.

**Marinsky, J.A. Wolf, A. & Bunzl, K. (1980).** The binding of trace amounts of lead (II), copper (II), cadmium (II) and calcium (II) to soil organic matter. *Talanta,* **27**, 461–468.

**Martin, M.H. & Coughtrey, P.J. (1976).** Comparisons between the levels of lead, zinc and cadmium within a contaminated environment. *Chemosphere,* **5**, 15–20.

**Martin, M.H. & Coughtrey, P.J. (1981).** Impact of heavy metals on ecosystem function and productivity. *The Effect of Heavy Metal Pollution on Plants,* Vol. 2 *Metals in the Environment* (Ed. by N.W. Lepp), pp. 119–158. Applied Science Publishers, London

**Martin, M.H., Duncan, E.M. & Coughtrey, P.J. (1982).** The distribution of heavy metals in a contaminated woodland ecosystem. *Environmental Pollution (Series B),* **3**, 147–157.

**Mayer, R. (1981).** Naturliche und anthropogene komponenten des schwermetallhaushalts von waldokosystemen. *Gottinger Bodenkundliche Berichte,* **70**, 1–152.

**Mayer, R. (1983).** Interaction of forest canopies with atmospheric constituents : aluminium and heavy metals. *Effects of Accumulation of Air Pollutants in Forest Ecosystems* (Ed. by B. Ulrich & J. Pankrath), pp. 47–55. Reidel Publishing Co., Dordrecht, Holland.

**Miller, W.P., McFee, W.W. & Kelly, J.M. (1983).** Mobility and retention of heavy metals in sandy soils. *Journal of Environmental Quality,* **12**, 579–584.

**Parker, G.R., McFee, W.W. & Kelly, J.M. (1978).** Metal distributions in forested ecosystems in urban and rural Northwestern Indiana. *Journal of Environmental Quality,* **7**, 337–342.

**Rutherford, G.K. & Bray, C.R. (1979).** Extent and distribution of soil heavy metal contamination near a nickel smelter at Coniston, Ontario. *Journal of Environmental Quality,* **8**, 219–222.

**Scokart, P.O., Meeus-Verdinne, K. & DeBorger, R. (1983).** Mobility of heavy metals in polluted soils near zinc smelters. *Water, Air, and Soil Pollution,* **20**, 451–463.

**Sharma, R.P. & Shupe, J.L. (1977).** Trace metals in ecosystems : relationships of the residues of copper, molybdenum, selenium and zinc in animal tissue to those in vegetation and soil in the surrounding environment. *Biological Implications of Metals in the Environment.* pp. 595–608. Hanford life sciences symposium 1975. CONF 750929. ERDA symposium series 42.

**Sidle, R.C. & Kardos, L.T. (1977).** Transport of heavy metals in a sludge-treated forested area. *Journal of Environmental Quality,* **6**, 431–437.

**Sidle, R.C. Hook, J.E. & Kardos, L.T. (1977).** Accumulation of heavy metals in soils from extended wastewater irrigation. *Journal of Water Pollution Control Federation,* **49**, 311–318.

**Sidle, R.C., Kardos, L.T. & Genuchten, M.Th. van, (1977).** Heavy metal transport model in a sludge-treated soil. *Journal of Environmental Quality,* **6**, 438–443.

**Smith, W.H. & Siccama, T.G. (1981).** The Hubbard Brook Ecosystem study: biogeochemistry of lead in the northern hardwood forest. *Journal of Environmental Quality,* **10**, 323–333.

**Sopper, W.E. & Kerr, S.N. (1980).** Cadmium in forest ecosystems. *Cadmium in the Environment,* Part I. *Ecological Cycling.* Chapter 17. (Ed. by J.O. Nriagu), pp. 655–667. Wiley, New York.

**Tyler, G. (1970).** Blyets fordelning in ett sydsvenskt barrskogsekosystem. *Grundforbattring,* **23**, 45– 49.

**Tyler, G. (1972).** Heavy metals pollute nature, may reduce productivity. *Ambio,* **1**, 52–59.

**Tyler, G. (1978).** Leaching rates of heavy metal ions in forest soil. *Water, Air and Soil Pollution,* **9**, 137–148.

**Van Hook, R.I., Blaylock, B.G., Bondietti, E.A., Francis, C.W., Huckabee, J.W. Reichle, D.E., Sweeton, F.H. & Witherspoon, J.P. (1976).** Radioisotope techniques to evaluate the environmental behaviour of cadmium. *Environmental Quality and Safety,* Vol. 5. *Global Aspects of Chemistry, Toxicology and Technology as Applied to the Environment.* (Ed. by F. Coulston & F. Korte), pp. 167–182. Theinne, Stuttgart and Academic Press, New York.

**Van Hook, R.I., Harris, W.F. & Henderson, G.S. (1977).** Cadmium, lead and zinc distributions and cycling in a mixed deciduous forest. *Ambio,* **6**, 281–286.

**Whicker, F.W. & Schultz, V. (1982).** *Radioecology: Nuclear Energy and the Environment.* Vol. 2. CRC Press, Florida.

**Williams, D.E., Vlamis, J. & Pukite, A.H. (1980).** Trace element accumulation, movement and distribution in the soil profile from massive applications of sewage sludge. *Soil Science,* **129**, 119–132.

**Zunino, H., Aquilera, M., Caiozzi, M., Peirano, P., Borie, F. & Martin, J.P. (1979).** Metal binding organic macromolecules in soil. 3. Competition of MgII and ZnII for metal binding sites in humic and fulvic type model polymers. *Soil Science,* **128**, 257–266.

# Shorebirds (S.Os Charadrii and Scolopaci) as agents of transfer of heavy metals within and between estuarine ecosystems

P. R. EVANS, J. D. UTTLEY, N. C. DAVIDSON
*Department of Zoology, Science Laboratories, University of Durham, South Road, Durham DH1 3LE*

P. WARD
*Institute of Terrestrial Ecology, Monk's Wood Experimental Station, Abbots Ripton, Huntingdon.*

## SUMMARY

**1** Shorebirds (waders, Sub-Orders Charadrii and Scolopaci) consume large proportions of the annual production of benthic intertidal invertebrates in temperate estuaries. They retain only a very small proportion (much less than 1%) of the quantities of heavy metals they ingest in the soft tissues of their invertebrate foods, even though they digest a high proportion of these tissues and perhaps their accompanying metals.
**2** Regular excretion of metals must occur through urine, and possibly through nasal salt glands and the preen gland. Seasonal excretion occurs through loss of feathers at moult, and through egg laying.
**3** Daily intake of metals by shorebirds should be highest in mid-winter, but amounts stored in liver and kidney rise sharply in late winter. These accumulations are lost by the following winter, except in the case of cadmium, which is found at higher levels in old than young birds of several species. Even when daily rates of storage are at their maximum, the percentage of daily metal intake that is retained is very small.
**4** Considerable proportions of some non-essential metals (but not cadmium) stored during late winter are probably transferred to the breeding plumage during the pre-breeding moult. Metal levels in liver (and kidney) differ amongst dunlin from different British estuaries, broadly in line with known sources of pollution, but differences in amounts stored are small.
**5** Shorebirds act as efficient recyclers of metals from benthic invertebrates to water and sediments within an estuary. Because most species move from their many overwintering sites to a few large European estuaries immediately before they moult into breeding plumage, they transfer small quantities of metals between estuaries. Their new plumage probably reflects metal levels on the wintering sites rather than the moulting site.

## INTRODUCTION

Estuaries are some of the most productive ecosystems in the temperate zones of the world. Amongst the macrofauna, the highest contributions to annual production arise from marine polychaete and oligochaete worms, small crustacea and bivalves that burrow into the soft and often muddy sediments. Harder substrates are usually colonized by mussels *Mytilus edulis* which also have high productivity.

Productivity of intertidal areas is higher than that of adjacent subtidal sites, even though they hold many of the same species. The main predators of the intertidal areas are shorebirds (waders, Charadii and Scolopaci) and some gulls when the areas are exposed; and flatfish, shore crabs *Carcinus maenas* and some ducks, notably shelduck *Tadorna tadorna,* as the areas become submerged. The invertebrates of subtidal sites escape predation by gulls and waders but are vulnerable to diving ducks, notably eider *Somateria mollissima.*

The aims of this review are to show that shorebirds are responsible for rapid rates of transfer of heavy metals, and a high proportion of the amounts transferred, from benthic intertidal invertebrates back to estuarine waters and sediments; and that seasonal and geographical differences in the quantities of metals stored in the birds' bodies have little effect on the transfer rates. Data to establish certain conclusions are fragmentary, but may soon be amplified by laboratory studies planned to commence in autumn 1986.

## SHOREBIRDS AS AGENTS OF TRANSFER OF HEAVY METALS

The proportion of the annual production of benthic invertebrates taken by predators is high (reviewed by Baird *et al.* 1985). In the Dutch Wadden Sea, more than half is taken in total, and 17% by shorebirds. On the Tees estuary in north-east England, where productivity is even higher, a similar total proportion is taken by predators, but almost all by shorebirds (Table 1), since few fish and crustacea appear able to tolerate the metal levels

TABLE 1. Production and consumption of intertidal macrobenthos (expressed as kJ $m^{-2} y^{-1}$). Adapted from Baird *et al.* (1985)

| | Dutch Wadden Sea | Seal Sands, Teesmouth |
|---|---|---|
| Production | 619 | 851 |
| Average standing crop | 556 | 200 |
| Consumption by: | | |
| shorebirds | 104 (17%) | 367 (44%) |
| fish | 105 | – |
| crabs | 62 | trace |
| shellfisheries | 42 | – |
| Total | 313 (51%) | 367+ (44+%) |

present in the waters of the estuary. Many estuaries are moderately enriched with organic materials such as sewage, and their invertebrate productivities may be higher than those of unpolluted sites (Gray 1976). Although the variety of benthic fauna in such enriched sites may be restricted, their high productivity (and consequent high densities of prey) makes them particularly attractive to shorebirds. If, in addition, the estuarine sediments and waters contain relatively high concentrations of heavy metals, then not only are the benthic invertebrates likely to contain high levels of metals, but their aquatic predators are likely to be scarce. In these circumstances, shorebirds are responsible for the majority of heavy metal transfer from intertidal invertebrates to other parts of the ecosystem.

It is possible to demonstrate that shorebirds retain only a very small proportion of the metal loads that they ingest in their food. As shown by Evans *et al.* (1979), a curlew *Numenius arquata* requires about 1200 polychaete worms of average dry mass 50 mg to satisfy its daily energy needs. From measurements of metal concentrations in purged ragworms *Nereis diversicolor* from Teesmouth, Evans & Moon (1981) calculated that such a bird would ingest (i.e. take in through its mouth, into its gut) about 12 mg zinc and 0·5 mg lead each day, and thus about 3 g zinc and 0·12 g lead during the period, August–March, which most curlews spend at Teesmouth. If none was excreted and each metal was evenly distributed throughout the body of an 850g bird, this would lead to an average concentration increase, during one winter, of 3·5 g $kg^{-1}$ of zinc and 140 mg $kg^{-1}$ of lead. Even if only 10% of the ingested metals were absorbed from the gut, and the remainder voided in undigested form, the average concentration increase during the winter should have been 350 mg $kg^{-1}$ of zinc and 14 mg $kg^{-1}$ of lead. Yet, in two adult curlews caught in March, the mean concentrations in the kidneys (which hold higher levels than the average for the whole body) were only 18 mg $kg^{-1}$ wet weight for zinc and 1·2 mg $kg^{-1}$ for lead. These storage levels had been acquired during several winters spent at Teesmouth, so that the total quantities of metals the birds had ingested during their lives were several times greater than the amounts eaten in a single winter and used for the calculation given above. This demonstrates that curlew retain much less than 1% of the metals they ingest in their food. Similar calculations can be made for other bird species.

Curlew, and most other shorebirds, digest most of the organic materials in each worm they ingest, except for chitinous parts such as jaw mandibles. Thus, it is probable that if metals are absorbed alongside the organic components of their food, the low retention of e.g. zinc and lead is a result chiefly of efficient excretion of digested materials, rather than of voiding in faeces and/or regurgitated pellets of undigested parts of their prey (though it is accepted that, e.g. jaws or exoskeletons of benthic invertebrates may contain metal levels higher than average for the whole prey animal). The possibility remains that any metal granules present in the soft parts of the prey may not, in fact, be able to cross the bird's gut wall and so will be voided in faeces. This is not easy to detect directly, since both faeces and urine are excreted from the bird's cloaca.

Although excretion via the kidney in urine is thought to be the major route of continuous loss of absorbed metals from a bird's body, other routes are available at certain times of year, notably during moult and egg laying. In most shorebirds, moult occurs twice each year: in late summer after breeding and in early spring before breeding. However, some metals appear not to be excreted in significant amounts in growing feathers, notably cadmium. For year-round excretion of metals, another possible route is via the preen gland, but this seems unimportant (Pannekoek, Kelsall & Burton 1974; Goede & de Bruin 1984). Birds do not have sweat glands. However, sea and shore birds have supra-orbital nasal (salt) glands which accumulate heavy metals such as lead and cadmium (Howarth,

Hulbeck & Horning 1981). It is not known whether they excrete these metals along with the concentrated salt solutions that they produce.

## SEASONAL CHANGES IN INTAKE AND STORAGE

Daily intake rates of heavy metals by shorebirds vary from month to month, for two reasons: first because their daily energy requirements vary seasonally and second because the metal concentrations in many intertidal invertebrates also vary seasonally.

Most shorebirds arrive on European estuaries in August, after breeding in arctic or sub-arctic regions. They stay in their non-breeding areas until April or May. Whilst shorebirds are on their wintering grounds, regulated changes take place in the level of fat reserves in their bodies (Dugan *et al.* 1981; Davidson 1981). These are related to seasonal changes in the likelihood of severe weather, which causes temporary reductions in the availability of their prey; these can be compensated for if birds draw upon their short-term fat reserves. In order to accumulate fat, birds in early autumn must eat more each day than they require to meet their energy demands. Conversely, in spring, they eat less each day than is needed to provide energy balance, making up the deficit as they draw upon and run down the fat reserves that are no longer needed. When temperatures are coldest and windspeeds highest, in mid-winter, convective heat loss is greatest and shorebirds' daily energy requirements are presumed to be at their highest. This is difficult to establish in the field, as most shorebirds feed by night as well as by day, and measurements of food intake rates at night are hard to quantify. Nevertheless, such a presumption has been confirmed for dunlin *Calidris alpina* by Worrall (see Chapter 2 in Evans, Goss-Custard & Hale 1984).

Once water temperatures fall in autumn, many estuarine invertebrates not only cease growth, but their mass (for a given size) slowly falls during the winter as they use up tissue glycogen and other materials (e.g. Phillips 1980). However, the amounts of metals carried by individuals do not fall appreciably, so that metal concentrations in their tissues rise. From the viewpoint of the predators, to gain a given biomass of prey of a particular size-class, shorebirds must eat correspondingly more individuals each day as the winter progresses, and hence ingest correspondingly larger quantities of metals. The combined effect of the two processes - seasonal changes in daily food requirements and progressively lighter prey of a given size - are illustrated in Fig. 1. This suggests that the daily rate of metal intake by shorebirds should rise sharply through the autumn to a probable mid-winter peak, before falling somewhat towards early spring. Is this matched by mid-winter peaks in metal loads and concentrations in shorebird tissues, or do excretion rates rise in compensation?

To examine this, seasonal changes in levels of heavy metals in dunlin *Calidris alpina* at the Tees estuary were studied by collecting samples of birds at monthly intervals during the winter of 1977–78 under licence from the Nature Conservancy Council. Metal levels in wet tissues were determined at the Institute of Terrestrial Ecology, Monk's Wood, by atomic absorption spectrometry (AAS) of diluted digests after oxidation with concentrated nitric acid. Mercury levels were measured by the Hatch & Ott technique (Ward 1979). Unfortunately, birds were held for some hours after capture, which is now known to result in losses in liver mass, though not to appreciable losses in kidney mass. Because metals are

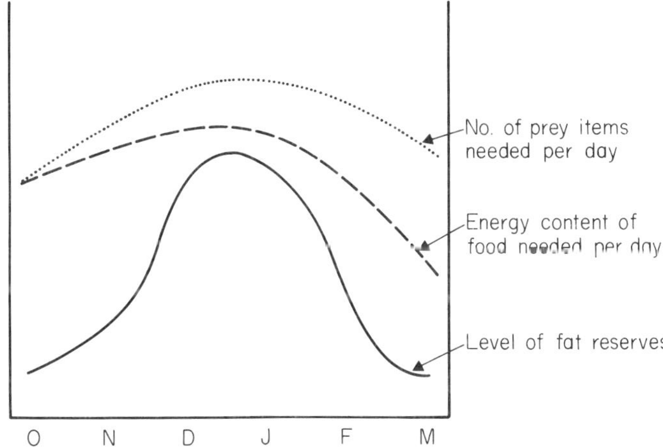


FIG. 1. Model of seasonal changes in daily food requirements of shorebirds, assuming no change in the size of items taken.

retained in the liver in the short-term, temporarily starved birds often show remarkably high *concentrations* of metals in their livers (but not correspondingly high kidney levels), even though the *loads* of metals in the livers are not elevated (Evans & Moon 1981). For this reason, results are shown here both as loads and as concentrations in each of the two tissues, liver and kidney. Median values are presented to avoid bias arising from the use of means as measures of central tendency in non-normal distributions. Data from adults and juveniles are presented separately only if this is justified statistically.

The results (Figs 2–7) show that, far from rising sharply in autumn, the concentrations and loads of mercury in dunlin livers actually fell significantly, and of zinc and cadmium showed little change, until December, after which all three metal levels increased considerably during the next 2 months. Although some dunlin are known to arrive in the Tees estuary from elsewhere, notably the Wash, in October and November, the wintering population remains unchanged throughout the rest of the winter, so the increases in metal levels in the samples should genuinely represent changes occurring in the whole population.

Knot arrive on the Tees from the Wash and the Dutch Wadden Sea in November (Dugan 1981). Though there is movement through the estuary in that month, the composition of the population from December onwards is thought to change little. Hence, once again, the differences between metal levels in the samples are believed to indicate real increases in levels of zinc, mercury and to a lesser extent cadmium, in all individuals. The differences in rates of increase of, e.g. mercury levels between dunlin and knot are presumed to result, at least in part, from differences in diet.

The conclusion from these data is that the seasonal changes in levels of metals stored in shorebird tissues do not reflect the predicted seasonal changes in intake rates of the different metals. Hence, there are seasonal changes in excretion rates.

Figs 2–7 illustrate two other points. The first is that cadmium is the only metal known to be present in significantly higher concentrations in adults than juveniles. This has been

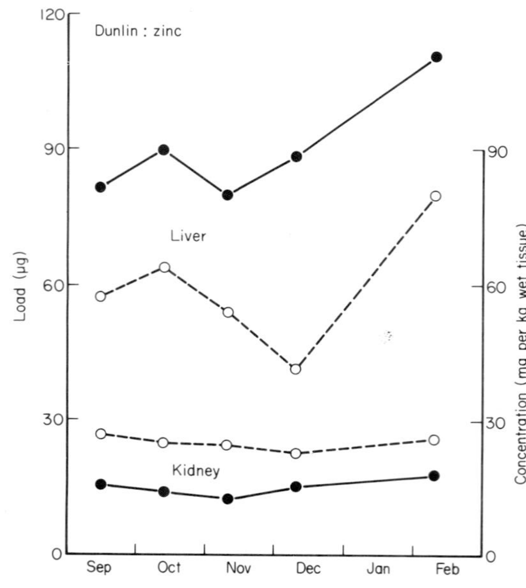


FIG. 2. Seasonal changes in wet tissue loads (●) and concentrations (○) of zinc in dunlin *Calidris alpina* from Teesmouth, north-east England.

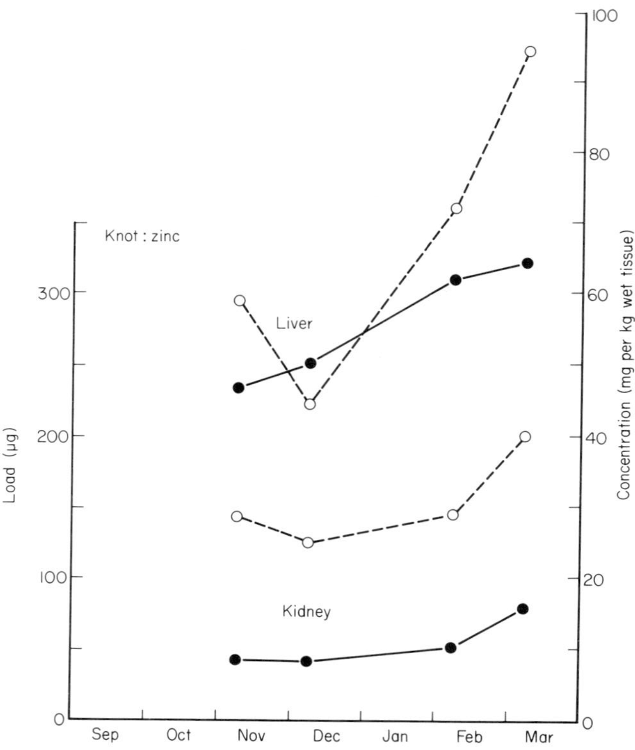


FIG. 3. Seasonal changes in wet tissue loads and concentrations of zinc in knot *Calidris canutus* from Teesmouth, north-east England. Key as in Fig. 2.

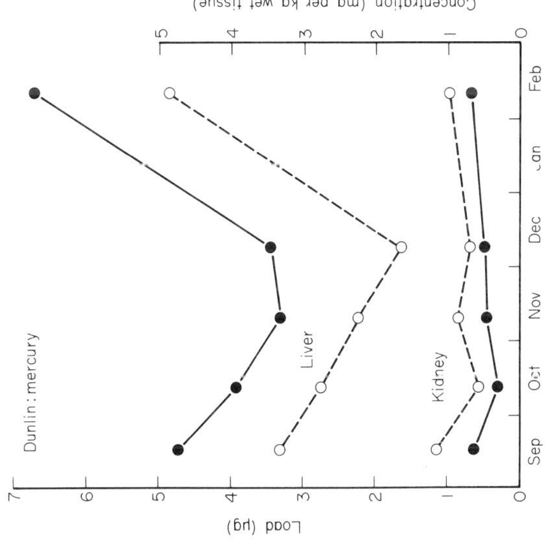


FIG. 4. Seasonal changes in wet tissue loads and concentrations of mercury in dunlin *Calidris alpina*. Key as in Fig. 2.

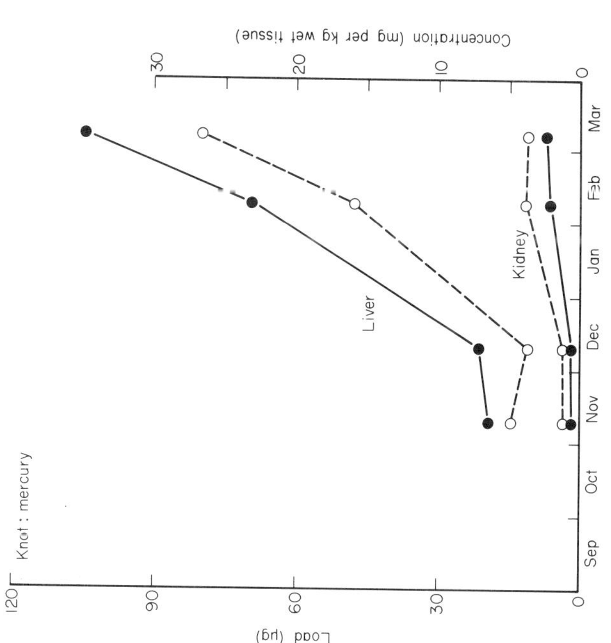


FIG. 5. Seasonal changes in wet tissue loads and concentrations of mercury in knot *Calidris canutus*. Key as in Fig. 2.

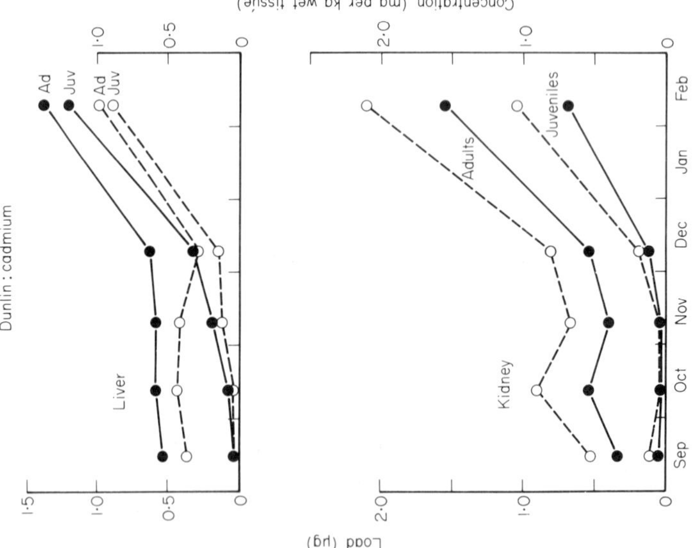


FIG. 6. Seasonal changes in wet tissue loads and concentrations of cadmium in juvenile and adult dunlin *Calidris alpina* from Teesmouth, north-east England. Key as in Fig. 2.

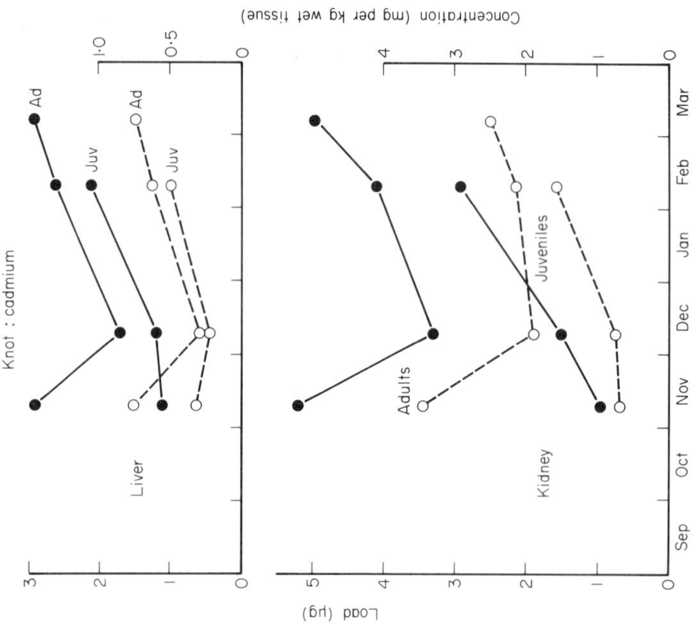


FIG. 7. Seasonal changes in wet tissue loads and concentrations of cadmium in juvenile and adult knot *Calidris canutus* from Teesmouth, north-east England. Key as in Fig. 2.

found also in oystercatchers *Haematopus ostralegus* (Hutton 1981), and in bar-tailed godwits *Limosa lapponica* (Evans & Moon 1981). The second is that a given metal is stored in different amounts and concentrations in different tissues, and that the tissue containing the highest concentration of one metal may not contain the highest of another. This is illustrated further in Table 2 which shows the relative distributions of four heavy metals in six parts of dunlins collected from the Mersey estuary in late summer. Such an order of 'affinity' of particular metals for particular tissues may vary according to the chemical form in which the metal is ingested (in particular, whether inorganic or organic) and according to age and season, depending on the periods of the animal's lifetime (or of the year) when metals can enter the different tissues. For example, metals can enter feathers from the body of the bird only once or twice a year, when they are growing. Thereafter, changes in metal levels in feathers are the result of leaching or external contamination (Goede & de Bruin 1984). At the other extreme, metals can enter (and leave) the liver over periods of days, if not hours, so that liver concentrations are much more labile.

Thus, the parallel increase in storage levels of metals in livers and kidneys of dunlin and knot in late winter do not necessarily reflect increased levels in other tissues. In particular, there may not be major increases of metals such as zinc stored in brain (Ward 1979) and pectoral muscle tissue (P.R. Evans, unpublished data).

TABLE 2. 'Affinities' of four heavy metals for different tissues of dunlin *Calidris alpina* from the R. Mersey*

Decreasing concentrations →

|  | | | | | | |
|---|---|---|---|---|---|---|
| Cadmium | Bone | Bill | Kidney | Feet | Liver | Pectoral muscle |
| Copper | Bone | Pectoral muscle | Liver | Kidney | Bill | Feet |
| Lead | Bill | Bone | Kidney | Liver | Feet | Pectoral muscle |
| Zinc | Bill | Bone | Feet | Liver | Kidney | Pectoral muscle |

*Calculated from the data of Anderson (1982).

## SEASONAL CHANGES IN METAL LOADS OF TISSUES

From measured rates of change of metal loads in liver and kidney of dunlin and knot it is possible to calculate daily storage rates for the two organs combined (Table 3). These may be compared, for zinc, with the estimated daily intake rate of this metal in the main foods taken by the two species.

Dunlin at Teesmouth feed chiefly upon small oligochaetes and polychaetes (Evans *et al.* 1979). Their daily food requirement totals about 140 kJ, i.e. about 8 g dry weight of worms. Small polychaetes such as *Nereis diversicolor* contain some 250 mg $kg^{-1}$ of zinc, so dunlin

TABLE 3. Maximum daily storage rates of heavy metals in liver and kidney of shorebirds at Teesmouth

| Metal | Shorebird | Storage rate µg | over days | Average daily storage (µg) |
|---|---|---|---|---|
| Zinc | Dunlin | 26 | 58 | 0·5 |
| | Knot | 88 | 30 | 3·0 |
| Mercury | Dunlin | 3·4 | 58 | 0·06 |
| | Knot | 36 | 30 | 1·2 |
| Cadmium | Dunlin (Adult) | 1·75 | 58 | 0·03 |
| | Knot (Juv.) | 1·5 | 58 | 0·025 |

must ingest about 2 mg of zinc each day. Of this, they store each day only 0·5 µg (0·025%) in kidney and liver. Hence, the role of the dunlin as a 'zinc transfer machine' – from benthic invertebrate tissues back to water and sediment – is hardly interrupted by the seasonal storage pattern of zinc in the birds' tissues, even if some is stored in tissues other than liver and kidney.

A similar picture emerges for knot, which feed chiefly on molluscs, notably mussels *Mytilus edulis* at Teesmouth (Evans *et al.* 1979; Dugan 1981). To obtain their daily energy requirements, they eat an average of 20 g dry weight of soft tissue per day, implying a daily intake of about 8 mg of zinc. Of this, only 3 µg (0·38%) is added to the kidney and liver load.

Unfortunately, no measurements are available of the mercury concentrations in these invertebrates during the winter in which the birds were studied. However, to achieve the same percentage storage rate the mercury concentrations in small polychaetes would need to have been about 35 mg $kg^{-1}$, and in mussels about 150 mg $kg^{-1}$ dry weight. These are probably unrealistically high. If so, then a slightly higher percentage of the mercury ingested in the diet of both birds is stored than of zinc.

Even if only 1% of the metal load ingested is actually digested (absorbed), the proportion stored is still extremely low, even at its maximum (i.e. 2·5% and 4% of zinc in dunlin and knot, respectively).

## GEOGRAPHICAL VARIATION IN STORAGE OF METALS IN SHOREBIRDS ON BRITISH ESTUARIES

The seasonal pattern of storage of metals in metabolically active tissues does not accord with what would have been expected on the basis of the seasonal pattern of daily quantities of metals ingested. Also, because such a high proportion is excreted, the possibility arises that changes in metal levels in sediments and benthic fauna are not reflected in changes in levels stored in shorebird tissues.

Levels of zinc, copper, cadmium and lead have been measured in dunlin from four areas: the southern side of the Severn estuary 15 miles downstream from Avonmouth; the Menai Straits (one sample from the Anglesey side at Beaumaris, another from the mainland side near Bangor); the Wash, at Snettisham (Norfolk); and the Firth of Forth, near Musselburgh. Dunlin were chosen because adults show great site fidelity during a winter and from year to year. Analyses of dry tissues were carried out at Durham University. Livers and kidneys were excised, dried in vacuum at 60 °C and completely digested with concentrated nitric acid. The digests were evaporated to dryness and redissolved in 3M HCl before metals were estimated using AAS. Results are shown in Figs 8 and 9, which are not directly comparable with Figs 2 to 7 because of the different method of analysis.

In view of the seasonal changes in metal levels found in dunlin at Teesmouth, the most appropriate comparisons are between birds from the Severn (sampled on 22 February 1981) and those from Bangor (sampled on 21 February 1981). Birds from Beaumaris (27 December 1978), the Wash (2 January 1983) and the Forth (15 January 1980) are also comparable on a seasonal basis, though sampled in different years. The birds from the two sides of the Menai Straits form a single wintering population which shifts roosting site from time to time. Hence, one might have expected higher levels of metals in the Bangor birds, sampled later in the winter, unless this affect is overridden by year-to-year changes.

Dunlin from the Severn contained significantly higher concentrations of cadmium and copper in their livers and kidneys than did those from Bangor, but they had significantly

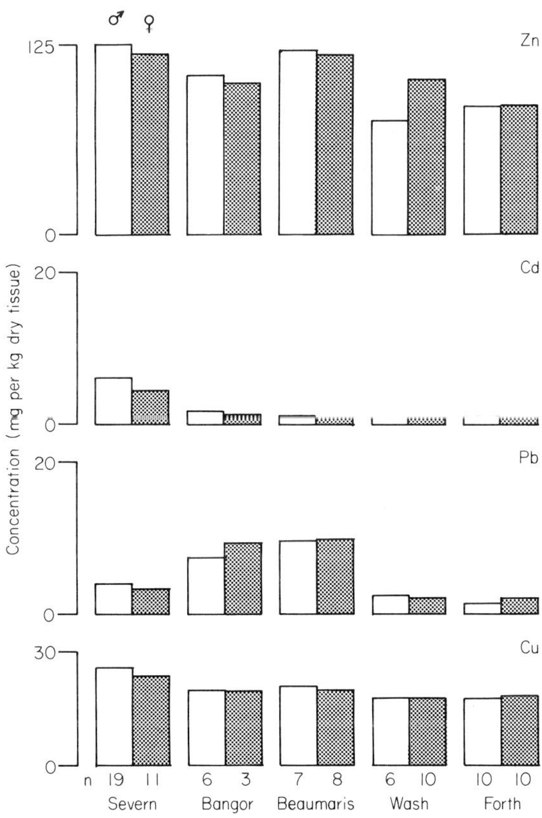


FIG. 8. Concentrations of four heavy metals in dry livers of dunlin *Calidris alpina* from four British estuaries.

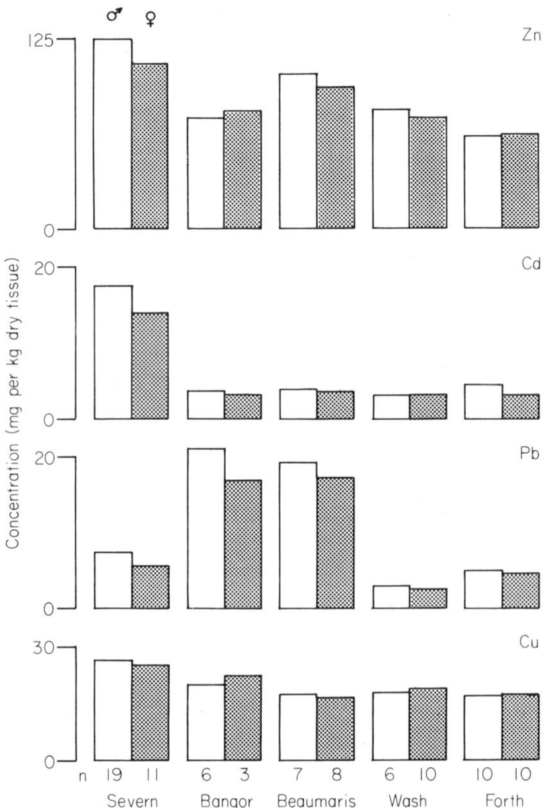


FIG. 9. Concentrations of four heavy metals in dry kidneys of dunlin *Calidris alpina* from four British estuaries.

smaller tissue masses, so that, although the tissue loads of cadmium were also significantly higher in Severn than Bangor birds, the tissue loads of copper were similar. Conversely, both concentrations and, more particularly, loads of lead were significantly higher in Bangor birds (Table 4).

Metal concentrations and loads in both livers and kidneys of birds from Bangor did not differ significantly from those in birds from Beaumaris, except for levels of copper in the kidney, which were significantly higher in the later winter sample (Bangor). Both samples showed higher variability amongst individuals in levels of lead, in both liver and kidney, than did samples from the other three estuaries.

Comparisons amongst birds from Beaumaris, the Wash and the Firth of Forth (Table 5) revealed significantly higher concentrations of zinc in the kidneys and livers of those from Beaumaris, but when their smaller size was taken into account (by calculating loads of zinc in these tissues) differences between sites became insignificant. Cadmium concentrations and loads were slightly, and in most cases significantly, higher in dunlin from the Forth. The North Wales birds had much higher levels of lead than those from the other two estuaries, but slightly, though significantly, lower loads of copper in both livers and kidneys than those from the Wash and Forth.

TABLE 4. Levels of significance of differences in metal loads and concentrations in livers and kidneys of dunlin from the Severn and Bangor, North Wales. The estuary named has the higher level of the relevant metal

| | Liver concentration | Liver load | Kidney concentration | Kidney load |
|---|---|---|---|---|
| *Males* | | | | |
| Zinc | N.S. | N.S. | Severn*** | Severn* |
| Cadmium | Severn*** | Severn*** | Severn*** | Severn*** |
| Lead | Bangor* | Bangor** | Bangor*** | Bangor*** |
| Copper | Severn*** | N.S. | Severn*** | N.S. |
| | | | | |
| *Females* | | | | |
| Zinc | N.S. | N.S. | N.S. | N.S. |
| Cadmium | Severn*** | Severn*** | Severn*** | Severn*** |
| Lead | Bangor* | Bangor* | Bangor* | Bangor* |
| Copper | Severn* | N.S. | Severn*** | N.S. |

Significance values (Mann – Whitney *U*-test):
N.S. = $P > 0{\cdot}05$,
* = $0{\cdot}05 > P > 0{\cdot}01$,
** = $0{\cdot}01 > P > 0{\cdot}002$,
*** = $P < 0{\cdot}002$.

These data indicate that levels of cadmium in the Severn sediments and biota, downstream from the Avonmouth smelter, are sufficiently high to cause elevated levels of cadmium in shorebirds feeding there, whilst levels of lead in the fauna of the Menai Straits are presumably also higher than normal (but perhaps as a result of natural processes acting on the metal-bearing rocks of North Wales), and cause elevated levels of lead in the birds. As we have no measurements of metal levels in the benthic invertebrates that form the foods of dunlin in these two areas or elsewhere, we cannot examine whether the ratio of metal levels in birds to levels in their foods are similar in all estuaries and for all metals. Nevertheless, the conclusion reached earlier, that almost all the amounts of each metal ingested as food are promptly excreted, remains unaffected.

## EXCRETION OF METALS INTO FEATHERS

Stickel *et al.* (1977) found that most of the total body burden of mercury (administered as methyl mercury) in ducks was eliminated into feathers at the moult, and Osborn (1979) noted that mercury was absent from the livers of starlings *Sturnus vulgaris* during moult. This raises the possibility that the seasonal accumulation of certain metals in livers and kidneys of shorebirds in late winter could be excreted into growing feathers when the birds moult into breeding plumage. Feathers form an appreciable proportion of the dry weight of a bird: some 10–15% depending on species. At least two-thirds of this refers to body feathers, which in most shorebirds are moulted twice each year, the remainder comprising flight feathers (primary, secondary, tertial and tail feathers) which are moulted annually in late summer. The use of feathers to monitor metal pollution has been reviewed by Goede & de Bruin (1984), who give some of the limited data available on levels of metals in shorebird

TABLE 5. Levels of significance of differences in metal loads and concentrations in livers and kidneys of dunlin from Beaumaris, the Wash and the Firth of Forth

| *Zinc* | Liver concentration | | Liver load | |
|---|---|---|---|---|
| | Beaumaris | Wash | Beaumaris | Wash |
| Wash | ** M | – | | – |
| Forth | *** M<br>** F | | | |
| | Kidney concentration | | Kidney load | |
| Wash | ** M<br>* F | – | | – |
| Forth | ** M | * M | ** M<br>* F | ** M<br>* F |

| *Cadmium* | Liver concentration | | Liver load | |
|---|---|---|---|---|
| | Forth | Beaumaris | Forth | Beaumaris |
| Beaumaris | | – | * M<br>** F | – |
| Wash | ** F | * F | | |
| | Kidney concentration | | Kidney load | |
| Beaumaris | * M | – | * | – |
| Wash | *** M | *M | ** M | |

| *Lead* | Liver concentration | | Liver load | |
|---|---|---|---|---|
| | Beaumaris | Forth | Beaumaris | Forth |
| Forth | *** M<br>*** F | – | *** M<br>*** F | – |
| Wash | *** M<br>*** F | | *** M<br>*** F | |
| | Kidney concentration | | Kidney load | |
| Forth | *** M<br>*** F | – | *** M<br>*** F | – |
| Wash | *** M<br>*** F | * M<br>* F | *** M<br>*** F | |

| *Copper* | Liver concentration | | Liver load | |
|---|---|---|---|---|
| | Wash | Forth | Wash | Forth |
| Forth | | – | | – |
| Beaumaris | | | *** M<br>*** F | *** M<br>*** F |
| | Kidney concentration | | Kidney load | |
| Forth | * F | – | * M<br>** F | – |
| Beaumaris | | | ** M<br>*** F | |

Significance levels (Mann – Whitney *U*-test): * $> 0{\cdot}05 > P = 0{\cdot}01$, ** $0{\cdot}01 > P > 0{\cdot}002$, *** $P < 0{\cdot}002$. Blanks in the table indicate $P > 0{\cdot}05$. M = male, F = female.

feathers. Unfortunately, these refer only to flight feathers in which metal concentrations differ in the vane and the shaft. Nevertheless, they allow order-of-magnitude calculations for body feathers, on the assumption that these, by weight, contain chiefly vane rather than shaft.

The body plumage of knots has a dry mass of about 3 g. Such feathers contain about 300 mg kg$^{-1}$ of zinc and perhaps 8 mg kg$^{-1}$ of mercury dry weight (Fig. 1 in Goede & de Bruin 1984). Thus, production of breeding plumage would lead to uptake of 900 $\mu$g of zinc and 24 $\mu$g mercury into the growing feathers. These quantities may be compared with the increases in kidney + liver loads of 88 $\mu$g zinc and 36 $\mu$g mercury, respectively, between February and March in knots at Teesmouth (Figs 3 & 5). It is clear that an appreciable proportion of the seasonally-accumulated mercury could be eliminated into the feathers, but that additional sources of zinc are required during moult. It is not known to what extent other non-essential heavy metals, besides mercury, are eliminated into feathers, but measurable concentrations of lead were recorded in flight feathers of almost all individual knot examined, and arsenic and cadmium in a few (Goede & de Bruin 1984).

## TRANSPORT OF METALS AMONGST ESTUARIES

Many of the shorebirds that over-winter on small British estuaries, such as the Tees, move to a few large European intertidal areas, such as the Wash (Norfolk/Lincolnshire) and the Dutch Wadden Sea (Boere 1976) in early spring, before moulting into breeding plumage. In these sites they shed the body feathers acquired at the time of their complete moult in the previous late summer and acquire new body feathers which, as shown earlier, may contain an appreciable proportion of the mercury (if not also of the lead and other non-essential metals) stored at the end of their stay on overwintering sites. The breeding plumage is retained until late summer, when it is shed, along with all the flight feathers, and replaced by winter plumage. The metal content of winter plumage, particularly of the earliest groups of feathers to be replaced, may well reflect the levels of non-essential metals acquired on the breeding grounds just before southward migration, rather than the levels present on the moulting grounds themselves, unless birds delay moult for some weeks after arrival there.

The major late-summer moulting grounds of shorebirds in Western Europe are the Dutch Wadden Sea, the west coast of Schleswig-Holstein, the Wash, the Ribble (Lancs) and Morecambe Bay (Lancs/Cumbria). These sites thus receive metal loads acquired on a variety of estuaries the previous winter and stored in feathers throughout the breeding season. For example, if 100 000 birds each shed, on average, 5 g of body feathers containing 8 mg kg$^{-1}$ of mercury on the Ribble estuary, this would represent an input of 4 g mercury, spread along the high water mark as feathers. If these were decomposed and dispersed into the top 1 mm of sediment over a total area of 10 ha, it would represent an increase in sediment levels of about 0·01 mg kg$^{-1}$ mercury. Natural levels in estuarine sediments lie in the range 0·1–0·4 mg kg$^{-1}$.

Thus, although shorebirds do transfer heavy metals from polluted to less polluted estuaries, the scale of transfer is relatively small and not nearly as important as their role in daily turnover of metals within an estuary.

## REFERENCES

**Anderson, J.L. (1982).** *Some heavy metal studies in an estuarine food chain.* Ph.D. thesis, University of Wales.

**Baird, D., Evans, P.R. Milne, H. & Pienkowski, M.W. (1985).** Utilization by shorebirds of benthic invertebrate production in intertidal areas. *Oceanography & Marine Biology Annual Review,* **23**,573–597.

**Boere, G.C. (1976).** The significance of the Dutch Wadden Sea in the annual life cycle of arctic, subarctic and boreal waders. *Ardea,* **64**,210–291.

**Davidson, N.C. (1981).** *Seasonal changes in nutritional condition of shorebirds during the non-breeding seasons.* Ph.D. thesis, University of Durham.

**Dugan, P.J. (1981).** *Seasonal movements of shorebirds in relation to spacing behaviour and prey availability.* Ph. D. thesis, University of Durham.

**Dugan, P.J., Evans, P.R., Goodyer, L.R. & Davidson, N.C. (1981).** Winter fat reserves in shorebirds: disturbance of regulated levels by severe weather conditions. *Ibis,* **123**,359–363.

**Evans, P.R. Goss-Custard, J.D. & Hale, W.G.** (Eds) **(1984).** *Coastal Waders and Wildfowl in Winter.* Cambridge University Press, Cambridge.

**Evans, P.R., Herdson, D.M., Knights, P.J. & Pienkowski, M.W. (1979).** Short-term effects of reclamation of part of Seal Sands, Teesmouth on wintering waders and Shelduck. *Oecologia (Berlin).* **41**, 183–306.

**Evans, P.R. & Moon, S.J. (1981).** Heavy metals in shorebirds and their prey in north-east England. *Heavy Metals in Northern England: Environmental and Biological Aspects* (Ed. by P.J. Say & B.A. Whitton), pp. 181–190. University of Durham.

**Goede, A.A. & de Bruin, M. (1984).** The use of bird feather parts as a monitor for metal pollution. *Environmental Pollution* (Series B), **8**, 281–298.

**Gray, J.S. (1976).** The fauna of the polluted River Tees. *Estuarine & Coastal Marine Science,* **4**, 653–676.

**Howarth, D.M. Hulbert, A.J. & Horning, D. (1981).** A comparative study of heavy metal accumulation in tissues of the Crested Tern *Sterna bergii* breeding near industrialized and non-industrialized areas. *Australian Wildlife Research.* **8**, 665–672.

**Hutton, M. (1981).** Accumulation of heavy metals and selenium in three seabird species from the UK. *Environmental Pollution (Series A),* **26**, 129–145.

**Osborn, D. (1979).** Seasonal changes in the fat, protein and metal contant of the liver of the Starling *(Sturnus vulgaris). Environmental Pollution,* **19**, 144–154.

**Pannekoek, W.J., Kelsall, J.P. & Burton, R. (1974).** Methods of analyzing feathers for elemental content. *Environment Canada Fish & Marine Service Technical Report,* **498**, 1–16.

**Phillips, D.J.H. (1980).** *Quantitative Aquatic Biological Indicators.* Applied Science Publishers, London.

**Stickel, L.F., Stickel, W.H., McLane, M.A.R. & Bruns, M. (1977).** Prolonged retention of methyl mercury by Mallard drakes. *Bulletin Environmental Contamination & Toxicology,* **18**, 393–400.

**Ward, P. (1979).** *Heavy metals in waders overwintering in a polluted estuary.* Report to the Nature Conservancy Council by Institute of Terrestrial Ecology (ITE Project No. 179).

# Pathways of radionuclides from soils to wheat

H. A. GROGAN[‡], N. G. MITCHELL*

M. J. MINSKI[†] AND J. N. B. BELL

*Department of Pure and Applied Biology and [†]The Reactor Centre, Imperial College at Silwood Park, Ascot, Berkshire SL5 7PY

## SUMMARY

**1** Soil–wheat transfer factors for a range of radionuclides were measured over a time-course between sowing and harvesting. The programme consisted of two parts: an initial set of experiments where the wheat was grown in large pots outdoors and the second in which lysimeters were used to simulate agricultural conditions more closely. Rates of uptake of the different radionuclides studied varied in the order $^{99}Tc > ^{137}Cs > ^{106}Ru > ^{144}Ce/^{125}Sb$.

**2** Results obtained are expressed as transfer factors (Bq $g^{-1}$ plant dry weight per Bq $g^{-1}$ air-dried soil at sowing). Those for the large pots are comparable with the results obtained using lysimeters. Transfer factors for $^{137}Cs$ are comparable with those of other European workers. In the case of $^{144}Ce$ and $^{106}Ru$ the transfer factors obtained are the first for wheat under European conditions; however, the values compare favourably with those given for North American conditions.

**3** The radionuclides showed considerable difference in their distribution within the plant, with concentration occurring in the straw, rather than the ears. This was most marked for spring wheat.

**4** Harvests carried out over a time-course indicate distinct trends in levels of activity for the straw and ears but there are major differences between the radionuclides. $^{137}Cs$ content decreased progressively in straw and with $^{106}Ru$, showed a decrease in transfer factors between 1981 and 1982. $^{99}Tc$ had an extremely high transfer factor at the first harvest reflecting its high level of availability.

**5** The results show that soil type influences the transfer factors. Sandy loam resulted in the highest uptake of $^{137}Cs$ and $^{106}Ru$ from the pots. Loam and silty clay produced highest uptakes for the lysimeters. A comparison of washed with unwashed samples showed that resuspension plays a major role in the contamination of the straw, but height of the plant organ above ground is a critical factor.

## INTRODUCTION

The responsibility for controlling operational releases of radioactive materials into the environment and for formulating emergency responses after an accidental release, rests with the national authorities in the UK. These are undertaken in accordance with the

*Present address: Associated Nuclear Services, Eastleigh House, 60 East Street, Epsom, Surrey KT17 1HA.
‡Present address: Eidgenössische Institut für Reaktorforschung. Würenlingen, CH-5303, Switzerland.

recommendations of the International Commission on Radiological Protection (ICRP), as interpreted by the National Radiological Protection Board. Currently, ICRP recommendations (ICRP 1977) impose limits on doses incurred by both radiation protection workers and members of the general public, the overall aim being to keep exposures as low as reasonably achievable, economic and social factors being taken into account (ALARA principle). Additionally, there is a requirement to evaluate the population detriment. This approach represents an advance on the earlier one recommended by the ICRP (ICRP 1959), where dose limitation was achieved solely by considering the critical nuclide, critical pathway, and critical organ concerned. The current approach thus requires doses to be estimated, considering all nuclides, pathways, and organs. Since many of the assessments are either predictive in nature or arise from the inability to measure extremely low environmental radiation levels, recourse has to be made to calculations, using validated computer models. One of the pathways neglected in previous assessments, which has subsequently been shown to be important in many cases (e.g. Nair, Hand & Merlin 1982; Kelly *et al.* 1983) is the food chain pathway. Initially, simple models were developed, applicable only to assessment of ingestion dose arising from equilibrium conditions (Nair, Hotson & Stacey 1980; Ng *et al.* 1978). However, recently a range of dynamic models have been developed in the UK (Nair 1984), which are capable of evaluating the time-course of radio nuclide transfer through food chains and are thus applicable to the assessment of ingestion dose from both accidental and routine releases.

Calculations of reactor accident release source-terms have generally indicated that these consist of a mixture of gaseous and other volatile fission products, together with a small amount of activation products. Nair (1983), for example, has listed a source-term for a national atmospheric release from a Magnox reactor, which indicates that $^{90}Sr$, $^{137}Cs$, $^{131}I$ and $^{106}Ru$ are among the key radionuclides. However, very little is known about the chemical form of these radionuclides on release, but for the earlier models (e.g. Clarke 1979), where only the inhalation pathway was considered, the physical properties were asssumed to be the important ones. Now that all pathways are being investigated, the ingestion pathway to man must be included and here the chemical characteristics of the radioactive material become important. Initially, following an atmospheric release to the environment, the various radionuclides contained in the plume will be deposited onto plants by wet or dry deposition depending on meteorological conditions, and the point of deposition will depend on a number of factors, including windspeed and radioactive half-life. At the point of deposition the chemical form of a particular radionuclide could be very different from that at the time of release, but the processes involved in uptake of the radionuclide into the plant (crop) and hence the subsequent ingestion by man will involve foliar absorption, root uptake, resuspension and translocation, all dependent on the chemical properties of the particular radionuclide.

Current dynamic models for prediction of radiation dose to man via the soil–crop pathway following nuclear reactor accidents until recently have been based largely upon data generated from either bomb test sites in arid lands of the USA, or from small-scale pot experiments in glasshouses. Consequently, their validity for western European field conditions is open to question, in view of the different crops, soils, and climatic conditions occurring in the USA, and the unrealistic nature of experiments using plants grown indoors in small containers. During the last few years, awareness of this problem has prompted a number of research programmes in which crops characteristic of western Europe are grown under quasi-field conditions (e.g. Haisch, Capriel & Stark 1982; Eriksson 1983; Steffens, Fuhr & Mittelstaedt 1982). In these investigations soils are spiked with radionuclides to simulate circumstances following hypothetical reactor accidents or operational releases, in order to obtain values for soil-plant transfer factors (Bq $g^{-1}$ plant dry weight per Bq $g^{-1}$ air-dried soil at sowing). These can then be used to refine the existing food-chain models, in order to make them more relevant to European agriculture (Grogan 1984).

In this paper we describe an experimental programme, which represents the first attempts in the UK to determine soil–crop transfer factors over a time-course between sowing and harvest, using controlled additions of radioactive contamination to the soil under conditions designed to represent the field as closely as possible. Wheat was selected as a crop grown over large areas of western Europe and representing a major portion of human diet. There is a considerable range of soil types on which wheat is grown in Europe, in the UK in particular, and thus it was decided that the effect of soil type should be investigated as a major variable influencing radionuclide uptake into the crop. Contamination of the aerial parts of plants by radio nuclides originating from the soil can arise by both root uptake and resuspension of soil particles due to wind and rain action. Soil characteristics, particularly particle size distribution, are likely to influence the magnitude of resuspension, consequently resuspension was also determined in the present work.

## METHODS

The programme consisted of two main parts: an initial set of experiments, where the crop was grown in large pots located outdoors; and a second investigation in which the pots were replaced with lysimeters in order to simulate agricultural conditions more closely (Grogan 1984).

### *Radionuclides*

For the pot experiments, four radionuclides were selected on the basis of the high probability of their release in significant amounts during a reactor accident. These were $^{144}$Ce, $^{137}$Cs, $^{106}$Ru, and $^{99}$Tc, which are all potentially hazardous via soil–crop pathways in view of their relatively long half-lives, which range between 284 days and $2{\cdot}14 \times 10^5$ years. $^{137}$Cs was included in the study because there is already a considerable body of knowledge on its behaviour in the environment, thereby providing a check on the validity of the results for the three less well studied radionuclides. For the lysimeter experiments, $^{99}$Tc was replaced by $^{125}$Sb. All the radionuclides were obtained from Amersham International as carrier-free chlorides, except for $^{99}$Tc which was only available as ammonium pertechnetate ($6{\cdot}29 \times 10^2$ MBq $g^{-1}$Tc).

### *Pot experiments*

Three soil types were selected for the pot experiments as being typical of the most productive agricultural regions of the UK. These were obtained from experimental husbandry farms of the Agricultural Development and Advisory Service of the Ministry of Agriculture, Fisheries and Food, and were classified as a clay, a sandy loam, and an organic soil; their characteristics are shown in Table 1.

TABLE 1. Soil characteristics

| | Soil type | | | | | | |
|---|---|---|---|---|---|---|---|
| | Pot experiments | | | Lysimeter experiments | | | |
| | Clay | Sandy loam | Organic | Sandy loam | Silty loam | Loam | Silty clay |
| Texture | | | | | | | |
| Sand (%) | 37·1 | 69·5 | 54·9 | 53·1 | 10·8 | 31·0 | 20·1 |
| fine | 75·3 | 97·4 | – | – | – | – | – |
| coarse | 24·7 | 2·6 | – | – | – | – | – |
| Silt (%) | 19·9 | 13·7 | 35·6 | 36·1 | 65·3 | 42·0 | 48·9 |
| Clay | 43·0 | 16·8 | 9·5 | 10·8 | 23·9 | 27·0 | 31·0 |
| pH | | | | | | | |
| $H_20$ (2:1) | 6·6 | 7·2 | 5·9 | 7·9 | 6·7 | 6·6 | 6·8 |
| N KC1 (2:1) | 5·7 | 6·9 | 5·2 | 7·2 | 6·4 | 5·6 | 6·1 |
| Loss-on-ignition (% of oven dry wt) | 6·9 | 4·0 | 50·4 | 7·9 | 4·0 | 6·5 | 5·3 |
| Bulk density (g dry wt $cm^{-3}$) | – | – | – | 1·08 | 1·42 | 1·14 | 1·38 |

In Experiment 1 the clay was used, with 10·5 kg of soil being placed in each of eighteen 25·4 cm diameter plastic pots (25 cm deep). Each pot was free-draining, supported on a plastic cylinder, enclosed within a double layer of polythene in order to retain leachate. The pots were placed in an open secure location, beneath a fruit cage and sown with *Triticum aestivum* L. cv. Timmo (spring wheat) on 1 April 1981 at the agricultural rate of 250 seeds $m^{-2}$. Aqueous solution (200 ml) containing 0·74 MBq each of $^{144}Ce$, $^{137}Cs$, and $^{106}Ru$ and 0·37 MBq $^{99}Tc$ were applied evenly over the surface of each of fifteen pots, 5 days later and before emergence of the seedlings. A smaller amount of $^{99}Tc$ was applied than in the case of the other radionuclides to avoid chemical toxicity, since this isotope was not available in a carrier-free form. The remaining three pots were used as non-radioactive controls. A general NPK fertilizer was applied at the start of the experiment and at intervals as necessary according to normal agricultural practice throughout this and subsequent experiments. The soil only received natural rainfall, with the exception of very dry periods in summer when hand watering was carried out.

Three harvests were performed in this first experiment, at 74, 103 and 132 days from sowing, with the crop being fully mature at the final harvest. On each occasion the plants from five pots (and one control) were cut off at ground level and divided into straw (stem and leaves) and ears (if present). Half of each sample was washed several times in tap water to remove surface contamination resulting from resuspension, while the remainder was left unwashed. All samples were subjected to freeze-drying prior to being ground to a fine powder. The activities of $^{137}Cs$, $^{106}Ru$, and $^{144}Ce$ were determined using $\gamma$-ray spectrometry. $^{99}Tc$ was activated to $^{100}Tc$ in the research reactor at Imperial College and this $\gamma$-emitter counted in the same manner as the other radionuclides.

After the final harvest of the first experiment, the soil was allowed to remain fallow for 68 days and then a winter wheat, *T. aestivum* cv. Armada was sown as before on 19 October 1981 (Experiment 2). In this case the effect of soil type on transfer factors was determined by sowing the wheat also on the sandy loam and organic soils which had been used for other crops over the previous summer, before which they had been contaminated on the same date and with the same amount of radioactivity as for the clay. The organic soil was held in larger containers, with twice the volume of the 25·4 cm pots. There were only five replicates (plus one control) in this experiment and consequently all the crop was allowed to grow to maturity, when a single final harvest was performed on 11 August 1982.

Experiment 3 overlapped with Experiment 2 and consisted of five replicates of *T. aestivum* cv. Timmo sown on 10 April 1982, with a single final harvest at maturity on 7 September 1982. A fourth experiment was carried out using *T. aestivum* cv. Armada on all three soil types. The crop was sown on 15 October 1982 and five replicates harvested in April 1983 and the remaining plants harvested at maturity after 167 days on 10 August 1983. On this occasion it was only possible to analyse the ears due to the shortage of time remaining for the investigation.

### *Lysimeter experiment*

The second part of the investigation commenced in 1982 when a system of lysimeters became available, thereby providing the opportunity for a more realistic simulation of field conditions. The use of pots in obtaining soil to plant transfer factors was considered to have three possible drawbacks viz. the rapid exhaustion of soil water during dry periods, limited insulation of the root zone against changes in air temperature, and the small amount of plant material produced.

The acquisition of eighteen lysimeters built in 1959, allowed for a larger and better insulated soil volume to be used. The lysimeters each measure 1·62 $m^2$with a soil volume of 1·46 $m^3$. They were dug out, cleaned, and rendered watertight, prior to a layer of gravel being placed in the bottom and a drainage pipe installed in each at a depth of 50 cm. The lysimeters were then refilled with four soil types characteristic of the major wheat producing areas of the EEC: four replicates each of a loam and a silty loam; and five replicates each of a silty clay and a sandy loam. Soil characteristics are described in Table 1. An attempt was made to preserve the soil profiles where possible by removing the soils from their original sites in two layers, corresponding to the plough layer and the next horizon. The soils were allowed to settle for 5 months, during which time a spring wheat crop was grown on them and the surrounding area. The fifth replicate of the silty clay and the sandy loam was a free-draining control, which was not treated with radionuclides.

Winter wheat was drilled in eight 10-cm spaced rows in October 1983, and a surface application of radionuclides made 1 week later at 3·7 MBq $m^{-2}$. The radionuclide mixture contained $^{137}Cs$, $^{144}Ce$, $^{106}Ru$ and $^{125}Sb$ as previously described.

The crops were subjected to normal agricultural practice with respect to fertilizer, herbicide and fungicide applications. No irrigation was given. The first harvest was completed on 9 January 1984, further harvests were made on 25 April 1984, 15 June 1984 and at maturity on 17 August 1984. Each lysimeter was divided into four blocks, and a 30 cm strip along the drill was taken from each of these at each harvest. The plants at all harvests were cut off at 1 cm above soil level.

The harvested plants were analysed both unwashed and washed as for the pot experiments.

### *Analysis of results*

For both sets of experiments, the arithmetic mean and standard error of the mean were calculated, and analyses of variance performed after logarithmic (pot experiments) and arcsin (lysimeter experiments) transformations. In some instances the radioactivity in a sample was below the detection limit for the instrument concerned. When this occurred, '< detection limit' is quoted, and zero values were calculated on the basis of 20 cm soil depth and, unless otherwise specified, for washed samples only, thereby eliminating contributions of radioactivity from resuspension.

## RESULTS

### *Pot experiments*

#### *Spring wheat (clay soil)*

In Experiment 1, the $^{137}Cs$ content of the straw decreased from an initially high level of 52·9 Bq $g^{-1}$ at the first harvest to 17·7 and 11·6 Bq $g^{-1}$ at the second and final harvests, respectively (Table 2); however, this decrease was only significant ($P < 0·01$) between the

TABLE 2. $^{137}Cs$ content of spring wheat growing on clay soil

| Harvest | Plant part | Activity (Bq $g^{-1}$ dry wt) | | | Transfer factor | |
|---|---|---|---|---|---|---|
| | | No wash | Wash | Wash/no wash | No wash | Wash |
| Experiment 1 | | | | | | |
| First (1981) | Straw | 58·7 | 52·9 | 0·90 | $6·75 \times 10^{-1}$ | $6·09 \times 10^{-1}$ |
| | | ±5·1 | ±3·0 | ±0·09 | $±5·91 \times 10^{-2}$ | $3·49 \times 10^{-2}$ |
| | Ears | not formed | | | | |
| Second (1981) | Straw | 26·4 | 17·7 | 0·67 | $3·03 \times 10^{-1}$ | $2·03 \times 10^{-1}$ |
| | | ±3·3 | ±5·9 | ±0·24 | $±3·78 \times 10^{-2}$ | $±6·83 \times 10^{-2}$ |
| | Ears | 1·8 | 1·1 | 0·61 | $2·11 \times 10^{-2}$ | $1·29 \times 10^{-2}$ |
| | | ±1·1 | ±0·6 | ±0·49 | $±1·29 \times 10^{-2}$ | $±6·72 \times 10^{-3}$ |
| Third (1981) | Straw | 26·7 | 11·6 | 0·43 | $3·07 \times 10^{-1}$ | $1·34 \times 10^{-1}$ |
| | | ±7·7 | ±2·1 | ±0·15 | $±8·86 \times 10^{-2}$ | $±2·42 \times 10^{-2}$ |
| (At maturity) | Ears | 2·4 | 2·3 | 0·96 | $2·72 \times 10^{-2}$ | $2·61 \times 10^{-2}$ |
| | | ±0·2 | ±0·5 | ±0·22 | $±2·72 \times 10^{-3}$ | $±5·45 \times 10^{-3}$ |
| Experiment 3 | Straw | 13·8 | 6·0 | 0·44 | $1·62 \times 10^{-1}$ | $7·08 \times 10^{-2}$ |
| | | ±3·9 | ±1·1 | ±0·15 | $±4·58 \times 10^{-2}$ | $±1·27 \times 10^{-2}$ |
| (1982) | Ears | 1·2 | 1·4 | 1·13 | $1·40 \times 10^{-2}$ | $1·59 \times 10^{-2}$ |
| (At maturity) | | ±0·2 | ±0·2 | ±0·25 | $±2·50 \times 10^{-3}$ | $±2·00 ± 10^{-3}$ |

first two harvests. Ears had not developed by the first harvest but at the second harvest their $^{137}Cs$ activity was considerably less than the straw and remained the same up to the final harvest. The results for 1982 (Experiment 3) showed a similar pattern, the ears containing significantly ($P < 0{\cdot}05$) less $^{137}Cs$ ($1{\cdot}4$ Bq $g^{-1}$) than the straw ($6{\cdot}0$ Bq $g^{-1}$). The transfer factors at the final harvest for two successive years for straw and grain, showed some indication of a decrease from $1{\cdot}34 \times 10^{-1}$ to $0{\cdot}71 \times 10^{-1}$ and from $2{\cdot}61 \times 10^{-2}$ to $1{\cdot}59 \times 10^{-2}$, respectively; however these were not statistically significant (Fig. 1).

Statistical analyses of the $^{106}Ru$ and $^{144}Ce$ data are of limited application, as many samples contained no detectable levels of these two radionuclides. General trends were as follows. The $^{106}Ru$ content of spring wheat in Experiment 1 increased considerably from the first to the second harvests (Table 3) and in washed straw decreased slightly by the final harvest. There was some indication of a lower $^{106}Ru$ content in washed straw for 1982 (Experiment 3) than at the final harvest in 1981; however, the two values were not significantly different. No $^{106}Ru$ was detected in the ears at any of the harvests. $^{144}Ce$ was not detected in any of the washed samples during 1981 and 1982.

The results for $^{99}Tc$ contrast sharply with those for the other three radionuclides. $^{99}Tc$ was taken up extremely rapidly, resulting in very high activities in the plants, with a maximum of $3{\cdot}1 \times 10^3$ Bq $g^{-1}$ in the straw at the first harvest in Experiment 1 (Table 4). The level in the straw subsequently declined significantly ($P < 0{\cdot}001$) from the first to the final harvest but the content at the second harvest was not significantly different from either of the other harvests. Even though the $^{99}Tc$ activity was $1{\cdot}7 \times 10^3$ Bq $g^{-1}$ and

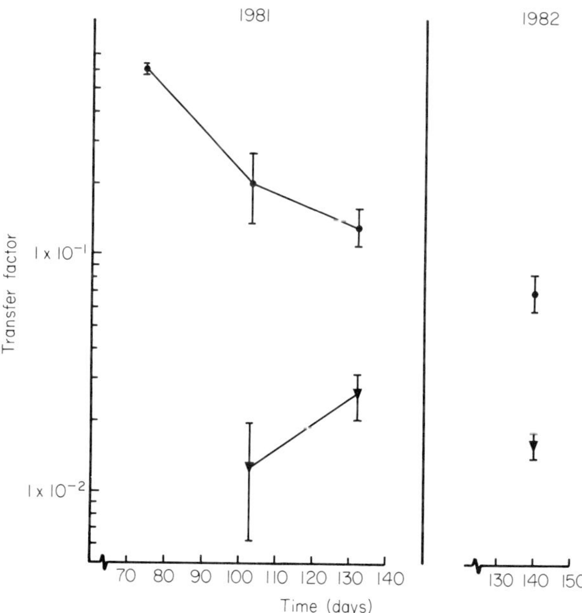


FIG. 1. Transfer factors (mean ± SEM) for $^{137}Cs$ from clay soil to washed spring wheat straw (•) and ears (▼) in Experiments 1 (1981) and 3 (1982).

TABLE 3. $^{106}Ru$ content of spring wheat growing on a clay soil

| Harvest | Plant part | Activity (Bq g$^{-1}$ dry wt) No wash | Wash | Wash/no wash | Transfer factor No wash | Wash |
|---|---|---|---|---|---|---|
| Experiment 1 | | | | | | |
| First (1981) | Straw | 1·6 | 1·6 | 1·00 | $1{\cdot}89 \times 10^{-2}$ | $1{\cdot}89 \times 10^{-2}$ |
| | | ±1·6 | ±1·6 | | $\pm 1{\cdot}89 \times 10^{-2}$ | $\pm 1{\cdot}89 \times 10^{-2}$ |
| | Ears | not formed | | | | |
| Second (1981) | Straw | 24·1 | 12·0 | 0·50 | $2{\cdot}77 \times 10^{-1}$ | $1{\cdot}37 \times 10^{-1}$ |
| | | ±3·6 | ±4·5 | ±0·20 | $\pm 4{\cdot}15 \times 10^{-2}$ | $\pm 5{\cdot}20 \times 10^{-2}$ |
| | Ears | <0·5 | <0·5 | | $< 5{\cdot}60 \times 10^{-3}$ | $< 5{\cdot}60 \times 10^{-3}$ |
| Third (1981) | Straw | 25·3 | 7·3 | 0·29 | $2{\cdot}91 \times 10^{-1}$ | $8{\cdot}43 \times 10^{-2}$ |
| (At maturity) | | ±3·8 | ±3·5 | ±0·15 | $\pm 0{\cdot}44 \times 10^{-1}$ | $\pm 4{\cdot}06 \times 10^{-2}$ |
| | Ears | <0·5 | <0·5 | | $< 5{\cdot}60 \times 10^{-3}$ | $< 5{\cdot}60 \times 10^{-3}$ |
| Experiment 2 | | | | | | |
| 1982 | Straw | 4·1 | 1·9 | 0·45 | $9{\cdot}47 \times 10^{-2}$ | $4{\cdot}28 \times 10^{-2}$ |
| (At maturity) | | ±1·7 | ±0·8 | ±0·27 | $\pm 3{\cdot}94 \times 10^{-2}$ | $\pm 1{\cdot}86 \times 10^{-2}$ |
| | Ears | <0·3 | <0·3 | | $< 6{\cdot}81 \times 10^{-3}$ | $< 6{\cdot}81 \times 10^{-3}$ |

TABLE 4. $^{99}Tc$ content of spring wheat growing on a clay soil (Experiment 1)

| Harvest | Plant part | Activity (Bq g$^{-1}$ dry wt) No wash | Wash | Wash/no wash | Transfer factor No wash | Wash |
|---|---|---|---|---|---|---|
| First (1981) | Straw | $2{\cdot}9 \times 10^{3}$ | $3{\cdot}1 \times 10^{3}$ | 1·04 | 67·1 | 69·9 |
| | | $\pm 0{\cdot}1 \times 10^{3}$ | $\pm 0{\cdot}4 \times 10^{3}$ | ±0·14 | ±3·1 | ±8·7 |
| | Ears | Not formed | | | | |
| Second (1981) | Straw | $1{\cdot}4 \times 10^{3}$ | $1{\cdot}7 \times 10^{3}$ | 1·23 | 32·2 | 39·7 |
| | | $\pm 0{\cdot}3 \times 10^{3}$ | $\pm 0{\cdot}3 \times 10^{3}$ | ±0·32 | ±6·0 | ±7·1 |
| | Ears | <33·3 | <33·3 | | <0·76 | <0·76 |
| Third (1981) | Straw | $8{\cdot}4 \times 10^{-2}$ | $9{\cdot}9 \times 10^{-2}$ | 1·18 | 19·3 | 22·7 |
| (At maturity) | | $\pm 0{\cdot}6 \times 10^{-2}$ | $\pm 0{\cdot}3 \times 19^{-2}$ | ±0·35 | ±1·5 | ±6·6 |
| | Ears | <33·3 | <33·3 | | <0·76 | <0·76 |

$9{\cdot}9 \times 10^{2}$ Bq g$^{-1}$ in the straw at the second and final harvests, respectively, no activity was detected in the corresponding ears. The transfer factor declined over the course of Experiment 1 on a log-linear basis (Fig. 2).

No $^{99}Tc$ was detected in the spring wheat in Experiment 3, or in any of the other plant samples analysed after Experiment 1.

### *Winter wheat (clay, sandy loam, and organic soils)*

The results from Experiment 2 showed almost twice as much $^{137}Cs$ in the straw from the sandy loam than the clay soils ($P < 0{\cdot}001$) (Table 5). The straw grown on the organic soil contained a similar level of $^{137}Cs$ to that grown on the clay. There was consistently less $^{137}Cs$

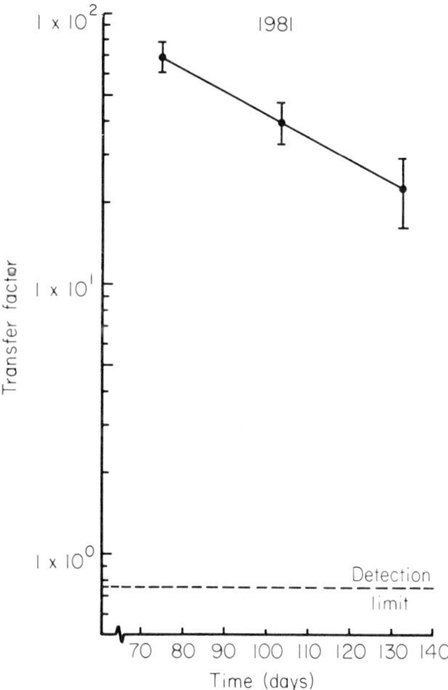


FIG. 2. Transfer factors (mean ± SEM) for $^{99}$Tc from clay soil to washed spring wheat straw (•) in Experiment 1 (1981). No $^{99}$Tc was detected in the ears.

TABLE 5. $^{137}$Cs content of winter wheat at maturity growing on clay, sandy loam, and organic soils in 1981–82 (Experiment 2)

| Soil | Plant part | Activity (Bq g$^{-1}$ dry wt) No wash | Activity (Bq g$^{-1}$ dry wt) Wash | Wash/no wash | Transfer factor No wash | Transfer factor Wash |
|---|---|---|---|---|---|---|
| Clay | Straw | 4·0 | 3·4 | 0·84 | $4{\cdot}64 \times 10^{-2}$ | $3{\cdot}91 \times 10^{-2}$ |
| | | ±0·6 | ±0·4 | ±0·15 | $\pm 7{\cdot}00 \times 10^{-3}$ | $\pm 4{\cdot}12 \times 10^{-3}$ |
| | Ears | 2·6 | 2·2 | 0·85 | $2{\cdot}97 \times 10^{-2}$ | $2{\cdot}54 \times 10^{-2}$ |
| | | ±0·3 | ±0·2 | ±0·12 | $\pm 3{\cdot}09 \times 10^{-3}$ | $\pm 2{\cdot}36 \times 10^{-3}$ |
| Sandy loam | Straw | 9·9 | 6·6 | 0·67 | $1{\cdot}15 \times 10^{-1}$ | $7{\cdot}70 \times 10^{-2}$ |
| | | ±1·6 | ±1·2 | ±0·16 | $\pm 1{\cdot}90 \times 10^{-2}$ | $\pm 1{\cdot}38 \times 10^{-2}$ |
| | Ears | 2·3 | 1·9 | 0·80 | $2{\cdot}72 \times 10^{-2}$ | $2{\cdot}16 \times 10^{-2}$ |
| | | ±0·5 | ±0·6 | ±0·31 | $\pm 5{\cdot}28 \times 10^{-3}$ | $\pm 7{\cdot}25 \times 10^{-3}$ |
| Organic | Straw | 3·7 | 3·9 | 1·07 | $4{\cdot}23 \times 10^{-2}$ | $4{\cdot}51 \times 10^{-2}$ |
| | | ±0·7 | ±1·1 | ±0·36 | $\pm 7{\cdot}64 \times 10^{-3}$ | $\pm 1{\cdot}29 \times 10^{-2}$ |
| | Ears | 2·2 | 2·4 | 1·09 | $2{\cdot}58 \times 10^{-2}$ | $2{\cdot}82 \times 10^{-2}$ |
| | | ±0·5 | ±0·5 | ±0·35 | $\pm 6{\cdot}22 \times 10^{-3}$ | $\pm 5{\cdot}92 \times 10^{-3}$ |

in the ears than the straw; however, soil type had no significant effect on the former. Results for the 1982–83 season (Experiment 4) showed no significant difference between straw grown on either the clay, organic or sandy loam (Table 6). At the final harvest, when

TABLE 6. $^{137}Cs$ content of winter wheat growing on clay, sandy loam, and organic soils in 1982–83 (Experiment 4)

| Soil | Plant part | Activity (Bq g$^{-1}$ dry wt) No wash | Activity (Bq g$^{-1}$ dry wt) Wash | Wash/no wash | Transfer factor No wash | Transfer factor Wash |
|---|---|---|---|---|---|---|
| Clay | Straw | 7·4 | 4·4 | 0·59 | $8{\cdot}80 \times 10^{-2}$ | $5{\cdot}23 \times 10^{-2}$ |
| | (1st harvest) | ±1·1 | ±0·4 | ±0·10 | $\pm 1{\cdot}27 \times 10^{-2}$ | $\pm 0{\cdot}50 \times 10^{-2}$ |
| | Ears* | | 3·0 | – | – | $3{\cdot}56 \times 10^{-2}$ |
| | (Final harvest) | | ±0·5 | | | $\pm 0{\cdot}60 \times 10^{-2}$ |
| Sandy | Straw | 6·3 | 4·3 | 0·67 | $7{\cdot}55 \times 10^{-2}$ | $5{\cdot}06 \times 10^{-2}$ |
| loam | (1st harvest) | ±2·1 | ±0·9 | ±0·26 | $\pm 2{\cdot}45 \times 10^{-2}$ | $\pm 1{\cdot}04 \times 10^{-2}$ |
| | Ears* | – | 12·3 | – | – | $1{\cdot}46 \times 10^{-1}$ |
| | (Final harvest) | | ±2·7 | | | $\pm 0{\cdot}32 \times 10^{-1}$ |
| Organic | Straw | 6·3 | 3·4 | 0·54 | $7{\cdot}54 \times 10^{-2}$ | $4{\cdot}08 \times 10^{-2}$ |
| | (1st harvest) | ±1·0 | ±0·2 | ±0·09 | $\pm 1{\cdot}17 \times 10^{-2}$ | $\pm 0{\cdot}27 \times 10^{-2}$ |
| | Ears* | – | 5·7 | – | – | $6{\cdot}81 \times 10^{-2}$ |
| | (Final harvest) | | ±0·6 | | | $\pm 0{\cdot}74 \times 10^{-2}$ |

*Only washed samples were analysed at the final harvest.

only the ears were analysed, those grown on the sandy loam had a significantly greater ($P < 0{\cdot}001$, $P < 0{\cdot}05$) $^{137}Cs$ content than those grown on the clay and organic soils, respectively. There was no significant difference between the $^{137}Cs$ in the ears from the clay and organic soils. The transfer factors for the ears derived in 1982–83 were the same as for in 1981–2 on both the clay and organic soils (Fig. 3). However, the transfer factor for the ears of wheat grown on the sandy loam was nearly seven times the value obtained for 1981–82, this being significant at $P < 0{\cdot}001$.

$^{106}Ru$ was taken up into winter wheat in the 1981–82 season to a lesser extent than the $^{137}Cs$ (Experiment 2). Trends between soils were similar to those for $^{137}Cs$, e.g. the plants grown on the sandy loam contained greater amounts of $^{106}Ru$ than those on either the clay

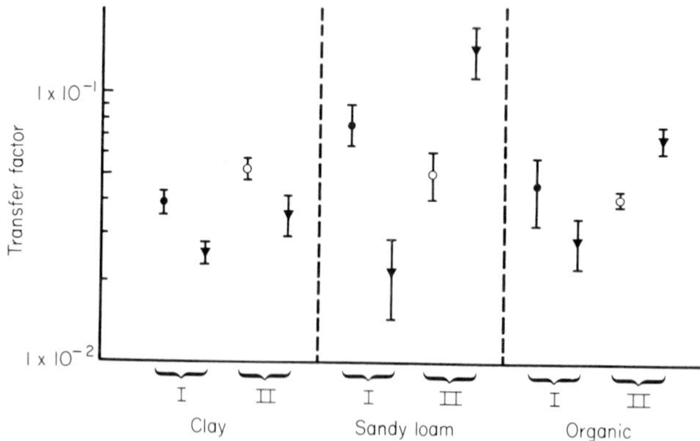


FIG. 3. Transfer factors (mean ± SEM) for $^{137}Cs$ from clay, sandy loam, and organic soils to washed winter wheat in Experiment 2 (1981–82) and 4 (1982–83). Straw at maturity, 1981–82 (●); straw at first harvest, 1982–83(○); ears at maturity, 1981–82 and 1982–83 (▼). I, 1981–82 season; II, 1982–83 season.

or organic soils (Table 7). In the 1982–83 season (Experiment 4) no $^{106}Ru$ was measured in either straw or ears at either harvest.

The winter wheat grown on clay and organic soils in Experiment 2 contained no detectable levels of $^{144}Ce$ in either the straw or ears. However, a small amount was detected in two straw samples and one ear sample grown on the sandy loam soil. No $^{144}Ce$ was detected in any winter wheat samples from the 1982–83 season (Experiment 4).

### *Lysimeter experiment*

Only the results for the first harvest of winter wheat, carried out 81 days after sowing, are available; at this point the ears had not appeared.

The $^{137}Cs$ content of the straw was significantly higher ($P < 0{\cdot}001$) on the loam and silty clay soils compared with the silty loam and sandy loam (Table 8), and this pattern is reflected in the transfer factors, despite differences in soil bulk density. $^{106}Ru$ showed a similar trend, although the only significant difference was between the loam and the other three soils (Table 9). Uptake of $^{144}Ce$ (Table 10) and $^{125}Sb$ (Table 11) was lower than for the other two radionuclides with about 30% of the values for the former on clay loam being less than the detection limit.

### *Resuspension*

The contribution of resuspended soil material to the contamination of both straw and ears is shown by the ratio of radioactivity in washed and unwashed samples (Tables 2–11).

The results indicate that washing generally removed a proportion of the $^{137}Cs$, $^{106}Ru$, $^{144}Ce$, and $^{125}Sb$ from the straw. For example, the washed samples contained 9·9%, 33·0%, and 56·5% less $^{137}Cs$ than the unwashed samples at the first, second, and final harvests, respectively, in Experiment 1, and 56·1% less in Experiment 3. The same samples contained 50·3% and 71·1% less $^{106}Ru$ at the second and final harvests respectively, in Experiment 1, and 54·9% less in Experiment 3. In contrast to the straw, washing had no apparent effect on the radionuclide content of the ears.

$^{144}Ce$ and $^{99}Tc$ showed a different pattern from the other radionuclides. In the pot experiments, washing in nearly all cases eliminated the $^{144}Ce$ from the straw of the spring wheat. The lysimeter experiment showed that 33–65% of contamination of the straw arose from resuspension. However, in this case winter wheat was used and in the pot experiments levels of radioactivity in the unwashed samples for this crop were below the detection limit and thus resuspension could not be calculated. For all four radionuclides used in the lysimeters, the silty clay apparently gave rise to the most resuspension. The $^{99}Tc$ results, which are only applicable to Experiment 1, gave no indication of any effect of washing.

## DISCUSSION

An integration of the results of the four pot experiments with those for the lysimeters demonstrates that there are major differences in the transfer factors for the different

TABLE 7. $^{106}$Ru content of winter wheat at maturity growin on clay, sandy loam, and organic soils in 1981–82 (Experiment 2)

| Soil | Plant part | Activity (Bq g$^{-1}$ dry wt) | | Transfer factor | |
|---|---|---|---|---|---|
| | | No wash | Wash | No wash | Wash |
| Clay | Straw | $2 \cdot 0 \times 10^{-1}$ | $2 \cdot 9 \times 10^{-1}$ | $3 \cdot 10 \times 10^{-3}$ | $4 \cdot 49 \times 10^{-3}$ |
| | | $\pm 1 \cdot 9 \times 10^{-1}$ | $\pm 1 \cdot 9 \times 10^{-1}$ | $\pm 2 \cdot 94 \times 10^{-3}$ | $\pm 2 \cdot 94 \times 10^{-3}$ |
| | Ears | $< 3 \cdot 0 \times 10^{-1}$ | $< 3 \cdot 0 \times 10^{-1}$ | $< 4 \cdot 65 \times 10^{-3}$ | $< 4 \cdot 65 \times 10^{-3}$ |
| Sandy | Straw | $1 \cdot 1$ | $9 \cdot 0 \times 10^{-1}$ | $2 \cdot 72 \times 10^{-2}$ | $1 \cdot 77 \times 10^{-2}$ |
| loam | | $\pm 3 \cdot 4 \times 10^{-1}$ | $\pm 2 \cdot 9 \times 10^{-1}$ | $\pm 0 \cdot 66 \times 10^{-2}$ | $\pm 0 \cdot 57 \times 10^{-2}$ |
| | Ears | $1 \cdot 13 \times 10^{-1}$ | $1 \cdot 17$ | $2 \cdot 03 \times 10^{-3}$ | $2 \cdot 31 \times 10^{-2}$ |
| | | $\pm 1 \cdot 3 \times 10^{-1}$ | $\pm 7 \cdot 7 \times 10^{-1}$ | $\pm 2 \cdot 03 \times 10^{-3}$ | $\pm 1 \cdot 52 \times 10^{-2}$ |
| Organic | Straw | $< 0 \cdot 30$ | $< 0 \cdot 30$ | $< 4 \cdot 65 \times 10^{-3}$ | $< 4 \cdot 65 \times 10^{-3}$ |
| | Ears | $1 \cdot 3 \times 10^{-1}$ | $1 \cdot 7 \times 10^{-1}$ | $2 \cdot 03 \times 10^{-3}$ | $2 \cdot 67 \times 10^{-3}$ |
| | | $\pm 1 \cdot 3 \times 10^{-1}$ | $\pm 1 \cdot 7 \times 10^{-1}$ | $\pm 2 \cdot 03 \times 10^{-3}$ | $\pm 2 \cdot 67 \times 10^{-3}$ |

TABLE 8. $^{137}$Cs content of winter wheat after 81 days growing on four soil types in lysimeters

| | Activity (Bq g$^{-1}$ dry wt) | | Wash/no wash | Transfer factor | |
|---|---|---|---|---|---|
| | No wash | Wash | | No wash | Wash |
| Sandy loam | 8·67 | 4·95 | 0·57 | $5 \cdot 2 \times 10^{-1}$ | $2 \cdot 9 \times 10^{-1}$ |
| | ±1·35 | ±0·34 | ±0·08 | $\pm 1 \cdot 6 \times 10^{-1}$ | $\pm 0 \cdot 8 \times 10^{-1}$ |
| Silty loam | 8·02 | 4·83 | 0·60 | $6 \cdot 3 \times 10^{-1}$ | $3 \cdot 8 \times 10^{-1}$ |
| | ±3·19 | ±1·56 | ±0·14 | $\pm 2 \cdot 0 \times 10^{-1}$ | $\pm 1 \cdot 1 \times 10^{-1}$ |
| Loam | 10·68 | 8·16 | 0·76 | $6 \cdot 7 \times 10^{-1}$ | $5 \cdot 13 \times 10^{-1}$ |
| | ±3·58 | ±0·53 | ±0·25 | $\pm 3 \cdot 2 \times 10^{-1}$ | $\pm 1 \cdot 8 \times 10^{-1}$ |
| Silty clay | 13·54 | 7·34 | 0·54 | 1·03 | $5 \cdot 56 \times 10^{-1}$ |
| | ±6·08 | ±2·59 | ±0·15 | $\pm 4 \cdot 25 \times 10^{-1}$ | $\pm 1 \cdot 7 \times 10^{-1}$ |

TABLE 9. $^{106}$Ru content of winter wheat after 81 days growing on four soil types in lysimeters

| | Activity (Bq g$^{-1}$ dry wt) | | Wash/no wash | Transfer factor | |
|---|---|---|---|---|---|
| | No wash | Wash | | No wash | Wash |
| Sandy loam | 6·57 | 4·52 | 0·688 | $4 \cdot 19 \times 10^{-1}$ | $2 \cdot 88 \times 10^{-1}$ |
| | ±1·21 | ±0·48 | ±0·103 | $\pm 1 \cdot 58 \times 10^{-1}$ | $\pm 1 \cdot 0 \times 10^{-1}$ |
| Silty loam | 9·37 | 4·80 | 0·512 | $7 \cdot 81 \times 10^{-1}$ | $4 \cdot 0 \times 10^{-1}$ |
| | ±1·99 | ±0·17 | ±0·107 | $\pm 2 \cdot 90 \times 10^{-1}$ | $\pm 1 \cdot 2 \times 10^{-1}$ |
| Loam | 9·73 | 8·36 | 0·859 | $6 \cdot 52 \times 10^{-1}$ | $5 \cdot 60 \times 10^{-1}$ |
| | ±1·20 | ±0·66 | ±0·081 | $\pm 2 \cdot 41 \times 10^{-1}$ | $\pm 2 \cdot 0 \times 10^{-1}$ |
| Silty clay | 15·69 | 5·57 | 0·355 | 1·27 | $4 \cdot 50 \times 10^{-1}$ |
| | ±4·00 | ±0·75 | ±0·355 | ±1·35 | $\pm 1 \cdot 6 \times 10^{-1}$ |

TABLE 10. $^{144}Ce$ content of winter wheat after 81 days growing on four soil types in lysimeters

| | Activity (Bq g$^{-1}$ dry wt) | | | Transfer factor | |
|---|---|---|---|---|---|
| | No wash | Wash | Wash/no wash | No wash | Wash |
| Sandy loam | 4·73 | 2·85 | 0·603 | $3·55 \times 10^{-1}$ | $2·14 \times 10^{-1}$ |
| | ±0·79 | ±0·28 | ±0·081 | $\pm 1·10 \times 10^{-1}$ | $\pm 6·0 \times 10^{-2}$ |
| Silty loam | 4·11 | 2·10 | 0·511 | $4·05 \times 10^{-1}$ | $2·07 \times 10^{-1}$ |
| | ±1·33 | ±0·22 | ±0·157 | $\pm 1·50 \times 10^{-1}$ | $\pm 4·0 \times 10^{-2}$ |
| Loam | 3·96 | 2·68 | 0·676 | $3·14 \times 10^{-1}$ | $2·12 \times 10^{-1}$ |
| | ±1·03 | ±0·39 | ±0·145 | $\pm 1·24 \times 10^{-1}$ | $\pm 7·0 \times 10^{-2}$ |
| Silty clay | 4·96 | 1·75 | 0·353 | $4·73 \times 10^{-1}$ | $1·67 \times 10^{-1}$ |
| | ±1·67 | ±0·39 | ±0·089 | $\pm 1·85 \times 10^{-1}$ | $\pm 5·0 \times 10^{-2}$ |

TABLE 11. $^{125}Sb$ content of winter wheat after 81 days growing on four soil types in lysimeters

| | Activity (Bq g$^{-1}$ dry wt) | | | Transfer factor | |
|---|---|---|---|---|---|
| | No wash | Wash | Wash/no wash | No wash | Wash |
| Sandy loam | 3·48 | 2·57 | 0·739 | $2·11 \times 10^{-1}$ | $1·56 \times 10^{-1}$ |
| | ±0·36 | ±0·15 | ±0·062 | $\pm 7·0 \times 10^{-2}$ | $\pm 5·0 \times 10^{-2}$ |
| Silty loam | 3·93 | 2·45 | 0·624 | $3·12 \times 10^{-1}$ | $1·95 \times 10^{-1}$ |
| | ±0·81 | ±0·19 | ±0·119 | $\pm 1·0 \times 10^{-1}$ | $\pm 5·0 \times 10^{-2}$ |
| Loam | 6·21 | 3·00 | 0·483 | $3·97 \times 10^{-1}$ | $1·92 \times 10^{-1}$ |
| | ±3·81 | ±0·24 | ±0·294 | $\pm 2·72 \times 10^{-1}$ | $\pm 6·0 \times 10^{-2}$ |
| Silty clay | 9·86 | 2·93 | 0·297 | $7·61 \times 10^{-1}$ | $2·26 \times 10^{-1}$ |
| | ±1·19 | ±0·22 | ±0·028 | $\pm 2·14 \times 10^{-1}$ | $\pm 6·0 \times 10^{-2}$ |

radionuclides with the following order: $^{99}Tc > ^{137}Cs \geq ^{106}Ru > ^{144}Ce/^{125}Sb$. The very high transfer factors (up to 69·9) measured for $^{99}Tc$ are in accord with those obtained in the field by previous workers (1·4–153) (Hoffman *et al.* 1982b; Eriksson 1983). Pot experiments have shown even higher transfer factors (95–1890) (Gast, Landa & Thorvig 1976), suggesting that the containers used in the present work give results which reflect reasonably well those obtained under field conditions. The high transfer factors obtained for $^{99}Tc$ compared with the other radionuclides undoubtedly reflect the much greater mobility of an anionic form. In general, the order of uptake of the five radionuclides studied in the present work confirms those found by earlier workers for various crops. Thus, Nishita, Romney & Larson (1961) showed a lower rate of uptake of $^{144}Ce$ than $^{106}Ru$ or $^{137}Cs$ from artificially contaminated soil; these latter two radionuclides were, however, taken up to a similar extent. Russell (1966) agreed with this order, but proposed that $^{137}Cs$ uptake was greater than $^{106}Ru$, as in the present work. The lysimeter experiment indicated similar transfer factors for $^{125}Sb$ and $^{144}Ce$. This is in conflict with the results of d'Souza & Mistry (1973) who, while finding the expected order of uptake of $^{137}Cs > ^{106}Ru > ^{144}Ce$ from soil into the aerial parts of plants, classified $^{125}Sb$ as intermediate between $^{137}Cs$ and $^{106}Ru$.

In the case of $^{137}Cs$ two major field investigations on its uptake into wheat have been carried out in western Europe in recent years. The results of these are compared in Table 12

TABLE 12. Comparison of $^{137}Cs$ transfer factors from soil to wheat with those determined by other workers under European field conditions

| Time from application of radioactivity (years) | Soil type | Plant part | Transfer factor | Reference |
|---|---|---|---|---|
| 0·5 | Clay | Spring wheat straw | $1 \cdot 3 \times 10^{-1}$ | Present work |
| 0·5 | Clay | Spring wheat ears | $2 \cdot 6 \times 10^{-2}$ | |
| 1·5 | Clay | Spring wheat straw | $7 \cdot 1 \times 10^{-2}$ | |
| 1·5 | Clay | Spring wheat grain | $1 \cdot 6 \times 10^{-2}$ | |
| 1·5 | Clay | Winter wheat straw | $3 \cdot 9 \times 10^{-2}$ | |
| 1·5 | Clay | Winter wheat ears | $2 \cdot 5 \times 10^{-2}$ | |
| 1·5 | Sandy loam | Winter wheat straw | $7 \cdot 7 \times 10^{-2}$ | |
| 1·5 | Sandy loam | Winter wheat ears | $2 \cdot 2 \times 10^{-2}$ | |
| 1·5 | Organic | Winter wheat straw | $4 \cdot 5 \times 10^{-2}$ | |
| 1·5 | Organic | Winter wheat ears | $2 \cdot 8 \times 10^{-2}$ | |
| 1·0 | Podsol | Winter wheat straw | $1 \cdot 2 \times 10^{-1}$ | Steffens, Fuhr & Mittelstaedt (1982) |
| 1·0 | Podsol | Winter wheat grain | $6 \cdot 6 \times 10^{-2}$ | |
| 2·0 | Podsol | Winter wheat straw | $8 \cdot 8 \times 10^{-2}$ | |
| 2·0 | Podsol | Winter wheat grain | $4 \cdot 3 \times 10^{-2}$ | |
| 1·0 | Para brown earth | Winter wheat straw | $5 \cdot 0 \times 10^{-3}$ | |
| 1·0 | Para brown earth | Winter wheat grain | $1 \cdot 0 \times 10^{-3}$ | |
| 2·0 | Para brown earth | Winter wheat straw | $1 \cdot 0 \times 10^{-3}$ | |
| 2·0 | Para brown earth | Winter wheat grain | $1 \cdot 0 \times 10^{-3}$ | |
| 0·5 | Sandy | Spring wheat straw | $3 \cdot 1 \times 10^{-2}$ | Haisch, Capriel & Stark (1982) |
| 0·5 | Sandy | Spring wheat grain | $6 \cdot 3 \times 10^{-3}$ | |
| 1·5 | Sandy | Spring wheat straw | $1 \cdot 4 \times 10^{-2}$ | |
| 1·5 | Sandy | Spring wheat grain | $3 \cdot 4 \times 10^{-3}$ | |
| 0·5 | Clay | Spring wheat straw | $1 \cdot 7 \times 10^{-5}$ | |
| 0·5 | Clay | Spring wheat grain | $3 \cdot 0 \times 10^{-6}$ | |
| 1·5 | Clay | Spring wheat straw | $1 \cdot 9 \times 10^{-6}$ | |
| 1·5 | Clay | Spring wheat grain | $8 \cdot 8 \times 10^{-6}$ | |

with those of the present work for harvests at maturity only (time-dependent data have not been determined elsewhere). Meaningful direct comparisons between the soil types used by different workers are not possible because of different experimental procedures, but the data generally agree with those of Steffens, Fuhr & Mittlestaedt (1982). However, Haisch, Capriel & Stark (1982) obtained transfer factors from a clay soil to spring wheat which were two to three orders of magnitude lower than those in the present work, although the values for a sandy soil were reasonably close to those for the sandy loam. The reasons for this discrepancy are not clear, but overall in the present work the results from the large pots are comparable with those obtained using lysimeters. This suggests that containers of the dimensions used in the present work are reasonably representative of field conditions, contrary to the views of Steffens *et al.* (1980) who consider that values obtained from pot experiments are artificially high. Some support for this is provided by the lysimeter experiment where transfer factors for winter wheat ranged between $2 \cdot 95 \times 10^{-1}$ and $5 \cdot 56 \times 10^{-1}$, 81 days after sowing: the values for the winter wheat pot experiment harvest which corresponded most closely in stage of development to this were reasonably comparable, lying between $3 \cdot 91 \times 10^{-2}$ and $7 \cdot 7 \times 10^{-2}$; however, this comparison is to some extent confounded by the radionuclides having been present in the soil for 1 year longer in the pot experiments, which may explain the transfer factors actually being lower than in the lysimeter experiment.

The comparability of the $^{137}$Cs data with those of other European workers, lends weight to the validity of the results for the other radionuclides. In the case of $^{144}$Ce and $^{106}$Ru, the transfer factors obtained from the present work are the first for wheat under European conditions. However, a compilation of American data by Ng, Colsher & Thompson (1982) gives transfer factors of $^{106}$Ru to wheat over the range $1 \cdot 98 \times 10^{-3}$ to $1 \cdot 02 \times 10^{-2}$, which compares favourably with the values obtained here of $< 4 \cdot 65 \times 10^{-3}$ to $5 \cdot 98 \times 10^{-2}$. In the case of $^{144}$Ce, measurements of washed samples were below the detection limit and thus it can merely be stated that the transfer factors were below $1 \cdot 49 \times 10^{-2}$, which is not inconsistent with the values of $1 \cdot 38 \times 10^{-4}$ to $1 \cdot 49 \times 10^{-3}$ quoted by Ng *et al.* (1982).

The radionuclides showed considerable differences in their distribution within the plant, with concentration occurring in the straw, rather than the ears. This distinction was most marked for spring wheat, where the transfer factor differed by a factor of ten for $^{137}$Cs. In the case of $^{106}$Ru, $^{144}$Ce, and $^{99}$Tc, no activity was detectable in the ears, despite the very high values of the latter in the straw. For winter wheat, a similar pattern was detected but the level of discrimination was less marked, which probably reflects the longer growing season involved. The reduced level of activity in the reproductive compared with the vegetative organs has, in fact, been reported by various workers for a range of radionuclides with different crops, e.g. Steffens *et al.*(1982) and Haisch *et al.* (1982) for $^{137}$Cs in spring and winter wheat (Table 12).

The consecutive harvests carried out in Experiment 1 indicate distinct trends in levels of activity with the straw and ears over a time-course, but there were some major differences between the behaviour of different radionuclides. The progressive fall in $^{137}$Cs content of the straw was caused by growth dilution during the more favourable climatic conditions prevailing in the summer months; this was demonstrated by comparing total plant uptake at consecutive harvests (concentration $\times$ biomass) and, in the case of transfer factors, by adjusting all values to the final harvest weight (Grogan *et al.* 1984). In addition, there may be a contribution from increasing fixation of $^{137}$Cs in the soil from its initially highly available state: support for this is provided by the fall in transfer factors to both ears and straw at maturity in 1982 compared with 1981. Fixation of $^{137}$Cs is particularly marked in clay soils, as demonstrated by Squire & Middleton (1966) who concluded that this process could be completed within 3 years of a surface application. The decrease in transfer factors between 1981 and 1982 was also apparent in the case of $^{106}$Ru. Less is known about the soil chemistry of this radionuclide but it appears to become strongly fixed to soil colloids with time, thereby reducing its availability for plant uptake (Brown 1976). However, $^{106}$Ru showed a different temporal trend in spring wheat straw from $^{137}$Cs, with an initially low level of uptake into the plant, followed by a substantially higher concentration at the two subsequent harvests. This clearly cannot be explained in terms of either growth dilution or soil fixation and warrants further investigation.

In the case of $^{99}$Tc the very high transfer factor recorded at the first harvest must be a reflection of its high level of availability, as the pertechnetate anion is highly soluble in water and poorly sorbed to soils (Hoffman *et al.* 1982). In fact, there is evidence that the plants exhausted the surface soil of this radionuclide at a very early stage of growth, as none could be detected in a resuspended form, or in the soil or leachate at the first harvest. Calculation of total uptake shows that the progressive fall in $^{99}$Tc in the straw is accounted for primarily by growth dilution. However, surprisingly for such a mobile radionuclide, the $^{99}$Tc became fixed in the straw with no subsequent translocation to the ears. A more detailed study of the movement of $^{99}$Tc within wheat plants is required in order to explain the mechanisms responsible for this phenomenon.

The results show that soil type influences the transfer factors to wheat. In the pot experiments, the sandy loam resulted in the highest transfer factors for $^{137}$Cs and $^{106}$Ru, which is probably a reflection of the low clay content which will minimize the degree of fixation. Although the organic soil also had a high sand content, it produced low transfer factors which may be attributed to absorption onto humic materials. The only set of results available from the lysimeter experiment does not, in general, confirm the relative availability of the radionuclides from the different soils in the pot experiments. In this case the soils with the greater percentage of small particle sizes (loam and silty clay) generally produced the higher transfer factors (Fig 4). However, the plants on the sandy and silt loams were at a slightly more advanced stage of growth, which may account for this apparent anomaly. Analyses of data from the later lysimeter experiments should resolve this.

The comparison of washed with unwashed samples reveals that resuspension can play a major role in the contamination of the straw. However, the absence of any measurable effect of washing on contamination of the ears suggests that the height of a plant organ

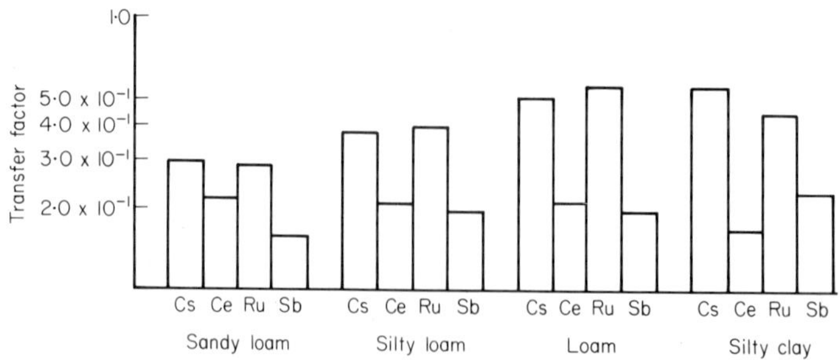


FIG. 4. Transfer factors for $^{137}$Cs, $^{144}$Ce, $^{106}$Ru and $^{125}$Sb for washed winter wheat straw (first harvest) in the lysimeter experiment.

above the soil is critical in determining the importance of resuspension compared with root uptake. It is difficult to make any other generalization concerning the contribution of resuspension at different stages of crop growth, because it is likely to be highly dependent on climatic conditions immediately prior to sampling. Nevertheless, the results suggest that the resuspension pathway is unlikely to make a major contribution to the dose received as a result of the consumption of wheat products contaminated with $^{137}Cs$ and $^{106}Ru$. In the case of $^{144}Ce$, where root uptake is extremely limited, resuspension accounted for effectively all contamination of the straw and must be considered as the potentially most important contributor to the contamination of wheat. In contrast, the absence of any indication of resuspension of $^{99}Tc$ shows that this highly mobile radionuclide only contributes to contamination of the aerial parts of the crop via uptake through the roots, except possibly at very early stages in the growth cycle.

## ACKNOWLEDGMENTS

We gratefully acknowledge the financial support of the Science and Engineering Research Council and the Central Electricity Generating Board in providing a CASE studentship for H.A.G. and the Commission of the European Communities for funding N.G.M. under the terms of the Radiation Protection Programme. Our thanks are also due to Dr S. Nair for his helpful comments on the manuscript.

## REFERENCES

**Brown, K.W. (1976).** *Ruthenium: Its behaviour in plant and soil systems.* EPA-600/3-76-019. USEPA, Las Vegas, Nevada.

**Clarke, R.H. (1979).** *A model for short and medium range dispersion of radionuclides released to the atmosphere.* National Radiological Protection Board Report, NRPB-R91. HMSO, London.

**Eriksson, A. (1983).** Behaviour of $^{99}Tc$ in the environment as indicated by a five-year field lysimeter experiment. *Transfer of Radioactive Materials in the Terrestrial Environment Subsequent to an Accidental Release to Atmosphere.* Vol. 1, pp. 353–374. Commission of the European Communities, Luxembourg.

**Gast, R.G., Landa, E.R. & Thorvig, L.J. (1976).** *The behaviour of technetium-99 in soils and plants.* Final report C00-2447-5. Nat. Tech. Info. Service, Springfield.

**Grogan, H.A. (1984)** *Pathways of Radionuclides from Soils Into Crops Under British Field Conditions.* Ph.D. thesis, University of London.

**Grogan, H.A. (1984).** *Pathways of Radionuclides from Soil Into Crops Under British Field Conditions.* Ph.D. Thesis, University of London.

**Grogan, H.A., Bell, J.N.B., Minski, M.J. & Nair, S. (1984).** *An experimental study of the time dependence of uptake from soil of* $^{137}Cs$, $^{106}Ru$, $^{144}Ce$ *and* $^{99}Tc$ *into green vegetables, wheat and potatoes.* CEGD/TPRD/B/0416/N84. Berkeley Nuclear Laboratory Report, Berkeley.

**Haisch, A., Capriel, P. & Stark, H. (1982).** Transfer factors determined at Munich, FRG. *Report on Workshop on Measurement of Soils to Plant Transfer Factors for Radionuclides. Part II,* pp. 134–136. International Union of Radioecologists (IUR). Euratom-Ital, Wageningen.

**ICRP (1959).** *Report of Committee II on Permissible Dose for Internal Radiation.* Pergamon Press, Oxford.

**ICRP (1977).** Recommendations of the International Commission on Radiological Protection. ICRP Publication No. 26. *Annals of the ICRP,* **1**. Pergamon Press, Oxford.

**Hoffman, F.O., Garten, C.T., Huckabee, J.W. & Lucas, D.M. (1982a).** Interception and retention of technetium by vegetation and soil. *Journal of Environmental Quality,* **11**, 134–141.

**Hoffman, F.O., Garten, C.T., Lucas, D.M. & Huckabee, J.W. (1982b).** Environmental behaviour of technetium in soil and vegetation. Implications for radiological assessment. *Environmental Science and Technology,* **16,** 214–217.

**Kelly, G.N., Charles, D., Broomfield, M. & Hemming, C.R. (1983).** *The radiological impact on the Greater London population of postulated accidental releases from the Sizewell PWR.* National Radiological Protection Board Report, NRPB–R146. HMSO, London.

**Nair, S. (1983).** A sensitivity study of ingestion population dose following a hypothetical atmospheric release from a nuclear power station. *Transfer of Radioactive Materials in the Terrestrial Environment Subsequent to an Accidental Release to Atmosphere* Vol. 2, pp. 523–552. Commission of the European Communities, Luxembourg.

**Nair, S. (1984).** *Models for the evaluation of ingestion doses from the consumption of terrestrial foods following an atmospheric radioactive release.* CEGB Report RD/B/5200/N8. Berkeley Nuclear Laboratory, Berkeley.

**Nair, S., Hand, T.D. & Merlin, J.H. (1982).** *Estimate of doses from routine atmospheric discharges at Sizewell A and B nuclear power stations.* CEGB Report TPRD/B/0166/N82. Berkeley Nuclear Laboratory, Berkeley.

**Nair, S., Hotson, J. & Stacey, A. (1980).** *Popfood – a computer code for calculating ingestion doses from continuous atmospheric releases.* CEGB RD/B/N4888. Berkeley Nuclear Laboratory, Berkeley.

**Ng, Y.C., Phillips, W.A. Ricker, Y.E. Tandy, R.K. & Thompson, S.E. (1978).** *Methodology for assessing dose commitment to individuals and to the population from ingestion of terrestrial foods contaminated by emissions from a nuclear fuel reprocessing plant at the Savannah River Plant.* UCID–17743. Lawrence Livermore National Laboratory.

**Ng, Y.C., Colsher, C.S. & Thompson, S.E. (1982).** *Soil-to-plant concentration factors for radiological assessments.* Final report. NUREG/RC–2975, UCID–19463, Lawrence Livermore National Laboratory.

**Nishita, H., Romney, E.M. & Larson, K.H. (961).** Uptake of radioactive fission products by crop plants. *Journal of Agricultural Food Chemistry,* **9,** 101–106.

**Russell, R.S. (1966).** Behaviour of radioactive materials in soil. *Radioactivity and Human Diet.* (Ed. by R.S. Russell), pp. 105–126. Pergamon Press, Oxford.

**d'Souza, T.J. & Mistry, K.B. (1973).** Comparative uptake of gamma-emitting fission product nuclides by plants. *Indian Society for Nuclear Technology and Agricultural Biology,* Newsletter, **2,** 17–20.

**Squire, H.M. & Middleton, L.J. (1966).** Behaviour of caesium-137 in soils and pastures in long term experiments. *Radiation Botany,* **6,** 413–423.

**Steffens, W., Fuhr, F. & Mittelstaedt, W. (1980).** Evaluation of small-scale laboratory and pot experiments to determine realistic transfer factors for the radionuclides $^{90}$Sr, $^{137}$Cs, $^{60}$Co, $^{54}$Mn. *A Systematic Approach to Safety.* Vol. 2, pp. 1135–1138. Proc. 5th Congress IRPA, Jerusalem, Pergamon Press.

**Steffens, W., Fuhr, F. & Mittelstaedt, W. (1982).** Transfer factors determined at Munich, FRG. *Report on Workshop on Measurement of Soils to Plant Transfer Factors for Radionuclides.* Part II, pp. 88–97. International Union of Radioecologists (IUR), Euratom-Ital. Wageningen.

# $^{137}$Cs uptake by sheep grazing tidally-inundated and inland pastures near the Sellafield reprocessing plant

B.J. HOWARD
*Institute of Terrestrial Ecology, Merlewood Research Station, Grange-over-Sands, Cumbria LA11 6JU*

## SUMMARY

**1** Field investigations into the transfer of $^{137}$Cs from pasture to sheep tissues have been made at two contrasting sites in west Cumbria, close to the nuclear fuel reprocessing plant at Sellafield. These were a saltmarsh bordering on the Esk estuary in 1982 and inland pastures close to the perimeter of the works in 1984. $^{137}$Cs concentrations of sampled vegetation from the saltmarsh were generally two orders of magnitude greater than from the inland pasture due to its inundation with $^{137}$Cs associated with silt. The relatively high $^{137}$Cs content of soil/silt compared to that of vegetation at each site meant that soil contamination of vegetation sometimes accounted for a substantial proportion of its $^{137}$Cs activity (up to 99% on the saltmarsh and 67% on the inland pastures). Considerable seasonal changes occurred in the extent of $^{137}$Cs contamination on the inland pastures, with late winter and early spring levels being up to 20-fold higher than those of the summer.
**2** $^{137}$Cs concentrations in tissues of lambs from the saltmarsh were consistently higher than for ewes; this was not true for sheep from the inland pastures. $^{137}$Cs concentrations in kidney were found to be higher than in all other tissues, both in the study flocks and in one of the controls.
**3** Transfer coefficients (calculated by dividing the $^{137}$Cs concentration of fresh tissue by the daily intake of $^{137}$Cs) were significantly ($P < 0 \cdot 05$) higher in saltmarsh sheep than in the inland pasture sheep; the difference in transfer coefficients was increased further if $^{137}$Cs activity due to soil/silt contamination was removed from the estimates of daily intake leaving $^{137}$Cs associated with the vegetation only to contribute to the transfer coefficients. Further studies are required to determine the availability of $^{137}$Cs associated with soil/silt particles which are ingested by animals.

## INTRODUCTION

It has recently become evident that the chemical form and valency state of a pollutant has a considerable influence on its biological availability (Bulman & Cooper 1986). It is therefore necessary to investigate the movement of different forms of pollutants in ecosystems so that differences in their behaviour in the environment can be taken into account when models and monitoring schemes are being devised.

In west Cumbria the nuclear fuel reprocessing works at Sellafield releases $^{137}$Cs into the environment, mostly via liquid discharges to the Irish Sea, but also in smaller quantities via aerosols from stacks. The discharges have resulted in enhanced concentrations of $^{137}$Cs in a variety of agricultural products in the vicinity of the works (Bradford, Curtis & Popplewell

1984). The transfer to sheep of $^{137}$Cs deposited on saltmarshes near Sellafield was studied during 1982 (Howard 1985; Howard & Lindley 1985). In 1984 a further study investigated the transfer of $^{137}$Cs to sheep grazing inland pasture close to Sellafield. This pasture is subject to wind-blown spray of a marine origin and aerosols emitted from the stack. In this paper initial results from the 1984 inland study are compared with those for the saltmarsh for 1982.

## STUDY SITES

Both of the study sites were located to the south of Sellafield (Fig. 1). The saltmarsh has developed behind sand dunes to the north of the mouth of the Esk estuary (National Grid Reference SD065975) and lies approximately 7 km south of Sellafield. The sheep have free access to both the saltmarsh and heathland beyond. The relative time spent grazing the different areas was estimated from observations conducted during the 1982 study (Howard 1985). The dominant vegetation on the saltmarsh consists of *Puccinellia maritima, Armeria maritima, Glaux maritima, Juncus gerardii* and *Salicornia* spp. The heathland beyond is slightly acidic and contains a wide variety of typical heath species including *Festuca rubra, Anthoxanthum odoratum, Trifolium repens, Juncus* spp, *Erica tetralix, Molinia caerulea* and *Sphagnum* spp.

The sheep flock maintained on the inland pasture grazed a number of different fields throughout the study period, ranging from 1·4 km to 3·6 km from the Sellafield stacks and 0·4 km (National Grid Reference NY035024) to 2·3 km (National Grid Reference NY055016) from the coast. The pastures used by this flock varied from permanent pastures, which had not been ploughed for more than 30 years, to rotational-land ploughed within the last 2 years. The flock grazed the unploughed land from April through to October 1984 and in January 1985. The dominant plant species in all the fields are *Lolium perenne* and *Trifolium repens* and the soil is a brown earth.

The sheep grazing the saltmarsh were a flock of Swaledale ewes which were brought from the Yorkshire Dales when about 3–4 years old. Grey-faced lambs were obtained from the ewes on the comparatively better lowland pasture on the coast for at least 3 years. The male lambs were sold for meat and the females for breeding. The inland flock consisted of grey-faced ewes, brought onto the farm at 0·5 years old and mated with Suffolk males to produce fat lambs.

## MATERIALS AND METHODS

$^{137}$Cs analysis was carried out for all samples using a pair of coaxial Ge(Li) detectors shielded by 10 cm of lead. The resolution at 1·33 MeV was 2·25 and 2·0 keV respectively, and the efficiencies were 27% and 29%. Tissue and control samples were counted for 170 K seconds, inland pasture samples for 80 K seconds and saltmarsh samples for 40 and 80 k seconds. The gamma spectra were analysed by the Canberra Spectran-F analysis system.

Three replicate vegetation samples were taken from the grazed areas each month using a sampling scheme devised to simulate the grazing behaviour of the sheep (Howard 1985). Daily intakes of $^{137}$Cs were estimated assuming that the ewes consumed the equivalent of

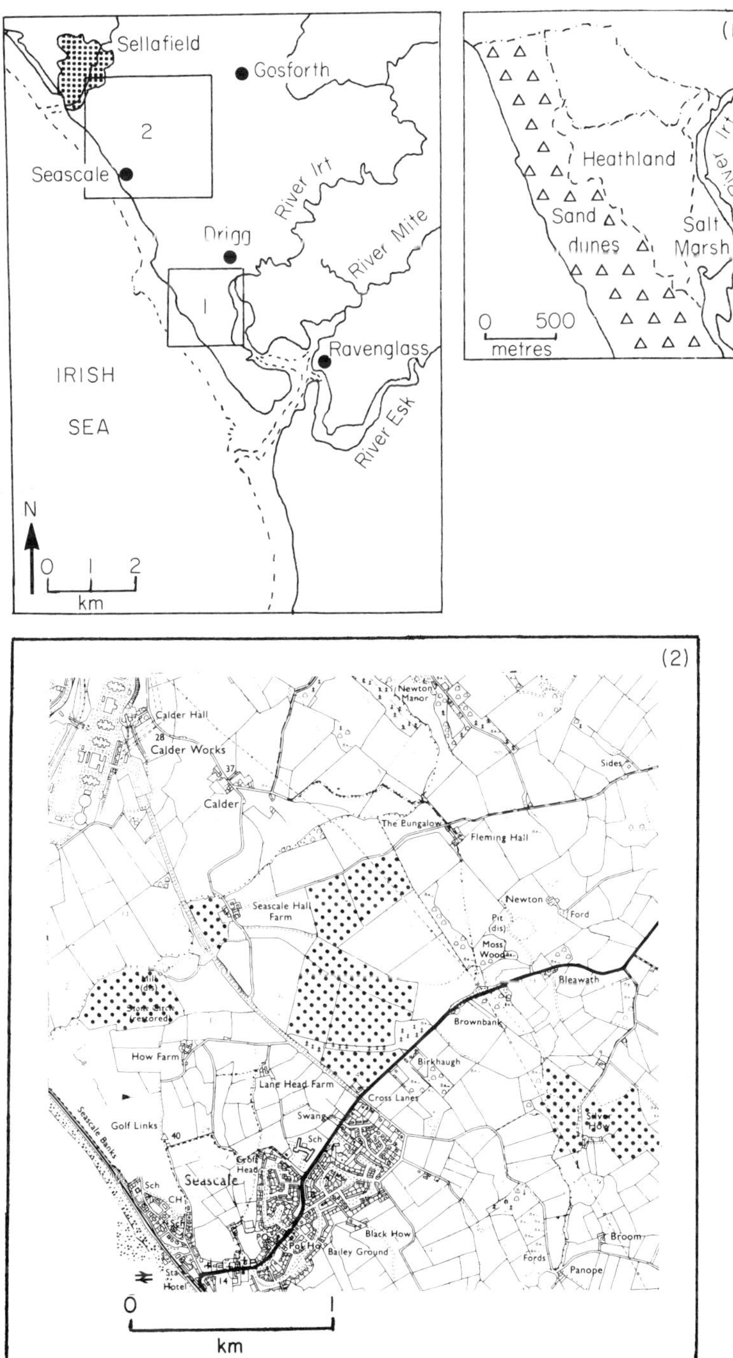


FIG. 1. Map showing location of the study sites in relation to Sellafield with insets of the pastures grazed by the two study flocks. (1) Saltmarsh site. (2) Inland pastures site: fields grazed by the inland flock are shaded.

1 kg dry weight of vegetation per day and the lambs 0·6 kg $d^{-1}$; additional feed supplements given during the lambing season were allowed for. Fresh faecal samples were taken (monthly on the saltmarsh; bi-monthly on the coastal pasture) 2 days after sampling the vegetation. Soil samples (0–15 cm in the ploughed fields and the top 2 cm on the permanent pasture and the saltmarsh) were also taken from each grazed area. The samples were dried at 105 °C for 24 hours, ground in a Tema mill and placed in plastic containers for analysis. All results are expressed on a dry weight basis.

Much of the $^{137}$Cs content of the vegetation may have been due to a surface deposit of soil or silt. Titanium is present in soils in reasonably large concentrations and the soil concentration would be expected to be about 10 000 times that of uncontaminated plants. Titanium analyses of the vegetation collected each month and of soil and silt were carried out, using the chemical fusion method of Sherman & Kanehiro in Black (1965) and X-ray fluorescence.

Control samples of vegetation (≃ 300 g dry weight), silt and soil (≃ 2 kg dry weight) were taken from a saltmarsh in the Humber estuary in September 1982 and the south of England in November 1984 (National Grid References TA235190 and SZ555089), respectively.

Three ewes and three lambs from both sites were slaughtered in September of the respective years just before the lambs were due to be sold. Muscle (hind leg), mineral bone (hind leg), liver, lung and kidney were removed and weighed both fresh and after drying in a freeze-drier. They were then ashed at 490 °C and analysed for $^{137}$Cs. Samples from a control ewe from the Yorkshire Dales from the farm supplying the Swaledale ewes (National Grid Reference SD982004) in 1982 and from the south of England (National Grid Reference SZ553089) in an area relatively free of nuclear installations in 1984 were also analysed.

## RESULTS

Saltmarsh vegetation and silt contained much more $_{137}$Cs than either the adjacent heathland or the inland pastures close to Sellafield (Fig. 2, Table 1). There was a pronounced seasonal variation in the $^{137}$Cs concentration in the saltmarsh and inland vegetation pasture which probably reflected differences in tidal inundation, rate of vegetation growth and grazing pressure. The $^{137}$Cs contents of the control vegetation were 0·52 and 13·06 Bq $kg^{-1}$ dry wt in the Humber estuary and southern England inland pasture, respectively.

The variation in the $^{137}$Cs content of faeces collected from the inland pastures followed similar seasonal patterns to the vegetation (Fig. 3). On the saltmarsh site the $^{137}$Cs content of the faeces collected in November was relatively low, reflecting the small amount of grazing that occurred on the saltmarsh at this time of year.

There was a significant correlation ($P<0{\cdot}001$) between the titanium content of the saltmarsh vegetation sampled monthly and the extent of $^{137}$Cs contamination (Fig. 4) which showed that the saltmarsh vegetation is covered by relatively more silt (with associated $^{137}$Cs) in the winter/spring than in the summer. The silt burden of the plants is obviously an important component of the total $^{137}$Cs intake of sheep grazing saltmarsh vegetation. The

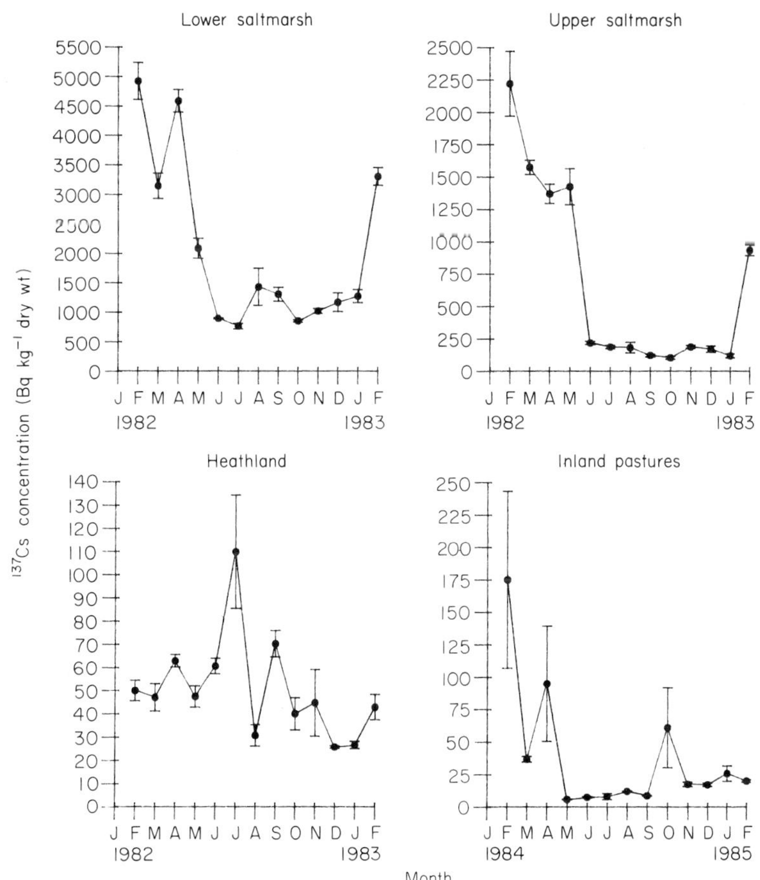


FIG. 2. $^{137}$Cs concentrations in the vegetation from the saltmarsh, heathland and inland pastures grazed by the sheep. Note the different scales for $^{137}$Cs (mean ± standard error).

TABLE 1. $^{137}$Cs concentrations in soil and silt from the study sites

| Sample | $^{137}$Cs concentration (Bq kg$^{-1}$ dry wt) |
|---|---|
| Saltmarsh site (1982) | |
| Lower marsh silt | 15580 |
| Upper marsh silt | 11210 |
| Heathland soil | 57·3 |
| Inland pastures site (1984) | 36·1 – 87·3 |
| Humber estuary control silt (1982) | 18·9 |
| Southern England control soil (1984) | 35·3 |

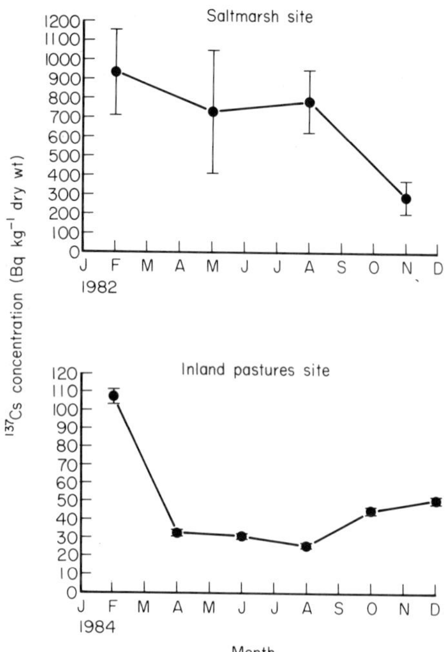


FIG. 3. $^{137}$Cs concentrations in faeces collected from the study sites (mean ± standard error).

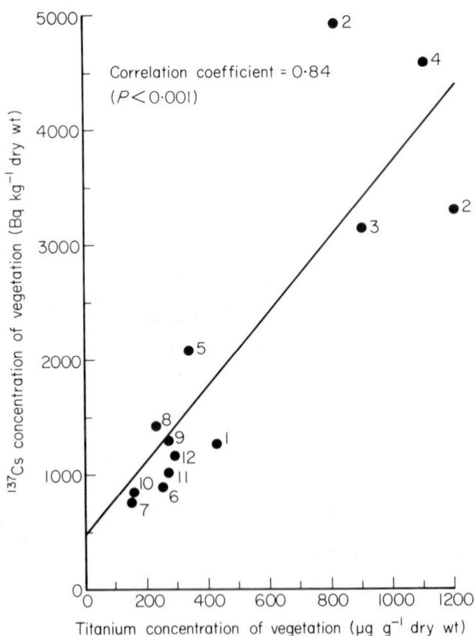


FIG. 4. Correlation between the concentrations of $^{137}$Cs and Ti of vegetation from the lower saltmarsh (according to Spearman's rank correlation coefficient). Numbers indicate the month of the year.

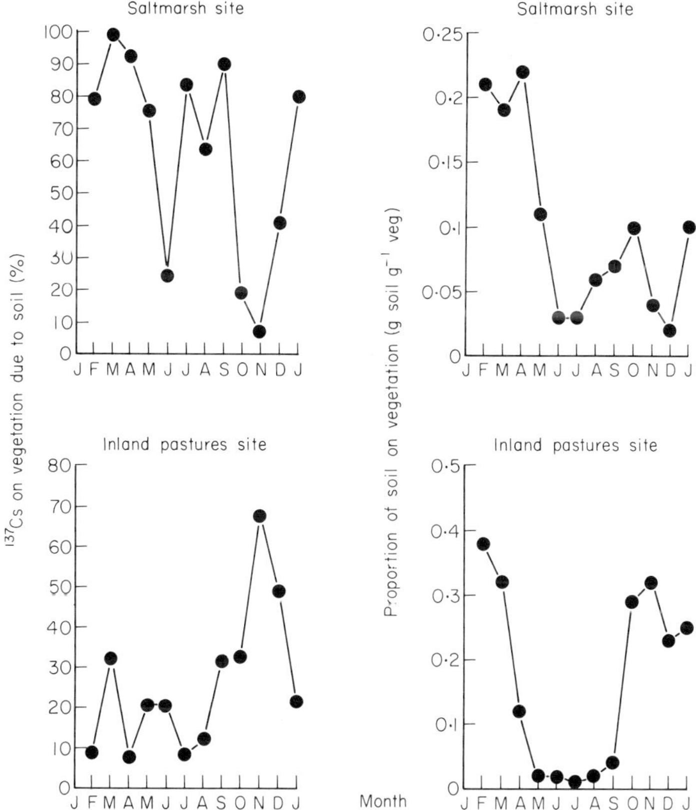


FIG. 5. Soil contamination on vegetation expressed as (i) the proportion of $^{137}$Cs activity of sampled vegetation due to soil and as (ii) the weight of soil on vegetation.

$^{137}$Cs activity on herbage due to silt or soil at the study sites (Fig. 5) was estimated by multiplying the $^{137}$Cs soil/silt concentration by its weight on the ingested vegetation (estimated using titanium determinations). The lower saltmarsh was inundated by the tide more frequently and the vegetation was covered with relatively more silt than the upper marsh, causing higher $^{137}$Cs concentrations on the sampled vegetation. The importance of silt contamination in influencing $^{137}$Cs levels on the saltmarsh was evident. In the summer months of July, August and September when much of the sheep grazing occurred on the saltmarsh, rather than the heathland, soil contamination of vegetation of 3–7% by weight accounted for 64–90% of the $^{137}$Cs activity by weight of vegetation. Over the sampled year the proportion of $^{137}$Cs in ingested vegetation due to silt/soil ranged from 7% (November) to 99% (March). On the inland pastures, the proportional weight of soil contamination varied markedly with season ranging from 1% (July) to 38% (February), accounting for 8–67% of $^{137}$Cs activity of the vegetation.

Sheep can take in $^{137}$Cs by both ingestion and inhalation, but from local measurements of the air the latter route probably accounts for less than 1% of the daily intake and can be discounted (Sumerling, Dodd & Green 1984). Similarly $^{137}$Cs concentrations measured in

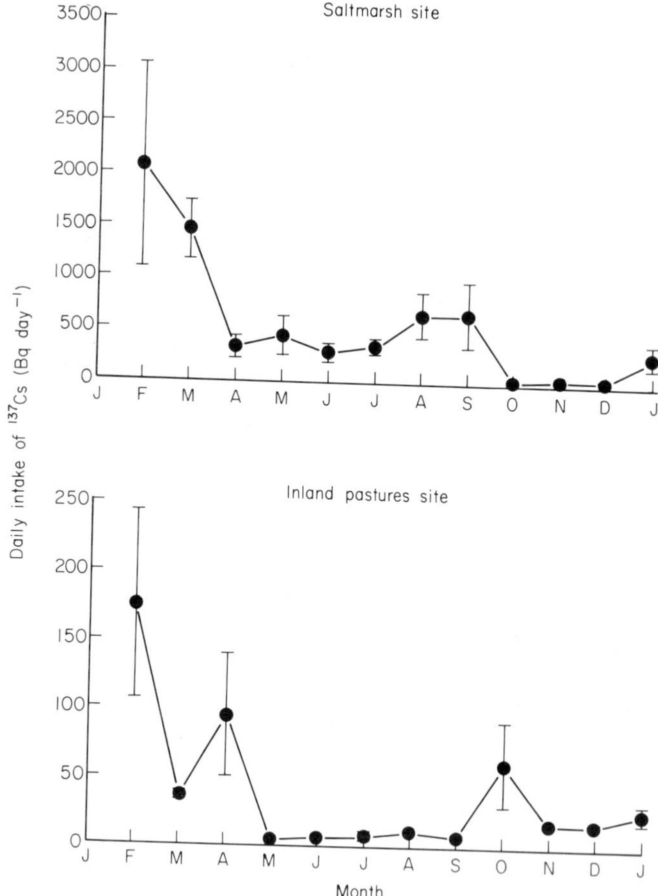


FIG. 6. Daily intake of $^{137}$Cs by ewes grazing the two study sites (mean ± standard error).

stream water in the pastures in this study were negligible. The daily intake of $^{137}$Cs by sheep on the saltmarsh was considerably greater than on the inland pasture (Fig. 6) and tended to decline throughout the sampled year. The daily intake of the inland pasture sheep was lower in the summer months than in the winter, spring and autumn.

The difference in the daily intake of $^{137}$Cs was reflected in the $^{137}$Cs concentration of the sheep tissues (Fig. 7). In the saltmarsh sheep, $^{137}$Cs concentration in the lambs' tissues was consistently higher than in the ewes, ranging from 274 to 756 Bq kg$^{-1}$ dry weight; the $^{137}$Cs content of saltmarsh ewes tissues ranged from 5 (lung) to 39 (kidney) times higher than the control ewe.

The $^{137}$Cs concentrations in soft tissues of lambs and ewes from the inland pasture were similar. Lamb bone had significantly more ($P<0{\cdot}05$) $^{137}$Cs than ewe bone. $^{137}$Cs tissue concentrations were considerably lower than in the saltmarsh sheep, ranging from 1·1 to 6·8 Bq kg$^{-1}$ dry weight in lamb tissue. The ewe tissues from inland pasture ranged from

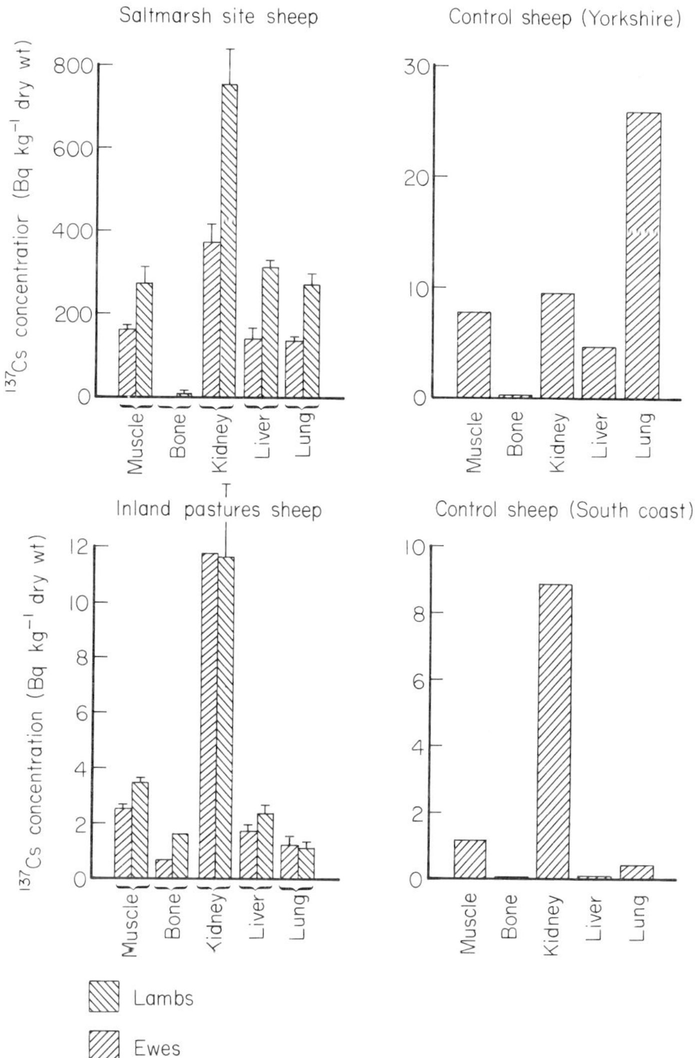


FIG. 7. $^{137}$Cs concentrations in the study and control sheep tissues (mean ± standard error). Where no error bar is shown $n \leqslant 2$.

only 1·0 (kidney) to 24·9 (liver) times higher than the control ewe tissues. Kidney $^{137}$Cs concentrations were similar in the 1984 control and the inland pasture ewes.

The transfer of $^{137}$Cs into sheep tissues can be expressed using transfer coefficients ($F_f$) calculated using the expression:

$$F_f \text{ (d kg}^{-1}\text{)} = \frac{^{137}\text{Cs concentration in tissues (Bq kg}^{-1}\text{ fresh wt)}}{\text{daily intake of }^{137}\text{Cs (Bq day}^{-1}\text{)}}$$

Because $^{137}$Cs is generally assumed to have a relatively short biological half-life of about

18 d (Coughtrey & Thorne 1983), the September tissue values were compared to the estimated daily intakes for the previous month (i.e. August).

Transfer coefficients have been calculated for two different types of daily intake (DI):

(a) Total daily intake, using the $^{137}$Cs content of the vegetation consumed ($DI_t$).

(b) Daily intake associated with the vegetation only i.e. excluding the fraction of $^{137}$Cs associated with the silt/soil component ($DI_v$).

There was a significant difference ($P<0·05$) in the transfer coefficients of the majority of the soft tissues of sheep at the two 'contaminated' sites (Table 2), the transfer coefficients

TABLE 2. Transfer coefficients for $^{137}$Cs of selected tissues of sheep at the two study sites, using daily intake estimates of (i) total intake ($DI_t$) and (ii) intake without soil component ($DI_v$)

| | Liver | Kidney | Muscle | Bone |
|---|---|---|---|---|
| Saltmarsh site | | | | |
| $DI_t$ | | | | |
| Ewe | $5·7 \times 10^{-2}$ | $1·5 \times 10^{-1}$ | $6·4 \times 10^{-2}$* | $1·4 \times 10^{-2}$ |
| CV(%) | 50·88 | 51·33 | 55·0 | 50·71 |
| Lamb | $2·0 \times 10^{-1}$* | $4·9 \times 10^{-1}$* | $1·8 \times 10^{-1}$* | $8·5 \times 10^{-2}$ |
| CV(%) | 55·00 | 53·06 | 51·67 | 52·94 |
| $DI_v$ | | | | |
| Ewe | $1·6 \times 10^{-1}$* | $4·1 \times 10^{-1}$* | $1·8 \times 10^{-1}$* | $3·9 \times 10^{-2}$ |
| Lamb | $5·5 \times 10^{-1}$* | 1·4* | $4·9 \times 10^{-1}$* | $2·3 \times 10^{-1}$ |
| Inland pastures site | | | | |
| $DI_t$ | | | | |
| Ewe | $3·7 \times 10^{-2}$ | $1·7 \times 10^{-1}$ | $5·3 \times 10^{-2}$* | $4·9 \times 10^{-2}$ |
| CV(%) | 27·78 | 24·29 | 23·56 | NA |
| Lamb | $8·0 \times 10^{-2}$* | $2·8 \times 10^{-1}$* | $1·2 \times 10^{-1}$* | $1·1 \times 10^{-1}$ |
| CV(%) | 19·18 | 55·41 | 17·83 | 112·15 |
| $DI_v$ | | | | |
| Ewe | $4·2 \times 10^{-2}$* | $1·9 \times 10^{-1}$* | $6·0 \times 10^{-2}$* | $5·6 \times 10^{-2}$ |
| Lamb | $9·2 \times 10^{-2}$* | $3·2 \times 10^{-1}$* | $1·4 \times 10^{-1}$* | $1·2 \times 10^{-1}$ |

NA, not available.

*Significant difference between saltmarsh and coastal site (Mann–Whitney *U*-test, $P < 0·05$).

for the inland pasture sheep being consistently lower than the saltmarsh site sheep. The difference was enhanced when soil contamination was excluded.

## DISCUSSION

Initially it was expected that sheep grazing inland pasture, subject to $^{137}$Cs input from aerial discharges, would take up more $^{137}$Cs from the gut than sheep grazing saltmarshes where $^{137}$Cs was adsorbed onto silt particles. However, this does not appear to be the case with much of the $^{137}$Cs of inland pasture vegetation grazed by the sheep being associated with the soil which might make it relatively unavailable for absorption. The addition of phyllosilicate clay minerals to fodder has been shown to greatly reduce the uptake of $^{134}Cs$ by sheep (Hoek Van den 1976).

The higher transfer coefficients for the saltmarsh site sheep are difficult to explain and it may be due to a number of different factors, such as the saline environment of the saltmarsh, the roughage content of the vegetation which Johnson *et al.* (1968) thought to affect $^{137}Cs$ uptake in cattle, or the availability of $^{137}Cs$ in saltmarsh vegetation, although the uptake of $^{137}Cs$ by *P. maritima,* the dominant saltmarsh grass, from silt was shown to be very low (Allen *et al.* 1984). An obvious omission in the work to date is the potential influence of potassium in $^{137}Cs$ uptake. Analysis of potassium in the samples is in progress and will be discussed in future publications.

Airborne discharges of $^{137}Cs$ from Sellafield have been reduced considerably since 1981 following the commissioning of new equipment (Table 3). Similarly, reductions have

TABLE 3. Recent annual airborne discharges of $^{137}Cs$ from the Sellafield reprocessing plant (BNFL 1980–84)

| Year | TBq |
|---|---|
| 1979 | 0·24 |
| 1980 | 0·45 |
| 1981 | 0·08 |
| 1982 | 0·01 |
| 1983 | 0·007 |

occurred in the liquid discharges of $^{137}Cs$ in recent years. This obviously limits direct comparison of the magnitude of the $^{137}Cs$ concentrations found on the different sites in different years; levels in 1984 samples would be expected to be less than those in 1982. Recent measurements of $^{137}Cs$ concentrations on the lower saltmarsh have been lower than those found in 1982 (e.g. 281 Bq $kg^{-1}$ and 209 Bq $kg^{-1}$ in July and August 1985, respectively). In the inland pastures the reduction is reflected in the relatively high $^{137}Cs$ activity in soil compared to vegetation around the reprocessing plant. In 1981 Sumerling, Dodd & Green (1984) estimated that only 30% of $^{137}Cs$ on cattle-grazed vegetation was being maintained by direct deposition and that the bulk of the activity of vegetation was being maintained by older deposits in the soil. They suggested that the mechanical transport of soil to the vegetation surface was more important than root uptake by plants, and accounted for approximately 60% of the $^{137}Cs$ activity consumed by the cattle. In the present study of sheep, between 7–99% and 8–67% (saltmarsh and inland sites respectively) of $^{137}Cs$ intake was calculated to arise from silt/soil contamination. The extent of tidal inundation dominated $^{137}Cs$ concentrations of saltmarsh vegetation due to the predominant influence of the silt in determining $^{137}Cs$ activity. However, the importance of soil contamination of vegetation ingested by the saltmarsh site sheep was also affected by the proportion of time that the sheep grazed on the heathland compared with the saltmarsh. Over 75% of grazing in June, July and August took place on the saltmarsh as opposed to less than 25% in December and January. In the inland pastures there was a marked seasonal aspect to the extent of soil contamination which would be influenced by factors such as stocking rates, pasture types, weather conditions and the rate of vegetation growth.

It is not clear whether the $^{137}$Cs activity associated with the soil/silt component is totally unavailable for uptake by the ruminant gut. Radionuclides artificially added to soils have been found to transfer to a variety of tissues in two tethered sheep (Healy, McCabe & Wilson 1970). $^{137}$Cs is associated with a number of different components in the soil, but the majority is bound to clay mineral particles. In soils collected close to Sellafield in 1983 F.R. Livens (pers. comm.) found, using sequential extraction techniques, described by McLaren & Crawford (1973) that 6–16% was labile, 5–13% was associated with organic material, 0–13% was bound to oxides and 63–82% present in the residue, presumably bound in clay-silicate lattices. Further experiments in controlled conditions are required to look at the absorption of $^{137}$Cs from different components of the pasture system, and more detailed investigations of the characteristics of the $^{137}$Cs present in faecal samples may help to identify vegetation components which retain $^{137}$Cs during passage through the gut. Titanium analyses of faeces may also provide more accurate estimates of soil ingestion especially if daily herbage intake measurements can be made.

Further laboratory work is also required in order to understand the differences in the relationship between $^{137}$Cs concentrations in ewes and lambs at the two sites. Trials, by the author in collaboration with the Hill Farming Research Organization, are being conducted feeding saltmarsh vegetation to 5 month old lambs and ewes (B.J. Howard & R. W. Mayes, unpublished). The management practices at the two sites might influence $^{137}$Cs levels in sheep since, in the inland pasture, ewes were bought at a much younger age (0·5 years old) than the saltmarsh ewes (3–4 years old). However, the short biological half-life of $^{137}$Cs in soft tissues ($<$ 1 month) should make the past history of the ewe relatively unimportant by the time they were sampled.

Concentrations of $^{137}$Cs in all tissues of inland sheep were lower than those measured by Popplewell *et al.* (1981) from a nearby site in 1979. $^{137}$Cs concentrations in kidneys were consistently higher than for the other soft tissues, and were comparable for the control and inland pasture samples.

## ACKNOWLEDGMENTS

I am indebted to both farmers for the use of their land and their help during this study. I would also like to thank colleagues at Merlewood, particularly A.D. Horrill and N.A. Beresford for practical help, constructive discussion and help with plotting figures.

This research was carried out under contract for the Department of the Environment, as part of its radioactive waste management research programme. The results will be used in the formulation of Government policy, but at this stage do not necessarily represent Government policy.

## REFERENCES

**Allen, S.E., Horrill, A.D., Howard, B.J., Lowe, V.P.W. & Parkinson, J.A. (1984).** *Radionuclides in terrestrial ecosystems.* NERC contract report to DOE (DOE/RW/84.013).

**Black, C.A.** (Ed). **(1965).** *Methods of Soil Analysis,* Vol. 2, American Society of Agronomy, Madison.

**Bradford, W.R. Curtis, E.J.C. & Popplewell, D.S. (1984).** Radioactivity in environmental samples taken in the Sellafield and Ravenglass areas of west Cumbria, 1977–1982. *Science of the Total Environment,* **35**, 267–283.

**BNFL (1980).** *Annual Report on Radioactive Discharges to the Environment 1979,* British Nuclear Fuels Limited, Risley.

**BNFL (1981).** *Annual Report on Radioactive Discharges to the Environment 1980,* British Nuclear Fuels Limited, Risley.

**BNFL (1982).** *Annual Report on Radioactive Discharges to the Environment 1981,* British Nuclear Fuels Limited, Risley.

**BNFL (1983).** *Annual Report on Radioactive Discharges to the Environment 1982,* British Nuclear Fuels Limited, Risley.

**BNFL (1984).** *Annual Report on Radioactive Discharges to the Environment 1983,* British Nuclear Fuels Limited, Risley.

**Bulman, R.A. & Cooper, J.R.** (Eds) **(1986).** *Speciation of fission and activation products in the environment.* Proceedings of Speciation 85 Seminar. Elsevier Applied Science Publishers, London.

**Coughtrey, P.J. & Thorne, M.C. (1983).** *Radionuclide Distribution and Transport in Terrestrial and Aquatic Ecosystems,* Vol. 1, A.A. Balkema, Rotterdam.

**Healy, W.B., McCabe, W.J. & Wilson, G. (1970).** Ingested soil as a source of micro elements for grazing animals. *New Zealand Journal of Agricultural Research,* **13**, 503–521.

**Hoek Van den, J. (1976).** Cesium metabolism in sheep and the influence of orally ingested bentonite on cesium absorption and metabolism. *Zeitschrift für Tierphysiologie Tierernahrung und Futtermittelkunde,* **37**, 315–321.

**Howard, B.J. (1985).** Aspects of the uptake of radionuclides by sheep grazing on an estuarine saltmarsh. 1. The influence of grazing behaviour and environmental variability on daily intake. *Journal of Environmental Radioactivity,* **2**, 183–198.

**Howard, B.J. & Lindley, D.K. (1985).** Aspects of the uptake of radionuclides by sheep grazing on an estuarine saltmarsh. 2. Radionuclides in sheep tissues. *Journal of Environmental Radioactivity,* **2**, 199–213.

**Johnson, J.E., Ward, G.M., Firestone, E. & Knox, K.L. (1968).** Metabolism of radioactive caesium ($^{134}$Cs and $^{137}$Cs) and potassium as influenced by high and low forage diets. *Journal of Nutrition,* **94**, 282–288.

**McLaren, R.G. & Crawford, D.V. (1973).** Studies on soil copper: I. The fractionation of Cu in soils. *Journal of Soil Science,* **24**, 172–181.

**Popplewell, D.S., Ham, G.J., Savory, T.E. & Bradford, W.R. (1981).** Radionuclide concentrations in some cattle and sheep from west Cumbria. *Radiological Protection Bulletin,* **41**, 20–23.

**Sumerling, T.J., Dodd, N.J. & Green, N. (1984).** The transfer of strontium-90 and caesium-137 to milk in a dairy herd grazing near a major nuclear installation. *Science of the Total Environment,* **34**, 37–72.

# Generalized models for the transfer and distribution of stable elements and their radionuclides in agricultural systems

D. JACKSON* AND A.D. SMITH
*Associated Nuclear Services, 60 East Street, Epsom, Surrey, KT17 1HA*

## SUMMARY

**1** Generalized equilibrium and time-dependent mathematical models for simulating the transfer and distribution of stable elements and radionuclides in food-producing systems are described. Dynamic models offer greater flexibility of use, but require a much more detailed database, than equilibrium models. Particular emphasis is placed on the development of sub models appropriate to animal components.
**2** A specific suite of computer codes (SPADE) developed on behalf of the Ministry of Agriculture, Fisheries and Food, describing transfer within the agricultural soil–plant–animal system, is described and illustrative results presented for sulphur. The models can be used in the predictive assessment of routine or accidental releases of radionuclides or stable elements.
**3** In order to validate predictions arising from such complex models it is necessary to obtain data sets independent of those used to construct and test the model. Data arising from monitoring studies may be appropriate, providing that adequate information describing environmental conditions are available.
**4** The use of 'sensitivity analysis', to determine parameter coefficients to which model results are sensitive, is demonstrated to be useful in providing inputs to programmes of research.

## INTRODUCTION

One aspect of the assessment of the impact of releases of radionuclides from the nuclear fuel cycle, or of stable elements from other sources, is the predictive calculation of distribution and transport in ecological systems. A number of models of varying form and complexity have been developed for this purpose. In the past, these models have tended to consider equilibrium end-points only, reflecting the 'concentration ratio' approach to data collation. More recently, available data have been reassessed and new data have been obtained allowing a time-dependent approach to modelling. Such data relate primarily to radionuclides, but stable analogues of various isotopes have also received some

*Present address: British Nuclear Fuels plc, Safety and Services Department, Sellafield, Cumbria, CA20 1PG.

consideration.

Several organizations within the U.K. and elsewhere have developed mathematical models describing the transport and distribution of radionuclides through terrestrial, agricultural, food-chain systems (e.g. Linsley, Simmonds & Haywood 1982; Nair *et al.* 1983; Roberts 1983; Thorne & Coughtrey 1983; Nair 1984; Jackson, Coughtrey & Crabtree 1985). Such models, which can be termed 'descriptive', allow the predictive assessment of the impact of unplanned releases, as well as providing an input into the definition of regulatory limits on discharge levels. Where the models are sufficiently sophisticated and mechanistic in approach (i.e. based on known mechanisms of transfer) they may also be used to identify important parameters and transfer coefficients for which further data are required.

This paper presents generalized equilibrium and time-dependent approaches to modelling the transfer to, and distribution of, stable elements and their radionuclides in domestic animals. In addition a specific model for soil–plant–animal systems (SPADE), developed for the Ministry of Agriculture, Fisheries and Food (MAFF), is described, and illustrative results presented. The structure and function of SPADE allows detailed analyses to be undertaken of the potential distribution of a wide range of elements in agricultural ecosystems, subsequent to deposition to soil or vegetation.

## METHODS FOR CALCULATING UPTAKE AND DISTRIBUTION OF RADIONUCLIDES AND STABLE ELEMENTS

Models currently in use to describe the uptake and distribution of radionuclides or stable elements in animals fall into two general categories: equilibrium models and dynamic models. Examples of equilibrium models for soil–plant–animal systems are presented by Nair, Hotson & Stacey (1980) and Coughtrey & Thorne (1983a). Such models tend to be conservative, in that no opportunity exists for changes to occur in activity concentrations with time from the input. Although widely used in the past, equilibrium models are very restricted in their scope and application. To overcome the inflexibility of equilibrium models, pseudo time-dependent models were developed. In such models, compartmental contents are taken to build up with time, but with equilibrium conditions being imposed rapidly, by incorporation of artificially high reflux terms (e.g. Simmonds, Linsley & Jones 1979; Simmonds & Linsley 1981). More recently, the continued development of time-dependent models has led to adoption of more realistic rate coefficients and reflux terms between compartments and such models are currently favoured for representing animal systems (e.g. Linsley *et al.* 1982; Nair *et al.* 1983; Thorne & Coughtrey 1983; Nair 1984; Jackson *et al.* 1985). The general approach to equilibrium and time-dependent modelling is described below.

### *Equilibrium models*

The advantages of using equilibrium models are that they are relatively simple in concept, and therefore easily understood, and they require only a set of static transfer coefficients specific to given radionuclides or elements. The concentration ($C$) of radionuclide or stable element per unit mass of animal product ($C_a$) is given by:

$$C_a = f_a \sum_g J_g C_g \tag{1}$$

where $J_g$ denotes the daily intake of food $g$ contaminated at level $C_g$ per unit mass; and $f_a$ denotes the equilibrium diet-to-animal product transfer factor. This last term has units of days per unit mass, and can be defined as:

$$f_a = \frac{\text{(activity concentration in animal product per unit mass)}}{\text{(total activity intake in food sources per day)}} \tag{2}$$

In cases where contamination by radionuclides is of interest, and food has been stored, the radioactive-decay corrected concentration, of single nuclides, in food sources ($C_g^*$) can be determined from:

$$C_g^* = C_g \exp(-\lambda_r t_s) \tag{3}$$

where $\lambda_r$ is the radioactive decay constant (time$^{-1}$); and $t_s$ is the time (expressed in the same units as $\lambda_r$) for which the food was stored. Where radionuclide production may occur within a food source, from the radioactive decay of 'parent' nuclide, both build-up and decay must be modelled in order to determine concentrations in food at any time.

*Dynamic models*

Mathematically, multi-compartment dynamic models can be described generally by sets of equations of the form:

$$\frac{\delta}{\delta t} q_i(t) = \sum_{i=1}^{n} \lambda_{ij} q_j(t) + I_i(t) \qquad \text{for } i = 1 \text{ to } n \tag{4}$$

where $q_i(t)$ is the content of compartment $i$ at time $t$; $I_i(t)$ is the rate of intake into compartment $i$ at time $t$; $\lambda_{ij}$ with $i \neq j$, is the constant (time$^{-1}$) for transfer from compartment $j$ to compartment $i$; $\lambda_{ij}$, with $i = j$, is $<0$ and is the total rate of loss from compartment $i$; and $n$ is the number of compartments.

For a multi-compartmental model of radionuclide transfer, it can be shown readily that the total rate of loss from compartment $i$ ($\lambda_{ii}$) is given by the following:

$$\lambda_{ii} = -\sum_{\substack{j=1 \\ j \neq i}}^{n} \lambda_{ji} - \lambda_r - \lambda_i^L \tag{5}$$

ANIMAL MODEL
Tracheobronchial region
ingestion
inhalation
inhalation
Upper gastro-intestinal tract
inhalation
inhalation
Pulmonary region
Lower gastro-intestinal tract
Tracheobronchial lymph nodes
Systemic circul-ation
Excretion to soil
Tissue comp. 1
Tissue comp. 2
Tissue comp. 3
Tissue comp. 4
Excretion to soil

PLANT MODEL
foliar deposition
External leaf
to soil as washoff
deposition of gases and vapours
Internal leaf
Grain
animal ingestion
Stem
Root Store
Root
from soil solution

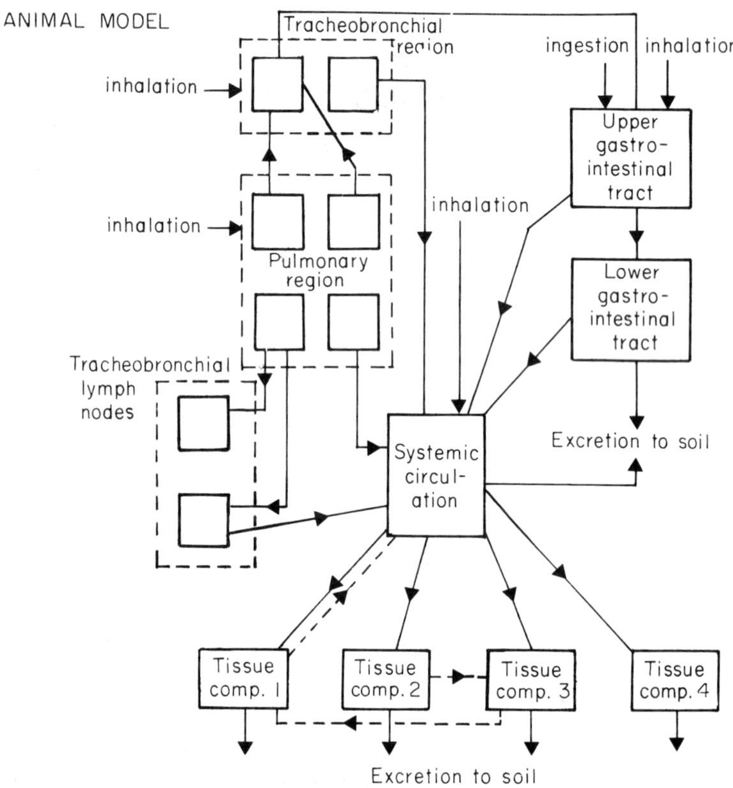

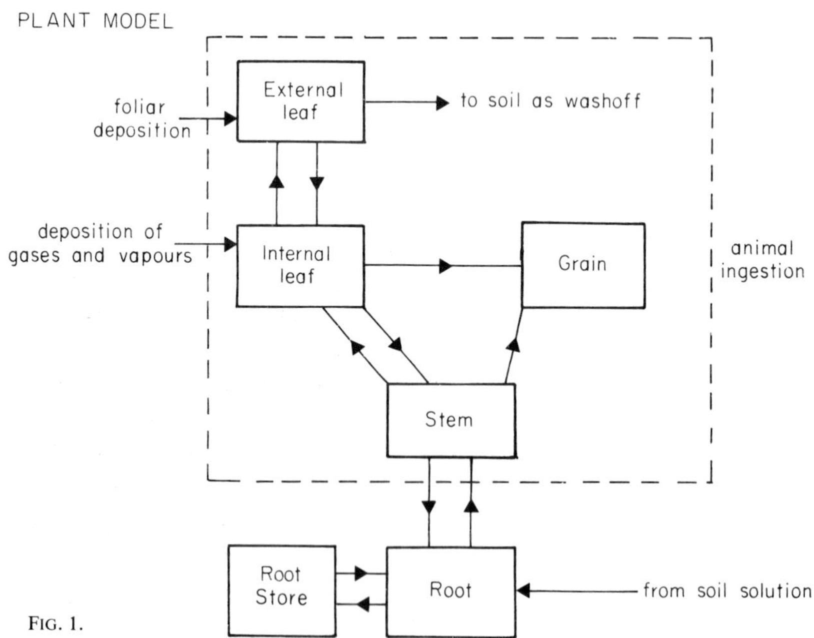

FIG. 1.

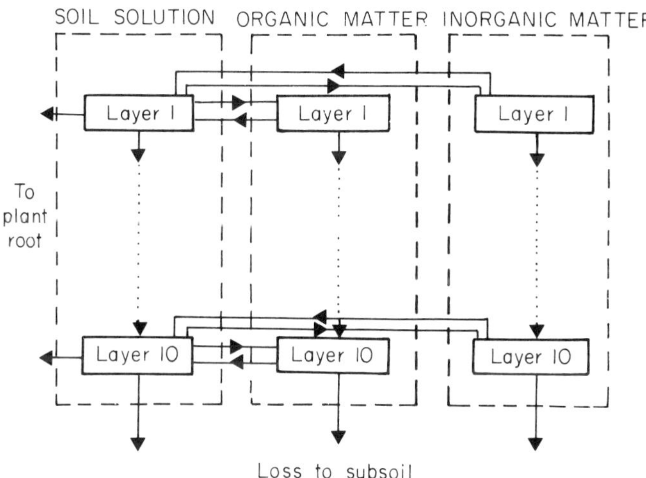


FIG. 1. Activation/fission product soil–plant–animal model.

where $\lambda_r$ is the radioactive decay constant of the radionuclide; and $\lambda_i^L$ is the rate of loss from compartment *i* to outside the system.

Such models are flexible in construction and are amenable to solution by various analytical and numerical methods (e.g. Harte 1976; Jacquez 1972; Thorne 1979). However, an extensive database is required in order to develop and test such models.

### *Dynamic models for soil–plant animal systems*

During the past 5 years, Associated Nuclear Services (ANS) has undertaken the development of several time-dependent mechanistic models to describe the uptake and distribution of various stable elements and their radioisotopes within animals. Computer codes currently available allow consideration of stable and radioisotopes of the more important elements arising as fission/activation products in nuclear power plants (ie. I, S, Sb, Ba/La, Zr/Nb, Cs, Sr, Ce, Tc, Ru and Te) and the actinides (i.e. Pu, Am, Cm and Np). Additional models under development will extend this list to include tritium and $^{14}C$ also.

The integrated suite of codes for application to soil–plant–animal systems (SPADE), commissioned by MAFF, has been outlined by Jackson *et al.* (1985). Specific models employed within SPADE are variable with element, but the general forms of model used for activation/fission product elements and the actinide elements are illustrated in Figs 1 and 2 respectively. It can be seen that whilst the soil–plant components of the models differ substantially, the animal component is essentially similar in both cases.

Parameters and coefficients for SPADE have been derived from extensive reviews of the literature, undertaken on behalf of the Commission of the European Communities (Coughtrey & Thorne 1983a, b; Coughtrey, Jackson & Thorne 1983a; Coughtrey *et al.*

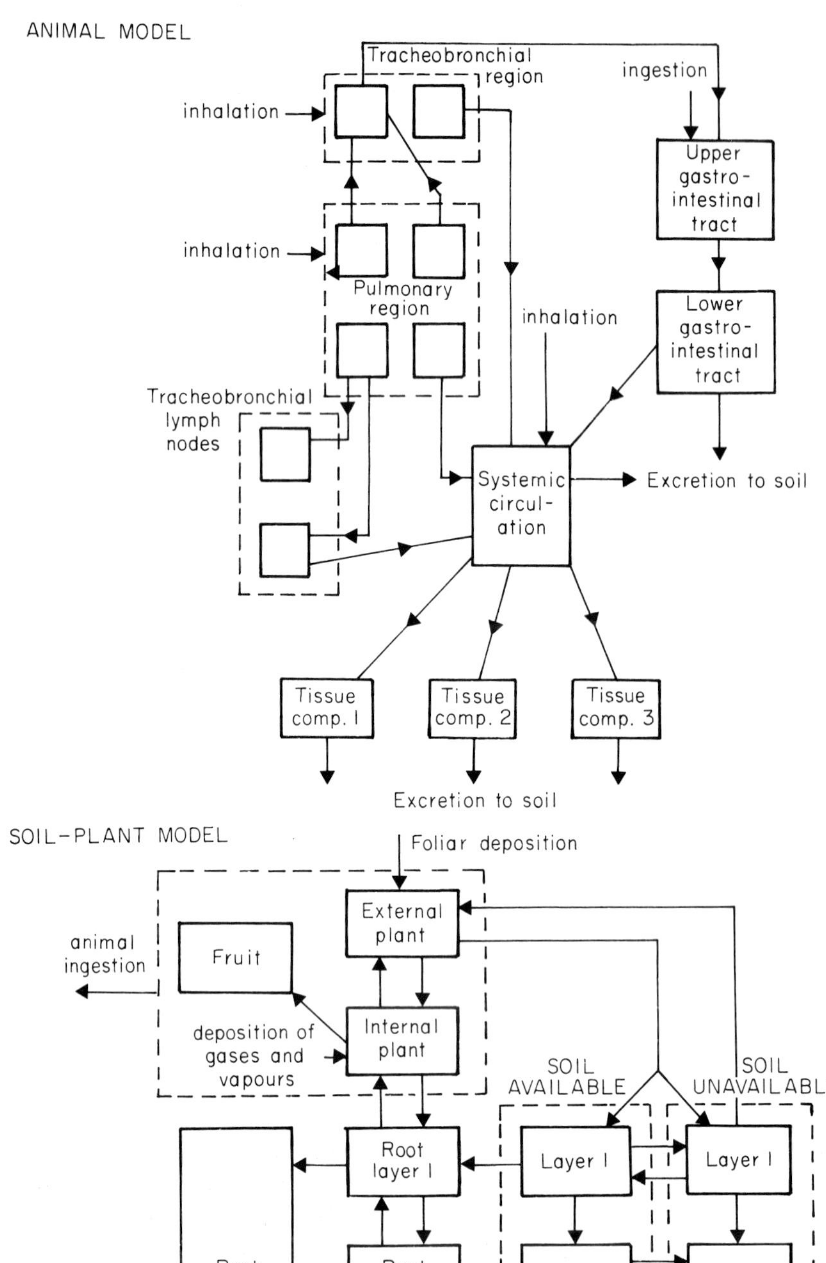


FIG. 2. Actinide soil–plant–animal model.

TABLE 1. Selected parameter coefficients used within the soil-plant-animal model SPADE to describe rates of transfer of inorganic compounds of sulphur between various compartments

| | |
|---|---|
| *Soil data* | |
| Soil solution to organic matter | 7·0E-9 $s^{-1}$ |
| Organic matter to soil solution | 7·0E-9 $s^{-1}$ |
| Soil solution to inorganic matter | 2·0E-9 $s^{-1}$ |
| Inorganic matter to soil solution | 2·0E-9 $s^{-1}$ |
| Soil solution to plant root | 6·0E-6 $s^{-1}$ |
| Depth modifiers: organic | 0·1 $cm^{-1}$ |
| (cut-off | 20·0 cm) |
| inorganic | 0·01 $cm^{-1}$ |
| *Plant data* | |
| Root to stem | 3·0E-6 $s^{-1}$ |
| Stem to root | 3·0E-6 $s^{-1}$ |
| Stem to internal leaf | 3·0E-6 $s^{-1}$ |
| Internal leaf to stem | 3·0E-6 $s^{-1}$ |
| External leaf to internal leaf | 1·3E-5 $s^{-1}$ |
| Internal leaf to external leaf | 6·0E-7 $s^{-1}$ |
| Internal leaf to grain | 3·0E-9 $s^{-1}$ |
| Stem to grain | 3·0E-9 $s^{-1}$ |
| Wash-off | 2·0E-5 $s^{-1}$ |
| Fraction of washoff to soil solution | 0·99 |
| Plant density | 1·25 kg $m^{-2}$ |
| *Animal data* | |
| Gastrointestinal uptake factor ($f_1$) | 0·8 |
| Grazing area per animal (cattle) | 500 $m^2$ |
| Forage consumption | 55 kg (*f*.w.) $d^{-1}$ |
| Soil consumption | 0·11 kg $d^{-1}$ |

Compartmental names do not represent strict equivalents to the physical or physiological terms used.

1984a, b; Coughtrey, Jackson & Thorne 1985), the Department of the Environment (Coughtrey *et al.* 1984c) and MAFF (Thorne, Coughtrey & Meekings 1983), supplemented by experimental studies commissioned by MAFF (Coughtrey *et al.* 1983b). The large number of parameters used in SPADE necessitate the use of a pre-processor which facilitates the creation of parameter files appropriate to specific applications. In order to allow site-specific calculations to be undertaken, a range of options are available on relevant parameters such as soil characteristics (e.g. structure, organic content, pH, drainage) and land use (e.g. pasture type, arable land, garden). Various season types can be specified, corresponding to agricultural practices and conditions and a number of input modes can be accommodated to reflect potential industrial practices.

Separate sub-models are available within the animal component of SPADE in order to simulate distribution and dynamics in broad groups such as goats, sheep or cattle (beef or dairy animals). For whichever animal type is specified, the model is structured so that intake by ingestion, or both inhalation and ingestion, may be considered. Default values are available with respect to breathing rate, grazing area, feeding rate, etc., although all coefficients can be modified in order to model particular situations or to undertake sensitivity analyses (e.g. Thorne *et al.* 1983; Jackson *et al.* 1985). As an example, selected default parameter coefficients employed within the code for modelling inorganic compounds of sulphur deposited to permanent pasture are listed in Table 1. Compartmental names referred to in Table 1 correspond to those given in Fig. 1. It should be noted that the terms used do not represent strict physical or physiological categories but are used for convenience of identifying transfer between compartments and compartmental contents.

Additional sub-models for the metabolism in man of several toxic chemicals have been developed, on behalf of the CEC, for use in conjunction with monitoring of exposure of workers to these substances (Thorne & Jackson 1983; Smith & Thorne 1984). The substances reviewed include metals of known toxicological importance such as lead, cadmium, arsenic and nickel. Extension of these models to include transport through the environment could yield an invaluable tool for the predictive assessment of the distribution, environmental levels and uptake by man resulting from routine releases of such materials from industrial plant, such as smelters.

At present, the sub-model used for lung in the soil–plant–animal model (Figs 1 and 2) is based on that developed for man by the International Commission on Radiological Protection (ICRP 1966), but the two rapidly cleared compartments normally used to represent the nasal passages have been removed, and activity which would have been deposited in them is transferred directly to the blood or gastrointestinal tract as appropriate. The sub-model for the gastrointestinal tract of ruminants (i.e. cattle) has been developed specifically for this work; the sub-model for monogastric species is based on that developed for man by Eve (1966). For systemic metabolism, the sub-models used are structurally similar to those developed by the ICRP for man (ICRP 1979). The applicability of these models to other species has not been tested rigorously.

## *Model applications and sensitivity analysis*

The soil–plant–animal model described is intended for use principally as an investigational tool, to determine the distribution of radionuclides at specific times following release to various components of the environment, and is suitable for evaluating the radiological implications of both 'spike' (single, instantaneous release) and continuous releases (uniform or time-variable). In addition, the systematic variation of selected rate coefficients singly or in combination ('sensitivity analysis') provides an opportunity to identify and define those parameters to which model results are most sensitive. The model can thus be envisaged as having three distinct applications. First, interpretation of routine monitoring data and assessing authorized limits on discharge. Second, predictively assessing the impact of accidental releases. Third, identification of requirements as a basis for definition of appropriate programmes of research and monitoring. Results obtained from the model for hypothetical releases are presented here to illustrate these three modes of operation.

*Assessment of routine releases*

It is commonly assumed, for the purposes of modelling radionuclide distribution, that routine releases are continuous over time and that equilibrium has been achieved between discharge and concentrations in the environment. However, consideration of effluents arising from factories and non-nuclear power-producing plants suggest that, whilst a continuous background emission may be maintained, this is likely to have periodic spike emissions superimposed. The periodicity of such spike emissions may be predictable, arising from normal factory-cleansing operations such as the refitting or refurbishing of stack 'scrubbers' in smelting plants, and do not necessarily imply exceeding short or long-term authorizations on discharge. Emissions of unpredictable periodicity may also occur, and where discharge does occur in this fashion, considerable divergence from equilibrium conditions may be apparent.

Results presented in Fig. 3 illustrate model predictions of radioactivity concentrations (Bq $l^{-1}$) in milk and contents in soil and external plant (Bq $m^{-2}$) arising from hypothetical input of $^{35}S$ ($T_{\frac{1}{2}} = 87{\cdot}4$ d) at a steady state level to soil and plants ($0{\cdot}5$ Bq $m^{-2}$ $d^{-1}$) with periodic (90 day) spike releases of 10 Bq $m^{-2}$. Deposition and transfer characteristics are taken to be appropriate to the release of $^{35}S$ in inorganic particulate form. In order to simplify interpretation of the data, results are presented over 450 days, assuming continuous growing agricultural conditions. The total activity content in the top $1{\cdot}5$ cm layer of soil rises rapidly over the first 200 days and tends towards an equilibrium value of 60 Bq $m^{-2}$ after 400 days. To some extent, the establishment of an equilibrium is dependent upon the relatively short half-life of $^{35}S$. Periods of spike release are visible as minor pertubations in the predicted activity levels. By contrast, total activity associated with external plant is generally low ($<0{\cdot}5$ Bq $m^{-2}$), but sixfold increases due to the spike releases are apparent. Due to the assumed rapid mobilization of sulphur from plant surfaces, these peaks in activity content are transient only. Milk produced by grazing cows, which effectively integrate activity recorded in plants spatially and temporally, exhibit less marked peaks in activity concentration than recorded in vegetation and a 'dynamic equilibrium' is evidenced from $\sim 200$ days. Nonetheless, considerable variation about the mean concentration of $1{\cdot}8$ Bq $l^{-1}$ is apparent.

By comparison, results are presented in Fig. 4 for predicted contents of stable sulphur in soil, plant and milk components, following a hypothetical release of the stable element corresponding to the release characteristics noted for $^{35}S$ above. It can be seen that the overall pattern of sulphur accumulation in milk and external plant resembles that predicted for $^{35}S$. The slight flattening of peaks in milk concentrations reflect the absence of physical decay of the element. A more substantial effect is noted for the concentration of sulphur in soil, with no equilibrium being apparent at 450 days. In this context, it is notable that the half-life of radioactive decay of $^{35}S$ is approximately 90 days, which is equivalent to the periodicity of the spike releases.

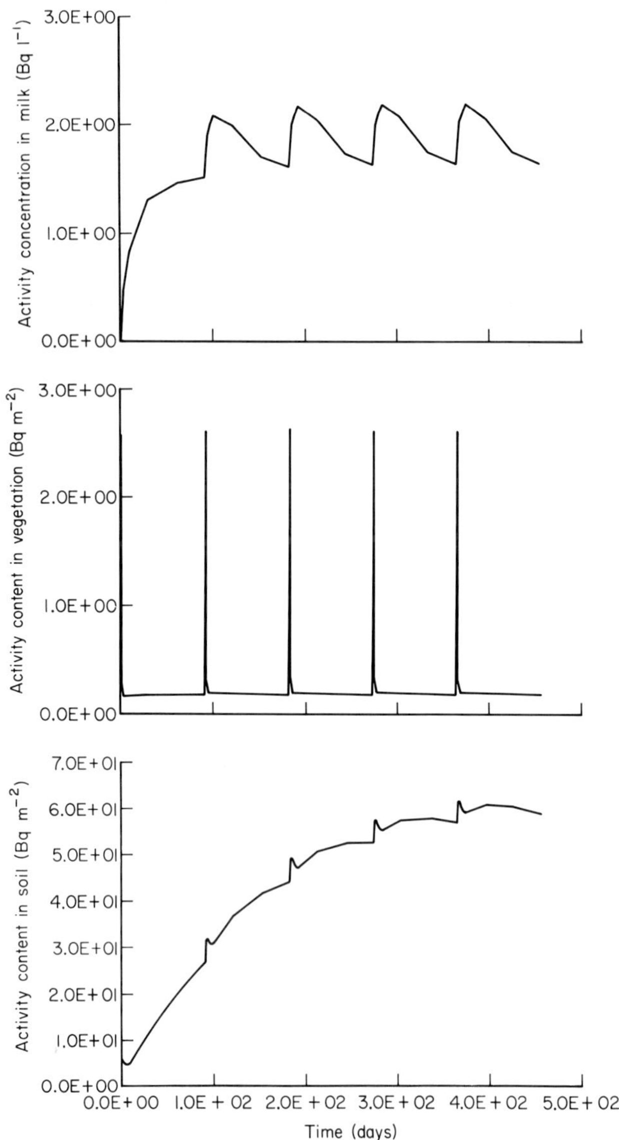


FIG. 3. Illustrative results following continuous uniform deposition of $^{35}S$ with multi-spike inputs to permanent pasture grazed by dairy cattle. A continuous background deposition to plant and soil of 0·5 Bq m$^{-2}$ d$^{-1}$ is assumed, with spike inputs of 10 Bq m$^{-2}$ at 90-day intervals. Deposited $^{35}S$ is assigned 0·2 to soil solution and 0·8 to external leaf.

### *Assessment of single spike releases*

Following single spike deposition of $^{35}S$ to dairy pasture, the activity content in plants declines rapidly, due to the swift assumed mobilization of sulphur and combined wash-off and grazing (Fig. 5). Activity content of the top 1·5 cm layer of soil declines more slowly. At the end of the 400 day period illustrated in Fig. 5, activity in soil falls to approximately one-tenth of its peak value, compared with a greater than five orders of magnitude

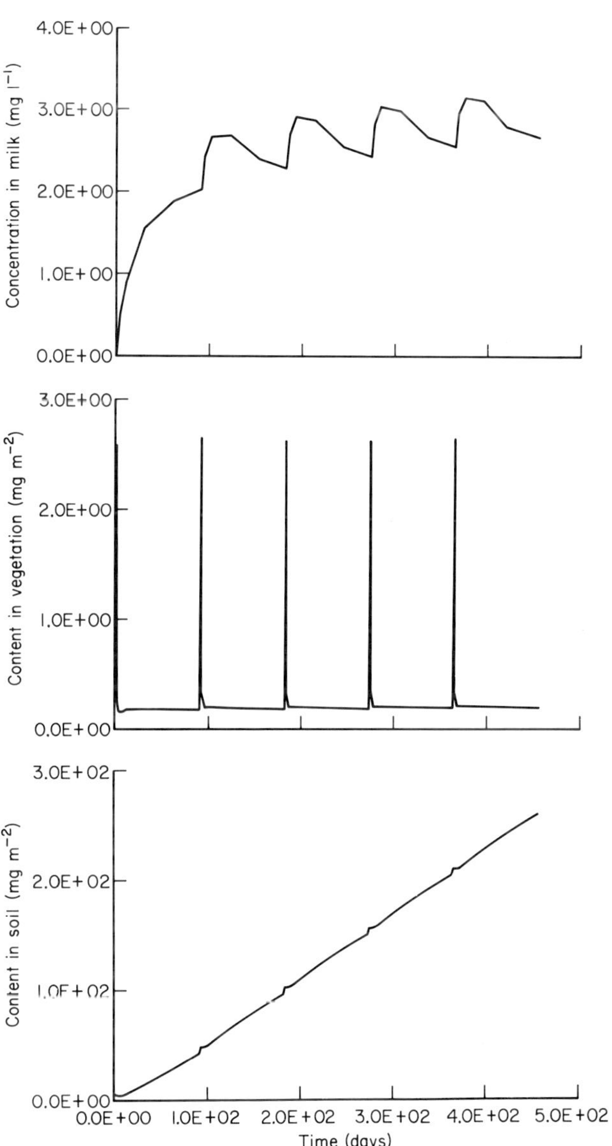


FIG. 4. Illustrative results following continuous uniform deposition of stable sulphur with multi-spike inputs tc permanent pasture grazed by dairy cattle. A continuous background deposition to plant and soil of 0·5 mg $m^{-2}$ $d^{-}$ is assumed with spike inputs of 10 mg $m^{-2}$ at 90-day intervals. Deposited sulphur is assigned 0·2 to soil solution anc 0·8 to external leaf.

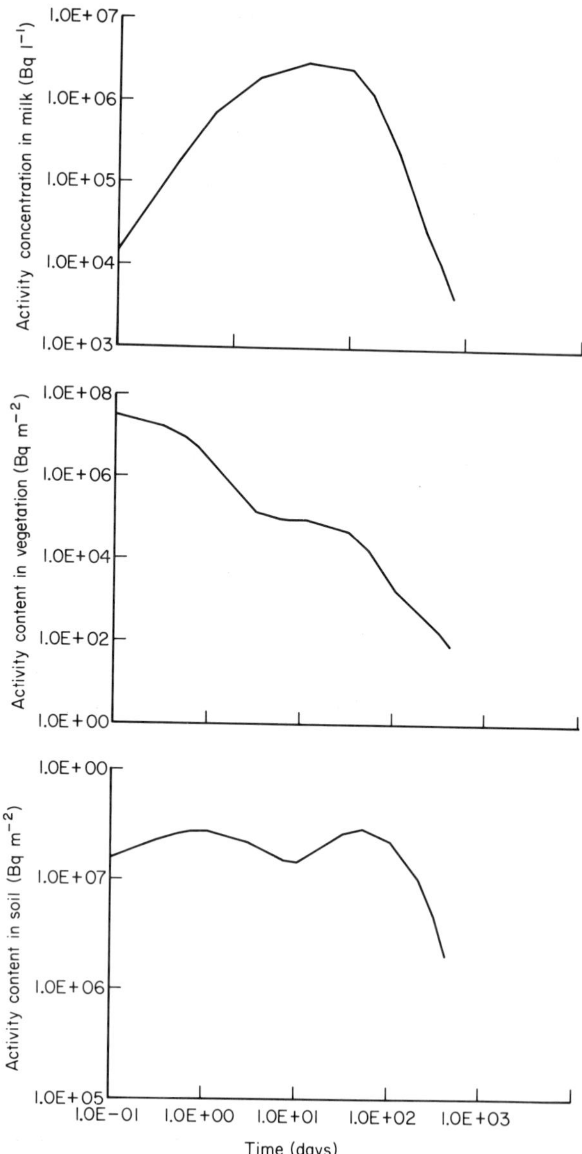


FIG. 5. Illustrative results following single spike deposition of $5 \times 10^7$ Bq $^{35}$S m$^{-2}$ to permanent pasture grazed by dairy cattle. Deposited $^{35}$S is assigned 0·2 to soil solution and 0·8 to external leaf.

variation in plant activity concentration. Activity in milk increases up to 10 days following spike input and declines thereafter over a period equivalent to 2–3 months. One hundred days after initial release, activity levels in milk are approximately one-tenth of those predicted at 10 days, and are equivalent to those predicted at 1 day after input. Thereafter, activity concentrations decline rapidly.

### *Sensitivity analysis*

Illustrative results are presented in Figs 6 and 7, assuming conditions of continuous uniform release in each case, but varying the parameter coefficients describing leaf wash-off and shoot-to-root transfer, and the values assumed for grazing area and the leaf interception fraction for depositing material. Results predicted assuming a grazing area of 1000 m$^2$ and a leaf interception fraction of 1·0 indicate an equilibrium activity concentration of $^{35}$S in milk approximately three times greater than predicted where a more typical grazing area of 500 m$^2$ and a leaf interception fraction of 0·5 is assumed (Fig. 6). Relative rates of accumulation of activity in milk are little affected by these varying assumptions and near equilibrium is reached in each case at 100 days. By contrast, varying the assumed rates of fractional leaf wash-off and shoot-to-root transfer by an order of magnitude either way results in a range of predicted equilibrium activity concentrations in milk of only 1–1·5 Bq l$^{-1}$, as illustrated in Fig. 7. A limited effect on the rate at which equilibrium is reached is observed, with equilibrium established at between 100 and 200 days.

## DISCUSSION

### *Implications for research and monitoring*

Consideration is given in the present study to the interpretation of predicted results in the context of differing input scenarios. Whilst none of the input scenarios modelled is based

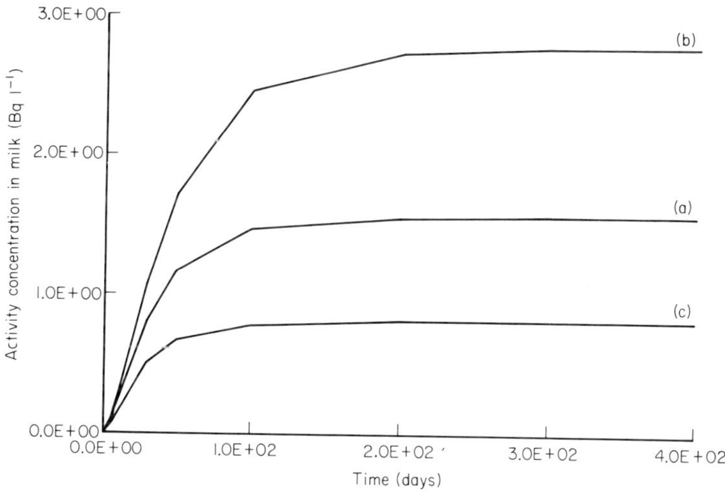


FIG. 6. Illustrative sensitivity analysis of cattle grazing area (GA) and leaf interception fraction ($r$) following continuous uniform deposition of $^{35}$S to permanent pasture grazed by dairy cattle. An input rate to soil and plants of 0·5 Bq m$^{-2}$ d$^{-1}$, is assumed throughout the 400-day period in all cases. Symbols: (a) GA = 500 m$^2$, $r$ = 0·8, remainder of activity to soil solution; (b) GA = 1000 m$^2$, $r$ = 1·0; (c) GA = 250 m$^2$, $r$ = 0·5, remainder of activity to soil solution.

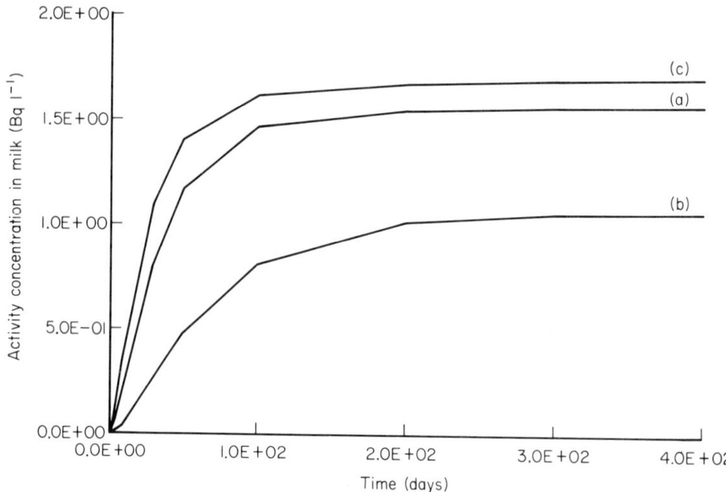


FIG. 7. Illustrative sensitivity analysis of leaf washoff rate (LWR) and shoot-to-root transfer coefficient (SRT), following continuous uniform deposition of $^{35}S$ to permanent pasture grazed by dairy cattle. An input rate to soil and plants of $0 \cdot 5$ Bq $m^{-2}$ $d^{-1}$ is assumed throughout the 400-day period in all cases. Symbols: (a) LWR $= 2 \times 10^{-5}$ $s^{-1}$, SRT $= 3 \times 10^{-6}$ $s^{-1}$; (b) LWR $= 2 \times 10^{-4}$ $s^{-1}$, SRT $= 3 \times 10^{-5}$ $s^{-1}$; (c) LWR $= 2 \times 10^{-6}$ $s^{-1}$, SRT $= 3 \times 10^{-7}$ $s^{-1}$.

on actual discharge data, they all correspond to assumptions regarding discharges applicable to modelling studies. Thus, single-spike discharges are typically considered as representative of accidental releases, whilst continuous uniform discharges may be taken to represent releases from continuously operating plant. The third scenario considered (i.e. continuous release with periodic minor spikes) conveniently describes the form of discharge which may be expected from actual operating plant. Results obtained are discussed below in the context of implications for programmes of research and monitoring, and for model validation and revision.

Results presented here are illustrative only; nonetheless, they do indicate areas for caution in the interpretation of monitoring data. Where periodic spike releases of activity occur, this may result in a dynamic equilibrium being established for milk concentrations. Correlation of contents in soil, plant and milk will not be apparent on the basis of spot samples (Figs 3 and 4). Similarly, long-term air sampling may not reflect adequately the transient rise in atmospheric concentrations. Consideration of releases of stable sulphur even greater care is required where correlation between concentrations in soil, plant and milk is required, particularly over periods short by comparison to periods of spike discharge. Analysis of single spike releases, as presented in Fig. 5, indicate that, even for this relatively simple case, an integrated approach to monitoring will be required, particularly at times soon after release.

Sensitivity analyses suggest that model predictions regarding concentrations of $^{35}S$ in milk are sensitive, at times prior to establishing equilibrium, to changes in parameter values of leaf wash-off and shoot-to-root transfer coefficients. Both of these parameters are poorly defined in the literature, especially in the context of their time dependence (Coughtrey *et al.* 1983a) and in any case may vary both with season and age of the plant. Variation in grazing area and leaf-interception fraction exhibit considerable influence on concentrations of sulphur in milk under equilibrium conditions. Grazing area is definable for individual sites of study. A rough approximation to the leaf interception factor ($r$) for depositing particulates can be derived from the model proposed by Chamberlain (1970) for grassland vegetation as follows:

$$r = 1 - \exp(\mu Y_V)$$

where $\mu$ is the proportionality constant (absorption coefficient, $m^2\ kg^{-1}$) and $Y_V$ is the dry weight density of vegetation ($kg\ m^{-2}$). Generally assumed values of $\mu$ range from 2·3 to 3·3 $m^2\ kg^{-1}$ (Miller 1980; Coughtrey & Thorne 1983a). Assuming a dry weight above-ground biomass of 0·1–0·5 $kg\ m^{-2}$ for permanent pasture, this range of values for $\mu$ corresponds to a range of values for $r$ between approximately 0·2 and 0·8. Given the uncertainty associated with mean estimates of $Y_V$ it is apparent that estimates of $r$ may vary between broad limits. In view of the sensitivity of the model to changes in this parameter, further definition of the leaf interception fraction for differing pasture types would appear to be of value, particularly for elements such as sulphur which may be released and deposited in a range of physical (e.g. gaseous or particulate) and chemical (e.g. $CS_2$, COS) forms.

### *Model validation and future code development*

Model construction, testing and validation are three distinct phases of development, requiring mutually-independent data bases. Parameters used during construction of the soil–plant–animal model were based on literature reviews. Some testing and refinement of the model has been undertaken on the basis of independent experimental studies. In order to validate the model, an independent and extensive database is required. Appropriate data may be available from various monitoring studies undertaken both in relation to the nuclear and non-nuclear industries. Nonetheless, it is apparent from Figs 6 and 7 that considerable care is required to match known environmental/agricultural conditions and model input data. Thus, not only grazing area, but pasture productivity, over-wintering practice, soil ingestion factors, and soil resuspension and inhalation factors all require consideration. Such data are rarely reported in monitoring studies and in their absence the alternative data presented are often of limited value for model validation.

Future model development will include the revision of the animal lung and gut sub-models, where this appears to represent an improvement in the physiological representation of processes of uptake. A preliminary review of the lung and gut models was undertaken as part of a recent study of the metabolism of selected toxic substances in man (Thorne & Jackson 1983; Smith & Thorne 1984). Additionally, the ICRP has set up a Task Group to review the current ICRP Lung Model, and a final report is expected in 1988 (Bair 1985). Revisions to the code, based on these reviews are likely to incorporate the following.

(a) Dissolution of particulate material deposited in the lungs.
(b) Translocation of particulate material within and from the deep lung, including clearance via the muco-ciliary escalator as well as movement in the lymphatic

system and within lung tissue.

(c) The effect of exposure conditions and physiological parameters on the deposition of material.

With respect to the gut model, Smith & Thorne (1984) concluded that the rate of transfer of material through the compartments of the gut was adequately described by the earlier model, but that the model structure should be modified to take into account the binding of material within the mucosal cells of the intestine. This was noted to be of particular importance because of the presence of metal-binding proteins in mucosal cells and also the fast rate of turnover of these cells. This may result in material absorbed into the mucosal layer of the gastrointestinal tract being excreted in the faeces before uptake into the systematic circulation can occur, due to intestinal-cell sloughing. The above factors were incorporated into the various substance-specific metabolic models proposed by Smith & Thorne (1984). Similar considerations have been noted by Crawford-Brown (1983).

In addition to amendments of sub-modules within SPADE, the existing model structure and database can be extended to allow the predictive assessment of the transfers of elements other than those considered currently. Models of this type could have particular uses in predicting the effects of pollutants from sites such as heavy industrial plant, including smelters and non-nuclear power producing plants, or assessing the consequences of heavy metal accumulation in sewage-sludge treated agricultural land. An application of the techniques discussed in this paper is presented by Martin & Coughtrey (these proceedings, p. 319) in respect of the migration of cadmium in contaminated soils.

Finally, the models need not be restricted to elements. If sufficient parameters can be derived, they could be extended to include a wide variety of industrial chemicals, allowing assessment of the impact of such chemicals and giving a basis for authorizations on discharge. This topic is currently under review.

## CONCLUSIONS

The principal conclusions from this study are summarized as follows.

(a) For predictive assessment of the transfer and distribution of radionuclides and their stable elements in domestic animals there is a clear requirement for the use of dynamic models.

(b) Use of such models can provide inputs to programmes of research and development.

(c) In order to validate these models on the basis of monitoring data there is a requirement for ancillary data describing environmental conditions.

In addition, data currently available suggest that some revision in commonly-accepted lung and gut models may be required.

## ACKNOWLEDGMENTS

Thanks are due to several colleagues within ANS, and in particular D. Hannant and T. Bobb for preparation of the figures and Drs M.C. Thorne and P.J. Coughtrey for helpful criticism. This study is based largely on work commissioned by MAFF and CEC, and thanks are due to Mr G.F Meekings and Dr D.F. Crabtree (MAFF) and Mr W. Hunter (CEC) for helpful comments at various stages of the work.

## REFERENCES

**Bair, W.J. (1985).** Task group to review models of the respiratory tract. *NRPB Radiological Protection Bulletin,* **63**, 5–6.

**Chamberlain, A.C. (1970).** Interception and retention of radioactive aerosols by vegetation. *Atmospheric Environment,* **4**, 57–58.

**Coughtrey, P.J., Jackson, D., Jones, C.H., Kane, P. & Thorne, M.C. (1984a).** *Radionuclide Distribution and Transport in Terrestrial and Aquatic Ecosystems: A Critical Review of Data,* Vol. 4. A.A. Balkema, Rotterdam.

**Coughtrey, P.J., Jackson, D., Jones, C.H. & Thorne, M.C. (1984b).** *Radionuclide Distribution and Transport in Terrestrial and Aquatic Ecosystems: A Critical Review of Data,* Vol. 5. A.A. Balkema, Rotterdam.

**Coughtrey, P.J., Jackson, D. & Thorne, M.C. (1983a).** *Radionuclide Distribution and Transport in Terrestrial and Aquatic Ecosystems: A Critical Review of the Literature,* Vol. 3. A.A. Balkema, Rotterdam.

**Coughtrey, P.J., Jackson, D. & Thorne, M.C. (1985).** *Radionuclide Distribution and Transport in Terrestrial and Aquatic Ecosystems: A Compendium of Data,* Vol. 6. A.A. Balkema, Rotterdam.

**Coughtrey, P.J., Jackson, D. & Thorne, M.C. (1984c).** *A Review of Current Research, Data and Theoretical Models.* ANS Report No. 374. Associated Nuclear Services, Epsom.

**Coughtrey, P.J., Jackson, D., Thorne, M.C. & Meekings, G.F. (1983b).** Studies on the mobility of radioisotopes of Ce, Te, Ru, Sr and Cs in soils and plants. *The Transfer of Radioactive Materials in the Terrestrial Environment Subsequent to an Accidental Release to Atmosphere,* Vol. 1, pp. 505–521. CEC V/3004/83, Luxembourg.

**Coughtrey, P.J. & Thorne, M.C. (1983a).** *Radionuclide Distribution and Transport in Terrestrial and Aquatic Ecosystems: A Critical Review of Data,* Vol. 1. A.A. Balkema, Rotterdam.

**Coughtrey, P.J. & Thorne, M.C. (1983b).** *Radionuclide Distribution and Transport in Terrestrial and Aquatic Ecosystems: A Critical Review of Data,* Vol. 2. A.A. Balkema, Rotterdam.

**Crawford-Brown, D.J. (1983).** An age-dependent model for the kinetics of uptake and removal of radionuclides from the G.I. tract. *Health Physics,* **44**, 609–622.

**Eve, I.S. (1966).** A review of the physiology of the gastrointestinal tract in relation to radiation doses from radioactive materials. *Health Physics,* **12**, 131–161.

**Harte, G.A. (1976).** *The Numerical Integration of a Stiff System of Linear Differential Equations.* CEGB Report RD/B/3693. Central Electricity Generating Board.

**ICRP (1966).** Deposition and retention models for the internal dosimetry of the human respiratory tract. *Health Physics,* **12**, 173–207.

**ICRP (1979).** *Publication 30. Limits for Intakes of Radionuclides by Workers. Part 1.* Annals of the ICRP, **2**, (3/4). Pergamon Press, Oxford.

**Jackson, D., Coughtrey, P.J. & Crabtree, D.F. (1985).** Dynamic models for soil–plant–animal systems. *Nuclear Europe,* **5**(4), 29–31.

**Jacquez, J.A. (1972).** *Compartmental Analysis in Biology and Medicine.* Elsevier, Amsterdam.

**Linsley, G.S., Simmonds, J.R. & Haywood, S.M. (1982).** *FOOD-MARC: The Foodchain Transfer Module in the Methodology for Assessing the Radiological Consequences of Accidental Releases.* National Radiological Protection Board, NRPB-M76. HMSO, London.

**Miller, C.W. (1980).** An analysis of measured values for the fraction of a radioactive aerosol intercepted by vegetation. *Health Physics,* **38**, 705–712.

**Nair, S. (1984).** *Models for the Evaluation of Ingestion Doses from the Consumption of Terrestrial Foods Following an Atmospheric Radioactive Release.* CEGB Report, RD/B/52004/N84. Berkeley Nuclear Laboratories, Berkeley.

**Nair, S., Grogan, H.A., Minski, M.J. & Bell, J.N.B. (1983).** Models for the prediction of doses from the ingestion of terrestrial foods. *Ecological Aspects of Radionuclide Release.* (Ed. by P.J. Coughtrey, J.N.B. Bell & T.M. Roberts), pp. 141–159. Blackwell Scientific Publications, Oxford.

**Nair, S., Hotson, J. & Stacey, A. (1980)).** *POPFOOD-A Computer Code for Calculating Ingestion Collective Doses from Continuous Atmospheric Releases.* CEGB Report, RD/B/N4888. Berkeley Nuclear Laboratories, Berkeley.

**Roberts, J.A. (1983).** Terrestrial pathways model for NUCRAC–SAI. *The Transfer of Radioactive Materials in the Terrestrial Environment Subsequent to an Accidental Release to Atmosphere,* Vol. 2, pp. 553–569. CEC V/3004/83, Luxembourg.

**Simmonds, J.R. & Linsley, G.S. (1981).** A dynamic modelling system for the transfer of radioactivity in terrestrial food chains. *Nuclear Safety,* **22**, 766–777.

**Simmonds, J.R., Linsley, G.S. & Jones, J.A. (1979).** *A General Model for the Transfer of Radioactive Materials in Terrestrial Foodchains.* National Radiological Protection Board, NRPB–R89. HMSO, London.

**Smith, A.D. & Thorne, M.C. (1984).** *Pharmacodynamic Models of Selected Toxic Chemicals in Man.* ANS Report No. 512–2. Associated Nuclear Services, Epsom.

**Thorne, M.C. (1979).** Analytical techniques for the analysis of multi-compartment systems. *Physics, Medicine and Biology,* **24**,815–817.

**Thorne, M.C. & Coughtrey, P.J. (1983).** Dynamic models for radionuclide transport in soils, plants and domestic animals. *Ecological Aspects of Radionuclide Release.* (Ed. by P.J. Coughtrey, J.N.B. Bell & T.M. Roberts), pp. 127–139.Blackwell Scientific Publications, Oxford.

**Thorne, M.C., Coughtrey, P.J. & Meekings, G.F. (1983).** A study of the sensitivity of a dynamic soil–plant–animal model to changes in selected parameter values. *The Transfer of Radioactive Materials in the Terrestrial Environment Subsequent to an Accidental Release to Atmosphere,* Vol. 2, pp. 505–521. CEC V/3004/83, Luxembourg.

**Thorne, M.C. & Jackson, D. (1983).** *Models for the Metabolism of Chemical Carcinogens in Man.* ANS Report No. 347. Associated Nuclear Services, Epsom.

# Subject index


*Abies alba,* 201
*Accipiter nisus,* 212–19
*Acer pseudoplatanus,* 184–6, 187, 188
Acetic acid, 143, 269, 270
Acid, defined, 139, 140
Acid neutralizing capacity (ANC) *see* Alkalinity
Acid rain, 1, 3, 139–40
  brown earth soil studies, 189–90
  pH definition, 148
  sulphur immobilization buffer, 186
Acidity
  and alkalinity, 144–50
  and altitude, 119, 175
  base neutralizing capacity (BNC), 139, 144, 148, 150
    at fixed pH values, 148–50
    measurement, 144–6, 147–8
    in natural waters, 146–7
    passage through ecosystem, 152
  and buffering, 139, 143
  defined, 139, 144
  ecosystem, internal production of, 140
    passage of BNC, 152
  equivalence points, 144
  fir tree ecosystem, 152–4, 176–8
  and humus, 153–4, 178–9
  neutralization, 143–4
  and proton donation, 139, 140, 141–3, 144
  and proton flux, 151–2
  soil, and heavy metal movement, 327, 328, 329, 331, 332, 333
    Micro-organism produced, 187
    and plants, 152, 178–9
  trees affected by, 152–4, 173, 176–8
    species effects, 177–8
Aerosol *see* Particulates
Agriculture
  cadmium inputs in, 225, 230, 231, 232, 235
  and element balance, 7, 95
  pollutants, 3
Alfalfa, 281
Algae, 24, 25, 250
Alkalinity, 139
  and acidity, 144–50
  budget, Douglas fir, 152–4
  components of, natural waters, 147
  defined, 144
  measurement, titration, 144–6, 147, 148
    at fixed pH values, 148–50
Alkyls, atmospheric, 241
*Allium ampeloprasum porrum,* 313
*Allium cepa,* 285
Aluminium, 101
  acidic properties, 141, 142, 143
  aerosol concentrations, 11, 93, 94, 95
    enrichment factors, 102, 108
  aerosol concentrations
    trends in, 101
    washout factors, 107
    winter/summer ratios, 100
  deposition trends, 108, 109
  in estuarine sediment, 64, 66–70, 71, 82
  -fluoride complexes, 203, 206
  human daily intake, 27
  -silicon ratio, human tissue and environment, 29
  smelting, and fluoride, 197, 198–9, 203
  and sulphate, 171, 180, 182
Ammonia, 3, 140
  and precipitation acidity, 119
Ammonium
  acid, $pK_a$ of, 143
  in rainfall, above and below canopy, 172, 174
Ammonium ion, cloud water, 133
Ammonium lead sulphate, 240, 244
Ammonium nitrate, as extractant, 269, 270, 272
Ammonium zinc sulphate, 240
Animals
  fluoride transport to, 193–4, 203, 204–7
  radionuclide transfer models, 386–400
  *see also* Birds; Cattle; Sheep
Antimony
  aerosol concentration, 11, 93, 94
    deposition, 107, 108, 109
    enrichment, 101, 102, 103, 104
    trends in, 101
    washout factors, 107
    winter and summer, 95, 100
  human daily intake, 27
  soil-wheat transfer, 355, 357, 363, 365
Aphids, 206, 207
Arsenic
  aerosol concentration, 11, 93, 94, 96–9
    deposition, 108, 109
    enrichment, 101, 102, 103, 104

trends in, 100, 101
washout factors, 107
winter and summer, 95, 100
in coal, 9
soil concentration, calculated 2003 AD, 110
Atmosphere
air particulate *see* Particulates
chemical composition, 7
element levels, rural Britain, 89–110
metal vapours and alkyls in, 241
pollutants, 1, 3
Avonmouth zinc smelter, 225–6, 233, 322, 333, 350

Barium, 27
Barley, *see Hordeum*
Base(s), 139, 140–1
neutralizing capacity (BNC), *see* under Acidity
and neutralization, 143–4
strength of, 141–3
Beans, *see Phaseolus*
Beech *see Fagus*
Beryllium, 27, 258
Beet, *see Beta vulgaris*
*Beta vulgaris,* 313
*Betula pendula,* 177
Bicarbonate, 143
Biosphere, and geosphere
elements, relationships, 5-31
Birch, *see Betula*
Birds
as heavy metal transfer agents, 337–52
feathers, excretion into, 344, 350–1
intake, seasonal changes in 340–4
and moulting grounds, 351–2
storage, geographical variation, 346–50
tissue metal loads, seasonal changes in, 344–6
predatory, bone fluoride in, 211–20
age-related, 214–16, 218
sex-related, 214, 218
Bisulphate, 143
Black acidic snow, 187
Blood, human 8
element abundance, correlation with non-blood groupings, 12, 13, 14, 15
v. aerosol (urban), 19, 23, 29
v. continental crust, 14, 15, 18, 23
v. seawater, 14, 18, 23
v. whole seal blood, 15, 21
K–Rb and Si–A1 ratio, 29

*Brassica rapa,* 279, 282, 285
Bristol Channel, Great Britain, cadmium discharge to, 225–6
Bromegrass, 281
Bromine
aerosol concentration, 11, 93, 94, 95, 96–9
deposition, 108, 109
enrichment, 102, 103, 104
trends, 101
winter/summer ratios, 100
human daily intake, 27
- lead ration, and petrol, 93–4
Bronsted-Lowry definition of acids, 139, 140–1
Bryophytes, 250, 258, 260, 264

Cadmium, 3
aerosol concentrations, 11, 93, 241
budget, 151
and cement manufacture, 229, 233
coal concentrations, 9
coastal water input, 225–6, 230, 232, 234
- containing products, 226–7, 233
cycling and fate, polluted woodland, 319–34
discharge sources, 223–35
and fossil fuel combustion, 228–9, 233
human daily intake, 27
human kidney, 10, 224
human liver, 10
and industrial emissions, 224, 225–6, 241
and iron and steel industries, 227–8, 233
liming affected levels, 271–3, 274
and non-ferrous metal production, 225–6
and phosphate fertilizers, 229–30, 233, 235
plant uptake, 267, 269–74
and refuse incineration, 223, 230–1, 233
in sediment and soil, 243
and sewage disposal, 231–2, 233
sludge additions, 303, 306–15
shorebirds, levels in, 343–9
soil residence times, 328–9
Caesium
aerosol concentration, 11, 93, 94
deposition, 107, 108, 109
trends in, 101
enrichment, 102, 104
trends in, 101
winter/summer ratio, 100
human daily intake, 27
soil-wheat pathways, 354–68
transfer to sheep, 371–82

Calcium
  aerosol concentrations, 11, 93, 94
    deposition, 108, 109
    enrichment, 102, 104
    trends in, 101
    winter/summer ratios, 100
  human daily intake, 27
  intra and extracellular, 26
  tree canopy affected, 172, 174
*Calidris alpina,* 340–50
*Calidris canutus,* 341–6, 351
Cancer, 113
Carbon, 7
  in cloud, 133
  soil content, and manure applications, 307–8
Carbonaceous chondrites, 16
Carbon dioxide, 128
Carbonic acid, 143
  and acidity, 139
  and alkalinity, 139, 147
  equivalence point, 144–5
*Carcinus maenas,* 338
Carnivores, 3, 4
Carrot, *see Daucus carota*
Cattle, 4
  radionuclide transfer, from pasture, 389, 393–9
Cells, element interactions, 26–8
Cement manufacture
  and cadmium discharge, 229, 233
Cerium
  aerosol concentrations, 11, 93, 94, 95, 100, 101
    deposition, 107, 108, 109
    enrichment, 102, 104
    washout factors, 107
  - europium ratio, 95
  - samarium ratio, 95
  soil-wheat transfer, 355–68
*Chameanerion angustifolium,* 196
Chloride, 174
Chlorine
  aerosol concentration, 11, 93, 94, 96–9
    deposition, 107, 108, 109
    enrichment, 102, 104
    trends in, 101
    washout factors, 107
    winter and summer, 95, 100
  human daily intake, 27
  - sodium ratio, 93
  tree canopy affected, 172
Chromium
  aerosol concentrations, 11, 93, 94
    enrichment, 102, 104
    trends in, 101
    washout factors, 107
    winter and summer, 95, 100
  in coal, 9
  deposition, total, 108, 109
  enrichment, 102, 104
  in estuarine sediment, 64, 66–70, 74, 76, 79, 83
  human daily intake, 27
  lichen, levels in, 258, 259, 260, 261
  in oil refinery effluent, 60
  and sewage sludge application, 302, 306–15, 332
  soil concentration, calculated 2003 AD, 110
Clay
  debris in organic evolution, 30
  oceanic pelagic, element abundance, 12, 13, 14, 16, 17, 29
  red, 8
  soils, and radionuclide uptake, 358–63, 364, 365, 366, 367
Cloud(s)
  acidic, 133
  droplet size, 126–7
  impaction efficiency, 127, 128
  insoluble particles, 133, 134, 135
  in precipitation processes, 118–22
  turbulence deposition, 128, 130
  water chemistry, 131–3, 134
  wind-driven, vegetation capture, 125–36
Coal
  burning, and element levels, 93, 100, 102, 133, 134
  as cadmium source, 223, 228–9
  element concentration, 9
    correlation with other groupings, 13, 14, 15
Cobalt
  aerosol concentration, 11, 93, 94
    deposition, 108, 109
    enrichment, 100, 102, 104
    trends in, 101
    winter/summer ratios, 100
  in coal, 9
  in estuarine sediment, 64, 66–70, 74, 76, 79, 83
  in lichen, levels, 258, 259, 260, 261, 262–3
  in soil
    concentration, calculated 2003 AD, 110
    mobility, 332
  washout factors, 107
*Coffea arabica,* copper levels, 289–99
  and mulching, 290, 291, 294, 295
  and wet processing method, 290, 291, 294

*Colletotrichum coffeanum,* 289
Contaminants, defined, 2
Copper
  accumulation, coffee stands, 289–95
  aerosol concentrations, 11, 93, 94
    trends in, 101
    winter/summer ratios, 100
  budget, coffee stands, 295–9
  bird ingested, 345, 347, 348, 349, 350
  deposition, 108, 109
  enrichment, 102, 103, 104
  in estuarine sediment, 64, 66–70, 74, 76, 79, 83
  forest ecosystems, 321
  fungicides, 289, 290, 299
  and $HNO_2$ formation, 110
  human daily intake, 27
  human plasma, 28
  in lichen, 258, 259, 260, 261
  in oil refinery effluent, 60
  plant availability, 267, 269–74
  residence time in soil, 328
  and sewage sludge application, 303, 306–15
  in soil and sediment, 243
  washout factors, 107
*Corylus avellana,* 322–34
Crown-leaching, 165–7, 171–2, 174
  accelerated, 179
  of fluoride, 197, 200
Crustal rocks, 8
  element abundances, correlation to non-crustal groupings, 12, 13, 14, 15, 16, 17, 18
    K–Rb and Si–Al ratio, 29
  and lichen element concentrations, 254–6, 262
*Cyperus esculentus,* 267, 268, 273–4

*Dactylis glomerata,* 206
*Daucus carota,* 313
Deposition
  dry, 89, 107–8
  interception, 165–7, 171–2, 174, 175–6
  occult, *see* Occult deposition
  soluble, and washout factors, 107
  total, trends in, 109–10
  velocity ($V_g$), 102, 104
  wet, modelling, 113–23
    and precipitation processes, 118–20
Diet, 28
  element abundance, correlation with other groupings, 13, 14, 15, 22
  element intake, national variations, 27–8
  K–Rb and Si–Al ratio, 29
Dihydrogen phosphate, 143
DNA, herring, 8, 13, 14, 15, 20
Douglas fir *see Pseudotsuga menziesii*
Dry deposition, 89, 107–8
  of fluoride, 198–9

Element(s)
  abundance, correlation between various groupings, 12–31
  atmosphere, rural Britain, 89–110
  biophile, 21–2, 24
  biophobe, 21–2, 24
  enrichment mechanism, 101
*Endymion nonscriptus,* 322
Enrichment of elements, 100–4, 105
  and air particulate size, 89, 101, 104
  lichen as indicators, 257–64
  mechanism, 101
  and solubility, 105
*Equisetum telmateia,* 196
Estuaries, 35
  flushing, 35, 36–7, 43
    time, calculation, 36
    and weather, 37
  internal hydraulic jump, 43, 51, 52
  metal turnover, by shorebirds, 337–52
  mixing, 43
    and neap tides, 40
    vertical, 48, 50–2
  pollutant transfer mechanisms, 47–53
  salinity distribution, 35, 43, 47, 48–53
    patchy distribution, 39–41
    and tributary flows, 42
    upper and lower estuary, 36
  secondary flows, 37–9, 44
  sediment contamination, oil refinery, 55–84
  spring tides, and mixing, 40, 43
  topographical features, and mixing, 43, 51
  tributary flows, 41–3
  *see also* under names of estuaries
*Eucalyptus,* 197, 200, 202
Europium
  aerosol concentration, 11, 93, 94
  – cerium ratio, 95
  – samarium ratio, 95
Evaporation
  and chemical concentrations, 135–6, 165, 167
  of plant intercepted water, *see* Interception loss
  of pollutants in precipitation, 119
Evolution, 7

*Fagus,* 321
*Falco sparverius,* 218
*Falco tinnunculus,* 212–19
Feathers, metal excretion into, 344, 350–1
Fertilizers, 109
 fluoride increase by, 203, 204
 and long-term metal transfers, 305, 306, 307, 309, 310, 311, 315
 phosphate, as cadmium source, 223, 229–30, 233, 235
Filter effect, vegetational *see* Interception deposition
Fires, 89, 193, 263
 and rapid fluoride uptake, 203
Fluoride, 3, 4
 air concentrations, rural and industrial, 193
 and aluminium, 197, 198–9, 203, 206, 217
 animals, transport to, 193–4, 203, 204–7
 bone, predatory birds, 211–20
 deposition, measurement limitations, 194–5
  rates of, 196–8
 dry deposition, 198–9
 gaseous, velocity, 198
 geographical occurrence, 217–18, 219
 particulate, velocity, 197
 rain scrubbing of, 199–200
 surface deposition, in animal uptake, 204–5, 207, 208
  as sink, 195–6
 vegetation, transport from, 200–2
 volatization to air, 201, 202, 207
 wet deposition, 199–200
Fluorosis, 193–4
Fog drift, *see* Occult wet deposition
Food supplies, daily, by geographical region, 28
Forests and woods
 contaminated, heavy metal cycling and fate, 319–34
  mathematical model, 322–5, 327–8, 329, 330, 331, 333
 decline, 3
 fires, 89, 263
 interception deposition, 165–7, 171–2, 174
 occult wet deposition, 128–31, 133
 sulphate deposition, 158, 160–4
  crown leaching and washoff factors, 165–7
 sulphur dioxide deposition, 164–7
 *see also* Trees
Formic acid, 143
Forth estuary, Great Britain, 347–50
Fossil fuel combustion, 89, 95, 102, 263
 ash, in cloud water, 133, 134
 and soil sulphur content, 182, 184, 185, 186, 187
 *see also* Coal
Foxes, 4
Froghoppers, 206, 207
Fungicides, copper, 289, 290, 299

Gallium levels, lichen, 258, 259, 260, 261
Gaussian plume model, 114–17, 123
 hidden assumptions demonstrated 115–17
Geosphere, and biosphere elemen relationships, 5–31
Germanium, 258
Gold, 93, 94
Grass, occult wet deposition to, 138, 130 132
Gypsum, 141, 223

*Haematomma ventosum,* 253
*Haematopas ostralegus,* 343
Hazel, *see Corylus avellana*
*Hemilea vastatrix,* 289
Herbivores, 3
 fluoride uptake, 4, 193–4, 204, 205–6
 *see also* Cattle; Sheep
Herring, 8, 13, 14, 15, 20
*Holcus lanatus,* 322
Homeostasis, 26
*Hordeum,* 313
Hydrochloric acid, 143
Hydrocarbons, 3
 estuarine sediment, 62, 65–71, 78–84
  GLC traces, 72–3
Hydrochloric acid, 143
Hydrogen fluoride, 193, 196
Hydrogen ion concentration, 139, 144
 cloud water levels, 133
 and microbial mineralization, 182
 natural waters, 146
 passage, in ecosystems, 150, 151–2
  of forest, 172, 173–4, 176–7, 179–80
Hydrogen peroxide, 119
Hydrogen phosphate, 143
Hydrogen sulphide, 181, 188

Ice core elements, 224, 263
Indium, 11, 93, 94
Industrial emissions, 93, 100, 104, 110, 185
 and cadmium, 224, 225–6, 227, 228–9,

231, 232–3, 241
and caesium, 371–2, 381
and fluoride, 193
metal aerosols, 240–1
spike emissions, 392–3
and sulphur, 184–5, 186, 187
Interception deposition, 171–2, 174
altitude variation, 175–6
chloride, 174
sulphur, 156, 165–7
Interception loss, 127, 129–31, 155, 167
Iodine, 119, 354
aerosol concentration, 11, 94
human daily intake, 27
Iron, 101
acidic cations of, 141
aerosol concentration, 11, 93, 94, 95 96–9, 100
deposition, 108, 109
enrichment, 102, 104
trends, 101
washout factors, 107
cell usage, 26–7
in estuarine sediment, 64, 66–70, 71, 82
Iron sulphide, 188
oxidation in soil, 190
Iron and steel industries, cadmium discharge, 227–8, 233, 234

Kidney, element concentration
human, 9–10
K-Rb and Si-A1 ratio, 29
shorebird, 342–3, 344–5, 346
*Lactuca sativa,* 279, 282, 285, 286
Landfill, and cadmium discharge, 223, 226, 228, 229, 230, 231, 232, 233
Lanthanum
aerosol concentrations, 11, 93, 94
coal concentrations, 9
*Larix kaempferi,* 177
Leaching
crown, *see* Crown leaching
of soil metals, 303–4, 311–12, 320
and sulphate levels, 155–6, 165–7, 182, 190
Lead
- acid battery works effluent, 246
aerosol concentrations, 11, 93, 94
and petrol, 94, 100, 107, 240
quarterly, rural locations, 96–9
speciation, 240–1
trends in, 100, 101
winter and summer, 95, 100
bird ingested, 339, 345, 347, 348, 349, 350
coal concentrations, 9
cycling and fate, contaminated woodland, 319–34
deposition, 106–9
enrichment, 101–4
gastrointestinal absorption, 239
global fluxes, human activities, 223–4
human daily intake, 27
inhaled, 239
lichen, levels in, 258, 259, 260, 261
peat core deposition rate, 107
plant uptake, from soil, 267–8, 269–74
from soil and air, 277–86
in sediment, speciation, 243
and sewage sludge application, 303, 306–15
soil, concentration, calculated 2003 AD, 110
residence time, 328
in street dust, 244
washout factors, 107
- zinc ores, cadmium source, 225
*Lecanora gangaleoides,* 253, 254, 255
Leeks, *see Allium ampeloprasum porrum*
Lepidoptera, 206, 207
Lettuce, *see Lactuca sativa*
Lichens, 217
trace element content, 249–64
and atmospheric deposits, 255–7, 262
and crustal abundance, 254–6, 262
trends in, 257–61, 263–4
Limestone washout, 109
Liming, and plant metal levels, 268, 271–3, 274
*Limosa lapponica,* 343
Lithium, 16, 27
cation selective processes, 30–1
Litter
effects on acidity, 153–4, 178–9
and fluoride transport, 197, 201, 202
as heavy metal sink, 320, 321, 324, 326, 327, 330, 331
and soil sulphur levels, 185–7
Liver, element concentration
human, 9–10, 13, 15, 22, 23
K-Rb and Si-A1 ratio, 29
shorebird, 342–3, 344–5, 346
*Lolium perenne,* 201, 313
Lung, human
element abundance, correlation to other groupings, 13, 22, 23
v. aerosol, 14, 19
v. human lymph nodes, 15, 20
K—Rb and Si—A1 ratio, 29–30
Lymph node, human

element abundance, correlation with various groupings, 13, 14, 15, 20, 22
K-Rb and Si-Al ratio, 29

Magnesium
aerosol concentration, 11, 93, 94
enrichment, 102, 104
trends in, 101
winter/summer ratio, 100
ground vegetation affected, 178-9
human daily intake, 27
rainfall, bulk, altitude affected, 175
seasonal effects, 177
tree canopy affected, 172, 174
tree species affected, 177
Manganese, 35, 172
acidic cations, 141
aerosol concentrations, 11, 93, 94
deposition, 107, 108, 109
enrichment, 102, 104
trends in, 101
winter/summer ratio, 100
in coal, 9
lichen levels, 258, 259, 260, 261, 262-3
washout factors, 107
Manure, applications and metal transfers, 304-15
Menai Straits, Wales, 347-50
Mercury
aerosol concentraton, 11, 93, 94
- potassium ratio, 95
shorebird levels, 342, 343, 346, 350-1, 352
vapour phase, 241
Metallothionein, 3
Micro organisms, and sulphur cycle, 181-91
Milford Haven, Wales, sediment contaminants, 55-84
geology and bathymetry, 57
oil spill statistics, 59
percentage mud, 58
Milk, radionuclide transfer models, 393-9
*Milium effusus,* 322
Mining, 95, 104, 263
cadmium, 225
Mist precipitation, *see* Occult wet deposition
Models, of radionuclide transfer
dynamic, 387
equilibrium, 386-7
release assessment, 391, 393
single spike, 393-4
sensitivity analysis, 391, 395-7
soil-plant-animal systems, 388-91, 392
validation, 399-400
Models, wet deposition, 113-25
Molybdenum
aerosol concentration, 11, 93, 94
coal concentrations, 9
and $HNO_2$ formation, 110
human daily intake, 27
lichen, levels in 258, 260
Monomers, 26
Monteith-Penman equation, 129, 131
Moon, element abundance, 13, 14
Muscle, human
element abundances, 13, 14, 15, 22, 23
K-Rb and Si-Al ratio, 29
Mussel, *see Mytilus*
*Mytilus,* 13, 14, 22, 23
*edulis,* 338, 346

Nickel
aerosol concentration, 11, 93, 94
deposition, 108, 109
enrichment, 100, 102, 104
trends in, 101
washout factors, 107
winter/summer, 100
- cadmium batteries, 226, 230, 233
in estuarine sediment, 64, 66-70, 74, 77, 79, 83
in lichen, levels, 258, 259, 260, 261
in oil refinery effluent, 60
and sewage sludge application, 303, 306-15, 332
smelting, 93, 100
soil, concentration, calculated 2003 AD, 110
mobility in, 332
*Nereis diversicolor,* 339, 346
Nitrate ion, 3, 4
cloud water levels, 133
Nitrates, 171
deposition, atmosphere, 108
bulk precipitation and altitude, 175
and ground vegetation, 178-9
total, trends, 109
tree canopy affected, 174
tree species affected, 177
Nitric acid
$pK_a$ of, 143
and rainfall acidity, 141
Nitic oxide, 3
Nitrogen, 7
aerosol concentration, 93
and base neutralizing concentration flux, 152
dioxide, 2, 110, 201

  microbial transformation, 191
  and occult deposition, 126, 133
Nitrogen oxide, 89, 133
Nitrous acid, 110
Nuclear weapons testing, 89
*Numenius arquata,* 339

Oak, *see Quercus*
Occult wet deposition, 119, 125–36
  chemical concentrations, 131–3
  drip rate, isolated tree, 127, 129
  evaporation, 129–31, 132
  to grass, 130
  impaction efficiencies, 127
  to *Pinus* forest, 130
  sedimentation, 127, 130
  shelter effect, 127
  turbulent deposition, 127–9, 130
Oceanic pelagic clay, 12, 13, 14, 16, 17, 29
Oil
  combustion, and cadmium discharge, 229
  effluents in estuaries, 55–84
Omnivores, 3
Onion, *see Allium cepa*
Organic soil, and radionuclide uptake, 360–3, 364, 365, 366, 368
Oxalic acid, 143
Oxygen, 7
Ozone, 3, 119
  phytotoxic, 2
*Parmelia,* trace elements in, 249–59
Particulates
  global, element transfer, 30
  K–Rb and Si–Al ratio, 22
  metal, speciation, 240–1
  precipitation scavenging, 89, 107, 119
  rural Britain, trace major elements, 89–102
    deposition, 102–10
  urban and rural, element abundance, 10–12, 14, 94
  whole blood v. urban aerosol, 19, 23, 25
Pea, *see Pisum sativum*
Peach damage, 193
Periodic Table of Elements
  associations, 5, 6, 7
  constancy, 7
  fluxes, 7
Pesticides, 3, 289
Petrol additive pollution, 94, 100, 104, 110, 240, 244, 263, 277
pH
  acid, 144, 147
  acid rain, defined, 148
  alkaline, 144, 147, 148
  and buffering, 139, 143, 147
  cloud water, 133, 134
  and deposition, 109
  of equivalence point, 145–6
  fixed value, base neutralizing capacity calculation to, 139, 148–9, 154
  and heavy metal movement, 327, 328, 329, 331, 332, 333
  and plant metal uptake, 271–3
  rain water, 134, 145–6
    throughfall, 173
  strong and weak acids, 139, 142–3
*Phaseolus,* 282, 313
  *vulgaris,* 198
Phenolphthalein alkalinity, 147
Phosphoric acid, 143
Phosphorus, 27, 172
Photoxidants, 2
*Pica pica,* 217
*Picea sitchensis,* 172, 176, 177, 321
*Pinus,* 130, 156, 177
*Pisum sativum,* 279, 282, 284, 285
$pK_a$, 142–3
  of weak acid, and BNC calculation, 149–50
  values, acids, 143
*Plantago lanceolata,* 196
Plants
  element distributions, 8
  heavy metal uptake, 267–74
  lead content, from soil and air, 277–86
  metal uptake, with sewage sludge, 312–14
  radionuclide transfer models, 286–400
  root copper uptake, coffee, 292, 296, 297
  root fluoride transport, 201
Plastics, and cadmium discharge, 226, 233
Platinum, 6
Podsol, transfer factors, 366
Pollutants, defined, 2
*Polygonum cuspidatum,* 196
Polymers, element interaction, 26
Potassium, 381
  aerosol concentration, 11, 93, 94
    enrichment, 102, 104
    trends in, 101
    winter/summer ratios, 100
  cell, 26
  in coal, 9
  deposition, 107, 108, 109
    tree canopy affected, 172, 174
  human daily intake, 27
  – mercury ratio, 95
  – rubidium ratio, human tissue and environment, 29

- sodium ratio, 95
Potatoes, *see Solanum tuberosum*
Primordial element abundances, 13, 14, 16
*Pseudomonas syringae,* 289
*Pseudotsuga menziesii,* 152
*Puccinellia maritima,* 372, 381

Quarrying, 104
*Quercus*
*coccinea,* 198
- hazel woods, 321, 322-34
*robur,* 322

Raddish, *see Raphanus sativus*
Radionuclides, 3, 6
in agricultural systems, models of transfer and distribution, 385-400
arctic lichen accumulation, 250
release predictions, 391, 393
single spike, 393-4
wet deposition modelling, 113-23
and resuspension of soil, 353, 355, 363, 368
soil-wheat pathways, 353-68
transfer to sheep from pasture, 371-82
Rainfall
acid, *see* Acid rain
convective storms, dynamical effects, 120-1
and element deposition, 105-7, 108, 109
processes of, 118-20
frontal systems, dynamical effects, 121-2
insoluble particles, 133, 134, 135
orthographic enhancement, 118, 119, 127
and stemflow, compared, 160, 161, 162
sulphate deposition, 155, 158, 161-3
and throughfall, compared, 157-8, 159, 160
and topography, 119
washout factor, 92, 107, 109
models, 114-17
*see also* Clouds
*Raphanus sativus,* 267, 268, 282, 283, 285
Redox potential, 320
of brown earth, acid rain affected, 189
of estuarine sediment, 61, 65, 66-70, 74
and manganese, 263
and plant metal levels, 268, 269, 273-4
Refuse incineration, 101, 263
as cadmium source, 223, 230-1, 233-4
Reindeer, 250
Renal dysfunction, 224
Resistance analogue, 3
*Rhizocarpon geographicum,* 253
Rice production, 273
Road
drainage water, trace metals, 243
- side dusts, 243-4
Rubidium, 11, 27, 29, 94
Ruthenium, soil-wheat pathways, 355-68
Ryegrass, *see Lolium*

Salinity
of cells, 26
estuarine
centre channel flux, 52, 53
decline, 47
distribution, 35, 36, 39-41, 42, 43
in vertical circulation, 48-50, 51
in vertical mixing, 51-2
Saltmarsh, $^{137}$Cs contaminated, 371-7, 380-1
Samarium
aerosol concentration, 11, 93, 94
- cerium ratio, 95
- europium ratio, 95
Sandy loam, and radionuclide uptake, 360-3, 364, 365, 366, 368
Savannah, 321
Scandium, 93, 100, 101
aerosol concentration, 11, 93, 94, 95
deposition, 108, 109
trends in, 101
winter/summer ratios, 100
enrichment, atmosphere, 102, 104
washout factors, 107
Scavenging
of aerosols by precipitation, 89, 107, 119
by suspended sediments, 56
Seawater
cadmium inputs to, 225-6, 228, 230-4
element composition, 7, 8
compared with non-seawater groups, 12, 13, 14, 15, 17, 18, 22, 23, 24, 25
in estuary flushing, 35, 36, 43
K-Rb and Si-Al ratio, 29
marine aerosol, 89, 95, 105
organisms of, 8
pollutants, 3
$SO_4$/Na ratio, 105
Seal blood and tissue
element abundance, 13, 14, 15, 21, 22, 23
Sediment
element abundance, correlation with non-sediment groups, 13, 14
estaurine, 35

contaminants, oil effluent related, 55–85
organic matter content, 63, 75
and % mud, 58, 80, 81
grain size, and contaminants, 63, 64, 66–70, 74–5, 82, 84
hydraulic conductivity, 63, 64, 66–70, 75, 78
metal speciation, 239, 242-3
suspended, oil refinery effluent, 60
Selenium, aerosol, 11, 93, 94, 96–9
deposition, 108, 109
enrichment, 101, 102, 104
trends in, 101
washout factors, 107
winter/summer ratios, 100
Sellafield nuclear fuel reprocessing plant, 119, 371-2, 381
Severn estuary, Great Britain, 347–50
Sewage, 3
cadmium discharge, 223, 231-2, 233, 246
cadmium inputs, UK, 234
sludge application, metal transfers, 301, 322
reversion hypothesis, 302
Sheep
$^{137}$Cs uptake by, 371–82
faecal, 376
tissue, 378–80, 382
soil ingestion, 206
Silicate, 143
Silicic acid, 143
Silicon, 7
- aluminium ratio, human tissue and environment, 29
Silty loam, and radionuclide uptake, 364, 365, 368
Silver
aerosol concentration, 11, 93, 94
content of lichens, 258, 260, 262
human daily intake, 27
Sitka spruce, *see Picea sitchensis*
Snow, acidic, 187
Snowdonia, Wales, lichen study, 250, 253–61, 264
Sodium, 30
aerosol concentration, 11, 93, 94, 96–9
seasonal changes, 95, 100
trends in, 101
bulk precipitation, altitude affected, 175
- chlorine ratio, 93
deposition, atmosphere, 107, 108, 109
rainwater, tree affected, 172, 174
enrichment, atmosphere, 102, 104
human daily intake, 27
organism control, 26
- potassium ratio, 95
washout factors, 107
Sodium bicarbonate, acidity and alkalinity measurement, 145, 147
Sodium carbonate, acidity and alkalinity measurement, 145, 147
Soil
acidity, and metal mobility, 327, 328, 329, 331, 332, 333
and micro-organisms, 187
and plants, 152, 178–9
bulk density, and organic manures, 307–9
copper fungicide affected, 292, 295, 296, 297, 298
element
atmospheric deposition, 110, 264
extractability, 302, 310-11
movement, 303, 319-34
relationships, 8
fluoride, 202–4, 207
as heavy metal sink, 320–34
lead, plant uptake, 277–86
metals, plant uptake, 267–74, 312–14
speciation, 242–3
movement, and metal loss, 314–15
radionuclide transfer models, 386–400
rare earth elements, 95
redox potential, *see* Redox potential
resistances, 3
and sewage sludge applications, 307–10
subsoil, re-sorbtion, 312
sulphur cycles, 181–91
anaerobic, 188–90
type, and radionuclide transfer, 353, 356–68
*Solanum tuberosum,* 313
Solar element abundances, 13, 14, 16
*Somateria mollissima,* 338
Soot deposits, and soil sulphur, 182, 187
SPADE model, 386, 388
Speciation of toxic metals, 239–46
atmosphere, 240–1
natural waters, 244–6
roadside dust, 243–4
sediment and soil, 243–3
X-ray powder diffraction, 240
Spinach, *see Spinacia*
*Spinacia,* 282
Spruce *see Picea*
Steel industries, cadmium discharge, 227–8, 233, 234
Stem-flow, 155
acidity, seasonal effects, 176–7
chemistry, and rainfall compared, 172–6

hydrogen ion deposition, and rainfall compared, 173
and stem cross-sectional area, 162
sulphate deposition, 161, 163–5
excess, sources, 167–9
tree species affected, 177–8
water amounts, and rainfall compared, 162
Straw
burning, 3
radionuclide concentration, 353, 358, 359, 360, 361, 362, 363, 367
*Strix aluco,* 212–19
Strontium, 27, 354
*Sturnus vulgaris,* 351
Subsoil, 312
Sugar beet, *see Beta vulgaris*
Sulphate ion, 3, 4
cloud water levels, 133
particle deposition, 163–4
in throughfall and stemflow, 165–7
Sulphates, 171, 172
atmosphere, deposition levels, 105, 108, 109
and leaching, 155–6, 165–7, 182
microbial reduction, 188–90
rainfall deposition, 160–3, 172, 175, 177
stemflow deposition, 161–7, 172, 177
thiosulphate, 185, 187
throughfall deposition, 161–2, 163, 165–7, 172, 177
tree species affected, 177–8, 179
Sulphide oxidation, 141
Sulphur
aerosol concentration, 93
global fluxes, human activities, 223
human daily intake, 27
microbial cycle, 181
immobilization, 186
leachate, liberation, 186
mineralization, 182, 183–6
oxidation–reduction subcycle, 187–90
and occult deposition, 126
radionuclide concentrations in milk, model, 393–7
Sulphur dioxide
and acidic cloud water, 133
aqueous phase oxidation, 110
forest deposition, 164–5, 166, 167, 333
lichens affected by, 250, 257
and nitrogen dioxide, 2
oxidation processes, 110, 119
and vegetation fluoride loss, 201
Sulphuric acid, 140
forested ecosystems, 180
$pK_a$ of, 143
and rainfall acidity, 141
Sycamore, *see Acer pseudoplatanus*

*Tadorna tadorna,* 338
Technetium, soil–wheat transfer, 355–68
Tees estuary, Great Britain
heavy metal transfer, shorebird, 338–46
tidal average dispersion, 47–53
Tetraalkylead compound analysis, 241
Tetrathionate, 185, 187
Thallium, 27
Thermal pollutants, 3
*Thiobacillus,* 187
Thiosulphate, 185, 187
Thorium
aerosol concentration, 93, 94
coal concentrations, 9
human daily intake, 27
Throughfall, 157
acidification, 172
seasonal effects, 176–7
in alkalinity budget, 153
canopy thickness affected, 157, 159
chemistry, and rainfall compared, 172–6
and fluoride transport, 197, 201, 202–3
hydrated aluminium ion, 142
hydrogen ion deposition, and rainfall compared, 173
species affected, 177–8
sulphate deposition, 161–2, 163
excess, sources, 165–7
water amount, and rainfall compared, 160
Tides, estuary
average dispersion mechanisms, 47–53
in flushing and dispersal, 38–43
neap, 40
spring, 40, 43
and residual drift, 48
and vertical circulation, 48–50
Tin, lichen levels, 258, 260, 261
Titanium
aerosol concentrations, 94
– caesium correlation, saltmarsh vegetation, 376
human daily intake, 27
lichen, levels in, 258, 259, 260, 261
Total suspended particulate matter (TSP), 89
rural areas, Great Britain, 93, 95
average annual, 94
trends in, 101
winter/summer ratios, 100

WHO guidelines, 93
Transfer factors, soil–wheat, 353, 355–68
Trees
acid neutralizing capacity budget, 152–4
age, and acidity fluxes, 178, 180
occult wet deposition rates, 130
species effects, rainwater chemistry, 177–9
stomatal $SO_2$ uptake, 166–7, 169
*Triticum aestivum,* radionuclide uptake, 356–63
Tungsten, *see* Wolfram
Turnip, *see Brassica rapa*
*Tyto alba,* 212–19

Uranium, 9, 27

Vanadium
aerosol concentration, 11, 93, 94, 96–9
deposition, 107, 108, 109
enrichment, 102, 103, 104
trends in, 101
washout factor, 107
winter and summer, 95, 100
in coal, 9
in estuarine sediment, 64, 66–70, 74, 77, 79, 83
lichen levels, 258, 259, 260, 261
and oil burning, 102
soil concentration, calculated 2003 AD, 110
Vapours, metal, in air, 241
Volcanic activity, 89, 95, 193, 263

The Wash, Great Britain, 347–50, 351
Wash-off, *see* Interception deposition
Washout factor, 92, 107, 109
frontal rain system, 121
models, 114–17
Gaussian plume, 114–15
Water
domestic, element abundance, 13, 15, 22
$K_w$, 144
natural, alkalinity, 146–7
metals speciation, 244–6
pollutants, 3
*see also* Seawater
Wet deposition
of fluoride, 199–200
modelling, 113–23
*see also* Occult deposition
Wheat, *see Triticum aestivum*
Windscale, *see* Sellafield nuclear fuel reprocessing plant
Wolfram, 11, 93, 94
Woodland, *see* Forests and woods

Yttrium, lichen levels, 258, 259, 260, 261
Ytterbium, lichen levels, 258, 259, 260

Zinc
aerosol concentration, 11, 93, 94, 96–9, 240–1
deposition, 105, 106, 108, 109
enrichment, 101, 102, 103, 104, 105
trends in, 100, 101
washout factors, 107
winter/summer ratios, 100
ammonium sulphate, 240
bird ingested, 339, 341, 342, 345, 346, 347, 348, 349, 351
cycling and fate, contaminated woodland, 319–34
as estuarine sediment contaminant, 64, 66–70, 74, 77, 79, 83
human daily intake, 27
- lead ores, cadmium source, 225
lichen, levels in, 258, 259, 260, 261, 262, 263
and nitrous acid formation, 110
in oil refinery effluent, 60
plant uptake, 267, 269–74
in sediment
estuarine contaminant, 66–7
speciation, 243
and sewage sludge, 303, 306–15
smelters, 225–6, 233, 241, 322, 332, 333, 350
soil, concentration calculated 2003 AD, 110
soil residence time, 328
Zirconium, lichen levels, 258, 259, 260, 261